A Comprehensive Guide to Solar Energy Systems

With Special Focus on Photovoltaic Systems

A Comprehensive Guide to Solar Energy Systems

With Special Focus on Photovoltaic Systems

Edited by

Trevor M. Letcher

University of KwaZulu-Natal, Durban, South Africa

Vasilis M. Fthenakis

Center for Life Cycle Analysis, Columbia University, New York, NY, United States

ELSEVIER

ACADEMIC PRESS

An imprint of Elsevier

Academic Press is an imprint of Elsevier
125 London Wall, London EC2Y 5AS, United Kingdom
525 B Street, Suite 1800, San Diego, CA 92101-4495, United States
50 Hampshire Street, 5th Floor, Cambridge, MA 02139, United States
The Boulevard, Langford Lane, Kidlington, Oxford OX5 1GB, United Kingdom

Library of Congress Cataloging-in-Publication Data
A catalog record for this book is available from the Library of Congress

British Library Cataloguing-in-Publication Data
A catalogue record for this book is available from the British Library

ISBN: 978-0-12-811479-7

For information on all Academic Press publications visit our website at
https://www.elsevier.com/books-and-journals

Publisher: Joe Hayton
Acquisition Editor: Lisa Reading
Editorial Project Manager: Serena Castelnovo
Production Project Manager: Sruthi Satheesh
Designer: Mark Rogers

Typeset by Thomson Digital

Contents

List of Contributors

Tom Baines

Stephenson Institute for Renewable Energy, The University of Liverpool, Liverpool, United Kingdom

Charles J. Barnhart

Western Washington University, Bellingham; Institute for Energy Studies, Western Washington University, Bellingham, WA, United States

Vítězslav Benda

Czech Technical University in Prague, Prague, Czech Republic

Khagendra P. Bhandari

Center for Photovoltaics Innovation and Commercialization, University of Toledo, Toledo, OH, United States

Rhys G. Charles

SPECIFIC-IKC, Swansea University, Swansea; Materials Research Centre, Swansea University, Swansea, United Kingdom

Fangliang Chen

Columbia University, New York, NY, United States

Matthew L. Davies

SPECIFIC-IKC, Swansea University, Swansea; Materials Research Centre, Swansea University, Swansea, United Kingdom

Thomas Döring

SolarPower Europe, Brussels, Belgium

Peter Douglas

Chemistry Group, Medical School, Swansea University, Swansea, United Kingdom; University of KwaZulu-Natal, Durban, South Africa

Beatrice Dower

MVGLA, Comrie, United Kingdom

Randy J. Ellingson

Center for Photovoltaics Innovation and Commercialization, University of Toledo, Toledo, OH, United States

Nesimi Ertugrul
University of Adelaide, Adelaide, Australia

Vasilis M. Fthenakis
Center for Life Cycle Analysis, Columbia University, New York, NY, United States

Michael Ginsberg
Center for Life Cycle Analysis, Columbia University, New York, NY, United States

Andrés Pinto-Bello Gómez
SolarPower Europe, Brussels, Belgium

Steven M. Grodsky
University of California, Davis; Wild Energy Initiative, John Muir Institute of the Environment, Davis, CA, United States

Ajay Gupta
EROI Energy Advisors Inc., Brampton, ON, Canada

Ingrid L. Hallin
Freelance Researcher, Edmonton, AB, Canada

Ross A. Hatton
Warwick University, Coventry, United Kingdom

Xiaoping He
China Center for Energy Economics and Research, The School of Economics, Xiamen University, Xiamen, China

Rebecca R. Hernandez
University of California, Davis; Wild Energy Initiative, John Muir Institute of the Environment, Davis, CA, United States

Zicong Huang
Guangdong University of Technology, Guangzhou, P.R. China

Aruna Ivaturi
University of Strathclyde, Glasgow, United Kingdom

Trevor M. Letcher
University of KwaZulu-Natal, Durban, South Africa; Laurel House, FosseWay, Stratton on the Fosse, United Kingdom

Jonathan D. Major

Stephenson Institute for Renewable Energy, The University of Liverpool, Liverpool, United Kingdom

Michelle Murphy-Mariscal

Mt. San Jacinto College, Menifee, CA, United States

Frank Pao

Columbia University, New York, NY, United States

Alyssa Pek

SolarPower Europe, Brussels, Belgium

Alexandre Roesch

SolarPower Europe, Brussels, Belgium

Michael Schmela

SolarPower Europe, Brussels, Belgium

Thomas P. Shalvey

Stephenson Institute for Renewable Energy, The University of Liverpool, Liverpool, United Kingdom

Katie Shanks

Environment and Sustainability Institute (ESI), University of Exeter, Penryn, United Kingdom

Graham Stein

National Grid, Warwick, United Kingdom

Senthilarasu Sundaram

Environment and Sustainability Institute (ESI), University of Exeter, Penryn, United Kingdom

Kristina Thoring

SolarPower Europe, Brussels, Belgium

David Timmons

University of Massachusetts Boston, Boston, MA, United States

Hari Upadhyaya

Wolfson Centre for Materials Processing, Institute of Materials and Manufacturing, Department of Mechanical, Aerospace and Civil Engineering, Brunel University, London, Uxbridge, United Kingdom

Zhangyuan Wang
Guangdong University of Technology, Guangzhou, P.R. China

James Watson
SolarPower Europe, Brussels, Belgium

Huiming Yin
Columbia University, New York, NY, United States

Eduardo Zarza-Moya
CIEMAT-PSA, Almería, Spain

Xudong Zhao
University of Hull, Hull, United Kingdom

Siming Zheng
Guangdong University of Technology, Guangzhou, P.R. China

Preface

Our book, *A Comprehensive Guide to Solar Energy Systems: With Special Focus on Photovoltaic Systems* is a companion volume to the recent book: *Wind Energy Engineering: A Handbook for Onshore and Offshore Wind Turbines* (Elsevier, 2017). It was felt that the solar energy industry like the wind turbine industry was developing so rapidly that it was now necessary to compile a collection of solar energy-related topics into one volume.

The use of renewable energy sources such as solar and wind for electricity generation is becoming commonplace in our society as we move away from fossil fuels to more sustainable forms of energy, free from carbon dioxide pollution. The move cannot come quickly enough as each month we hear that the previous month was the hottest month since records began and that CO_2 levels are increasing every year and have now passed the 410 ppm level.

Our book gives an all round view of solar energy with a special focus on technical issues surrounding photovoltaic cells. The 25 chapters are divided into the following six sections: Introduction; Solar Energy Resource and Worldwide Development; Thermal Solar Energy Technology; Photovoltaic Solar Energy—Generation of Electricity; Environmental Impacts of Solar Energy; Economics, Financial Modeling, and Investment in PVs, Growth Trends, and the Future of Solar Energy. In more detail, the book includes chapters on the following areas:

- Scientific aspects (basic theory of photovoltaic solar energy, global potential for producing electricity from the sun's energy);
- Wind energy in China, Europe, Africa, and the USA, to give a flavor of developments in very different countries but all with the same aim of reducing global warming while providing affordable, abundant, and sustainable energy;
- Thermal solar power in solar heaters, concentrated solar systems.
- Photovoltaics in all its different forms—crystalline silicon cells, cadmium telluride cells, perovskite cells, and organic cells;
- Large scale PV Integrated technologies (buildings);
- Integration into national grids;
- Small scale PV systems;
- Storing energy from PVs;
- Environmental issues and comparisons;
- Materials' abundance, purification, and energy cost for Si, CdTe, and CIGS photovoltaics;
- Life Cycle Analysis and Energy Return on Investment;

- Growth trends and the future of solar power;
- Minimizing the cost of resolving variability and energy storage.

It is hoped that the book will act as a springboard for new developments and perhaps lead to synergistic advances by linking ideas from different chapters. Another way that this book can help in serving the solar energy industry is through contact between readers and authors and to this effect addresses of the authors have been included.

Each topic is covered at the highest level with the very latest research and information, each chapter of this book has been written by an expert scientist or engineer, working in the field. Authors have been chosen for their expertise in their respective fields and come from ten countries: Australia, Belgium, Canada, Czechoslovakia, China, India, Spain, South Africa, United Kingdom, and the United States. Most of the authors come from developed countries as most of the research and development in this relatively new field is based in these countries. However, we look forward to the future when new approaches to solar energy, focusing on local conditions in emerging countries, are developed by scientists and engineers working in those countries. We are sure this new book will aid in this endeavor.

The chapters in this book can be considered as snapshots, taken in 2017, of the state of the solar PV industry. Our book goes hand in hand with four other books we have recently published: *Climate Change: Observed Impacts on Planet Earth*, 2nd edition, (Elsevier 2015); *Storing Energy: With Special Reference to Renewable Energy Sources* (Elsevier, 2016); *Wind Energy Engineering: A Handbook for Onshore and Offshore Wind Turbines* (Elsevier, 2017); and *Electricity From Sunlight: Photovoltaics Systems Integration and Sustainability* (Wiley, 2017).

For consistency and to appeal to an international audience, the International System of Units and Quantities is reflected in the book with the use of the Système International d'Unités (SI) throughout. Other units such as Imperial units are written in parenthesis. The index notation is used to remove any ambiguities; for example, billion and trillion are written as 10^9 and 10^{12} respectively. To avoid further ambiguities the concept of the quantity calculus is used. It is based on the equation: physical quantity = number × unit. To give an example: power = 200 W and hence: 200 = power/W. This is of particular importance in the headings of tables and the labeling of graph axes.

A vital concern related to development and use of renewable and sustainable forms of energy, especially solar, is the question of what can be done when it appears that politicians misunderstand or ignore, and corporations overlook the realities of climate change and the importance of renewable energy sources. The solution lies in sound scientific data and education. As educators we believe that only a sustained grassroots movement to educate citizens, politicians, and corporate leaders of the world has any hope of success. Our book is part of this aim. It gives an insight into the subject, which we hope readers will consider and discuss. The book is written not only for students, teachers, professors, and researchers into renewable energy, but also for politicians, government decision-makers, captains of industry, corporate leaders, journalists, editors, and all other interested people.

We wish to thank all 42 authors and coauthors for their cooperation, help, and especially, for writing their chapters. It has been a pleasure working with each and every one of the authors. Trevor thanks his wife, Valerie and Vasilis his wife Christina for their help, support, and encouragement they gave us over these long months of putting the book together. We also wish to thank Elsevier editors and staff for their professionalism and help in producing this well-presented volume.

Trevor M. Letcher
Stratton on the Fosse, Somerset
Vasilis M. Fthenakis
Columbia University, New York

PART 1

Introduction

1

Why Solar Energy?

Trevor M. Letcher
UNIVERSITY OF KWAZULU-NATAL, DURBAN, SOUTH AFRICA
trevor@letcher.eclipsr.co.uk

1.1 Introduction

The importance of the sun in sustaining life has probably been known to humans in all ancient societies, and many of these people, including the Babylonians, ancient Hindus, Persians, and Egyptians worshipped the sun. From written records, the ancient Greeks were the first to use passive solar designs in their homes and no doubt experimented with harnessing the sun's energy in many different ways. There is a story that, Archimedes in the 2nd century BC reflected the sun's rays from shiny bronze shields to a focal point and was thus able to set fire to enemy ships. The Romans continued the tradition of using the sun in their homes and introduced glass, which allowed the sun's heat to be trapped. The Romans even introduced a law that made it an offence to obscure a neighbor's access to sunlight.

By contrast, PV technology (the creation of a voltage by shining light on a substance) and the main focus of this book, is a very recent application. Scientists, as early as 1818, noticed that the electrical conductivity of some materials, such as selenium, increased by a few orders of magnitude when exposed to sunlight; however, it was not until the 1950s that scientists working on transistors at the Bell Telephone Laboratories showed that silicon could be used as an effective solar cell. This very soon led to the use of silicon solar cells in spacecraft; and in 1958, Vanguard 1 was the first satellite to use this new invention. This application paved the way for more research into better and cheaper solar cells. The work was further encouraged after the rapid oil price rise in the 1970s. In 1977, the US Government created the National Renewable Energy Laboratory. A further indication of the rapid rise of silicon solar cell technology was the building of the first solar park in 1982 in California, which could generate 1 MW; this was followed a year later by a larger Californian solar park, which could generate, at full capacity, 5.2 MW. The United States has now built several PV power plants in the range of 250–550 MW. It is amazing to think that just 34 years after the first solar farm was built in California, China has built a solar farm of 850 MW. Furthermore, the solar PV worldwide generating capacity, at the end of 2016, was in excess of 300 GW. To put this into perspective, 1000 MW (1 GW) is the power generated by a traditional fossil-fueled power station.

A Comprehensive Guide to Solar Energy Systems. http://dx.doi.org/10.1016/B978-0-12-811479-7.00001-4

In January 2017, it was reported that Chinese companies plan to spend US$1 billion building a giant solar farm (of 1 GW) on 2500 ha in the Ukraine, on the exclusion zone south of the land contaminated by the 1986 nuclear explosion.

The amount of solar energy shining on the earth (with wavelengths ranging from 0.38 to 250 μm) is vast. It heats our atmosphere and everything on the Earth and provides the energy for our climate and ecosystem. At night, much of this heat energy is radiated back into space but at different wavelengths, which are in the infrared range from 5 to 50 μm [1]. This energy heats the greenhouse gas molecules (such as carbon dioxide and methane) and water molecules in the atmosphere. The explanation is as follows. Using CO_2 and H_2O as examples, this heating process takes place because the radiated IR frequency is in sync (resonates) with the natural frequency of the carbon—oxygen bond of CO_2 and the oxygen—hydrogen bond of H_2O. The increased vibration of the bonds effectively heats the CO_2 and H_2O molecules. These heated molecules then pass the heat to the other molecules in the atmosphere (N_2, O_2) and this keeps the Earth at an equitable temperature. The vibrating frequencies of the O—O bond in oxygen and the N—N bond in nitrogen molecules are very different from these radiation frequencies and so are relatively unaffected. As there are many more water molecules than CO_2 or CH_4 molecules in the atmosphere, the overall contribution of the H_2O molecules to the greenhouse effect is larger than the contribution by CO_2 or CH_4 or the other minor greenhouse gases (GHGs), such as chlorinated hydrocarbons. However, as the CO_2 concentration has increased from 280 ppm (280 parts per million or 280 molecules per million molecules) before the industrial revolution, to 410 ppm (observed at Mauna Loa Observatory on April 21, 2017), and as the H_2O concentration in the atmosphere remains relatively constant, it is the CO_2 (together with other GHGs) that is largely responsible for present-day global warming.

Sunlight can be harnessed in a number of ever-evolving and ingenious ways, which include solar heating (usually water, Chapter 6), photovoltaics (for electricity production and the main focus of this volume), concentrated solar thermal energy (Chapter 7) and also solar ponds [2], space heating [3], molten salt power plants [4], and even artificial photosynthesis. Some of these technologies have been developed only in the past 30 years as ways of mitigating climate change and the build-up of atmospheric carbon dioxide from the burning of fossil fuel. The strength of solar energy lies in its inexhaustibility and also in the wide variety of ways that it can be harnessed ranging from small scale to large-scale applications.

In 2016, renewable energy supplied less than a quarter of electricity in the world. The renewable energy total of 23.7% is made up of: pumped hydroelectricity being the most prevalent, with 16.6%; wind 4%; and solar only 1.5% (Section 1.7). In spite of the relatively low values for wind and solar energy, their rate of implementation is amazingly rapid and the predictions for the future are promising. As an indication of things to come, we note that on May 15, 2017 Germany received almost all of its electricity from renewable and for 4 days (May 7–10, 2017) Portugal ran on renewable energy (wind, solar, and hydro) alone [5].

1.2 How Much Solar Energy Falls on the Earth and How Much is Used to Make Electricity?

There are many ways of expressing how much solar energy falls on the earth. Chris Goodall writes in *The Switch* that the sun supplies enough power in 90 min to meet the world's total energy needs for a year [6]. In more scientific language, the Earth receives 174×10^{15} W [174 PW (petawatts)] of incoming solar radiation (insolation) at the upper atmosphere. Approximately 30% of this is reflected back to space, while the rest is absorbed by the oceans and landmasses and things on the earth. At night this 70% absorbed energy is radiated back into space keeping the earth at a constant temperature.

The total solar energy absorbed by the Earth's atmosphere, oceans, and land masses is approximately 3.85×10^{24} J a^{-1} [3.85 YJ a^{-1} (yottajoules per annum)] [7]. Photosynthesis captures less than 0.1% of this, approximately 3.0×10^{21} ZJ a^{-1} (zettajoules per annum), in biomass [8]. The total energy consumption in the world today is less than 0.02% of the total solar energy shining on the earth.

Most people in the world live in areas with insolation levels of 150–300 W m^{-2} or 3.5–7.0 kW h m^{-2} d^{-1}, where d refers to day [9]. This magnitude of solar energy available makes it an appealing source of electricity. The United Nations Development Programme in its 2000 World Energy Assessment found that the annual potential of solar energy was between 16 000 and 50 000 $\times 10^{18}$ J (16 000–50 000 EJ). This is many times larger than the total world energy consumption, which was 559.8 EJ in 2012 [10].

Solar energy supplied only 0.45% of the total primary energy consumption in 2015. This is far below traditional forms of energy or other renewable forms of energy (Table 1.1) and reference [11]. Additionally, as mentioned earlier, solar energy produces 1.5% of all the electricity used globally. Therefore, much work has to be done to realize the suggestion of the International Energy Agency (IEA) that the sun could be the largest source of electricity by 2050, ahead of fossil fuels, wind, hydro, and nuclear. According to a recent report by IEA, solar PV systems could generate up to 16% of the world's electricity by 2050, while solar thermal electricity (STE) from concentrated solar power (CSP) plants could provide an additional 11%; this will require an early and sustained investment in existing and future solar technologies [12].

Table 1.1 World Energy Primary Consumption, 2015, Percentages [11]

Energy Types	Percentages (%)
Oil	32.9
Coal	29.2
Gas	23.9
Nuclear	4.4
Hydro	6.8
Wind	1.4
Solar	0.45

Note this table is not referring to electricity production. For the breakdown of electricity production see Table 1.3 in Section 1.7.

1.3 Types of Technology That Can Harness Solar Energy

There are two main types of solar power:solar thermal and solar PV.

Solar thermal includes domestic hot water systems (Chapter 6), cooking [13], solar-disinfecting water [14], energy storage—molten salts [4], solar power transport [15], fuel production [16], and CSP (Chapter 7). The latter involves focusing and tracking the sun's rays using mirrors (usually parabolic troughs or dishes) onto a working fluid, which vaporizes and expands and is used to drive a turbine. The temperature of the working fluid can reach 800°C. The great advantage of CSP is that the sun's energy is converted into heat, which can be readily stored. This is not true for PV systems, because electricity is more difficult to store, although battery technology is rapidly improving.

It has been estimated that solar energy could be used to supply up to 70% of household hot water in the United Kingdom and in sunnier climates, providing almost all domestic hot water. Today worldwide solar water heaters are responsible for 435 GW_{th} [17]. CSP supplies 5.01 GW electricity globally, this being less than 2% of all electricity supplied by solar energy; Spain is the CSP world leader with 2.5 GW investment followed by the United States (1.9 GW) [17].

Solar PV panels (Chapters 8 to 12) produce electricity directly and can be effective in both, direct, or diffuse cloudy solar radiation, although the systems are obviously more efficient in direct sunlight. Electricity is produced as a result of the sun's energy striking a solar panel (at present usually pure silicon), which causes electrons to be released; these in turn then travel through wires (Chapter 8). Until recently, the only solar panels (wafers) available were made of pure silicon (99.9999 purity), which is both costly and energy-intensive to manufacture (Chapters 9 and 21). Recent research into wafer technology has produced a range of new solar wafers, which include materials, such as cadmium telluride (Chapter 10) interesting alloys of copper indium and gallium (Chapter 21) and more recently perovskites (Chapter 11). Some of these involve elements, which are in short supply; and some involve elements, which are toxic, for example, cadmium (Chapter 21). Silicon wafers have improved significantly over the past 2 decades and the efficiency is of the order of 20%. Furthermore, with mass production, the price of silicon wafers has decreased enormously.

A recent report by Fraunhofer stated that in Germany, in 1990, the price for a typical rooftop system of 10–100 kW_p PV, was around 14 € $(kW_p)^{-1}$. At the end of 2016, such systems cost about 1.3 € $(kW_p)^{-1}$. This is a net-price regression of about 90% over a period of 26 years [18]. Solar panels suitable for use on roofs are now manufactured in such quantities that the electricity generated in several favorable locations, according to the World Economic Forum (WEF), has reached grid parity; that is the point where the direct, unsubsidized, cost of PV generated electricity is equal to that of fossil fuel generated power [19]. The growth in PV manufacturing has been driven by government incentives where, for example, in countries, such as the UK, Germany, Spain, and Australia the cost of electricity and technological innovation is subsidized. Under such schemes a premium tariff is paid for PV-generated electricity that is fed into the grid. This premium can be several times higher than the normal tariff paid for fossil-fuel-generated electricity. This has led to the establishment of a large number of wind farms, as well as many rooftop PV systems for individual houses.

In spite of its intermittent nature, solar power from PV panels has many advantages:

- The wafer panels are manufactured in modular form and can be retrofitted to roofs anywhere the sun shines.
- PV panels can be installed where the power is needed thus eliminating the need to integrate into grid systems. This is particularly important in areas, which do not have grid electricity.
- Often, particularly in hot countries, which have a high demand for air conditioning, the generation of PV electricity coincides with the greatest need for electricity during the day.
- PV electricity is useful over a wide range from the charging of mobile phones, street lighting using LEDs, telecommunications [20], space vehicles, solar pumps [21], and grid electricity.

Concentrated photovoltaics (CPV) use optical lenses or curved mirrors to concentrate light onto small but highly efficient solar cells. Often these systems are fitted with cooling systems because the efficiency of PV decreases with cell temperature [18].

Most (99%) of European solar cells are connected to the grid while off-grid systems are more common in Australia, South America, Africa, and South Korea [22].

PV systems are found in three marketing area: residential rooftop, commercial rooftop, and ground-mounted utility-scale systems (solar farms). In 2013, rooftop systems accounted for 60% of the global installations; this is changing rapidly with a shift toward utility-scale systems and as of 2017 utility systems in the United States have a higher installed capacity than the sum of residential and commercial. Residential systems are typically around 10 kW while commercial systems reach megawatt scale. The utility-scale power plants are in the range of 100–500 MW and moving to the 1 GW capacity, and are becoming more common especially in hot regions of the world. Three years ago, California's 550 MW (Topaz Solar farm) was the world's largest solar project. A year later, another large Californian solar farm (the 579 MW Star Solar farm) was built followed in 2016 by India's 648 MW Kamuthi Solar Power Project. This was surpassed in 2017 by China's Longyangxia Dam Solar Park of 850 MW [23].

A solar farm PV system connected to the grid, consists of the solar array and additional components usually called "balance of system" (BOS), which includes power conditioning equipment, and DC to AC power converters (called inverters) (Chapter 15).

The efficiency of commercial PV modules is about 16% and the modules are expected to have a life-time of 25 years. Higher efficiencies have been recorded [24].

1.4 Why We Need to Develop Solar Energy

Fossil fuels are the largest contributor to climate change and global warming and the only hope of cutting greenhouse gas emissions is to find cleaner methods of generating electricity and powering our vehicles [25]. This is perhaps the most important reason for developing renewable forms of energy, such as solar energy [26]. The agreement to hold the increase in the global average temperature to less than 2°C compared to preindustrial levels, and to pursue efforts to remain within a 1.5°C rise, which was made at the Paris Conference of the United Nations Framework Convention on Climate Change in December 2015, was a major triumph in the quest to manage global warming and climate change. Unfortunately, emissions continue to rise, with the average concentration of CO_2 in the Earth's atmosphere now at 410 ppm. It has been estimated that the 2°C limit roughly corresponds to a concentration of 450 ppm. To avert a temperature rise of above 2°C will require the almost total decarbonization of energy supply over the coming decades. In spite of these warnings, the world continues to burn ever-greater amounts of fossil fuel annually, especially for electricity production. Much of this is due to the massive exports of coal from Australia and the United States to Japan and China [26]. Another reason for reducing our dependence on imported fossil fuel is that it improves our energy security. After the oil supply disruptions of the early 1970s, many nations have increased their dependence on foreign oil supplies instead of decreasing it. This increased dependence affects more than just national energy policies; it leads to instability in world politics and many recent wars have been fought over oil [27].

Job creation is an important issue and instead of spending money on costly fuel imports, money can be spent on local development, materials, workmanship, and investment into solar energy and other renewable technologies. Furthermore, new developments and research into solar energy and other renewable energy technologies can provide a boost to international trade.

The development of solar energy technology with its infrastructure leads to a more stable and permanent industry than the present oil industry, which will someday end. The development of a solar energy industry can be considered as a positive legacy for our children's children as it will never run out.

With almost one-third of the world's population (2 billion people, mainly in Asia, Pacific, and sub-Saharan Africa) living without access to grid electricity, solar energy offers great promise to improve living standards and reduce greenhouse gas emissions [28]. Solar PV is particularly suitable for remote regions in warm climates where there is usually a suitable surface for the installation of panels. With the recent development of low-cost PV panels and efficient LED lighting, the technology can now displace traditional kerosene lamps as a cost-effective and safer alternative. There is no need to supply fuel to produce the electricity; and moreover, PV panels generate electricity that can be used on-site and there is no need for expensive transmission lines.

In many developing countries, there is a surprisingly high level of cell/mobile phone usage, despite a limited infrastructure for recharging the battery. Solar PV panels offer an ideal way of solving the problem.

Little has been said of solar thermal energy in this chapter but it too can be used to raise the standard of living for the almost one-third of the world's population living without electricity. Solar thermal energy offers a great way of heating water without the need for burning cow-dung (a major cause of eye disease, especially in rural India) or wood, which is often in short supply.

Although solar energy is a clean source of energy and has a much lower environmental impact than any of the conventional fossil fuels, coal, oil, and gas its replacement of fossil fuels is not going to be easy. First, the fossil fuel industry is well entrenched in our society and we have become totally dependent on it. Second, fossil fuel, in all its manifestations, is a wonderfully concentrated form of energy that packs a lot of energy into a small space, when compared to the major renewable forms of energy. The concentrated nature of fossil fuel, which has taken 2 billion years to accumulate, means the renewable forms of energy are at a significant disadvantage. However, humans are being forced to adapt, albeit slowly. There are three main driving forces for this. First, the burning of coal in huge quantities in power stations is causing major smog problems with its accompanying respiratory and other health issues (especially in China) and this is forcing governments around the world (especially Europe and China), to consider seriously reducing their reliance on coal-powered electricity generation and develop renewable energy sources. In Beijing, schools often have to close because of smog and protective masks are a common day-to-day sight.

Second, pumping oil or gas from both offshore or onshore wells is becoming more and more expensive, and this is particularly true in the case of oil wells where new sources are found only in increasingly inhospitable areas, making mining difficult, expensive, and in many cases environmentally disastrous.

Third, the price of renewable energy and, in particular, solar PV energy has decreased from \$10 W^{-1} in 2007 to less than \$1.56 W^{-1} for rooftop installations and \$0.86 W^{-1} for utility solar farm installations in 2017. Solar energy is now beginning to offer competitive prices with regard to energy from coal, gas, and oil [29,30]. This rapid reduction in price over just a decade is far better than the equivalent rate of change for any other renewable source of energy. This has led many to predict that soon solar PV will be the dominant renewable form of energy. However, on a commercial basis, solar energy must do more to out-compete traditional fossil fuels. Fossil fuels with their high energy density are just too convenient and wind and solar power is variable, delivering power only when the wind blows and the sun shines. Fossil fuel can deliver power when it is needed although their extraction and combustion pollutes the environment. The saving grace of solar and wind energy in scenarios of large penetration (e.g., $>30\%$) will be linked to their integration over large balancing zones and to the development of affordable energy storage technologies [31]. At the moment, battery technologies appear to be the most convenient storage method with the lithium-ion battery leading the way.

1.5 The Difficulties With Harnessing Solar Energy

The two main issues with harnessing solar PV energy are the following:

- Solar energy is perceived to be a dilute form of energy. As an example, in order to produce an average 1 GW of electricity (the size of a large fossil fuel power-station) from PV cells, in a hot tropical part of the world, peak mid-day sun energy level of (1200 W m^{-2}), the solar farm would be about 20–25 km^2. This includes the solar panels, spacing between them, access roads and corridors often left for wildlife mitigation. In a temperate country, such as the United Kingdom, each solar farm would probably have to more than twice this size to achieve a power of 1 GW. Although this seems a lot of land, if we account for the area occupied from coal surface mining, the coal life cycle can use more land than photovoltaics [32]. Of course when PV are installed in roof-tops they do not use any additional land.
- It also happens that in many countries, especially in the temperate parts of the world, when the sun is shining, the public demand for electricity is low. Peak demand on some national grids is usually in the early morning and evening when the sun is either not shining or its energy is weak. It is estimated that in temperate climates on average a solar farm produces useful electricity about 20% of the time, while in regions of very strong solar irradiation (e.g., the Atacama desert in Chile, the PV capacity is in excess of 35%).

These two issues make it difficult to feed solar PV electricity into a national grid (Chapter 15). This is largely because electricity generation and electricity consumption must be in balance at all times, even over such short times as seconds. However, large PV power plants that are utility-friendly, providing reactive power and ancillary services to the grid, have been developed [33]. Nevertheless, for supporting large penetrations of solar, storage facilities will be needed. This can take many forms [31]; at the moment the less expensive energy storage systems are pumped hydro and compressed energy storage, while the price of batteries (especially Li-ion ones) are being reduced catalyzed by increasing markets for electric vehicles.

1.6 Is Harnessing Solar Energy Cost Effective?

The efficiency of solar energy systems is rated according to their performance under a standard test irradiance of 1000 W m^{-2}, which corresponds to the maximum irradiance expected on a clear day in summer at moderate latitudes. The actual level of solar irradiance will depend on the latitude and local climatic conditions, but the annual average solar energy density lies in a range from 100 to 350 W m^{-2} for most locations. The capacity factor for solar collectors (actual output power/rated output DC power) therefore lies at 10%–35% depending on location. The cost of producing electricity from solar PV energy compared to the cost of producing electricity from coal has been estimated by Lazard [34] and the results are in Table 1.2.

Table 1.2 Estimation of Electricity Production Costs by Lazard [34]: Unsubsidized Levelized Cost of Energy

Energy Form	**Cost/$ (MW h)$^{-1}$**
Coal	60–143
Solar PV utility scale	49–61
Solar rooftop residential	138–222
Wind	32–62
Gas combined cycle	48–78
Biomass direct	77–110
Solar tower with thermal storage	119–182
Nuclear	97–136

The analysis shows that solar farming costs (estimated to be as low as $0.86 W^{-1} in 2017 [29]) are lower than coal (estimated to be in the range of $1.5–2.5 W^{-1}) in the same range as gas fired power stations but rooftop residential PV (estimated to be about $1.56 W^{-1} in 2017 [29]) is not really economical for feed-in to national grids without government subsidies.

1.7 A Comparison of Solar PV Installed Capacity With Other Renewable Forms of Energy

At present renewable energy is responsible for nearly a quarter of global electricity production, with pumped hydroelectricity and wind energy being the most successful of the renewable forms of energy. However, with all the new advances in solar PV, together with new developments in battery technology, solar PV is rapidly closing the gap. The breakdown of electricity production, as of 2016, is given in Table 1.3.

The installed capacity for wind and solar is, as expected, higher than the actual energy produced. The breakdown of global installed capacity of the renewable forms of energy is given in Table 1.4.

To put the data from Table 1.4 into perspective, one should consider the other major nonfossil fuel form of energy: nuclear energy. In 2016, the global net capacity of nuclear power was 391 GW from 449 nuclear operable reactors [36].

1.8 The Future of Solar Energy

Solar energy is the fastest growing renewable energy source. Table 1.5 compares worldwide wind and solar PV power capacity over the past 10 years and highlights the rapid growth of solar PV. Over the 5 years from 2011 to 2016, wind energy increases by about twofold, while solar PV increase over fourfold. This rate of increase is set to continue.

Table 1.3 A Breakdown of Energy Share of Global Electricity Production for 2016 [35]

Energy Types	Percentages (%)
Nonrenewable electricity	75.5
Hydro	16.6
Wind	4
Bio power	2
Solar PV	1.5
Ocean, CSP, geothermal	0.4

Table 1.4 Total Global Installed Renewable Power Capacity in 2016 [35]

Energy Types	Power/(GW)
Bioenergy	112
Solar PV energy	303
Wind energy	487
Hydroelectric	1243
Geothermal	13.5
CSP	4.8
Ocean	0.5

Table 1.5 Comparison of Wind and Solar Installed Capacity Over the Past 10 Years [35,37]

	Power/(GW)	
Years	Wind	Solar PV
2006	74	6.7
2011	238	70
2016	487	303

A Mckinsey report of November 2016 predicted that that by 2050, nonhydro renewables will account for more than a third of global power generation; a huge increase from the 2014 level of 6%. To put it another way, between now and 2050, wind and solar are expected to grow 4–5 times faster than every other source of power [38]. In 2008, Fthenakis, Mason, and Zweibel, published papers on the feasibility of solar with other renewables to provide 70% of the electricity of the United States by 2050 [39,40]. In 2011, a report by the International Energy Agency [41] predicted that just solar energy technologies, such as photovoltaics, solar hot water, and concentrated solar power could provide a third of the world's energy by 2060 if politicians commit to limiting global warming.

China is the leading country as far as installed solar PV capacity is concerned and is followed by Germany, Japan, and the United States. The figures for 2016 are 77, 42, 36, and 30 GW, respectively. The United Kingdom was in sixth position with a capacity of 11 GW; this is amazing when considering the United Kingdom's latitude and temperate climate (Table 1.6 [42]).

Table 1.6 Installed Solar PV Power Capacity by Country [42]

Countries	Solar Power/(%)
China	23
USA	14
Japan	14
Germany	13
Italy	6
Rest of world	30

Looking at the list in Table 1.6, one is struck by the fact that most solar installations have been in regions with relatively poor solar resources (Europe, Japan, and China) while the potential in high resource regions (Africa and Middle East) remains relatively untapped. Tax credits through Government policies have contributed to the development of the most mature solar markets (Europe and the United States). With solar power costs falling rapidly and with grid parity having been achieved in many countries, the solar PV industry is fast changing as emerging and developing nations enter the solar PV age. A start has been made in Dubai with the announcement on June 5, 2017 by the Dubai Electricity and Water Authority for four 200 MW installations at the newly created Mohammed bin Rashid Al Maktoum Solar Park. It will also include a 1 GW CSP tower for generating steam to produce electricity. The plan is to have a total solar PV power generating capacity of 5 GW by 2030, which is in line with Dubai's clean-energy strategy of generating 75% of its electricity needs from renewable energy by 2050 [43].

Earlier this year (February 2017), there was a report that Morocco was building a solar farm (Quarzazate Solar Complex) of 160 MW, increasing to 580 MW capacity with plans for solar developments of up to 15 GW in the future. The plan is that much of the electricity produced will be exported north to Europe and east to Mecca [44]. Other countries in the Middle East (Egypt and the United Arab Emirates) are now joining with Morocco through the Middle East Solar Industry Association (MESIA) with plans for their own new solar PV developments [45].

The rapid development of new and improved solar cell technologies (Chapters 9–12); the fallings cost of solar cells; and the surge in development of storing energy from renewable energy sources, will undoubtedly hasten the aforementioned predictions.

One new development that could have a major impact on the future growth within the solar PV industry is the building of floating solar farms (see the cover photograph of this book). One of the world's biggest floating solar farm is in China [46]; it is a 40 MW power plant with 160 000 panels resting on a lake. Aside from producing green energy, an added advantage to a floating solar farm is the reduction of evaporation from the lake's surface. It is of some interest that news broke of this new development on the day that Donald Trump considered pulled the United States out of the Paris Accord on climate change. A smaller floating solar farm (6.3 MW capacity) is currently being built on the Queen Elizabeth II reservoir near Heathrow airport [47]. While China remains the world's biggest emitter of CO_2, with two-third of its electricity still fuelled by coal, it is making great strides to wean itself off a fossil fuel dependency (Chapter 2).

1.9 Conclusions

In this chapter, the pros and cons of solar PV have been discussed together with its importance in the energy mix today. The development of solar PV over the past decade has been illustrated and the positive and encouraging picture of the present industry has been highlighted with references to recent developments.

Acknowledgment

My thanks to my coeditor, Vasilis Fthenakis for helpful suggestions.

References

[1] Tuckett RP: *Climate change: observed impacts on planet earth*. In Letcher TM, editor: 2nd ed., Oxford, 2016, Elsevier [chapter 24].

[2] Valderrama C, et al: *Storing energy: with special reference to renewable energy sources*. In Letcher TM, editor: Oxford, 2016, Elsevier [chapter 14].

[3] Cruickshank CA, Baldwin C: *Storing energy: with special reference to renewable energy sources*. In Letcher TM, editor: Oxford, 2016, Elsevier [chapter 15].

[4] Noël JA, et al: *Storing energy: with special reference to renewable energy sources*. In Letcher TM, editor: Oxford, 2016, Elsevier [chapter 13].

[5] Available from: https://www.theguardian.com/environment/2016/may/18/portugal-runs-for-four-days-straight-on-renewable-energy-alone, pp. 1. [Accessed February 8, 2018].

[6] Goodall C. The switch. London, United Kingdom: Profile Books; 2016.

[7] Morton O, A New Day Dawning? Silicon Valley Sunrise, Nature 2006; 443; 19–22. doi:10.1038/443019a.

[8] Energy Conversion by photosynthesis: Food and Agriculture Organization of the United Nation. Available from: http://www.fao.org/docrep/w7241e/w7241e06.htm.

[9] Available from: http://www.solar-facts.com/world-solar/world-insolation.php.

[10] United Nations Development Programme, World Energy Assessment, Bureau for Development Policy, One United Nations Plaza, New York, NY 10017.

[11] World Energy resources 2016. World Energy Council 2016 report, 62-64 Cornhill, London, EC3V 3NH, UK, p. 2–6.

[12] Available from: https://www.iea.org/newsroom/news/2014/september/how-solar-energy-could-be-the-largest-source-of-electricity-by-mid-century.html, pp. 1. [Accessed February 8, 2018].

[13] Available from: https://solar-aid.org/wp-content/uploads/2016/06/solar-cooking-pack.pdf, p. 1–10. [Accessed 8 February 2018].

[14] Chong L, Po-Chun H, Hongtao Y, et al. Rapid water disinfection using vertically aligned MoS_2 nanofilms and visible light. Nat Nanotechnol 2016;11:1098–1104. doi:10.1038/nnano.2016.138.

[15] Available from: http://www.wired.co.uk/article/france-solar-panel-roads, pp. 1. [Accessed February 8, 2018].

[16] Available from: www.hypersolar.com/technology.php, p. 1–3. [Accessed February 8, 2018].

[17] REN21 2016 report, REN201 Secretariat, c/o UNEP, 1 Rue Miollis Building VII, 75015 Paris, France, pp. 21, 70.

[18] Available from: https://www.ise.fraunhofer.de/content/dam/ise/de/documents/publications/studies/Photovoltaics-Report.pdf, pp. 9. [Accessed February 8, 2018].

[19] Available from: https://www.ecowatch.com/solar-cheaper-than-fossil-fuels-2167117599.html, p. 1–3. [Accessed February 8, 2018].

[20] Paudel S, Shrestha J N, Neto F J, Ferreira J A F, Adhikari M, Optimization of Hybrid PV/Wind Power System for Remote Telecom Station, IEEE International Conference on Power and Energy Systems (ICPS), Madras; 2011. p. 22–24.

[21] Khan SI, Sarkar MR, Islam MQ: Design and Analysis of low cost water pump for irrigation in Bangladesh, *Journal of mechanical Engineering* 43(2):98–102, 2013.

[22] Available from: http://www.greenpeace.org/international/Global/international/publications/climate/2016/Solar-Thermal-Electricity-Global-Outlook-2016.pdf, p. 18–114. [Accessed February 9, 2018].

[23] Available from: https://www.engadget.com/2017/03/08/the-worlds-largest-solar-farm-from-space/, p. 1–3. [Accessed February 9, 2018].

[24] Available from: https://www.ise.fraunhofer.de/content/dam/ise/de/documents/publications/studies/Photovoltaics-Report.pdf, pp. 7. [Accessed February 9, 2018].

[25] Letcher TM, editor. Climate change: observed impacts on planet earth. 2nd ed. Kidlington, Oxford, United Kingdom: Elsevier; 2016 [33 chapters].p. 608.

[26] Letcher TM, editor. Future Energy: Improved, Sustainable and Clean Options for our Planet, 2nd edition, London, United Kingdom, Elsevier, 2014, [31 chapters], 716 pages, ISBN:978-0-80994246; and Letcher TM editor. Wind Energy Engineering: A Handbook for Onshore and Offshore Wind Turbines.25 London Wall,EC2Y 5AS Elsevier (Academic Press), [26 chapters], 600 pp., ISBN 978-0-12-809451-8.

[27] Yergin D. The Prize. London, United Kingdom. Simon and Schuster Ltd; 1993, [36 chapters]. p. 887.

[28] Imperial College London Grantham Institute for Climate Change Briefing Paper No 1 June 2009. Solar energy for heat and electricity: the potential for mitigating climate change consumer retail prices.

[29] Jäger-Waldran A. Snapshot of photovoltaics—March 2017. Sustainability 2017;9:783. doi:10.3390/su9050783.

[30] PV Insights. Available from: http://pvinsights.com/.

[31] Letcher TM, editor. Storing energy: with special reference to renewable energy sources. New York, NY: Elsevier; 2016 [25 chapters]. p. 565.

[32] Fthenakis VM, Kim HC: Land use and electricity generation: a life cycle analysis, *Renew Sustain Energ Rev* 13:1465–1474, 2009.

[33] Morjaria M, Anichkov D, Chadliev V, Soni S: A grid-friendly plant the role of utility-scale photovoltaic plants in grid stability and reliability, *IEEE Power Energy Mag* 12:88–95, 2014.

[34] Available from: energyinnovation.org/2015/02/07/levelized-cost-of-energy/, p. 2–3. [Accessed February 9, 2018].

[35] Available from: http://www.ren21.net/wp-content/uploads/2016/06/GSR2016_Full-Report.pdf, pp. 32, Figure 3. [Accessed February 9, 2018].

[36] Available from: https://www.nei.org/Knowledge-Center/Nuclear-Statistics/World-Statistics/World-Nuclear-Generation-and-Capacity, pp. 1. [Accessed February 9, 2018].

[37] Available from: http://www.ren21.net/wp-content/uploads/2016/06/GSR_2016_Full_Report.pdf, pp. 62 figure 15; pp. 77 figure 23. [Accessed February 9, 2018].

[38] Available from: http://www.mckinsey.com/industries/oil-and-gas/our-insights/energy-2050-insights-from-the-ground-up, p. 1–6. [Accessed February 9, 2018].

[39] Zweibel K, Mason J, Fthenakis V: A solar grand plan, *Sci Am* 298(1):64–73, 2008.

[40] Fthenakis V, Mason J, Zweibel K: The technical, geographical and economic feasibility for solar energy to supply the energy needs of the United States, *Energ Policy* 37:387–399, 2009.

[41] Available from: https://www.iea.org/publications/freepublications/publication/Solar_Energy_Perspectives2011.pdf, pp. 21. [Accessed February 10, 2018].

[42] World Energy resources 2016 World Energy Council Report 2016. Available from: https://www.worldenergy.org/wp-content/uploads2017/03/WEResources_Solar_2016.pdf, pp. 4 figure 2. [Accessed February 10, 2018].

[43] Available from: https://www.thenational.ae/business/dubai-set-for-world-s-cheapest-night-time-solar-power-1.35494, p. 1–3. [Accessed February 10, 2018].

[44] Available from: https://www.theguardian.com/environment/2016/feb/04/morocco-to-switch-on-first-phase-of-worlds-largest-solar-plant: http://fortune.com/2016/11/05/moroccan-solar-plant-africa/, p. 1–3. [Accessed February 10, 2018].

[45] Available from: https://cleantechnica.com/2017/02/22/middle-east-north-africa-region-5-7-gigawatts-solar-pipeline/, p. 1–2. [Accessed February 10, 2018].

[46] Available from: https://www.smithsonianmag.com/smart-news/china-launches-largest-floating-solar-farm-180963587/, p. 1–2. [Accessed February 10, 2018].

[47] Available from: https://www.lightsource-re.com/uk/stories/qe2/, p. 1–2. [Accessed February 10, 2018].

PART 2

Solar Energy Resource and World Wide

2 Solar Power Development in China

Xiaoping He
CHINA CENTER FOR ENERGY ECONOMICS AND RESEARCH, THE SCHOOL OF ECONOMICS, XIAMEN UNIVERSITY, XIAMEN, CHINA
xphe@xmu.edu.cn

2.1 Introduction

China is one of the fortunate countries in the world blessed with abundant solar energy. Its annual horizontal solar irradiation is equivalent to 2.4×10^{12} t (2.4 trillion metric tonnes) of standard coal, which could correspond to the total electricity output by tens of thousands of the Three Gorges Hydropower Station [1]. In over two-thirds of China, the annual sunshine duration ranges between 2200 and 3300 h, while the solar radiation intensity varies between 5016 and 8400 MJ m^{-2}, equivalent to 170–285 kg of standard coal per square meter [2]. Early development of solar technologies in China was driven by space applications in the 1970s, and ground applications were limited to household small systems in remote area without access to grid electricity. In 2005, there were only six photovoltaic (PV) manufacturers, producing silicon cells, with a total capacity of 40.6 MW [3]. During the pre-2013 years, more than 90% of China's PV products were exported to Europe and North America. This was due to a lack of incentive policy and insufficient regulation in China to promote the deployment of PVs in the domestic market.

In 2013, the European Union and the United States both imposed antidumping (AD) and countervailing duties (CVDs) on imports of silicon PV modules and cells originating from China, seriously frustrating China's PV export market resulting in an overcapacity in the Chinese PV industry. To activate the PV market at home, the Chinese government progressively put in place a number of incentive policies and special investment schemes for the development of solar power. "Opinions on promotion of healthy development of PV industry," issued by the State Council in July 2013, specified for the first time the key polices of PV power, with regard to subsidy duration, electricity billing and settlement methods, and electricity feedback to grid. The feed-in-tariff (FiT) incentive policy for PV applications and various direct financial subsides (e.g., Golden-sun Demonstration Program) resulted in a growth of the domestic PV market (Fig. 2.1). At the end of 2015, China was the world's leader (ahead of Germany for the first time) in terms of solar installations.

China has now, by far, the world's largest PV industry, either in terms of PV manufacturing or application. The PV generation capacity increased from a small capacity of 0.26 GW in 2010 to 77.42 GW (including 10.32 GW distributed PV), currently accounting for 4.7% of China's total installed capacity (Fig. 2.1) and translating into 1.1% of the total electricity produced (Fig. 2.2). Yanchi PV Power Plant in Qinghai, with a total planned capacity of

A Comprehensive Guide to Solar Energy Systems. http://dx.doi.org/10.1016/B978-0-12-811479-7.00002-6

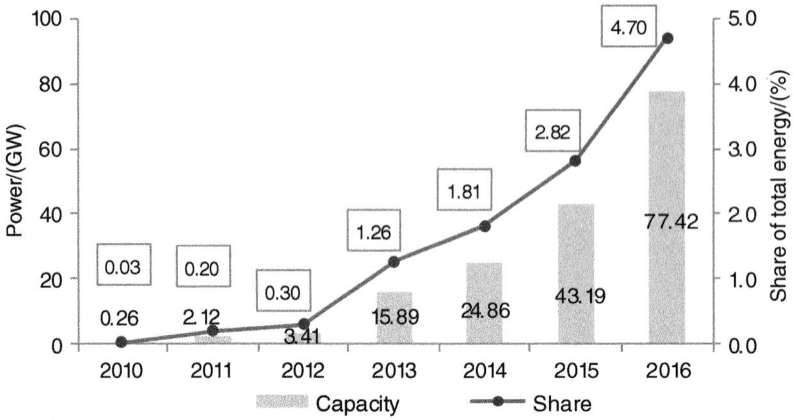

FIGURE 2.1 PV power installations and share in China total. *Author's compilation, on statistics by China Electricity Council (CEC): http://www.cec.org.cn/guihuayutongji/tongjxinxi/*

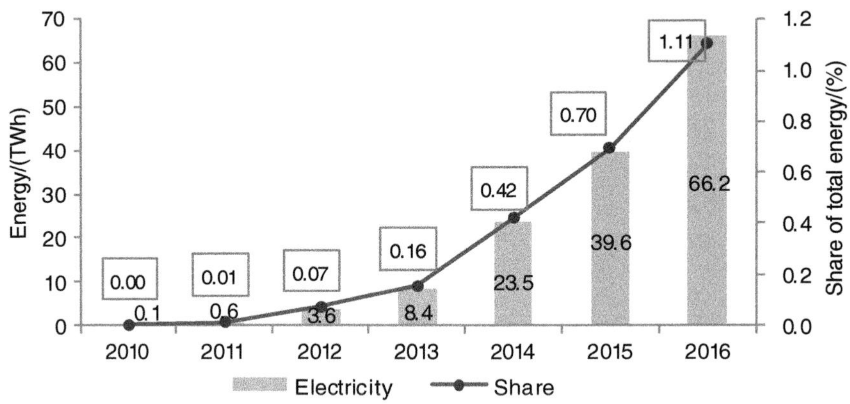

FIGURE 2.2 Solar power output in China. *Author's compilation, on statistics by CEC: http://www.cec.org.cn/guihuayutongji/tongjxinxi/*

2 GW, integrated into grid and put into operation in June 2016, is the largest PV power station in China and in the world.

The aim of this chapter is to provide a detailed description of solar power development in China, with regard to the present status, dynamics, as well as the policies of the Chinese Government. This chapter is organized as follows: Section 2.2 reviews the development of China's PV manufacturing industry over the last 5 years and the achievements; Section 2.3 discusses China's policy for solar power and the changes; Section 2.4 presents China's development plan on solar PV industry over the next 5 years; and Section 2.5 concludes.

2.2 Photovoltaic Manufacture

After a decade of development, a complete PV industrial chain has been established in China, composed of poly-Si materials, wafers, cells and modules, as well as various PV special equipment and application products (Fig. 2.3).

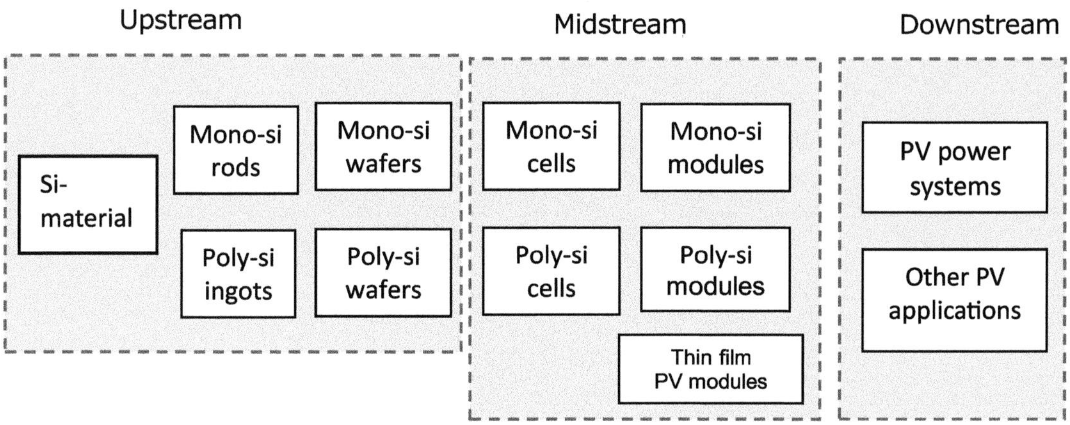

FIGURE 2.3 PV industry chain in China. *Drawn by the author, according to the development of China's PV industry.*

China's PV manufactures are concentrated in the east areas of the country. Production of the top 10 PV manufacturers accounts for more than 70% in its total PV product output, and the combined output of the top five accounts for more than half. In response to the ADs and CVDs following trade conflicts, major Chinese companies have established overseas production capacities in more than 20 countries, and the overseas capacity is now over 5 GW [4]. This has led, to a certain extent, to a diversification of Chinese PV production bases.

The number of the PV manufacturers in China, above a designated size, is estimated to be over 700, with a combined output value of RMB yuan 336 $\times$ 10^9 (336 billion) [4]. China has held the position of the world's largest producer of PV cells and modules since 2007. In terms of product, growth has concentrated on modules and wafers; and in terms of technology, passivated emitter rear cells (PERCs), black silicon, and N-type cell have advanced the most (Fig. 2.3).

2.2.1 Production

Despite the success of PV industry in China, the industry has long been troubled by a shortage of silicon feedstock, China accounts for over 50% of the world's total output of PV equipment. At present, the Chinese PV manufacturing is overproducing, and the average utilization rate was no more than 80% in 2016.

China has been the largest producer and consumer of polysilicon in the world for many years (Fig. 2.4). The top 10 solar-grade polysilicon producers together account for more than 90% of the total output in the country, and the top five account for over 70%. In 2016, the production of polysilicon reached to 194 000 t, accounting for 52.4% of global output. Approximately 40% of China's domestic polysilicon consumption for solar cells was imported, mainly from Germany, Korea, and Malaysia.

The output capacity of silicon wafer in China was estimated to be over 70 GW by the end of 2016, while production reached 63 GW. The capacity of various solar cells reached 76.8 GW [7], including 49 GW of crystalline silicon cells (41 GW were poly-Si cells). The poly-Si cell capacity of China accounted for approximately 66% of the global total (Fig. 2.5).

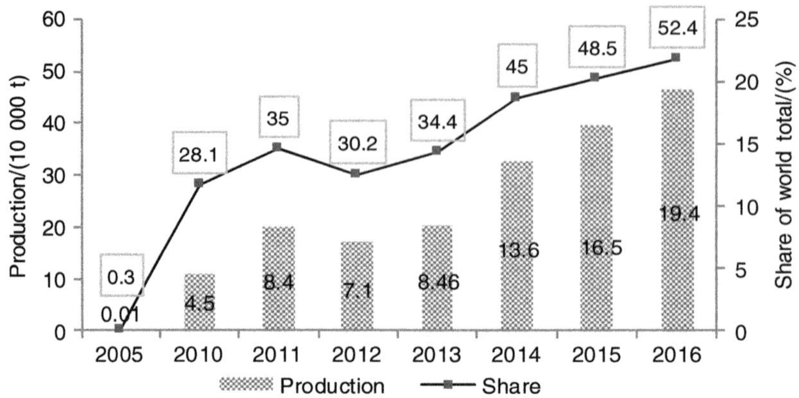

FIGURE 2.4 Polysilicon production in China. *Author's compilation, with data from MIIT. Brief report on photovoltaic industry in 2016. Available at: http://www.miit.gov.cn/n1146285/n1146352/n3054355/n3057643/n3057654/c5505791/content.html; 2017 [accessed March 2, 2017]; IEA. National survey report of PV power applications in China 2015. IEA PVPS. Available at: http://www.iea-pvps.org/index.php?id=93 [accessed June 9, 2016]; PHEI. Yearbook of China information industry 2015*(in Chinese). Beijing: PHEI; December 2015 *[4–6].*

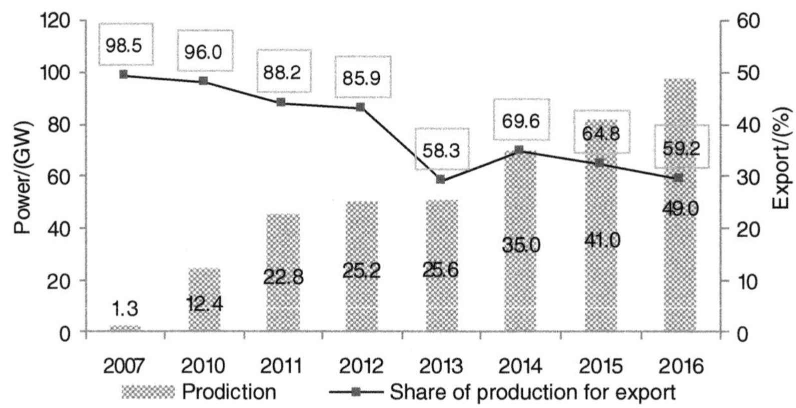

FIGURE 2.5 Production and export of poly-Si PV cells. *Author's compilation, with data from IEA. National survey report of PV power applications in China 2015. IEA PVPS. Available at: http://www.iea-pvps.org/index.php?id=93 [accessed June 9, 2016]; CPIA and CCID. The 2016 edition of the roadmap of China's PV industry development (in Chinese). China Photovoltaic Industry Association and CCID Group; 2017 [5,8].*

The capacity in China of high-efficient crystalline silicon cells, such as b-Si textured solar cell, PERC, and N-type double-sided cells, was estimated to be over 10 GW.

Crystalline silicon has long dominated the production of PV modules and cells in China, and thin film manufacturing activities remained relatively small in China. A small production of silicon-based thin film cells amounting to 420 MW was reported in 2015 [9]. Despite some production lines of a-Si cells having been established, no output in scale was reported. Other types of PV cells, in terms of material, such as polymer cell and nanocrystal cell remained at the research and development stage.

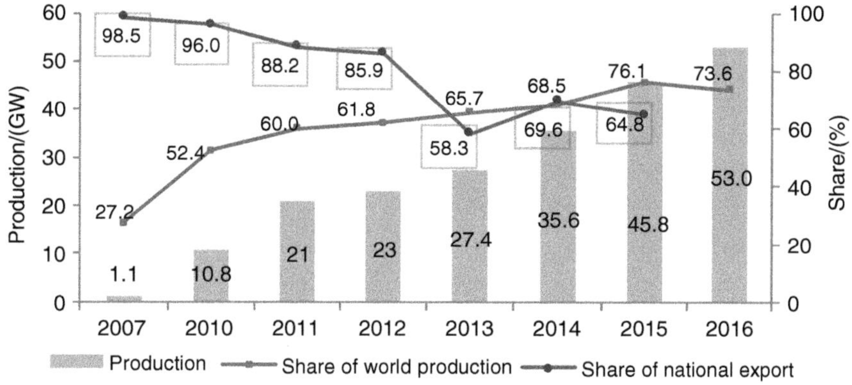

FIGURE 2.6 Production of Solar modules. *Author's compilation, with data from MIIT. Brief report on photovoltaic industry in 2016. Available at: http://www.miit.gov.cn/n1146285/n1146352/n3054355/n3057643/n3057654/c5505791/content.html; 2017 [accessed March 2, 2017]; Wang SC. China's PV policy review and outlook. Solar Energy 2016;6:19–26 [4,10].*

Production output of solar modules in China reached 53 GW in 2016, which implies an annual average growth of 20% in output over the period of 2011–16, and currently accounts for approximately 73.6% of the world total production (Fig. 2.6). The vast majority of module production has been silicon based, as can be inferred from the output volume of Si-based modules which amount to roughly 99% of China's total module production.

2.2.2 Photovoltaic Technology

After a few years of following technology advancements from other countries, China's PV industry is now able to keep pace with international advanced technology, and in some areas has gone beyond the level of its international competitors. The average efficiencies of China produced high-performance poly-Si and mono-Si cells and some certain types of thin film cells have now reached the international advanced level. Furthermore, due to the technological progress and the significant reduction in the cost of polysilicon material, the economics of PV power generation has improved significantly.

2.2.2.1 Technical Efficiency

Currently, China's PV industry has completely mastered the key production technologies of crystalline silicon cell. In 2016, the conversion efficiency of regular monocrystal Si PV cells has, on average, approached 20.5% and polycrystal 19.1% [4]. This is equivalent to the efficiency levels of PERC solar cells produced in China in the previous year. These values were 20.5% and 19%, respectively, for mono-Si and poly-Si PERC [8,11]. This implies that in China the efficiencies of regular cells on average have gone beyond the technical efficiency standards specified by the National Energy Administration (NEA). This is significant progress, when compared with the average efficiency of 17.5% and 16.5% [11], respectively, for mono-Si and poly-Si cells in 2010. The energy consumption for production of major

Table 2.1 Conversion Efficiency of Conventional Crystalline Silicon Cells/(%)

	2010	2011	2012	2013	2014	2015	2016
Poly-Si	16.3–16.6	16.6–16.8	16.8–17.3	17.3–18.4	17.6–18.3	18–18.4	18.3–19.2
Mono-Si	17.8–18	18–18.5	18.5–19	19–19.3	19.1–19.5	19.3–19.8	19.8–20.8
Si-based thin film	6	NA	10	NA	NA	NA	NA

Source: Author's compilation, with data from MIIT. Brief report on photovoltaic industry in 2016. Available at: http://www.miit.gov.cn/n1146285/n1146352/n3054355/n3057643/n3057654/c5505791/content.html; 2017 [accessed March 2, 2017]; CPIA. China PV industry review and outlook (report by Wang Bohua, the CPIA Secretary General). Available at: http://solar.ofweek.com/2017-02/ART-260009-8420-30104440.html; 2017 [accessed February 17, 2017]; MIIT. Brief report on photovoltaic industry in 2013. Ministry of Industry and Information Technology (MIIT). Available at: http://www.miit.gov.cn/n1146312/n1146904/n1648373/c3336480/content.html; 2014 [accessed April 23, 2014][4,12,13].

poly-Si cells dropped to 80 kW h kg^{-1}, the same as the international advanced level [4]. This recent progress has pushed China to the global frontier in cell technology (Table 2.1)

China's 13th Five-year Plan for Solar Energy Development set a target of conversion efficiency of 23% for advanced crystalline silicon cells produced on an industry scale by the year 2020. China has the resource advantage of developing CIGS thin film cells due to its abundance of indium and gallium. Hanergy, one of the leading manufactures of solar cells in China and globally, has acquired the intellectual property rights on CIGS and GaAs (gallium arsenide) technologies from other leading international cell players such as Solibro, MiaSolé, Global Solar Energy, and Alta Devices. The company presently produces CIGS and GaAs PVs in the laboratory with high efficiencies of 21% and 31.6%, respectively.

More than 85% of the solar cell market in China is taken up by back-surface field (BSF) PVs. This situation is expected to be maintained for many years despite BSF cells facing increased competition from other types of cells with new technology such as the PERC PVs which currently make up, roughly, 10% of the market in China.

"China Efficiency" of various types of solar inverters on average reached 98.3%. The China Efficiency is an average weighted efficiency value based on maximum power point tracking efficiency at different input voltage conditions and on the climate conditions of the location. The related technical standards were issued by the Ministry of Industry and Information Technology in March 2015. The primary types of inverters used for large utility-scale PV systems in China have been 1 MW, while 2 MW and even 3MW units are increasingly popular. The inverter equipment for distributed PV systems involves switching from 40 kW/50 kW to more advanced grades. Meanwhile, the share of microinverters in the market at home is increasing with more applications required for household PV systems and other civilian PV products. Of the total shipment of China's inverters, roughly 15% is exported.

2.2.2.2 Economics

As the industry scale continues to grow, PV power generation in China has become more economical than before, due to technological progress and reductions in the price of PV components. Investment costs of 10 000 t-grade poly-Si production line dropped to RMB

Table 2.2 Changes in PV Prices (RMB yuan W^{-1})

	2010	2011	2012	2013	2014	2015	2016
Poly-Si module	13	9	4.5	4	3.8	3.6	2.5[a]
Typical on-grid system	25	17.5	10	8.5	8.0	7.0	6.2[a]

RMB 1 yuan = EURs 0.136.
[a]For advanced enterprises.
Source: Author's compilation, with data from MIIT. Brief report on photovoltaic industry in 2016. Available at: http://www.miit.gov.cn/n1146285/n1146352/n3054355/n3057643/n3057654/c5505791/content.html; 2017 [accessed March 2, 2017]; Wang SC. China's PV policy review and outlook. Solar Energy 2016;6:19–26 [4,10].

yuan 150×10^6 per 1000 t and is expected drop further to below RMB yuan 100×10^6 per 1000 t within the next 5 years [8] (Table 2.2).

The electricity consumption for advanced manufacturers in China to produce polysilicon has dropped to <70 kW h kg^{-1} [4], and the production cost of polysilicon to \$10 kg^{-1}, by the end of 2016 [11]. Furthermore, the production cost of a PV module has dropped to below RMB yuan 2.5 W^{-1} for major manufacturers [4].

The financial investment per kilowatt and cost per kilowatt hour has also dropped significantly. Currently, the investment cost of PV power system is RMB yuan 7–8 W^{-1}. For concentrated PV power systems, the construction cost has decreased by 80% compared with 2010. For the most efficient investors, the construction cost has dropped to RMB yuan 6.2 W^{-1}. The generation cost ranges between RMB yuan 0.6 and 0.9 $(kW\,h)^{-1}$ [9]. Since 2010 China has achieved a decrease of 60% in terms of average PV generation cost. In the areas that are rich in solar resources, the PV generation cost has dropped down to RMB yuan 0.65 $(kW\,h)^{-1}$; this is approximately 70% higher than the national average of on-grid electricity prices.

Generally speaking, concentrated PVs in China produce a financial return rate of 10%–15%, and hence are attractive to investors. However, additional costs not directly related to the PV costs, such as land rents, taxes and fees, financing difficulty for utility-scale PV systems, and rents for roof-top PV sites, have shown an increasing trend. Assuming that these nontechnical costs can be excluded, investment cost per kilowatt is expected to further decline to RMB yuan 5 W^{-1} by 2020.

2.2.3 Photovolttaic Export

The export destination of PV equipment produced in China is shifting to emerging PV markets such as Asian and Latin American countries (Fig. 2.7). China's largest export market of PV products used to be Europe. In terms of value, Europe accounted for approximately 60% of China's export in 2011, followed by Asia and North America; these markets represented approximately 21.2% and 16.5%, respectively, of the overseas markets. Germany was once the largest buyer of the PV products originating from China.

Europe reductions in the subsides for solar power, combined with the AD and CVDs, have resulted in a slump in China's export to its traditional PV markets, such as Germany, Holland, Italy, and the United States. In 2016, exports to Europe and the United States fell to below 30% of China's total PV export. Meanwhile, exports to the emerging markets such as India, Korea, Turkey, Chile, and Pakistan have significantly improved. Japan is currently the largest overseas PV market, followed by India.

In pre-2009 years, the PV manufacturing in China was almost entirely an export industry, with over 90% of the modules and cells shipped overseas. With Chinese manufacturers establishing factories overseas, its export of silicon wafers is increasing, while exports of module and cell are decreasing. In terms of value, China's PV cell exports had held the upper hand for a long time prior to 2015; however, its dominance has been replaced by modules. The share of module exports with regard to total PV export value increased to 75%, from a value of 6% in 2011 (Table 2.3).

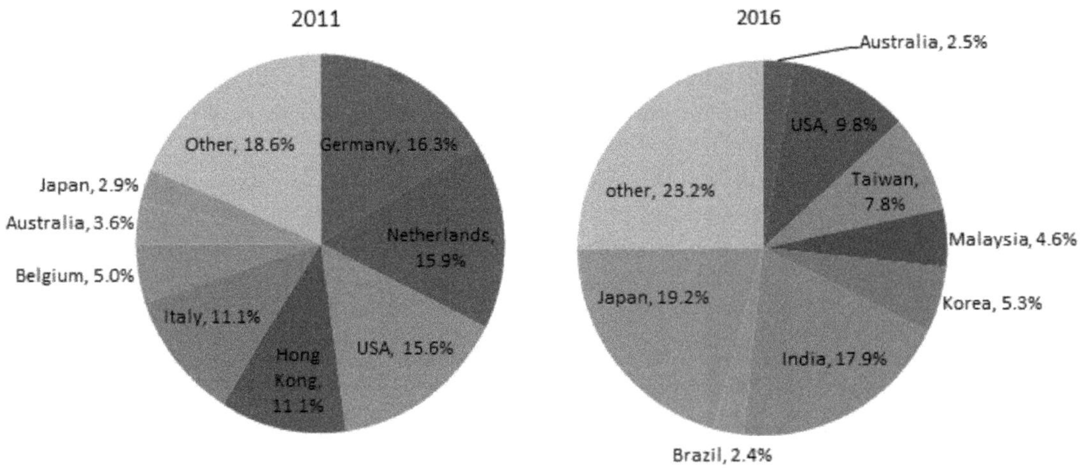

FIGURE 2.7 Changes in China's PV export destination. *Author's compilation, on statistics by General Administration of Customs (GAC).*

Table 2.3 China's PV Trade

	2011				2016			
	Volume		Value/($100 × 10⁶)		Volume		Value/($100 × 10⁶)	
	Import	Export	Import	Export	Import	Export	Import	Export
Poly-Si/(1000 t)	64.6	1.25	381.4	0.475	136		205	
Wafers/(10⁹ piece)						3.45[a]		26.90
Cells/(GW)				226.75		2.90		8.10
Modules/(GW)		15.30		21.50		21.30		105
Total/ ($100 × 10⁶)				358.21				140

[a]Here the volume of wafers is measured by number of pieces, not by physical capacity such as GW.
Source: Author's compilation, based on GAC statistics, and data from MIIT. Brief report on photovoltaic industry in 2016. Available at: http://www.miit.gov.cn/n1146285/n1146352/n3054355/n3057643/n3057654/c5505791/content.html; 2017 [accessed March 2, 2017] [4].

2.3 Industrial Policy

The development path for the Chinese PV industry has been clear over the last two 5-year plans. The priority over the period of 12th five-year plan was to establish a complete industrial chain and during the 13th five-year plan, the aim is to promote technological progress, cost reduction, and diversification of application. A number of polices have been put in place to incentivize the development of solar energy and are discussed in the following subsections.

2.3.1 Laws and Regulations

The Renewable Energy Act came into effect on January 1, 2006. It specified the implementation of FiT on renewable energy and full purchase of renewable energy, and the difference between the FiT and price of the electricity derived by conventional energy be apportioned among the customers with access to the grid. The Act of 2005 version placed an obligation on grid companies to purchase all the renewable energy added to the grid, regardless of the cost.

In the Renewable Energy Act Amendment issued in December 2009, the provision regarding full purchase of renewable energy was modified to "guaranteed purchase of renewable energy"; hence, providing the grid companies with the justification to decline the access of renewable energy when an electricity surplus occurs or when the price of conventional energy falls.

According to the amendment, the Renewable Energy Development Fund (REDF) was established and financed by special financial funds for renewable energy and revenue from the "Electricity Surcharge Due to Renewable Energy" (ESRE). The ESRE, levied on industrial and commercial users of electricity (approximately 80% of the national total consumption), has been exclusively used for the price subsidies of renewable energy and investment subsidies to projects that integrate renewable energy into the grid (including wind power, biomass power, and PV power). The special financial funds for renewable energy were used for the renewable power projects in remote unelectrified areas, and renewable energy pilot projects.

2.3.2 Government Funds Available for Solar Energy

The economic incentives for PV power have been obtained from price subsidies and direct financial subsidies of the initial investments. The investment subsidies were used for the Golden Sun Demonstration projects, and distributed PV projects such as Photovoltaic Architecture. Since 2014, the economic incentives for PV power have focused more on price rather than utility investment and favored more distributed PVs.

The ESRE, the primary revenue for the government to subsidize renewables, has been levied since August 2006. Being raised every 2 years, it has increased to RMB 1.9 cents (kW h)$^{-1}$ from the initial level of 0.1 cents (kW h)$^{-1}$. In the next few years, the annual revenue from the ESRE is expected to be about RMB yuan 70–90 $\times$ 10^9, based on the volume of industrial and commercial electricity consumption. However, the actual ESRE revenues in the past were far below the expected (Table 2.4).

As of the end of 2016, seven batches of solar and wind power projects, totaled 89.14 GW, were granted subsidies by the Ministry of Finance. At the same time, the installed capacity of solar and wind power totaled 226 GW, which implies only approximately 40% of these projects gained subsidies. Moreover, the subsidies to renewable energy were often in arrears by months. Obviously, there has been a huge imbalance between the revenue available and the funds demanded for renewable energy. Currently, the estimated gap between the needed subsidies and available funds for renewable energy is about RMB yuan 3050×10^9. The problem of funding deficit might become more intense with the rapidly growing size of renewable energy.

2.3.3 Price Policy for Photovoltaic Power

Pricing for PV power in China has gone through four regimes: pricing by specific negotiations, concession bidding, unified FiT, and region classified FiT (Table 2.5).

Specific negotiation pricing was applied in 2008–09. The negotiated price, confirmed by NEA for the first bench of on-grid PV stations, was as high as RMB yuan 4.0 $(\mathrm{kW\ h})^{-1}$. This price has been applied to Chongming Island PV Plant (1 MW) in Shanghai, Erdos PV Plant (205 kW) in Inner Mongolia, and Yang Bajing PV Plant (100 kW) in Tibet. The PV Concession Bidding program was introduced from 2009 to 2011, with a clear aim of reducing the generation cost. The prices of the bid-winning plants in 2010 averaged RMB yuan 0.847 $(\mathrm{kW\ h})^{-1}$. Next, in July 2011, a nationally unified feed-in benchmark price for PV power was introduced, initially fixed at RMB yuan 1.15 $(\mathrm{kW\ h})^{-1}$.

On January 1, 2014, the unified tariff was replaced by the region-classified FiT, according to the Notification on Promoting the Healthy Development of PV Industry by Price Leverage issued by the National Development and Reform Commission (NDRC) [14]. The notification clearly specified the benchmark prices for concentrated and distributed PV

Table 2.4 Final Accounts of ESRE Revenue and Expenditure

	2010	2011	2012	2013	2014	2015	2016
Expenditure/ (RMB $\times 10^9$)	0.55	0.58	14.61	28.23	44.84	57.96	59.51
Revenue/ (RMB $\times 10^9$)	NA	NA	19.62	29.8	49.14	51.49	64.78

RMB 1 yuan = EURs 0.136.
Source: Finance Yearbook of China (2011 and 2012); Ministry of Finance website: http://www.mof.gov.cn/zhengwuxinxi/caizhengshuju/index.htm.

Table 2.5 On-Grid Prices for PV Power Over 2008–16 (RMB yuan $(\mathrm{kW\ h})^{-1}$)

Negotiation	Bidding		Unified FiT		Region Classified FiT	
2008–09	**2009**	**2011**	**2011**	**2012–13**	**2014**	**2015–16**
4.00	0.73–1.09		1.15	1.00	0.90–1.00	0.80–0.98

The prices include tax breaks. RMB 1 yuan = EURs 0.136.
Source: Author's compilation.

power systems, respectively. The tariff for concentrated PV ranged in RMB yuan 0.90–1.0 $(kW\ h)^{-1}$ depending on the regional condition in solar resources. Meanwhile, distributed PV (e.g., rooftop PV) was allowed a pattern of "self-generation self-use and surplus feed in grid"; the tariff employed for electricity feed to grid was RMB yuan 0.42 $(kW\ h)^{-1}$ plus the local feed-in benchmark price for coal-fired electricity. The coal-fired electricity prices at the time were in a range of RMB yuan 0.35–0.45 $(kW\ h)^{-1}$, so the subsided price of distributed PV was RMB 0.7–0.8 yuan $(kW\ h)^{-1}$.

The FiT scheme for PV power, since its adoption in 2013, has strongly driven the development of the domestic PV market. Meanwhile, due to the low returns on investment, coupled with lack of transparency in electricity dispatching and other operations directly related to PV power (for instance, cash flows and capital transferring), the domestic distributed PV market was inactive. Here, distributed PV refers to electricity that is produced at or near the point where it is used. To turn the status around, in September 2014 NEA adjusted the policy for distributed PV systems, allowing the owners to choose between a self-consumption pattern and a pure feed-in pattern, with limited possibilities to switch the remuneration pattern during the system lifetime. The pure feed-in model applied to the feed-in benchmark price for utility-scale PV, while the other model applied to the subsided price for distributed PV.

The FiT scheme for PV has been entirely financed by the revenue from the ESRE. NDRC has successively twice lowered the feed-in price for concentrated utility-scale PV while keeping the subsidy of 0.42 yuan $(kW\ h)^{-1}$ for distributed PV unchanged. Currently, FiT for concentrated PV has dropped to a level of RMB yuan 0.65–0.85 $(kW\ h)^{-1}$ (Table 2.6). Hence, the prices for distributed PV and concentrated PV are roughly equal; moreover, the price differences between solar resource regions increased from the initial RMB yuan 0.05 $(kW\ h)^{-1}$ to 0.10 $(kW\ h)^{-1}$. These changes in FiT scheme were obviously more favorable for distributed PV, and were intend to curb the growth of traditional grid-connected PV systems in the Northern and Western regions and boost the development of distributed PV in other regions.

Table 2.6 2017 Updated FiT Schedule for PV Generation (RMB yuan $(kW\ h)^{-1}$)[a]

Concentrated and "All Feed-to-Grid" Distributed		Distributed[b]	
Solar Region	**FiT**	**Self-Consumed**	**Feed-to-Grid Surplus**
I	0.65	Retail price +0.42	On-grid coal-fired price +0.42
II	0.75		
III	0.85		

The prices include tax breaks. RMB 1 yuan = EURs 0.136.

[a]The schedule was published at the end of 2016 and put into effect on January 1, 2017.

[b]According to the stipulations by NEA and the State Grid, the distributed PV refers to (1) the power source that is connected to 35 kV-grade grid (or integrated to 10 kV-grade grid, with the on-grid capacity no larger than 6 MW), while the proportion of self-use electricity is more than 50%; (2) the power source that is connected to l0 kV-below grid, with the on-grid capacity no larger than 6 MW.

Source: Author's compilation based on NDRC. Notification on adjustment of PV power and onshore wind power feed-in benchmark price (no. 2729). NDRC; 2016 [15].

NDRC has stated the principle that TiT levels will be decreased over time. The likelihood of lowering the price subsidy for distributed PV in the future cannot be ruled out. Furthermore, being pegged to the feed-in benchmark prices for coal-fired power implies a downward trend embodied in the prices of distributed PV power because the benchmark prices of coal-fired electricity have witnessed reductions over the past few years.

2.3.4 Special Projects

A few schemes had been implemented in the pre-2015 years to motivate the development of utility-scale PV, rooftop PV in urban areas, and other types of distributed PV for microgrid and off-grid applications in the unelectrified areas. The schemes launched by the central government mainly included Golden Sun Demonstration Scheme and Photovoltaic Architecture schemes.

The Golden Sun Demonstration Scheme and Photovoltaic Architecture program were carried out over 2009–14, by providing subsidies for the initial investment of PV projects, using funds from the REDF. The Photovoltaic Architecture projects could obtain a subsidy of RMB yuan 15–20 W^{-1}. For Golden Sun Demonstration projects, the subsidy for on-grid PV systems accounted for approximately 50% in the total system investment (including the matching transmission and distribution project), and 70% for off-grid systems.

Besides, the Three-year Action Plan for Non-electrical Population was implemented over the period from 2013 to 2015, providing electricity for the last 2.73 million people who were living in areas not connected to the electricity grid, and serving 43.6% of these people by independent PV systems. The independent PV systems were invested by private investors, and the local government provided subsidies for the system maintenance and cell replacement.

2.4 Future Solar Energy in China

China's climate change mitigation targets are to achieve a share of at least 15% for nonfossil energy with regard to primary energy consumption by 2020, and 20% by the year 2030. This has pushed China to boost its development of renewable forms of energy.

2.4.1 Development Target

The 13th version of China's five-year energy plan sets targets of 335 GW installations (not including hydropower) for renewable power and 105 GW for PV power (concentrated 45 GW + distributed 60 GW) by the year 2020 (Table 2.7). The targeted PV capacity was expected to produce 124.5 TW h of energy (or 36.73×10^6 t coal equivalent). The proportion of solar PV power will be 5.3% with regard to total installed generation capacity by 2020, increased from a level of 2.8% in 2015, and to 1.8% from 0.7% in electricity output.

The target for solar energy requires China installing approximately 13 GW of PV systems every year over the period of 2015–20. This seems achievable by 2020 and may even be exceeded because the cumulated capacity of PV installations in China had already reached

Table 2.7 China's Target for Power Development in 2020

	2015		2020		
	Capacity/(GW)	**Electricity/(TW h)**	**Capacity/(GW)**	**Electricity/(TW h)**	**Coal equivalent/ (Mt)**
Total power	1530	5550	2000	6800–7200	
Nonhydrorenew-able	172.18	276	335	655	193.13
Solar power	43.19	39.6	110	144.5	42.63
Concentrated PV	37.12	39.2	60	124.5	36.73
Distributed PV	6.06		45		
Photothermal	0.014	0.4	5	20	5.90

Source: Author's compilation, based on NEA. Thirteenth five-year plan for solar energy development (no. 354). NEA; 2016; NDRC. Thirteenth five-year plan for renewable energy development (no. 2619). NDRC; 2016; NDRC and NEA. Thirteenth five-year plan for energy development (no. 2744). NDRC, National Energy Administration (NEA); 2016; NEA. PV statistics 2015. NEA. Available at: http://www.nea.gov.cn/2016-02/05/c_135076636.htm; 2016 [11,16–18].

77.42 GW by the end of 2016, an increment of 34.54 GW within the year. As NEA once publicly stated [19], the target for PV installations was derived according to China's structural objective with regard to primary energy (15% for nonfossil energy), and it should therefore be treated as the minimum rather than the binding value.

Despite the fact that the market is concentrated in the traditional grid connected system, the 13th Five-year Plan for Solar Energy Development proposes that the priority of solar PV in the future would be "application diversifying," by promoting applications of distributed PV (to establish 100 rooftop PV demonstration areas by the year 2020), and "PV+" (PV integration with other industries) such as PV for agricultural greenhouses, ad-hoc PV installations for fisheries, and hydro-PV hybrid plants.

With regard to price, NEA [11] has set a target for the price of PV electricity to decrease by more than 50%, compared with the 2015 level; moreover, on-grid electricity prices of PV systems should be roughly equivalent to the average retail price, and furthermore the generation cost of solar thermal power should decline to less than RMB yuan 0.8 $(\mathrm{kW\,h})^{-1}$.

2.4.2 Development Orientation

Dominance of crystalline silicon cell in the domestic PV market of China will be hard to change in the short term. Although poly-Si has captured over 80% of the domestic market, the demand for mono-Si is increasing. Price difference between mono-Si and poly-Si wafers is narrowing, due to improvement in monocrystalline-Si wafer quality and processing technology of PERC PVs. The generation cost of mono-Si PV power systems might be reduced so as to be competitive with poly-Si PVs in the near future, by taking advantage of the high conversion efficiency of mono-Si cell.

In terms of technology orientation, as suggested by NEA [11], the supportive policy intends to focus on PERC, n-mono, and other types of high-efficient silicon cells, industrialization of thin film cell, and R&D activities on key PV equipment. A large number of demonstration projects will be implemented to investigate the feasibility of new PV technology with regard to application in various regions, climate, and power network conditions. To ensure that advanced products have access to the market, the government-sponsored PV projects will give priority to advanced PV technologies and products.

With the share of PV electricity growing in the power system, the integration of large utility-scale PV generations to the grid has become an acute question. Currently, the solar energy utilized in China almost all come from solar PV, and 80% of the PV solar power are utility-scale systems. Here, distributed PV refers to a small PV system that is proximate to the user and the electricity is mainly used on site; utility scale PV refers to large solar stations that are integrated to grid. Despite that application of utility-scale PV is relatively more mature in China, in comparison to distributed PV, insufficient capability of the grid pushes the government to put priority on distributed PV. In view of this, in addition to the scarcity of land resources and high electricity load in Eastern and central China, PV installations would be more deployed outside the western regions in the next few years.

There were signs that PV deployment is shifting to the east-central region. According to NEA [20], 72% (24.8 GW) of the capacity increment during the year 2016 was deployed in the east. Furthermore, the capacity of distributed PV (10.32 GW) increased by 70% in 2016 when compared with the previous year.

2.4.3 Special Schemes

2.4.3.1 Photovoltaic Pioneer Program

In June 2015, to ensure a wider application of advanced PV technologies, the Chinese government issued the Opinions on Promoting Application of PV Products with Advanced Technologies and Industrial Upgrading [21], proposing the implementation of the PV Pioneer Program. The program is being carried out by constructing large PV demonstration bases and pilot engineering projects, and all the projects will be compulsorily put out to tender. The aim is to unify the deployment of large utility-scale PV stations as well as providing market support for advanced PV products and technologies.

For the PV demonstration bases and projects under this program, the adoption of the advanced PV products and technologies that have been chosen by the government is compulsory. The program will encourage the use of innovative products that lead to a high ratio of renewable forms of energy which can be integrated into the power system. A comparison of the technical standard of the general PV power projects and PV Pioneer Program power projects shows that the standard for the Pioneer Program products is noticeably higher (Table 2.8).

Of the solar PV projects approved by NEA for construction in 2016, the Pioneer Program projects, with a combined capacity of 5.5 GW, accounted for more than one third. NEA stipulated that the individual scale of PV power project should, in principle, be larger than 100 MW. It is possible that in future the NEA might launch an updated version of PV Pioneer Program, constructing 2–3 demonstration bases of at least 1 GW each.

Looking forward, it will be increasingly difficult to get the green light for the construction of PV systems if the project cannot be included in either of the national special schemes mentioned here and in the following section.

Table 2.8 Conversion Efficiency Standards by Module Type

	2015 PV Pioneer/(%)[a]	General PV/(%)
Poly-Si	>16.5	≥15.5
Mono-Si	>17	≥16
HCPV	>30	≥28
Silicon based	>12	≥8
CIGS	>13	≥11
CdTe	>13	≥11
Other thin film	>12	≥10

[a]Technical standards for the Pioneer Program projects approved by NEA in 2015.
Source: Author's compilation on NEA, et al. Opinions on promoting application of PV products with advanced technologies and industrial upgrading (no. 194). NEA, MIIT, Certification & Accreditation Administration; 2015 [21].

2.4.3.2 Other Programs

Other special supporting programs for PV include the Micro-Grid pilot program and the PV poverty alleviation program.

The Micro-Grid Demonstration program aims to establish 30–50 microgrid demonstration projects (including off-grid and on-grid projects) over the next 3–5 years These projects will be used to test the operation feasibility of the local power systems to integrate the generation, transmission, distribution, and storage of the PV energy systems, taking into account the high ratio of renewable power.

PV Poverty Alleviation Program was initially launched in October 2014, jointly by NEA and the Poverty Relieve Office of the State Council, providing financial support to PV projects in poor areas. The 13th version of five-year plan for solar energy expanded the size of the program to 15 GW (distributed PV 5 GW and ground-mounted system 10 GW) by the end of 2020 [22]. The program is expected to benefit 3 million poor households by raising their average annual income to a level of over 3000 yuan by 2020. By implementing the program, the capacity increment would account for 20% of total solar PV increment over the period of 2015–20.

2.5 Conclusions

Benefiting from FiT incentives and various direct financial subsides, China's solar energy has developed remarkably well within a few years. The PV industry currently dominates the solar energy development in China. Excluding hydropower, PV power is second only to the wind power in terms of renewable capacity. Domestic installed capacity of PV systems is now 2.3 times the nuclear power capacity and has approached 5% of the national total generation capacity. China not only has the world's largest PV manufacturing capacity but also the largest PV market. In that sense, one can expect that by the year 2020, a cumulative capacity in excess of 150 GW is possible.

Despite NEA making it clear that the priority of PV power in the future will be distributed PV, there are difficulties such as project location and financing issues, both of which could obstruct the commercial development of distributed PV in China. The rapid growth of renewable installations puts huge financial pressure on the Chinese government. For example, the subsidies to renewable energy are often in arrears and additional costs that are not directly related to the PV system are increasing. Although the reforming of the power market in China is progressing well, it is uncertain how more competitive PV industry might fare in the market place.

In terms of production, China needs to improve the availability of PV feedstock, improve the capacity of key technologies, improve the equipment on the domestic front, and promote the diversification in application patterns. The core issue facing China's solar energy industry in the future is to develop a mature commercial business in as short period as possible. Therefore, with the increasing capacity and the decreasing costs, the adjustments and the reduction of the subsidies to the industry, at the right time, are crucial.

References

[1] China low carbon yearbook 2010 (in Chinese). Beijing: China Financial & Economics Publishing House; 2010. p. 714-715.

[2] Shen WZ, editor. Solar PV technologies and applications (in Chinese).Shanghai: Shanghai Jiaotong University Press; 2013. p. 47.

[3] China electronic information industry statistics yearbook 2005 (in Chinese). Beijing: Publishing House of Electronics Industry (PHEI); 2005.

[4] MIIT. Brief report on photovoltaic industry in 2016 (in Chinese).Ministry of Industry & Information Technology of the PRC (MIIT).Available at: http://www.miit.gov.cn/n1146285/n1146352/n3054355/n3057643/n3057654/c5505791/content.html; 2017 [accessed March 2, 2017].

[5] IEA. National survey report of PV power applications in China 2015.IEA PVPS, International Energy Agency. Available at: http://www.iea-pvps.org/index.php?id=93 [accessed June 9, 2016].

[6] Yearbook of China information industry 2015 (in Chinese). Beijing: PHEI, China Industry and Information Publishing Group; December 2015.

[7] MIIT. Brief report on electronic and information manufacturing industry in 2016 (in Chinese).MIIT. Available at: http://www.miit.gov.cn/n1146312/n1146904/n1648373/c5507751/content.html; 2017 [March 3, 2017].

[8] CPIA and CCID. The 2016 edition of the roadmap of China's PV industry development (in Chinese). China Photovoltaic Industry Association (CPIA),China Center for Information Industry Development Group (CCID); February 2017.

[9] China electronic information industry statistics yearbook 2015 (in Chinese). Beijing: MIIT; June 2015.

[10] Wang SC. China's PV policy review and outlook. Solar Energy [J].2016;6:19-26.

[11] NEA. Thirteenth five-year plan for solar energy development (no. 354).National Energy Administration (NEA); 2016.

[12] CPIA. China PV industry review and outlook (reportby Wang Bohua, the CPIA SecretaryGeneral). Available at: http://solar.ofweek.com/2017-02/ART-260009-8420-30104440.html; 2017 [accessed February 17, 2017].

[13] MIIT. Brief report on photovoltaic industry in 2013 (in Chinese).MIIT. Available at: http://www.miit.gov.cn/n1146312/n1146904/n1648373/c3336480/content.html; 2014 [accessed April 23, 2014].

[14] NDRC. Promoting the healthy development of PV industry by price leverage (no. 1638).National Development and Reform Commission (NDRC); 2013.

[15] NDRC. Notification on adjustment of PV power and onshore wind power feed-in benchmark price (no. 2729). NDRC; 2016.

[16] NDRC. Thirteenth five-year plan for renewable energy development (no. 2619).NDRC; 2016.

[17] NDRC and NEA. Thirteenth five-year plan for energy development (no. 2744).NDRC, NEA; 2016.

[18] NEA. PV statistics 2015 (in Chinese).NEA. Available at: http://www.nea.gov.cn/2016-02/05/c_135076636.htm; 2016.

[19] NEA. NEA press conference on the 13th five-year plan for energy (in Chinese). Available at: http://www.gov.cn/xinwen/2017-01/05/content_5156795.htm#allContent; 2017.

[20] NEA. PV statistic information 2016 (in Chinese). Available at: http://www.nea.gov.cn/2017-02/04/c_136030860.htm; 2017.

[21] NEA, et al. Opinions on promoting application of PV products with advanced technologies and industrial upgrading (no. 194). NEA, MIIT, Certification & Accreditation Administration; 2015.

[22] PRO. Opinions on implementation of PV poverty alleviation work (no. 621). Poverty Relieve Office of the State Council; 2016.

3

Solar Power in Europe: Status and Outlook

Michael Schmela, Thomas Döring, Andrés Pinto-Bello Gómez, Alexandre Roesch

SOLARPOWER EUROPE, BRUSSELS, BELGIUM

m.schmela@solarpowereurope.org

3.1 The Past: Solar Developments in Europe (2000–16)

In spite of Europe leading global solar developments for many years, it is currently in a transition phase, trailing far behind the rapidly growing solar markets in Asia and in the United States. Last year was another disappointing year for solar in Europe. With only 6.7 GW of newly installed PV capacity in 2016, the European solar power market shrank by 22% year-on-year. This drop comes after a small increase in 2015 that followed several years of market contraction, which started in 2012. This contraction began as a result of the European solar pioneering countries slashing their previous lucrative feed-in tariff incentive programs.

Last year, Europe's installation volume dropped by 1.9 to 6.7 GW from 8.6 GW in 2015. The 2016 PV additions are basically level with the annual PV capacity that was added 7 years earlier in 2009, Europe installed 6.6 GW on its way up to its 22.5 GW climax in 2011. The European solar PV annual grid connections from 2000 to 2016 for selected countries are given in Fig. 3.1.

3.1.1 Leading European Solar Markets

The European leading solar market remained the same in 2016 as it was in 2015: the United Kingdom achieved that title for the 3rd year in a row. However, with only 1.97 GW grid connected in the United Kingdom, newly installed capacity decreased by 52% from the 4.1 GW, added the year before. The UK government's abandoning solar support is the main reason for Europe's demand drop in 2016. There was only one short spike in the United Kingdom in 2016 when 1.2 GW was grid-connected by March in response to the Renewable Obligation Certificate Scheme's termination for larger solar systems at the end of the first quarter; for the rest of the year, monthly PV additions remained mostly below 50 MW. A cut of the country's other solar support mechanism, a feed-in tariff (FIT) for small installations, had been announced a few days after the Climate Summit in Paris in December 2015. The UK's Solar Trade Association published a study in June 2016 stating that one out

A Comprehensive Guide to Solar Energy Systems. http://dx.doi.org/10.1016/B978-0-12-811479-7.00003-8

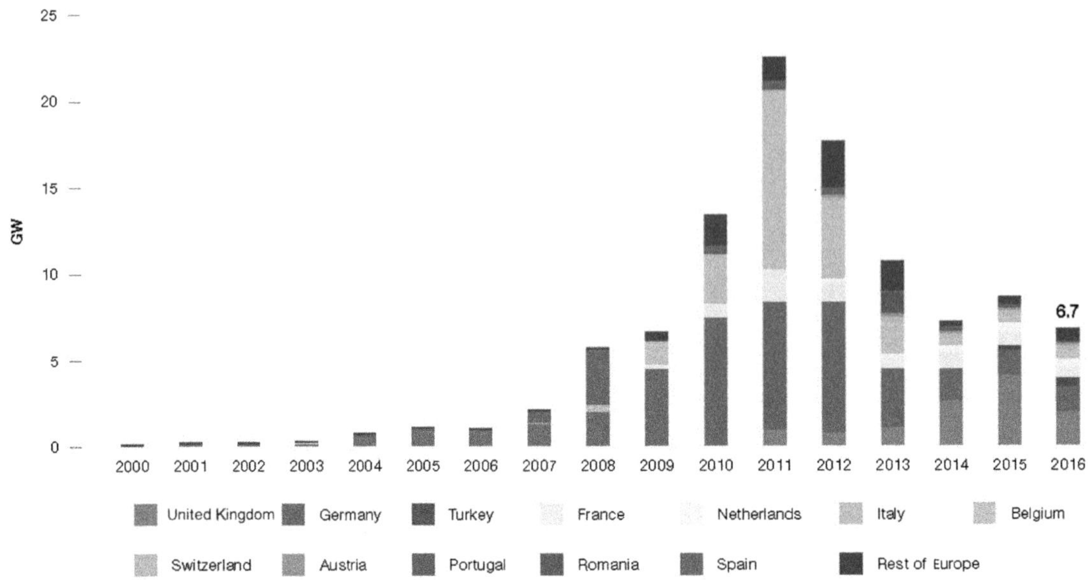

FIGURE 3.1 The European solar PV annual grid connections from 2000 to 2016 for selected countries. *Copyright of SolarPower Europe 2017.*

of every three solar jobs was lost over the past year. An impact assessment study by the country's Department of Energy and Climate Change found that the incentive cuts could wipe out up to 18 700 of the UK solar industry's 32 000 jobs.

The second largest European market was once again, Germany, adding almost exactly the same capacity as the year before: 1.42 GW in 2016, compared to 1.45 GW in 2015. For most of 2016, it seemed as if Germany would barely manage to reach the 1 GW mark, but a year-end rally, triggered by the termination of the FIT for ground-mounted systems up to 10 MW, resulted in over 400 MW of installations in December 2016. Still, in both years, 2016 and 2015, Germany clearly missed the government's solar installation target range of 2.4–2.5 GW. Apart from the United Kingdom and Germany, no other European country has installed anything like 1 GW of solar PV capacity in 2016.

Turkey, the new number three on the European solar map was able to increase PV installations by nearly 200% to 571 MW, from 191 MW in 2015, finally beginning to deliver what investors had been hoping for, for years. It is noteworthy that this growth rate took place despite the massive political turmoil, but the solar return promise is high and local financing is now available. Turkey had already passed a feed-in tariff law in 2010, but the 13.3 US cents $(\mathrm{KW\,h})^{-1}$ level was too low at the time. The first 600 MW tender round for "licensed" solar projects between 1 and 50 MW in 2013 was about 15 times oversubscribed, but it took until 2016, before the first two systems awarded in 2014 were actually built. The high license fee has been the main obstacle for these systems. Most of the PV installations in Turkey belong to the "unlicensed" systems category up to 1 MW, though it is possible to bundle these projects—the largest PV park in Turkey today, the Kayseri OSB power plant, started with 6 MW in 2014 and has been expanded to 51 MW in April 2017. By the end of September 2016, the Turkish authorities had received applications for an amount of 7 GW solar PV power of unlicensed projects; of this a total of 4.1 GW has been approved.

The solar additions for France were the most surprising disappointment in 2016. The country added only 559 MW, down 38% from 895 MW the year before. This strong drop should be considered as an exception. The market will pick up in 2017 as last June the government announced a detailed agenda to augment total installed solar capacity to around 20 GW by 2023 through regular tenders for building integrated PV (BIPV), rooftop PV and ground-mounted solar power plants. At the end of 2016, the cumulative PV capacity in France was 7.1 GW. Moreover, in early 2017, the French parliament passed a law that will facilitate investment in self-consumption systems, which had been hindered by levies and complicated regulatory frameworks. The first set of solar tenders have already been issued and awarded.

Apart from Turkey, there were other positive solar developments in Europe, though only the Netherlands installed a similar volume (500 MW) that was driven mostly by its net-metering scheme, with ground-mounted solar starting to play a big role through its SDE+ (an operating feed-in-tariff subsidy) program. In 2016, the Netherlands saw its first PV system above 10 MW being built—a 31 MW installation in the northern part of the country, while developments have begun for a 103 MW system. While PV additions in Italy increased by 23% in 2016, the absolute amount in the once world leading solar market stayed rather low (369 MW). A good sign is Belgium's new PV capacity (170 MW), which means a 70% growth year-on-year; this is based, to a large extent, on residential and commercial systems.

Strong growth signals were seen primarily in Eastern Europe. Poland doubled its new PV capacities in 2016, adding around 100 MW. The non-EU countries—Ukraine, Russia, and Belarus—went from almost no new solar in 2015 to between 50 and 80 MW each in 2016. The total European grid capacity for solar energy is given in Fig. 3.2.

Europe was the first region worldwide that reached the level of 100 GW of cumulative installed solar before the spring of 2016. This record was short lived as the Asia-Pacific countries ended the year with a total PV capacity of 147.2 GW, with China installing about one-third of Europe's cumulative solar capacity in a single year. The bulk of Europe's total installed solar power capacity is still carried by two countries: Germany (39.4%) and Italy (18.2%). Some distance behind trails the United Kingdom, where a short 3-year solar boom has resulted in an 11.1% share by the end of 2016. France remains fourth (6.8%), Spain keeps its fifth spot (5.3%) although it has added less than 600 MW over the last 5 years (Fig. 3.2).

3.1.2 Market Segmentation

Solar PV installations in Europe remain geographically scattered. Usually, emerging markets start their solar engagement with utility-scale solar plants, which are relatively easy to build, as there is no need for setting up sophisticated sales and installers networks or educating customers to quickly install large volumes.

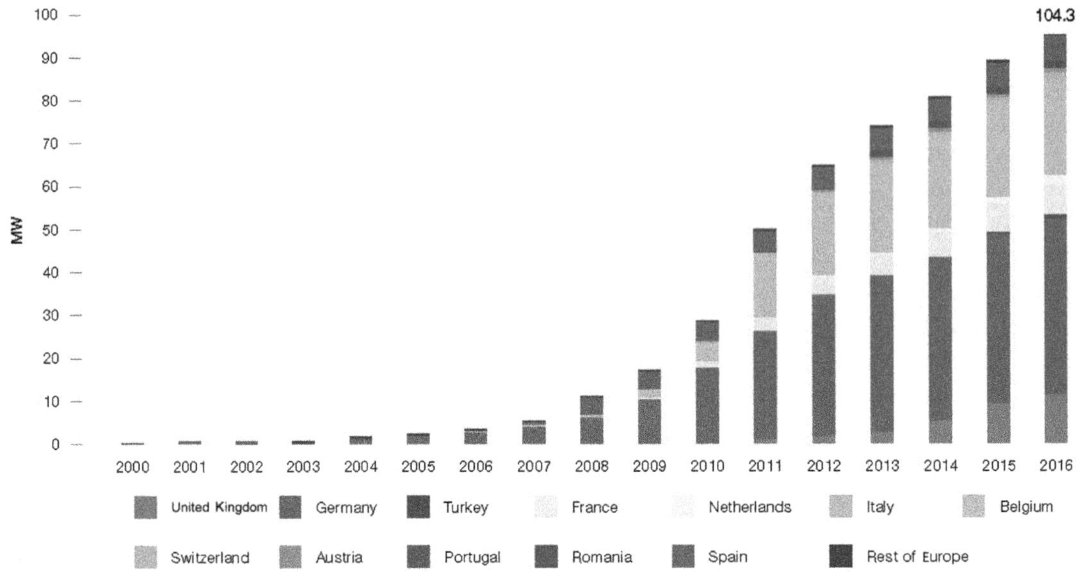

FIGURE 3.2 The total European grid connected capacity for solar PV power. *Copyright of SolarPower Europe 2017.*

This has also been true for several European markets. The rule of thumb is still valid that the less active a once prospering solar market is in Europe, the larger is the share of total installed ground-mounted solar utility plants. After its short feed-in tariff-driven solar boom periods based on utility-scale plants, Romania, Bulgaria, or Spain have never managed to build up notable markets for rooftop installations. Additionally, the latest European solar highflyer, the United Kingdom, grew primarily on subsidies for utility-scale systems, before the program has been terminated. The newest emerging solar markets on the continent—the Ukraine, Russia, and Belarus—have completely focus their attention on utility-scale installations, as has Turkey. Even the Netherlands, traditionally a fully-fledged rooftop market, has started to take advantage of low-cost ground-mounted systems, while Europe's largest solar market Germany tenders 600 MW ground-mounted systems between 750 kW and 10 MW per year.

More than two-thirds of solar systems in Europe are on the roofs of buildings: residential, commercial, and industrial in 2016. This dominance is expected to continue. The breakdown of the European solar PV capacity (residential, commercial, industrial, and utility) for the period up to 2016 for selected countries is given in Fig. 3.3.

3.2 The Future: 5-Year Market Outlook (2017–21)

It is predicted that after 2016, the several yearlong European solar PV market will come to an end. As of 2017, it is very likely that a new growth cycle will start for solar power in Europe.

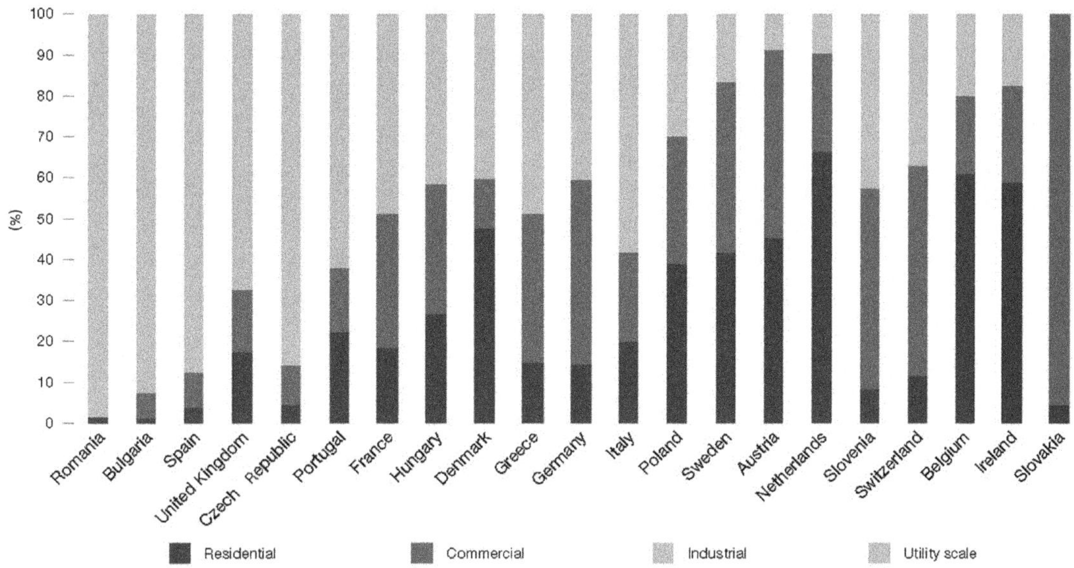

FIGURE 3.3 The breakdown of the European solar PV capacity (residential, commercial, industrial, and utility) for the period up to 2016 for selected countries. *Copyright of SolarPower Europe 2017.*

3.2.1 Main Reasons for Solar Market Growth in Europe

Despite a further decline of solar demand in the United Kingdom by more than half in 2017, the European PV market is forecasted to grow to 8.8 GW. The continent's solar sector association, SolarPower Europe, anticipates that other European countries will experience a market growth in solar PV in 2017, which will continue in the coming years.

There are several reasons for this solar growth phase in Europe:

- *Economic benefits of self-consumption*: Solar is already much cheaper than retail electricity in most European markets today and will continue to decrease in cost. This will be a key driver for people to invest in on-site power generation. At the same time, a stage has been reached in the European PV markets when consumers are increasingly starting to understand that solar PV energy often makes economic sense even without high feed-in tariff incentive programs. The quickly falling cost of battery energy storage together with the benefits of digital and smart energy products, supports the sales case for solar PV, as most consumers prefer to reach a higher "energy autonomy" status and fully control their energy bills.
- *Tenders*: Tenders or auction tools have fully disclosed the low cost of solar power and have been embraced by several European countries, substituting traditional uncapped feed-in tariff schemes. France has recently announced and already started a massive solar tender program for the next few years; Turkey has awarded a 1 GW tender and in Germany the 2015 started tender pilot was turned into a regular program and expanded in size. In Spain, solar PV was basically not awarded anything

in a first technology neutral renewables tender in 2017, but after the rules were changed, at the second 5 GW auction, solar outcompeted wind and won 3.9 GW.

- *EU 2020 targets*: A number of EU governments still have some way to go to meet their national binding EU 2000 renewable energy sources (RES) targets, are or will be, strengthening their support for solar, as they have realized that the technology is both very popular and low-cost and is a means to increase their renewables share.
- *Low solar cost triggering demand in new and dormant markets*: The low cost of solar PV is attracting European countries to embrace solar power that haven't been very active in field, such as Belarus and Russia. In other European markets, where the solar development stopped with the termination of early subsidy programs, direct bilateral power purchase agreements (PPAs) will compete with wholesale power markets. Finally, cheap solar is now increasingly able to outcompete other renewables in technology open support programs, like in the Netherlands, where in April 2017 solar PV accounted for 2.65 GW, or nearly 70% of the preassigned capacity of the first round of the SDE+ Program for large scale renewable energy projects.
- *Regulation*: The European Commission, as well as national governments, has begun addressing the needs of a flexible renewable energy system, working on a new electricity market design and implementing new tools and regulations to overcome barriers that have inhibited solar power's growth possibilities in recent years. These include rules for self-consumption in France or lacking guidelines for solar power sharing and on-site community solar power in Germany.

3.2.2 Solar Markets' Growth Scenarios

SolarPower Europe expects in its most probable medium scenario a strong 31% growth jump in 2017 that will somewhat flatten over the coming years, with annual installations to increase up to 15.7 GW in 2021. There are also models for high and low scenarios of solar developments in Europe. The results of the models shows very wide spreads for the coming 5 years—and over time it is increasingly widening. Why? Because the way solar will grow in Europe fully depends on policy makers in Brussels and the European countries.

If Europe fully embraces the enticing business case of low-cost solar, in 2020 the market could be nearly as big as in the record year of 2012, reaching 22.4 GW. This high scenario would require elimination of any trade barriers (as in Turkey or in the EU) and any taxes on self-consumption. If the European Commission's Clean Energy for All Europeans Proposal is quickly passed but with slight improvements, including a higher renewable energy target for 2030, as called for by SolarPower Europe, and adapted accordingly by the EU member states, these measures could even propel the European market to an annual level of over 27 GW in 2021. Conversely, if the bulk of European governments completely disregard solar's potential and the benefits to their citizens, and do not abandon coal quickly and continues to build new nuclear plants the low scenario will pay out. While on this topic it makes no economic sense to propose the building of expensive nuclear plants as has recently been in the United Kingdom, where a 35-year power supply contract was signed

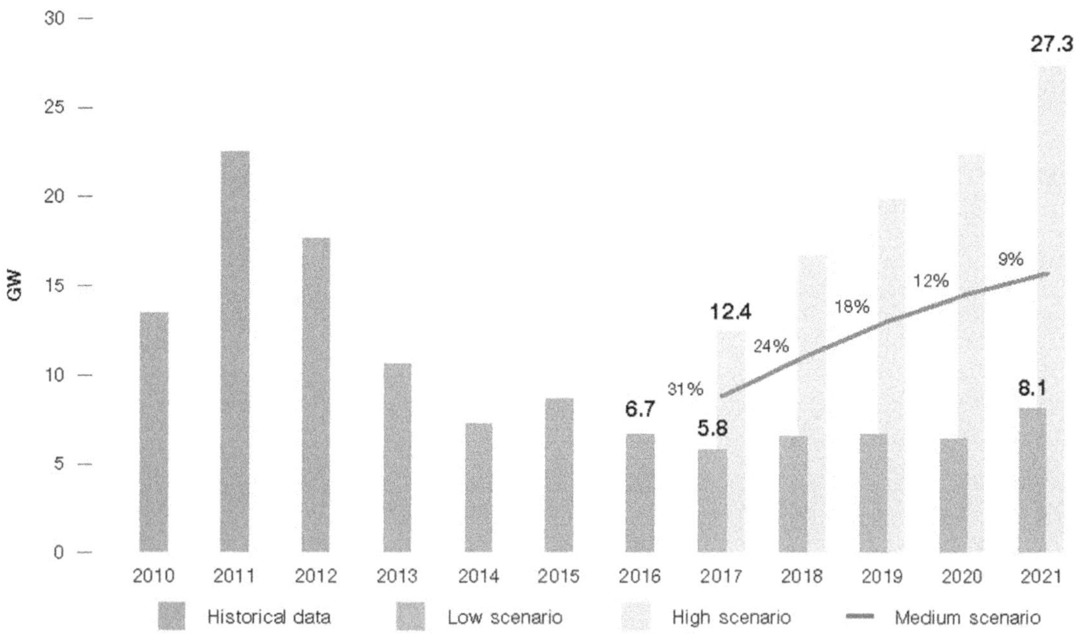

FIGURE 3.4 The European annual solar PV market scenario for 2017–21. *Copyright of SolarPower Europe 2017.*

based on the energy cost of £92.50 (MW h)$^{-1}$, which is around twice the level utility-scale solar costs in Central Europe.

The European annual solar PV market scenario for 2017–21 is given in Fig. 3.4.

If SolarPower Europe's high scenario comes to fruition, Europe's cumulative PV capacity could nearly double to 202.9 GW by the end of 2021 from the 104.3 GW installed by 2016. The medium scenario forecast is 167.2 GW. If the low scenario case plays out, Europe's solar power capacity would only grow by 33.6 to 137.9 GW—that's even less than the PV capacity added by China in the first 7 month of 2017 (35 GW). The European total solar PV market scenario for 2017–21 is given in Fig. 3.5.

3.2.3 European Countries' Solar Prospects

In Europe, the political support prospects for solar are not as bright as elsewhere in the world for the coming years. The solar "weather forecast" for European countries between 2017 and 2020 is still mostly rather cloudy but shows increasingly sunny areas and just one rainy spot. The United Kingdom is the only European country expected to add less new solar power year on year until 2019.

The top three markets to contribute the largest shares of new solar capacity until 2021 will be likely Germany, France, and Turkey. In Germany, an updated renewables law (EEG 2017) has been promulgated, which sets a stable regulatory framework and should provide the basis for a new growth phase, in particular, because utilities are increasingly engaging in engineering, procurement, and construction (EPC) and operations of solar power technology.

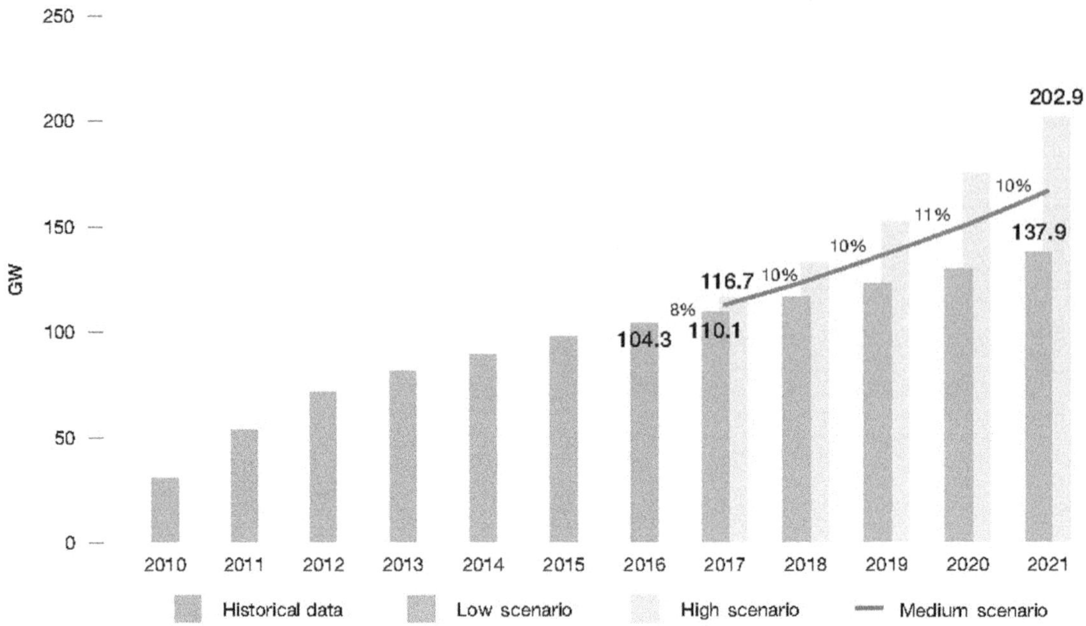

FIGURE 3.5 The European total solar PV market scenario for 2017–21. *Copyright of SolarPower Europe 2017.*

Regular solar tenders are now providing a steady "base" demand. France has published new solar targets with a concrete tendering schedule and passed a law to overcome regulatory obstacles for small solar installations. While the French market was disappointing in the first half year of 2017, adding only 264 MW, it is anticipated that the legislative improvements will show positive results as of 2018. Although the solar sector in Turkey seems to have mastered the limitations of the political turmoil so far, which has made it more difficult to access financing, its recent protectionist turn could make solar more expensive and negatively impact growth prospects in the country. Still, the Turkish solar business case is better than it is in the EU: the country's population and its need for power is quickly growing, there is plenty of space, excellent irradiation, and no power plant overcapacities.

The medium scenario anticipates the 15 largest European markets to install each a minimum of 1 GW until 2021, with Germany as the largest one adding 12.5 GW and France over 8 GW. In the first round in 2017 of the Netherlands SDE+ Program for large scale renewable energy projects, the relatively low-cost solar PV was very successful; it is predicted that the country will be one of the top four solar markets in the coming years.

A new European market in the list of top 15 is Russia; it is expected to add around 1.5 GW before the end of 2021. Another new European market is Ireland and though only installing 17 MW in 2016, is predicted to add 3.2 GW before the end of 2021. In total, SolarPower Europe anticipates Europe to add 62.9 GW (up from 52 GW) from 2017 to 2021, based on our most probable medium scenario.

The top European solar PV market prospects are given in Table 3.1.

Table 3.1 The Top European Solar PV Market Prospects

	2016 Total Capacity (MW)	2021 Total Capacity Medium Scenario by 2020 (MW)	2017–2021 New Capacity (MW)	2017–2021 Compound Annual Growth Rate (%)	Political Support Prospects
Germany	41 111	53 611	12 500	5	
France	7134	15 229	8095	16	
Turkey	820	7380	6560	55	
Netherlands	1911	7691	5980	32	
United Kingdom	11 547	15 822	4275	7	
Italy	18 983	22 525	3542	3	
Ireland	17	3233	3216	187	
Austria	1077	3377	2300	26	
Poland	182	2262	2080	66	
Switzerland	1681	3367	1686	15	
Russia	94	1559	1465	75	
Spain	5491	6771	1280	4	
Belgium	3423	4503	1080	6	
Greece	2611	3650	1039	7	
Sweden	182	1217	1035	46	
Rest of Europe	8060	14 289	6228	12	

In 2021, Germany is expected to be Europe's largest solar market, according to the medium scenario, followed by France, Turkey, the Netherlands, and Italy. The United Kingdom, which led the European solar sector from 2014 to 2016, is not expected to be among the top 5 European markets. The capacity additions and share of top European solar PV markets in 2016 and 2021 are given in Fig. 3.6.

3.3 Solar in the European Electricity System

3.3.1 Inflexible Energy Generation Needs to be Strongly Reduced Between Now and 2030

In 2016, solar produced enough electricity to cover nearly 4% of the 28 EU members (EU-28) electricity demand. This corresponds to the annual power consumption of the Netherlands alone or the demand for Ireland, Bulgaria, and Portugal combined. An increasing share of this electricity is produced locally by citizens and businesses, increasingly using solar

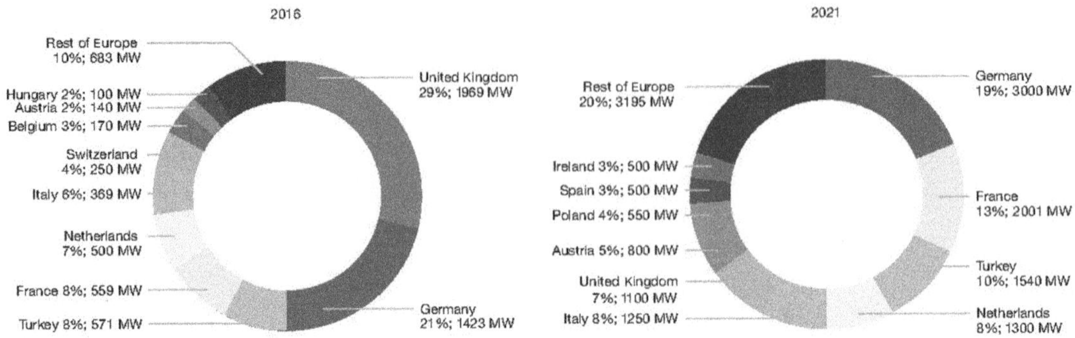

FIGURE 3.6 The capacity additions and share of top European solar PV markets in 2016 and 2021. *Copyright of SolarPower Europe 2017.*

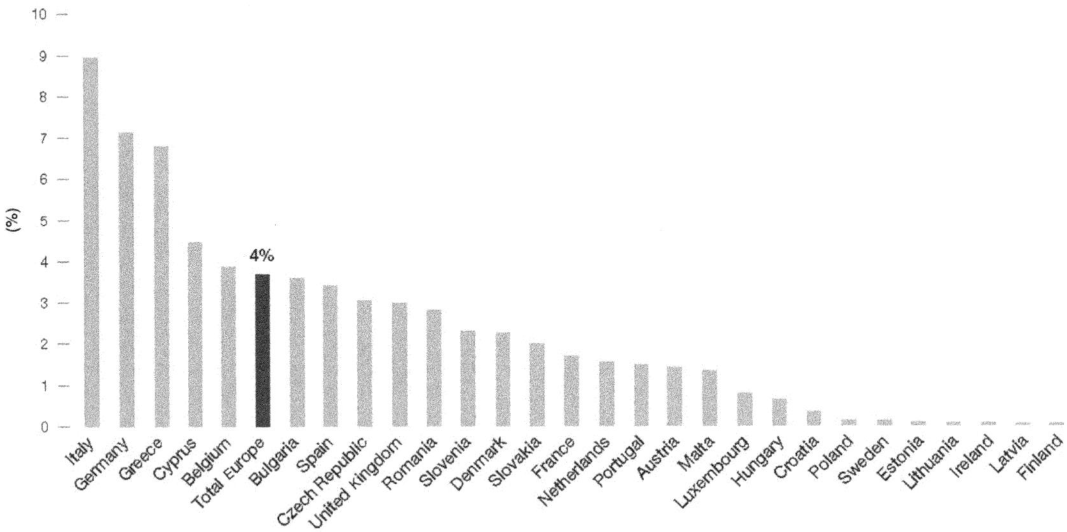

FIGURE 3.7 The share of electricity demand covered by solar PV in EU-28 in 2016. *Copyright of SolarPower Europe 2017.*

energy and storage. At the end of 2015, Germany had around 35 000 battery storage systems installed (>200 MW h) in combination with residential solar systems, an increase of around 25 000, or more than 70%, over the course of 2016. The share of electricity demand covered by solar PV in EU-28 in 2016 is given in Fig. 3.7.

Looking ahead, it is expected that the share of renewable electricity in the EU power mix will grow from 28.3% at the end of 2015 [1] to at least 46% by 2030 [2]. This would correspond to a renewable energy share in the final gross energy consumption of at least 27% in 2030.

The contribution of low-cost solar will depend on the future policy frameworks in the different EU member states. In any case, an energy market with a large share of renewables calls for a much a more flexible energy system: both on the demand and supply side. While the power output of wind, and to a lesser extent solar, has constantly increased since 2010,

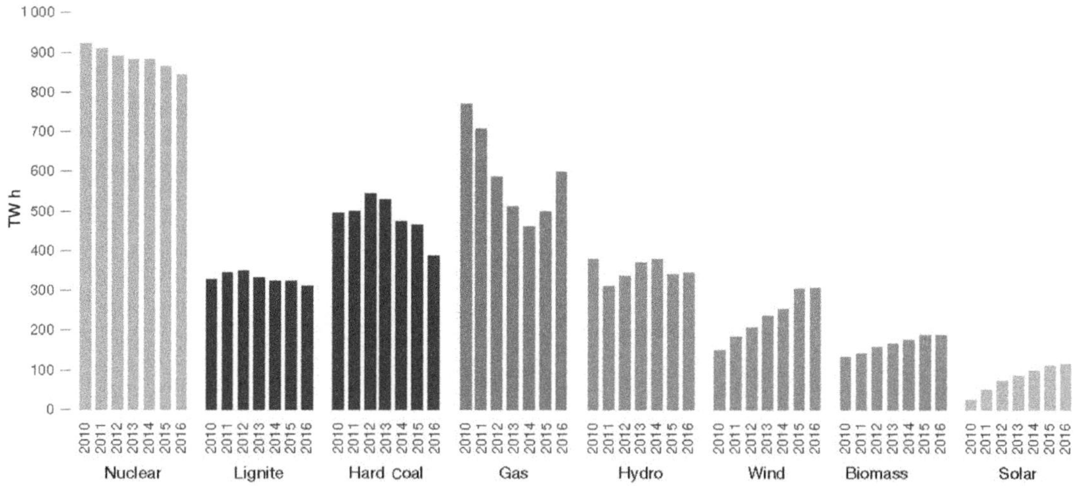

FIGURE 3.8 The development of EU-28 power output by different technologies from 2010 to 2016. *Agora Energiewende and Sandbag (2017); Copyright of SolarPower Europe 2017.*

it has not been mirrored by a corresponding reduction of inflexible generation in the European power mix. Hard coal and lignite still represented 22% of the total power generation in 2016, while nuclear produced 26% of the electricity in the same year. The development of EU-28 power output by different technologies from 2010 to 2016 is given in Fig. 3.8.

3.3.2 Accelerate the Energy Transition via Reliable and Ambitious Long-Term Signals

Once the European solar market passed the symbolic mark of 100 GW of installed capacity in the spring of 2016, solar become one of the lowest cost power generation technologies in Europe. In the German solar tender in February 2017, the lowest awarded bid was 6 euro cents $(\mathrm{KW\,h})^{-1}$. Assuming the same system prices and financing conditions but much better irradiation in southern countries, solar could generate power between 3 and 4 euro cents $(\mathrm{KW\,h})^{-1}$. The theoretical levelized cost of electricity (LCOE) in different European countries based on German Q2/2017 tender results is given in Fig. 3.9.

If Europe would build on its solar experience and utilize low cost solar, SolarPower Europe estimates that the solar capacity in Europe could easily be expanded to 375 GW by 2030 and cover up to 15% of Europe's power needs. Assuming SolarPower Europe's medium market growth scenario takes place, as expected up to 2021, it would require annual additions of 23 GW in each of the following years until 2030 to meet the aim of 375 GW. To provide the necessary signals to the investment and financing community long-term targets are needed.

The 27% renewable energy target by 2030 as suggested by the European Commission falls short of ambition, as it would translate into a fivefold market contraction for renewables between 2020 and 2030, as compared to the volumes expected to be deployed in the previous decade 2010–20.

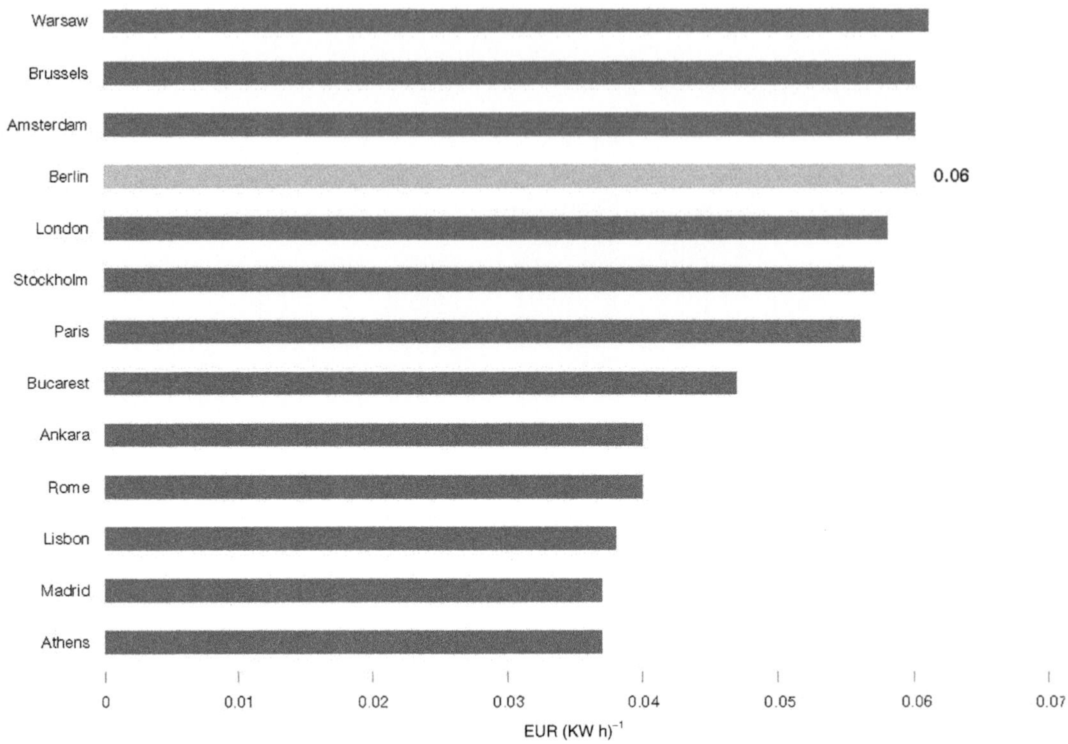

FIGURE 3.9 The theoretical levelized cost of electricity (LCOE) in different European countries based on German Q2/2017 tender results. *Copyright of SolarPower Europe 2017.*

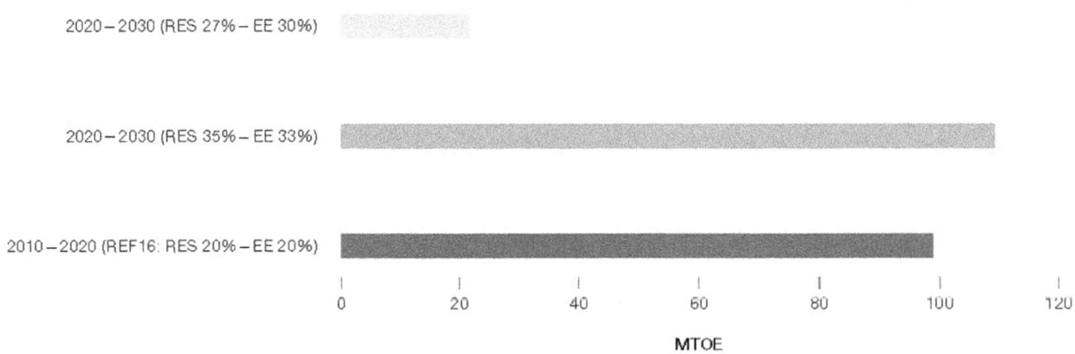

FIGURE 3.10 Renewable energy sources (RES) additions to the gross final energy consumption [millions of tonnes of oil equivalent (MTOE)] in EU-28 under different scenarios. *Data based on European Commission projection performed in 2013; Copyright of SolarPower Europe 2017.*

The renewable energy, RES, additions to the gross final energy consumption [millions of tons of oil equivalent (MTOE)] in EU-28 under different scenarios is given in Fig. 3.10.

The analysis of the various scenarios shows that only a RES target of "at least 35%" as asked for by SolarPower Europe would maintain the impetus for the renewable industry and be in line with the 2050 decarbonization objectives of the EU.

The projected RES penetration (aforementioned) and green house gas (GHG) emission reductions (later) in the EU-28 under different scenarios are given in Fig. 3.11.

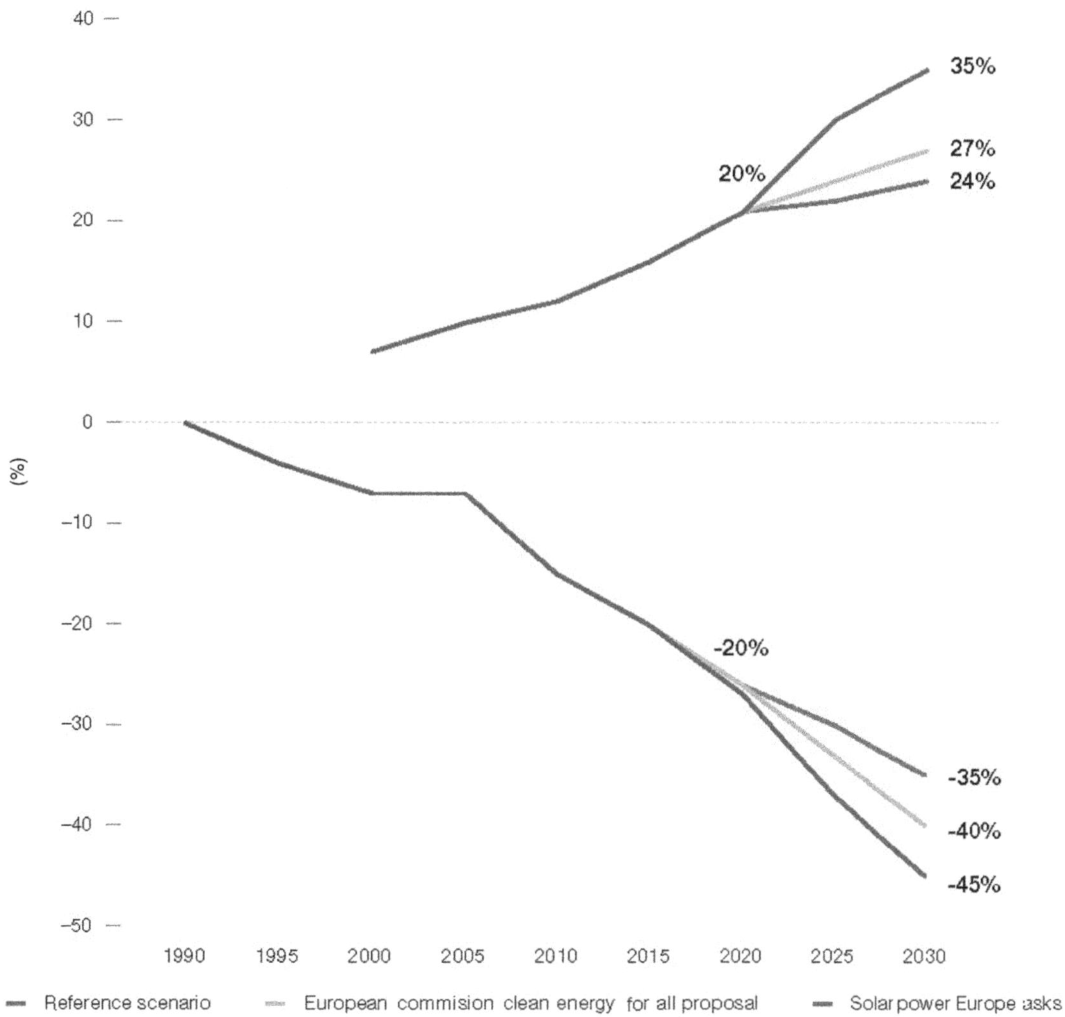

FIGURE 3.11 The projected RES penetration (above) and green house gas (GHG) emission reductions (below) in the EU-28 under different scenarios. *EC Reference Scenario 2016; Copyright of SolarPower Europe 2017.*

3.4 Policy Recommendation for Solar in Europe

The European Commission presented a set of new legislative proposals in November 2016 known as the "Clean Energy for All Europeans" package, which sets out the post-2020 framework for the power sector in the EU.

SolarPower Europe sees three strategic reforms that need to be pursued in parallel in order for solar PV to further penetrate the EU's energy mix:

- *A reliable governance framework to steer investments in renewable energy* is needed, and Europe needs to organize the orderly retreat from inflexible polluting plants. Considering the long lead-time of investments in the energy sector and the fact that literally all the investments compatible with the energy transition are capital intensive, attracting low-cost finance will be key. Besides, the need for an ambitious EU binding target (see previous section), clear national binding targets will signal to the investment community the direction to take. Such long-term visibility is crucial and needs to be made credible via the introduction of clear enforcement mechanisms in case the national 2020 targets are not met, or in case the trajectory toward the 2030 target is jeopardized by a lack of renewable deployment.
  While it is expected that around half of the power in Europe will be generated by renewable energy sources by 2030, with solar playing a key role, it will be important to organize an orderly retreat from inflexible and polluting generation capacities over the next decade to create the space in the electricity market for solar plants. In parallel and during the same period, flexible assets need to be ramped up. The so-called "Governance" Regulation presented by the European Commission in 2016 provides a unique opportunity to precisely steer the development of flexibility, and at the same time, organize the corresponding exit of inflexible generation. This can be achieved by the introduction of "national flexibility roadmaps," which will provide visibility on the evolution of the energy system at the national level.
- *Market rules are needed, which allow for a market-based energy transition* and enable a flexible system to harness renewable energies. Compared to just 5 years ago, there is now a broad recognition that market rules and products need to be redesigned to reflect the specific characteristics of variable solar and wind generation. More liquid intraday and balancing markets are crucial, alongside appropriate rules for the development of storage, demand response, and aggregation—three key enablers for further solar penetration.
  To ensure the energy transition will be market-based, we need a system that is able to make the best of variable renewable energies when they are abundantly available. This requires a cross-sectoral approach where electricity will play an increasing role in mobility and heating and cooling, but also a specific regime to reduce the occurrence of periods during which solar electricity is curtailed to let more polluting, less flexible plants run. The current debate around capacity mechanisms is a perfect occasion to consider the potential lock-in effect of subsidizing existing—or even worse—new generation capacities in high carbon emitting technologies or inflexible nuclear

assets, which will not be in line with the EU decarbonization objectives and might further distort market price signals.

- *A modernised framework for renewable energies is needed,* which supports the uptake of new business models and puts prosumers (because they both consume and produce electricity) at the core of the energy transition. Large-scale solar generation is increasingly being deployed via tendering mechanisms across the globe and also in Europe. The extensive experience gathered over the last years in tenders should be used now to enshrine high-level principles on the design of such mechanisms in the upcoming Renewable Energy Directive. The right for EU member states to run technology-specific tenders needs to be recognized.

  In the small-scale solar segments, self-consumption business models emerge in an increasing number of European countries. Self-generation and consumption is a very concrete lever for households and businesses, who want to control their energy costs. The proposal for a revised Renewable Energy Directive recognizes for the first time at the EU level the right to self-generate, store, and consume, either individually or collectively. All the models, which will make solar accessible to a larger number of citizens (e.g., joint purchasing, cooperatives, leasing) need to be promoted as well, notably by building on the mechanisms for collective self-consumption very recently adopted in France, Germany, and Austria.

  Long-term signals for ensuring a vibrant home market, adjusted market rules for unlocking new business models and an enabling framework for renewables: these are three prerequisites for ensuring a strong industrial basis for solar in Europe. Going forward, it will be important to ensure that any policy intervention will benefit the full solar European value chain, which is currently diversifying into new areas, such as storage, buildings, or digitalisation. An *industrial policy for solar in Europe* should be developed with the support of the European Commission, the European Parliament, and the Member States to capture, in a dynamic perspective, the future growth, and job potential.

3.5 Conclusions

As a solar pioneer, Europe has been experiencing the winds of sudden subsidy changes for years, the latest example was the United Kingdom slashing its solar incentive programs, which was the main reason for the continent's more than 20% market contraction in 2016. Now, the bottom seems to have been reached, and it looks like a new growth phase is beginning. Brussels and several EU member states are progressing in their efforts to address the challenges of the energy transition from a large centralized system to a distributed one based on a flexible energy market with a high penetration of renewables. In addition, there are many new markets in Europe and many other regions that finally expand into solar to profit from its attractive price offering.

What Europe and other early solar markets are experiencing is something that most emerging solar countries won't be spared: At a relatively early point in the development of a renewables-based system (solar in Germany produces on average only 7% of total annual electricity needs), daily demand peaks are mostly shaved, wholesale prices not only plummet but increasingly turn negative, and current transmission/distribution networks face challenges. If you want a relatively smoothly energy transition, it is key to think about intelligent new electricity market designs in a timely manner—and quickly put this into practice with a concerted effort.

A good example for a region that has reacted late to the challenges of the energy transition is the European Union. The EU is suffering from power generation overcapacities, as it has not been able to implement efficient measures for its member states to orderly retreat from dirty and inflexible coal. However, the recent "Clean Energy for All" proposal of the European Commission, though it needs some improvements, is a step in the right direction, offering an appropriate toolset to master many barriers toward a renewables-based economy.

In summary, SolarPower Europe sees three key topics for the EU to move to a clean and low-cost energy economy, which are similarly applicable for other countries:

- The EU needs a *reliable governance framework* to navigate the economy toward renewable energy with ambitious and binding targets for renewables. Subsidies for dirty and inflexible power technologies must be eliminated and phase out plans set up.
- *Electricity market design* needs to enable profitable investments and operation of variable renewable energy sources, taking into account rules for storage, demand response, and aggregation to provide new services. A cross-sectoral-approach for the power, heating, and transport sectors is required that will increasingly be based on renewables-generated electricity.
- *Modern renewable energy frameworks* are needed to enable new business models for solar and storage that put active consumers in the heart of the energy transition, allowing self-consumption without the burden of prohibitive taxes, or other barriers. While tenders are good mechanisms for efficient planning and deployment of utility-scale solar plants, their design is crucial to guarantee long-lasting, high-quality power generation.

If policy makers in Brussels and European countries would really take the characteristics for flexible solar and its renewable peers into consideration, as they strive for clean energy security of their economies, the energy transition could proceed much faster and at a much lower cost.

References

[1] Renewable Energy Progress Report, European Commission, Brussels (1.2.2017) 2017.

[2] Impact Assessment accompanying the draft revised Renewable Energy Directive. European Commission, Brussels.

4

Solar Power in the USA—Status and Outlook

Michael Ginsberg, Vasilis M. Fthenakis
CENTER FOR LIFE CYCLE ANALYSIS, COLUMBIA UNIVERSITY, NEW YORK, NY, UNITED STATES
vmf5@columbia.edu

4.1 Overall US Market Indicators

Presently, solar energy remains a minority contributor to the United States (US) energy production at 1.4% of the 2016 energy mix [1]. In recent years, the US energy market has shifted from coal and nuclear energy to increased reliance on natural gas in the ß US and solar and wind renewable energy in the Western US between 2009 and 2014. Natural gas based power generation increased by 204.6 TW h, a 22% increase, and non-hydrorenewable energy by 130.8 TWh, an 85 % increase (Table 4.1) [2].

US Energy Demand: US energy consumption in 2016 totaled 102.3 EJ (97.4 quadrillion Btu), a slight increase from 2015 (Fig. 4.1). The 9% decrease in coal use was more than offset by rising natural gas, petroleum, and renewable (mostly solar and wind) energy generation [3] (Fig. 4.2).

In the US, the world's second largest energy producer and consumer, the demand for electricity has historically been determined by heating, ventilation, and air conditioning (HVAC) needs, or heating and cooling degree days, and the state of the economy, or Gross Domestic Product (GDP).

The Decoupling of Economic Growth from Energy Use: The US National Oceanic and Atmospheric Administration (NOAA) showed the total number of cooling degree days (CDDs), which is tied to air conditioning and electricity usage, have increased over the past 100 years. CDDs for the summer of 2016 surpassed the 2011 record [3].

Historically, US energy and electricity use has been tied to economic growth. Over the past 25 years, however, GDP has risen by 80% while primary energy consumption rose by only 14%. Over the past 10 years, GDP rose by 12%, and energy consumption fell by 3.6%. This points to a new factor, energy efficiency, with some estimates that efficiency gains account for 60% of the energy intensity improvements since 1980 (Fig. 4.3) [4].

Utility decoupling, the separation of utility revenues from the amount of electricity generated and sold, has been key to this trend. In addition, energy efficiency regulations have incentivized utilities to implement energy efficiency measures among customers. In 2015, electric utilities spent over $6 billion on efficiency programs and natural gas utilities

A Comprehensive Guide to Solar Energy Systems. http://dx.doi.org/10.1016/B978-0-12-811479-7.00004-X

Table 4.1 Change in US Generation From Major Fuel Type Across Markets, 2009–2014 [2]

	Coal		Natural Gas		Nuclear		Nonhydro Renewable		Total	
	Absolute Change (TWh)	Percent Change	Absolute Change (TWh)	Percent Change	Absolute Change (TWh)	Percent Change	Absolute Change (TWh)	Percent Change	Absolute Change (TWh)	Percent Change
US	−171.3	−10	204.6	22	−1.7	0	130.8	85	132	3
WECC	−13.8	−6	−4.3	−2	−10.3	−15	43.4	92	11.9	2
SERC	−53.9	−11	94.8	51	3.8	1	12.7	52	49.8	5
RFC	−83	−15	65.1	85	12.1	5	17.5	102	13.5	1
NPCC	−17.4	−62	11.8	12	0.2	0	14.5	148	−6.4	−2
SPP	−0.8	−1	−5.7	−10	−0.2	−2	4	29	3.4	2
MRO	−9.6	−6	2.7	31	−3.9	−11	19.2	105	12.2	6
FRCC	−4.1	−7	30.6	29	−1.2	−4	0	−1	9.7	4
TRE	11.4	10	9.7	6	−2.2	−5	19.4	105	37.8	12
Alaska	−0.1	−11	−0.3	−8	0	0	0.2	1,484	−0.7	−10
Hawaii	0	1	0	0	0	0	0.5	74	−1.3	12

FRCC, Florida Reliability Coordinating Council; *MRO*, Midwest Reliability Organization; *NPCC*, Northeast Power Coordinating Council; *RFC*, ReliabilityFirst Corporation; *SERC*, SERC Reliability Corporation; *SPP*, Southwest Power Pool; *TRE*, Texas Reliability Entity; *TWh*, terawatt-hours; *WECC*, Western Electricity Coordinating Council.

In recent years, the electricity generation mix in the western United States has shifted from fossil fuels and nuclear power to nonhydro renewables. In the eastern part of the United States, generation has shifted primarily from coal to natural gas. Texas has seen a growth in generation from both coal and nonhydro renewables.

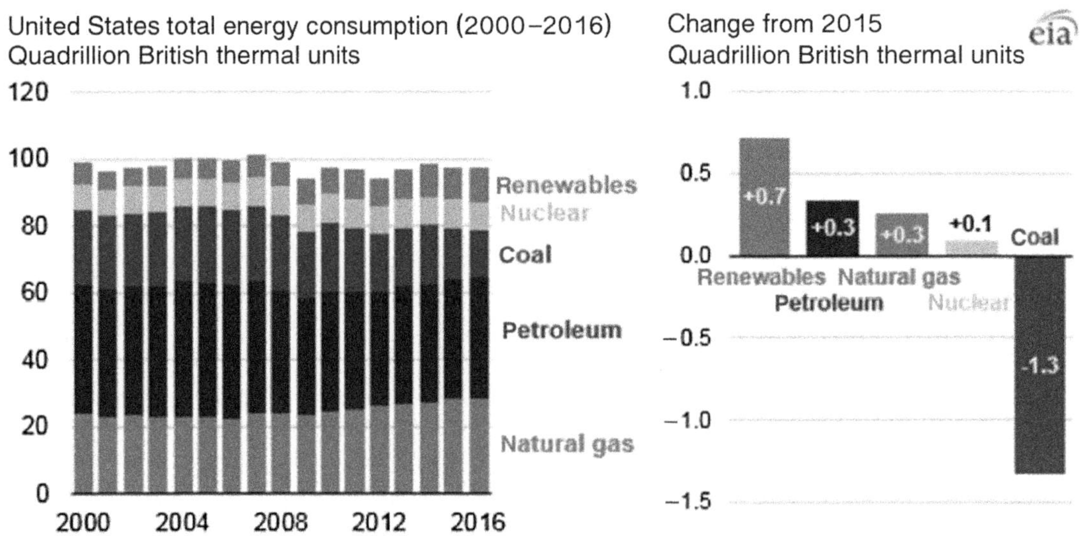

FIGURE 4.1 US total energy consumption (2000–2016) and change from 2015 to 2016 in Btu [3]. Here quadrillion refers to 10^{15} and 1 Btu = 1044.06 J.

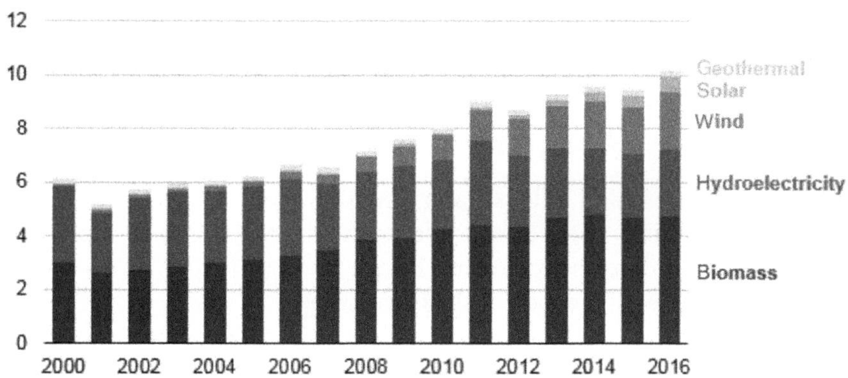

FIGURE 4.2 US renewable energy consumption (2000–2016) [3].

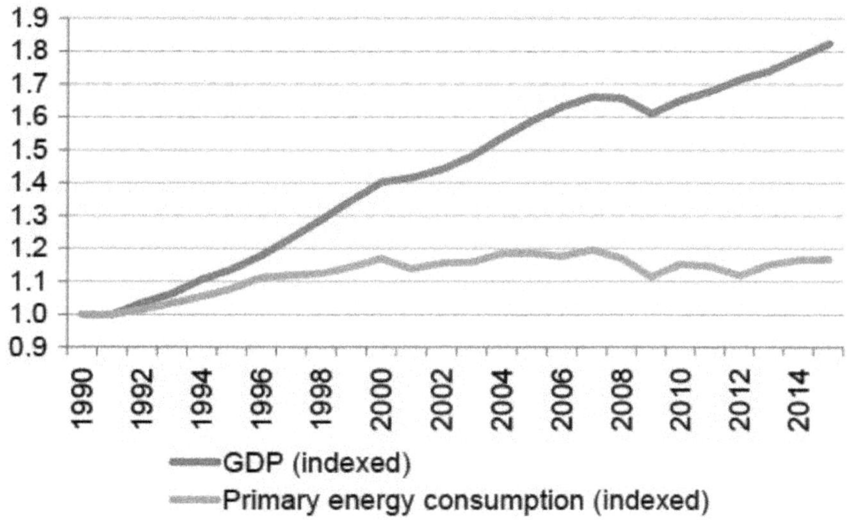

FIGURE 4.3 US GDP and primary energy consumption (1980–2014) [4].

$1.4 billion [5,4]. The decoupling is likely to persist as technology and smart control systems continue to improve energy efficiency.

Concentrating Solar Power (CSP) Indicators: Globally, the US and Spain have the greatest amount of installed CSP capacity. Growth in CSP has been largely due to the ITC, state renewable portfolio standards, and federal loan guarantees in conjunction with decreasing tower costs. From 2012 to 2016, CSP capacity in the US increased threefold to 1650 MW. Rapidly declining photovoltaic (PV) costs have negatively impacted the rate of CSP installation in the US but such displacement is somewhat mitigated by the operational and capacity benefits of CSP with thermal energy storage. This will be of particular importance during high renewable energy (RE) penetration, discussed in Section 4.3 of this chapter [6].

Solar PV Indicators: Cost declines in solar PV have enabled significant gains in the global and US solar markets. From \$75 W^{-1} in the 1970s to \$1 W^{-1} for an installed utility-scale system by the end of 2016, solar energy has achieved grid parity in many states, already nearing the Department of Energy (DOE) Sunshot Initiative's goal of \$.06 $(kW\ h)^{-1}$ (\$.06 per kilowatt-hour) by 2020 [7].

Residential median installed prices have declined from \$9 $(W_{DC})^{-1}$ in 2006 to \$2.75 $(W_{DC})^{-1}$ in 2016 (Figs. 4.4 and 4.5), a 69% decrease over 10 years or 6.9% per year. Commercial sector (over 500 kW) prices have declined from \$7.50 $(W_{DC})^{-1}$ in 2006 to \$1.60 $(W_{DC})^{-1}$ in 2016, representing a 79% decrease within 10 years [8].

The reductions in installed prices are driven by gains in the design and manufacturing of modules and balance of system (inverters, and module level power electronics, MPLE).

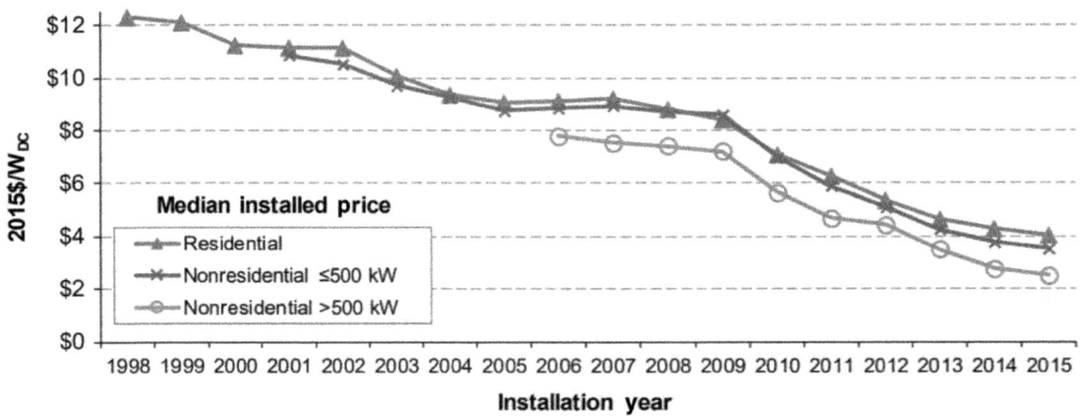

FIGURE 4.4 Median installed price of PV systems by sector from 1998 to 2015 [8].

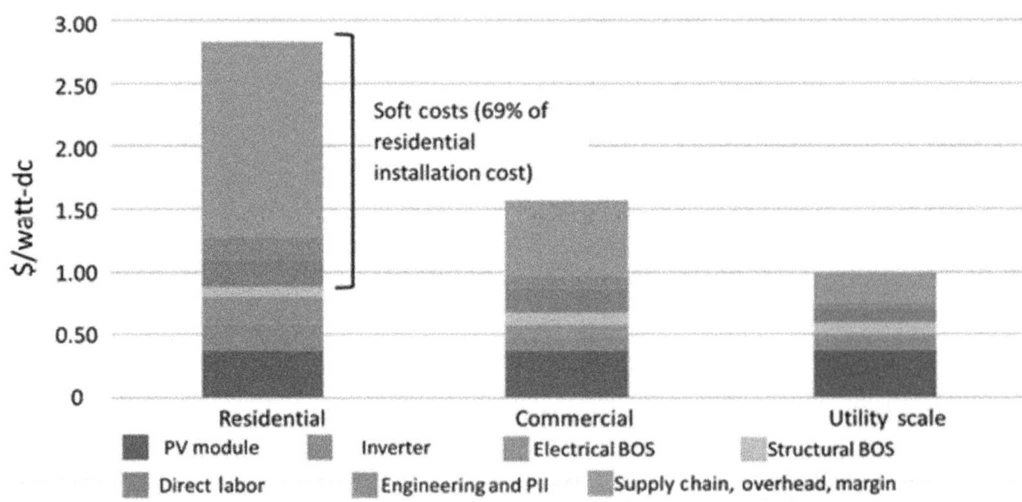

FIGURE 4.5 Q4 2016 quoted PV prices [9]. *GTMResearch and SEIA.*

These efficiencies enabled the installation of 14.8 GW_{DC} of PV in the US in 2016, a 97% increase over 2015, bringing cumulative installed capacity to 40.4 GW_{DC}, of which 2016 contributions were an astounding 37%. Significant private investment, $30 × 10^9$ ($30 billion) in deployed capital, was leveraged in 2016 to install over 1.25×10^6 (1.25 million) PV systems, making the US the second highest in annual installations behind China [8].

This spread is especially pronounced in the residential market with the largest sample size and where, in the 2015 residential market, 20% of installed prices were below $3.30 W^{-1} and 20% above $5.0 W^{-1}, with the remaining 60% in between (Figs. 4.6 and 4.7). This indicates differences among installers and regional markets, and reinforces the need to reduce soft costs.

Utility-scale: While residential PV is also increasing, utility-scale PV continues to dominate new capacity (Fig. 4.8).

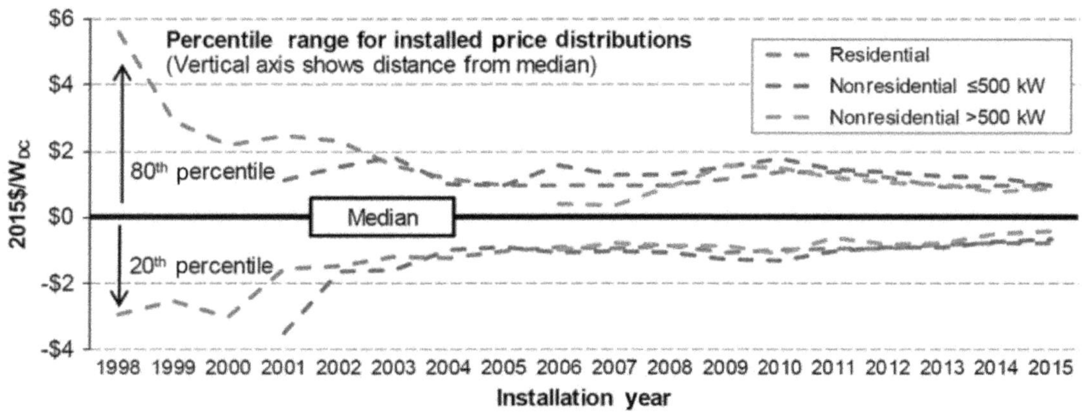

FIGURE 4.6 Installed price percentile ranges over time [8].

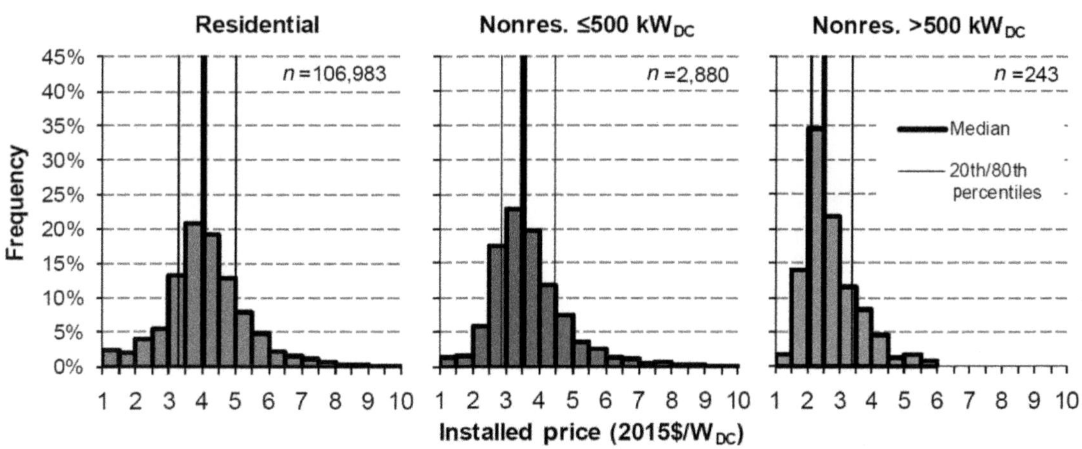

FIGURE 4.7 Installed price distributions for systems installed in 2015 [8].

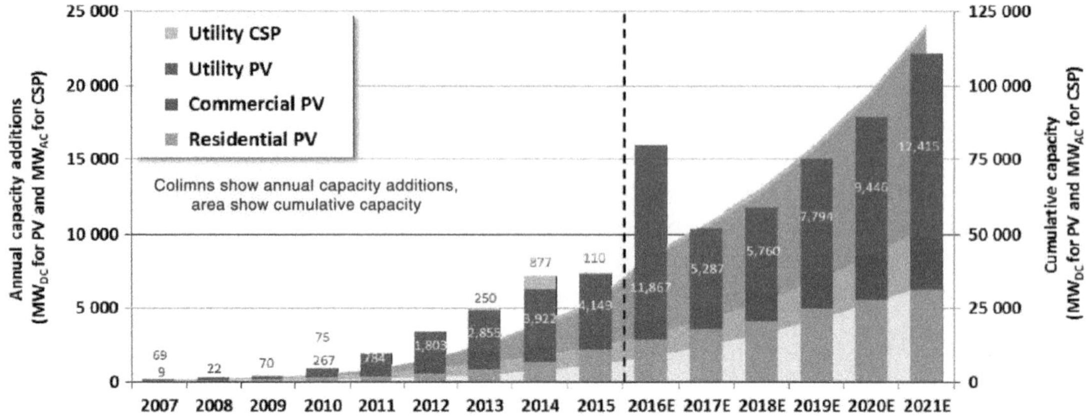

FIGURE 4.8 Historical and projected PV and CSP capacity by sector in the United States [10].

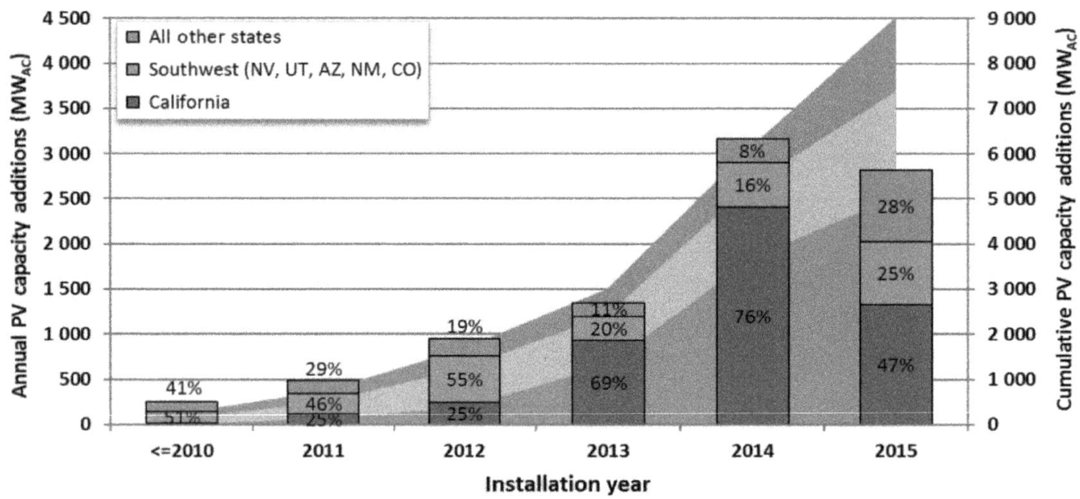

FIGURE 4.9 Annual and cumulative capacity additions in California, other Southwest states and other states [10].

Notable trends in the US utility-scale market include increased use of one-axis tracking, from 58% of installs in 2014 to 65% in 2015, and the growth of installations beyond the high-insolation zones of California and the rest of the Southwest, reflecting the impact of strong state policies despite moderate resources, such as in New Jersey, North Carolina, and Massachusetts. In 2015, 28% of newly added utility capacity came from areas outside of California and the Southwest (Figs. 4.9 and 4.10).

While California comprised nearly half of installed utility-scale PV capacity in 2016, large-scale solar is expanding to areas with less irradiance, such as North Carolina, Georgia, New Jersey, and Massachusetts (Fig. 4.11).

In addition, decreasing module costs have led to higher inverter loading ratios (ILRs) and more clipping (curtailment) losses. Despite the clipping losses of oversizing a DC array

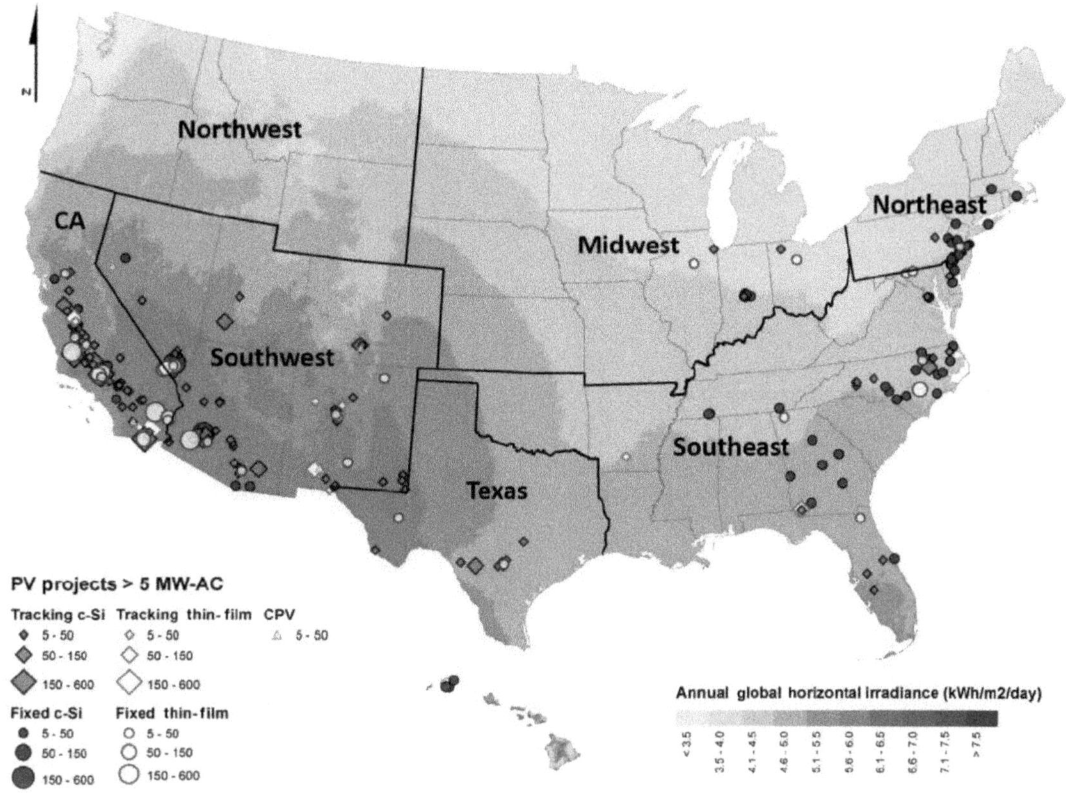

FIGURE 4.10 Map of global horizontal irradiance (GHI) and major utility-scale PV power plant locations [10].

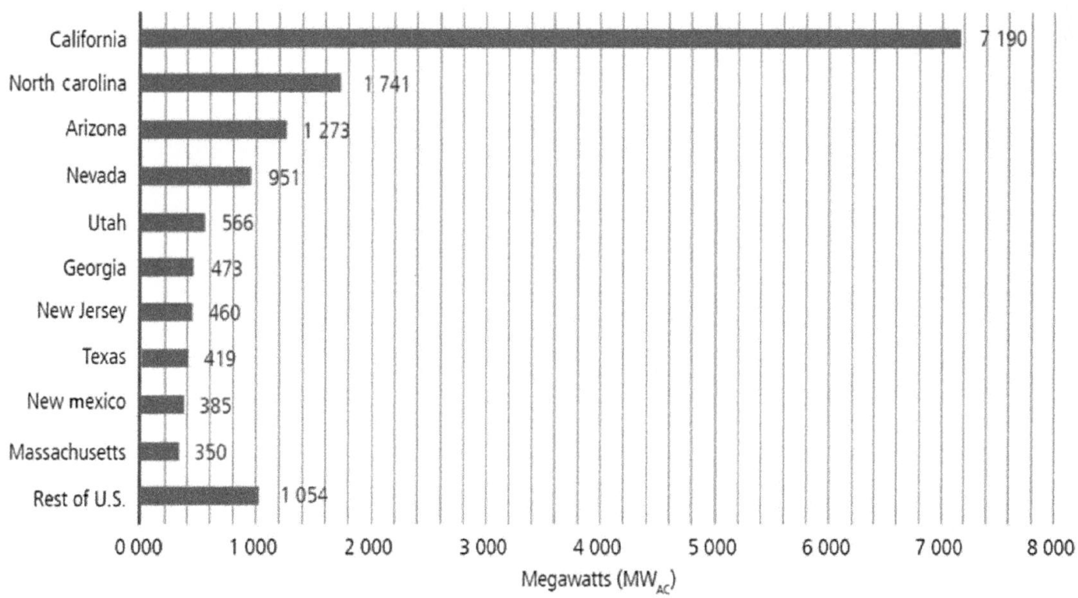

FIGURE 4.11 Utility-scale PV installed capacity, top 10 states, August 2016 [2].

relative to the AC capacity of the inverter during peak production, the use of more modules enables the inverter to operate closer to its peak rating more often, particularly during morning and afternoon hours when solar insolation is lowest. The increased revenue generated during the morning and afternoon hours has outweighed the clipping losses during the hours of greatest insolation, 9 AM to 3 PM, in summer months. This is increasingly valuable as grid operators require high ramp rates from other generators during these times. In 2015, average ILRs fixed-tilt and tracking PV increased to 1.31% from 1.2% in 2010 [10].

Another indication of the grid parity of solar is decreasing Power Purchase Agreement (PPA) prices due to falling hard costs, narrowing the gap between solar PPAs and existing gas-fired generators. Prices are down from \$100 $(\mathrm{MW\,h})^{-1}$ in 2011 to as low as \$30 $(\mathrm{MW\,h})^{-1}$ in 2015 [10].

In addition, PV may provide significant hedging benefits. While natural gas fuel cost projects are highly uncertain, the PV PPA prices are fixed. According to a recent Lawrence Berkeley National Laboratory study, 2017 natural gas PPAs were in the range of \$25 $(\mathrm{MW\,h})^{-1}$ to \$31 $(\mathrm{MW\,h})^{-1}$, and a wide spread is projected through 2040. As shown in Fig. 4.12, generation-weighted average PV prices are expected to decline gradually, converging with current (historically low) natural gas price projections around 2028 [10].

Despite that solar comprised only 1.4% of US electricity in 2016, it represented 3% of installed capacity and 40% of new added capacity. This is remarkable when viewed in a global context. Germany's Energiewende (Energy Transition), for instance, relied on significant government support, while federal policy in the US, aside from the 30% ITC, has been largely fragmented.

In a distinctly American fashion, growth in the US PV market has been largely driven by private investment. As solar comprises an increasing share of US electricity, the need

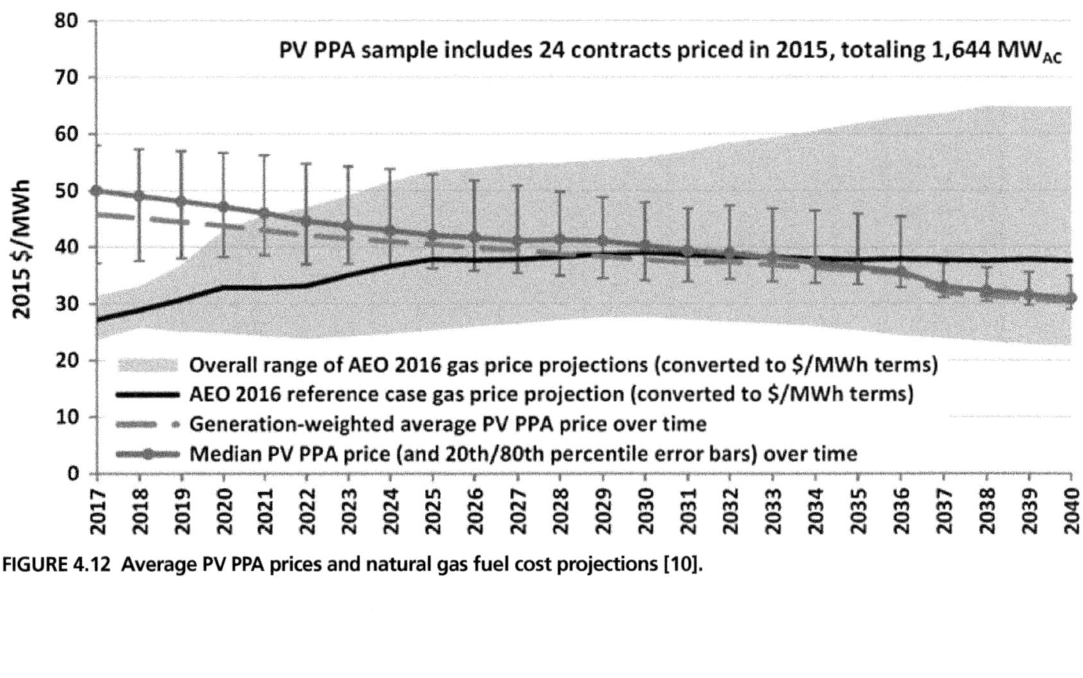

FIGURE 4.12 Average PV PPA prices and natural gas fuel cost projections [10].

arises for modernization of the grid and new models for energy valuation and transaction between producers and consumers, as discussed in Section 4.3 of this chapter.

4.1.1 Reducing Soft Costs

Basic research and improvements in the manufacturing process have greatly accelerated PV adoption. In the rapid decline of material costs, the US PV industry has encountered an arguably more complex obstacle—disorganization and fragmentation. Q4 2016 quoted PV prices showed that soft costs (direct labor, engineering and PII, supply chain, overhead, and margin) comprised over half of installed prices in both the residential and commercial sectors, and an astounding 69% of quoted residential prices (see Figure 5 in reference) [9].

As shown in Fig. 4.13, material prices declined much more steeply than nonmodule costs, or soft costs, from \$5 $(W_{DC})^{-1}$ in 1998 to \$1 $(W_{DC})^{-1}$ in 2015.

Streamlining the PV permitting process: One approach to reducing soft costs is to standardize the permitting process. As the US permitting requirements vary by region, state and local authority, the DOE, in collaboration with state agencies and industry bodies, have been working to simplify the permitting process. One result of these efforts is Bill Brook's "Expedited Permit Process for PV Systems: A Standardized Process for the Review of Small-Scale PV Systems," sponsored by the Department of Energy through the Solar America Board for Codes and Standards [11]. The document includes templates for a variety of system types, including those that use module level power electronics, such as microinverters. Several local jurisdictions now accept the templates as part of permitting packages for residential systems.

Data-Based Tools: Another area of extensive R&D toward soft cost reduction is data-based tools for designers, installers, and owners. One such public private partnership is NYSolar Smart, an initiative led by the City University of New York (CUNY), with support from the DOE, the New York State Energy and Research Development Authority (NYSERDA) NY-Sun, and One City, Built to Last. The team has been working to create a Solar Map and

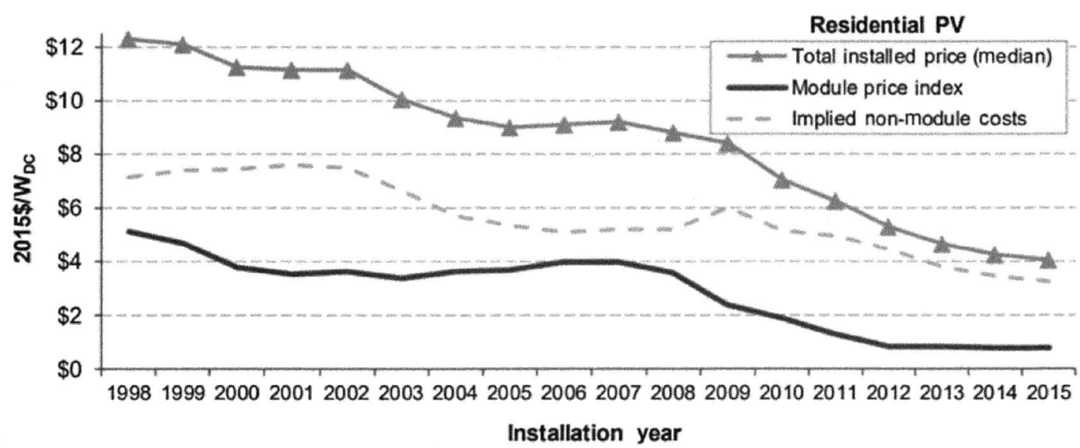

FIGURE 4.13 Installed price, module price index, and non-module costs over time for residential PV systems [8].

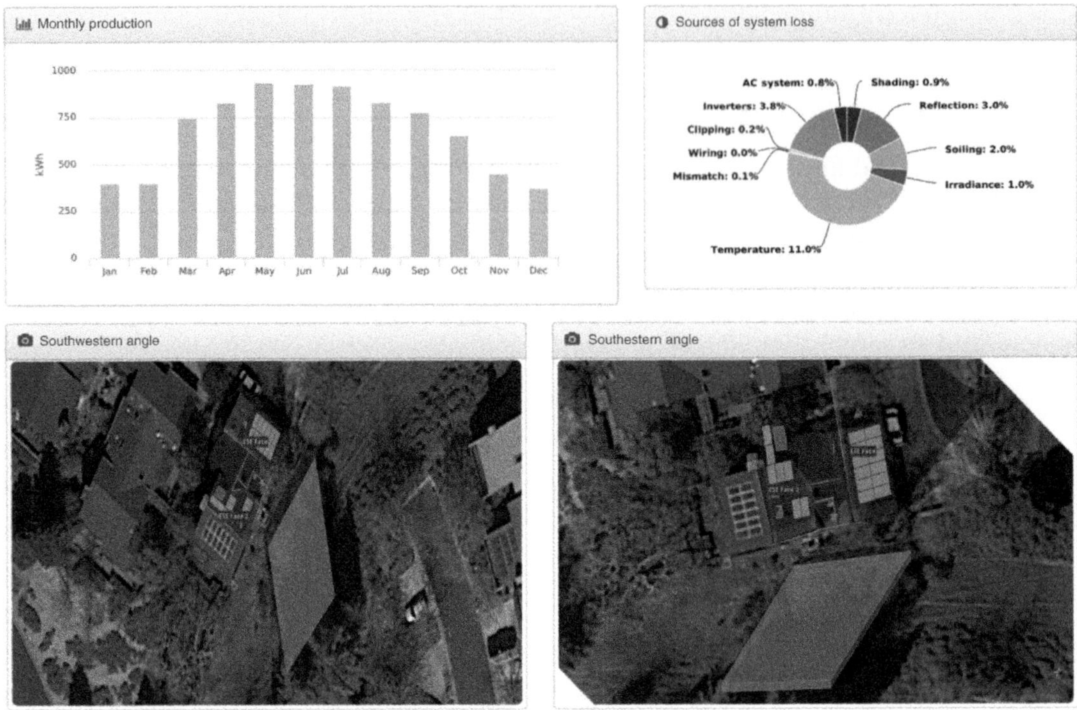

FIGURE 4.14 Sample shading report from heliscope of residential system [13]. Mastering Green.

Portal that provides installers and owners with data for installed capacity, cost per watt, and incentive levels, among other information [12].

Other noteworthy tools funded by the DOE include Helioscope and Aurora Solar, PV design software that enables overlay of structural and electrical designs onto geospatial data (roof area, LIDAR, shading) to quickly estimate system performance and financial returns without the need for AutoCAD and PVsyst. A sample of a Heliostat application by Ginsberg is shown in Fig. 4.14.

Skills-building: Improvement of installer skills is another critical component to soft cost reduction. The North American Board of Certified Energy Practitioners (NABCEP), with its PV Installer certification, is one of the primary educational bodies that accredits trainers, and will be discussed in Section 4.3 on Workforce Development.

4.1.2 Federal Policy

Politics has been a critical determinant of the US energy industry since the advent of electricity. Aside from the 30% ITC for PV, which was extended through 2021, congressional action has been largely absent. Politics continues to dominate the debate on RE, most recently from aspirational to pessimistic analyses on the feasibility of rapid RE penetration to the grid. The spectrum of opinions is wide, from Jacobson and Delucchi's "50-State Roadmap" to 100% renewable energy by 2050, and Fthenakis, Zweibel and Mason's 69% RE by 2050, to the study on "Protecting the Long-Term Reliability of the Electric Grid" commissioned by the Secretary of Energy Rick Perry that focuses on the risks of increased RE penetration in the shift away from coal, oil, and natural gas. SunShot estimates 80–230 GW of installed PV capacity by 2030 and 150–530 GW by 2040 [14–16].

The Obama Administration viewed renewable energy as critical to the decarbonization of the economy. In 2014, President Obama tasked the DOE by Executive Order with analyzing the overall US energy market and providing an update every four years. The Quadrennial Energy Review (QER) Report, includes an "integrated view of, and recommendations for, Federal energy policy in the context of economic, economic, environmental, occupational, security, and health and safety priorities" [2].

Primary obstacles for increasing RE penetration in the US, include the need for system flexibility, demand response, fast ramping of natural gas generation, storage, and enhancements in data modeling and analysis to improve our understanding of the grid operations (Fig. 4.15). These are further discussed in Section 4.3 of this chapter. Also, the QER noted a strong connection between states with a renewable portfolio standard (RPS) and renewable energy growth [2].

The QER noted four primary requirements for increased solar PV penetration. The integration of variable PV increases the need for *system flexibility*, which can be achieved through access to a more diverse portfolio of generation sources, demand response, fast-ramping of conventional generation sources, and storage. *Data modeling and analysis* refers to the need to understand the electricity system to improve system flexibility. Timely data on how utilities are changing operations to integrate increasing RE generation is pivotal to policymaking and investment. *Transmission system upgrades* are required to accept new RE sources, particularly those far from urban centers. Finally, *metrics for valuing solar* are critical to understanding the costs and benefits of increased PV integration compared to conventional fuels, particularly in light of environmental externalities, such as GHG emissions and water usage. Each of these requirements will be detailed in Sections 4.3 and 4.4 of this chapter.

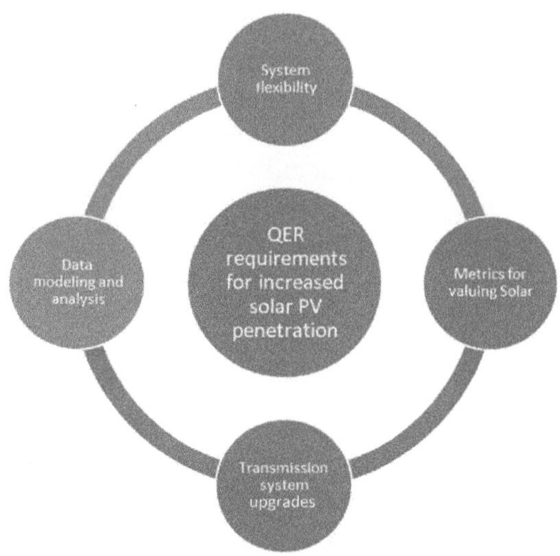

FIGURE 4.15 Requirements for increased solar PV penetration to the grid [2].

President Obama placed additional emphasis on energy and clean technology in foreign relations, forming the Bureau of Energy and Natural Resources within the Department of State, and making the US a founding member of the Paris Climate Accord. In contrast, the Trump Administration has proposed to cut funding by 71.8% to the Department of Energy's Office of Energy Efficiency and Renewable Energy [17].

The Administration's energy policy can be characterized by reduced reliance on foreign fuel. As evidenced by the recent US withdrawal from the Paris Climate Accord, renewable energy is not viewed as an environmental issue tied to climate change mitigation. Rather, it is viewed as a means of achieving energy independence, only if it can be sustained by the private sector. An example of this logic is President Trump's adoption of the idea for installing PV along the US-Mexico border is clear. President Trump does not view solar energy as political in and of itself, but rather a market from which US businesses should and could benefit. As Fthenakis and Zweibel wrote in the WSJ article, "A Shiny Border Wall That Pays for Itself: Forget a traditional barrier and build a 2000 mile solar field along the border," such a grandiose idea may make sense, since the cool, dry climate, low latitude, and minimal shade would be an optimal environment for solar PV production [18].

In January 2018 the Trump Administration imposed tariffs of up to 30% on imported solar equipment (President Donald Trump Imposes 30% Tariffs on Solar Panels (n.d.). Retrieved January 24, 2018, from: http://time.com/5113472/donald-trump-solar-panel-tariff/), in response to a World Trade Organization (WTO) ruling in September 2016 that China was illegally discriminating against US solar exports. According to the filing, "the petition alleges that increasing imports have taken market share from domestic producers and have led to bankruptcies, plant shutdowns, layoffs, and a severe deterioration of the financial performance of the domestic industry" [19]. Imported PV panels from China have been key to reducing domestic prices. This effort is clearly in line with President Trump's America First policy, but, if achieved, it would increase installed PV prices and may slow the advance of domestic solar diffusion.

While the Obama Administration's energy policy can be summed up as an analytical yet cautious "all of the above" prescription and the Trump Administration's doctrine politically potent and decidedly protectionist, neither should be mistaken for a coordinated federal approach to the promotion of solar energy in the US. While not a stated priority of the Trump Administration, increased solar energy penetration to the grid may benefit from progress on grid modernization, a decidedly less partisan issue. Nonetheless, impending cuts in Federal funding "reflects an increased reliance on the private sector to fund later-stage research, development and commercialization of energy technologies" and continued dependence on state policies" [20].

4.2 The United States as a Patchwork of States

The US is comprised of numerous state markets that form an often conflicting and dissimilar tapestry of regulations. Energy policy in the US has long been driven by states, particularly since the end of the Carter Administration and the OPEC oil embargo. In each of the 51 states, communities are governed by local bodies, over 18 000 in total, known as

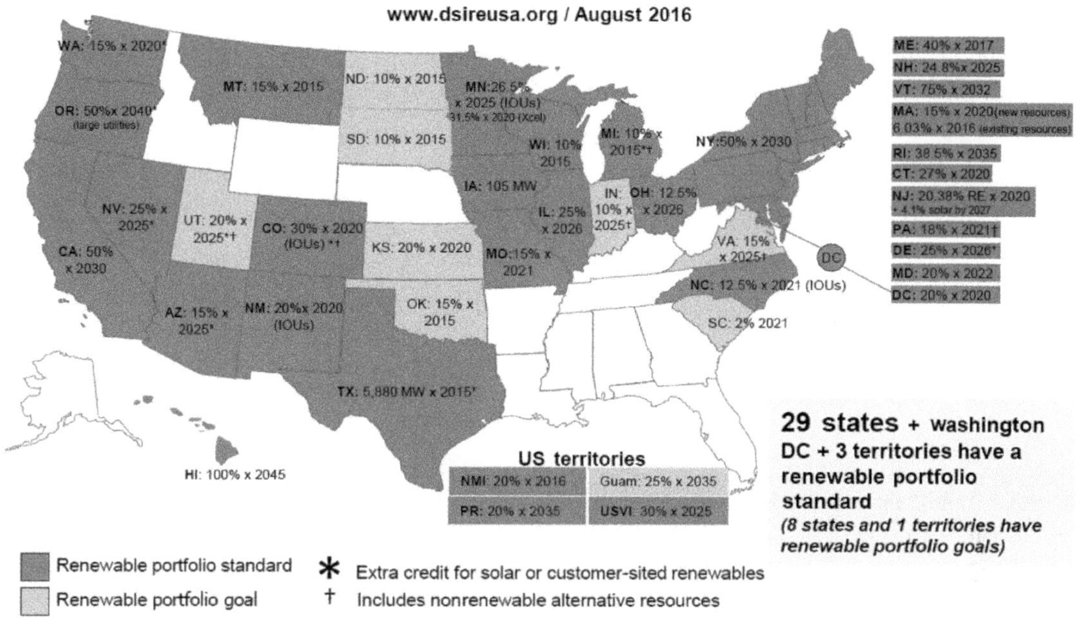

FIGURE 4.16 State renewable portfolio standards [22]. *dsireusa.org accurate as of August 2016.*

authorities having jurisdiction (AHJs). This has led to an abundance of design, installation and permitting requirements, imposing a heavy burden on solar firms, which has been reflected in high soft costs.

State Renewable Portfolio Standards (RPS) have led to increased renewable energy capacity, and those with solar carve-outs to more solar installations. In total, 29 states have RPS obligations. In 2013, 5600 MW of renewable electrical capacity was added to meet the objectives and 200 000 gross domestic renewable energy jobs created; see Fig. 4.16 [2,21].

4.2.1 Leading States

With the exception of Georgia, with almost exclusively utility-scale solar, the top 10 states for installed PV have a renewable portfolio standard or goal. Many of the top 10 states, such as California and Arizona, have an excellent resource, while some do not, such as Massachusetts and New York. California is the clear frontrunner with 9.4% more installed capacity than all of the other leading states combined in 2016 (18 296 MW in California and 16 730 MW combined in other top states). California benefits from an excellent resource, strong RPS, and robust incentive programs. In 2016, PV supplied 35.7% of the state's electricity demand; see Figs. 4.17 and 4.18 [23].

State Variability: There is a high degree of price variability across and within states. Median 2015 residential prices ranged from \$3.2 W^{-1} in Nevada to \$4.8 W^{-1} in Minnesota, reflecting substantial differences in regional labor, permitting and engineering costs [8]. Other contributing factors to price variability are dissimilar incentive levels and electricity rates.

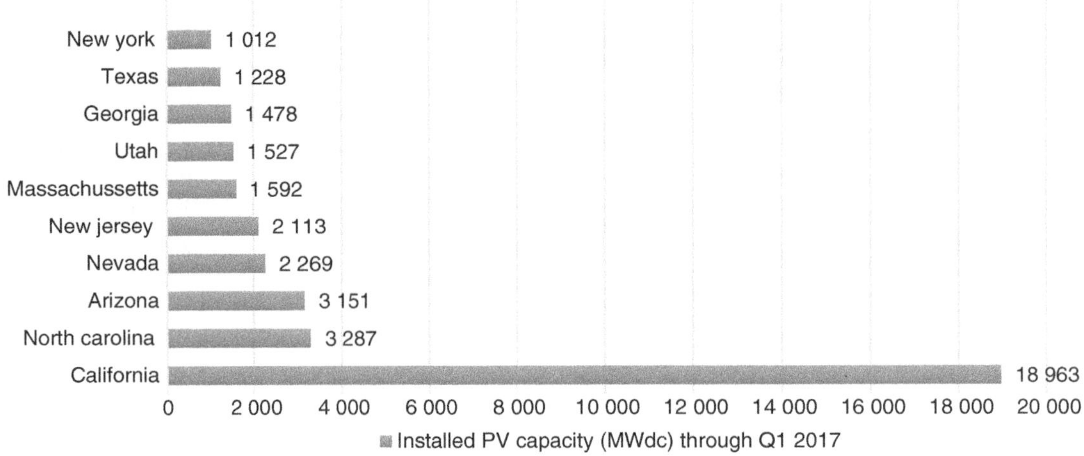

FIGURE 4.17 Top 10 solar states based on cumulative amount of PV capacity installed through Q1 2017 for all sectors [9]. GTM Research, SEIA.

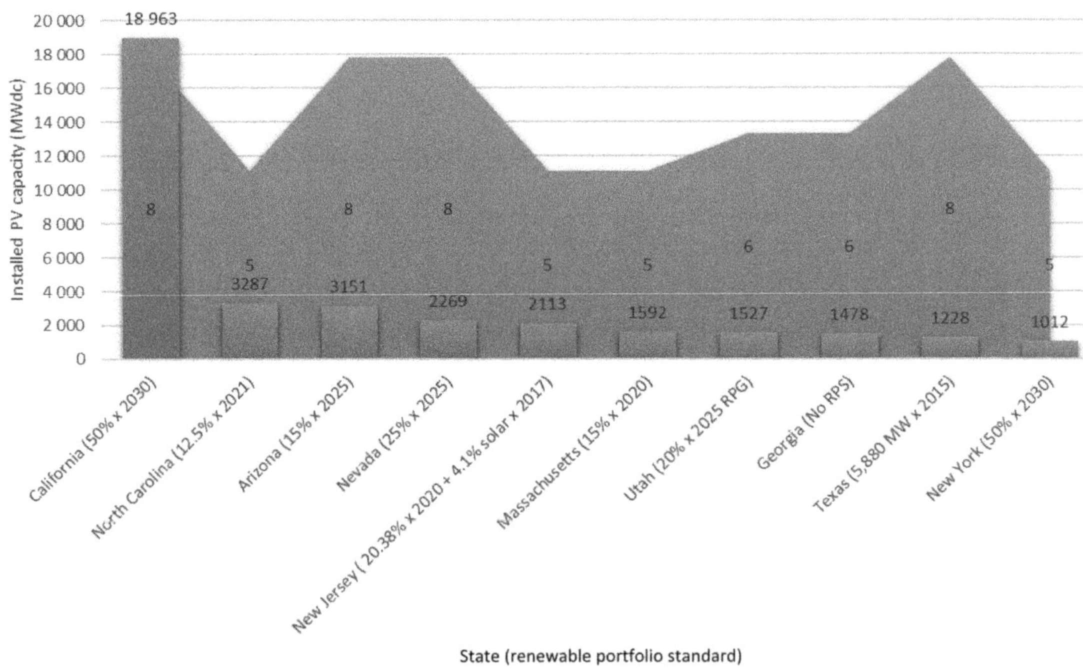

FIGURE 4.18 Cumulative installed PV capacity, solar radiation, and RPS' in top 10 states [9,22,24]. Note on solar radiation data: Data from NREL Flat Plate Collector Map facing south at fixed tilt equal to the latitude of the site. Maps of average values are produced by averaging all 30 years of data for each site. "Highest average daily radiation per month" indicates that when a state had more than one range of averages (intra-state solar resource variations), the highest possible kW h m^{-2} yr^{-1} was taken for consistency in comparison. National Renewable Energy Laboratory (NREL) Flat Plate Collector Map, dsireusa, SEIA.

State Spotlight: California is the world's eighth largest economy. The state recently increased its RPS to 50% by 2030, and on May 31, 2017, Senate Bill 100, which mandates 100% renewable energy by 2045, was passed and is likely to become law [25].

California currently generates about 35.66% of its electricity from renewable energy, of which PV supplied 8.11%, large hydropower 10.21%, wind 9.06%, and biomass, geothermal, and small hydro the remaining 8.28% [23]. Coupled with its high average residential electricity price of 15.34 cents $(\mathrm{kW\,h})^{-1}$, the state's "Go Solar California!" Campaign has laid the foundation for a thriving solar market [26].

Go Solar is a joint campaign of the California Energy Commission and Public Utilities Commission, and includes various incentive programs to "encourage Californians to install 3000 MW of solar energy systems on homes and businesses by the end of 2016." The campaign provides online resources for installers, homeowners and other stakeholders.

Go Solar is comprised of the California Solar Initiative (CSI), which is a rebate program for customers of investor-owned utilities for PV, as well as thermal, installs. Incentive levels are tied to anticipated and verified performance of solar systems, requiring careful energy modeling pre-installation, and measurement and verification post-installation. A subset of the CSI, the New Solar Homes Partnership (NSHP) provides incentives for new construction and also requires detailed energy efficiency modeling and verification [27].

In addition to leading the nation in installed RE, California is helping fill the need identified in the Second Installment of the DOE QER for an improved understanding of the electricity system through providing data points for modeling and analysis.

4.3 US Solar Energy Market Outlook

Note on Modeling—In the NREL SunShot, ITC and Renewable Electricity Futures Study discussed in this section, two primary tools were employed for grid modeling and analysis, ReEDS and dSolar. The Regional Energy Deployment System Model (ReEDS), "a capacity expansion and dispatch model for the contiguous U.S. that relies on system-wide least cost optimization to estimate the type and location of future generation and transmission capacity," and Distributed Solar (dSolar), "a consumer adoption model for the U.S. rooftop PV market that simulates future rooftop PV deployment in the industrial, commercial buildings, and residential buildings sectors" [28].

ReEDS accounts for location-dependence, variability, and uncertainty of wind and solar resources through high spatial resolution and statistical methods, and dSolar combines a detailed representation of rooftop PV consumer segments with a county-level representation to model geographic differences in rooftop PV resource and electricity markets [28]. In addition, in the 2012 Renewable Electricity Futures Study, NREL used ABB GridView to model hourly operation of the grid with high levels of variable wind and solar generation [29]. The synergy of wind and PV in increasing the total renewable energy penetration in the state of NY was shown by another study by Nikolakakis and Fthenakis [30].

Achieving 80 Percent Renewable Energy Penetration by 2050: In the 2012 Renewable Electricity Futures Study, NREL modeled 80% RE penetration by 2050 under low and high electricity demand scenarios. The study found that under the low-demand scenario, there was greater PV than CSP installed capacity in low RE penetration (up to 30%), and more CSP in the high RE penetration scenario since CSP with thermal storage provides added dispatchability [29].

Under the high demand scenario, a much greater amount of solar and wind capacity were needed compared to the low-demand scenario. Under the low-demand scenario, 2.5 to 10% (100 to 290 GW) of PV were required and under the high-demand scenario 13% (420 GW) were required; see Fig. 4.19 [29].

In its analysis, NREL used ReEDS, dSolar, and ABB GridView to create a map of optimal energy mixes by region to achieve 80% RE by 2050 under the low-demand scenario [29].

Economic Carrying Capacity: The concept of economic carrying capacity (ECC) is particularly relevant to understanding current limitations to renewable energy penetration. Achieving greater RE penetration is more of a financial issue than a technological one as fixes exist, but at varying costs. The ECC is a cost benefit analysis that finds the "economically desirable limit associated with adding variable renewable energy" [31]. This limit is dependent on a number of factors, including existing transmission capacity, and regional resources and electricity prices. An NREL report shows that 30% variable RE penetration could be achieved by expanding transmission capacity and changes to system operations, yet beyond that greater investment in grid modernization and flexible market mechanisms would be required [31].

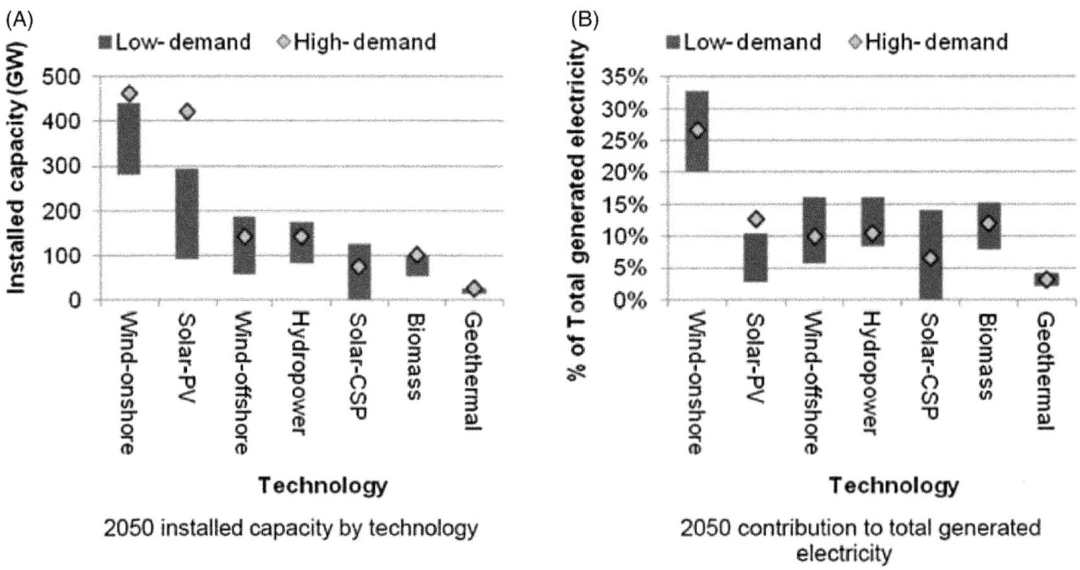

FIGURE 4.19 Range of 2050 installed capacity and annual generated electricity by technology for the low-demand core 80% RE scenarios and the high-demand 80% RE scenario [29].

Federal Investment Tax Credit (ITC) versus Natural Gas Prices: The 30% ITC for solar energy has been a strong driver of the US solar energy market. Under the Consolidated Appropriations Act of 2016, the solar ITC is set to gradually decline per sector in 2020. Utility-scale and commercial PV will receive a 26% ITC in 2020 and 22% in 2021, with 10% into the future, while residential PV will decline to 26% in 2020, 22% in 2021, and expire at the end of 2021 (Table 4.2) [28].

The decline and expiration (for the residential sector) of the ITC has introduced uncertainty into the US solar market. However, a 2016 NREL report that modeled several scenarios with and without the extension, found that while the ITC would accelerate RE deployment through the early 2020s, it did not result in significantly greater cumulative renewable energy deployment by 2030. Other factors, including rising natural gas prices, decreasing PV costs and the US Environmental Protection Agency's Clean Power Plan, appeared to propel continued growth through 2030 under both extension and no extension scenarios.

In the study, NREL compared installed PV capacity to natural gas prices. Using the US Energy Information Administration (USEIA's) 2015 Annual Energy Outlook (AEO), the authors created scenarios with limited oil and gas resources (*Base Gas Price* scenario) and high oil and gas resources (*Low Gas Prices*). NREL entered the AEO 2015 Reference and High Oil & Gas Resource cases into ReEDS, which took into account regionally differentiated fossil fuel prices, availabilities, and seasonal adjustments for natural gas prices. Base gas prices increase from 2015 beginning at $5.01(PJ)^{-1}$ ($4.75/MMBtu) to about $6.33(PJ)^{-1}$ ($6/MMBtu) in 2030, while base gas prices decreased from 2015 beginning at $5.01(PJ)^{-1}$ ($4.75/MMBtu) to about $4.75(PJ)^{-1}$ ($4.50/MMBtu) in 2030 (see Fig. 4.20) [28].

The study found that RE penetration is more sensitive to gas price than to the ITC, especially when natural gas prices are as low as $4.75(PJ)^{-1}$ ($4.50/MMBtu) (see Table 4.3) [28].

Utility-scale PV is shown to accelerate more rapidly in the early 2020s under the extension scenario, but reach the no extension scenario by 2030. The curve is steeper in the Base

Table 4.2 Schedule of Wind and Solar Tax Credits After the Consolidated Appropriations Act of 2016 [28]

New Policy		**2015**	**2016**	**2017**	**2018**	**2019**	**2020**	**2021**	**Future**
Wind PTC		Full	Full	80%	60%	40%	0%	0%	0%
Solar ITC	Utility	30%	30%	30%	30%	30%	26%	22%	10%
	Commercial/Third-party-owned	30%	30%	30%	30%	30%	26%	22%	10%
	Residential host-owned	30%	30%	30%	30%	30%	26%	22%	0%
Prior policy		2015	2016	2017	2018	2019	2020	2021	2022
Wind PTC		0%	0%	0%	0%	0%	0%	0%	0%
Solar ITC	Utility	30%	30%	10%	10%	10%	10%	10%	10%
	Commercial/Third-party-owned	30%	30%	10%	10%	10%	10%	10%	10%
	Residential host-owned	30%	30%	0%	0%	0%	0%	0%	0%

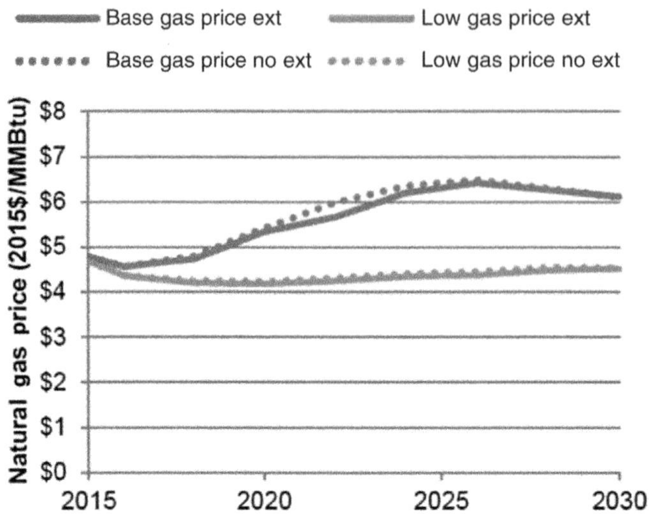

FIGURE 4.20 Model-estimated electric sector natural gas price (excluding "direct use" natural gas consumption from combined heat and power, on-site generating systems, and similar facilities) [28].

Table 4.3 The Impact of the ITC and Gas Price on RE Penetration in 2030 (RE Penetration Shown as Percentage of Total 2030 US Energy Mix) [28]

	Base Gas Price $6.33(PJ)^{-1}$ ($6/MMBtu)	**Low Gas Price $4.75(PJ)^{-1}$ ($4.50/MMBtu)**
ITC extension	33%	25%
No ITC extension[a]	33%	22%
Increase in RE penetration with extension	0%	3%

[a]"No ITC Extension" scenarios are based on the Prior Policy in Table 4.2. The scenario excludes the recently passed extension and assumes the solar ITC is set to decline to 10% for the utility and commercial sectors after 2016, and expire for the residential sector at the end of 2016.
Impacts of Federal Tax Credit Extensions on Renewable Deployment and Power Sector Emissions.

Gas Price scenario, with about 72 GW of installed capacity in 2030 compared to the low gas price scenario with about 45 GW of installed capacity in 2030; see Fig. 4.21 [28].

NREL modeled rooftop PV capacity with and without the ITC extension, but did not consider natural gas price scenarios, and found a greater sensitivity in this sector to the ITC; see Fig. 4.22 [28].

Federal Financing—The DOE Loan Guarantee Program, created to advance innovative energy and transportation projects, has been particularly useful for commercial and utility-scale PV installations. Created in 2005, the program has supported 28 energy projects by guaranteeing the debt of privately held energy generation and manufacturing projects in the event that the company defaults on a loan. The program has spurred over 25×10^9 ($25 billion) in private investment[32]. The program, which has 41×10^9 in existing spending authority, has been a target of Congress, particularly over some of the high-profile loan defaults, such as Solyndra. However, as of December 2016 the portfolio has been performing well with 36×10^9 in loan guarantees, 6.65×10^9 in repaid loan

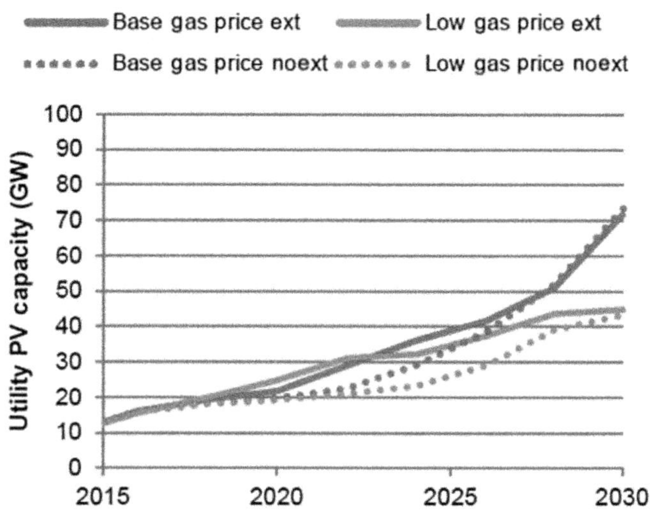

FIGURE 4.21 Installed utility PV capacity by scenario [28].

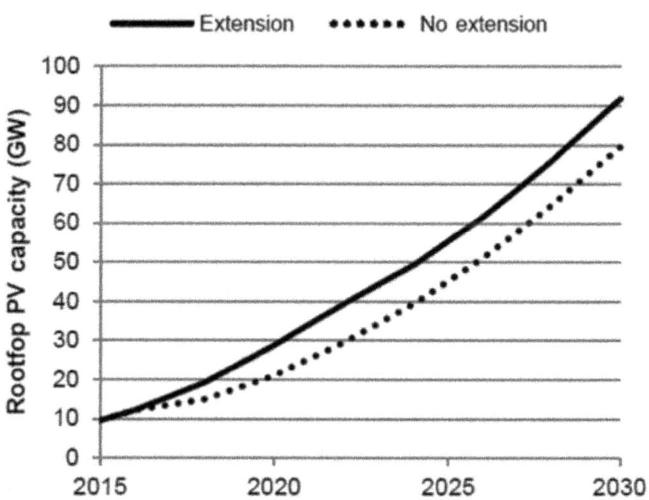

FIGURE 4.22 Installed rooftop PV capacity by scenario [28].

principal, and 1.79×10^9 in interest. Losses have been barely half of the interest paid so far, or just over 2% of commitments [33].

4.4 The United States as a Driver of Innovation

4.4.1 "Profoundly Disconnected": The Need for Workforce Development and Educator Training

In his book "Profoundly Disconnected," Former CNN Host Mike Rowe wrote of the widening skills gap in the US and the lack of technical education required for today's jobs [34].

The rapid growth of the US solar market has created an immediate demand for solar energy specialists and a dearth of competent professionals.

In 2016, the industry added workers 17 times faster than the overall economy, 2% of all jobs created in the US, with 178% growth since 2010. The Solar Foundation's Solar Jobs Census revealed 260,077 solar jobs were created in 2016, with the most in those states with RPS obligations.

Solar is the second highest employer in the energy industry, behind oil, represents one out of 50 new jobs in the US, and provides more employment than natural gas, twice as much as coal, three times as much as wind, and five times as much as nuclear. The median wage is $26 per hour and includes a wide range of positions, from installers to project developers and analysts (Fig. 4.23) [35].

Spotlight on Workforce Development: One of the initiatives designed to make US society less disconnected in the renewable energy sector is The Center for Renewable Energy Advanced Technological Education (CREATE). CREATE is comprised of community colleges with robust renewable energy programs, including the Madison Area Technical College (MATC) in Wisconsin, the College of the Canyons in California, and the Lane Community College in Oregon. With funding from the National Science Foundation, CREATE provides hands-on training for community college and high school teachers in STEM. Its mission is to "advance renewable energy educational programs nationwide by serving as a source of teacher support, business and industry networking, professional development, and of educational materials" [35].

CREATE was formed in 2016 from the success of the Renewable Energy Academies (REA). Beginning in 2010, MATC, in coordination with Solar Energy International, launched the REA, a five-day training in PV system installation for community college and high school teachers. Since then, MATC has provided a three-day professional development training for high school teachers, known as the STEM Solar Institute, providing classroom activities using PVWatts and Solar Pathfinder for participants' immediate use.

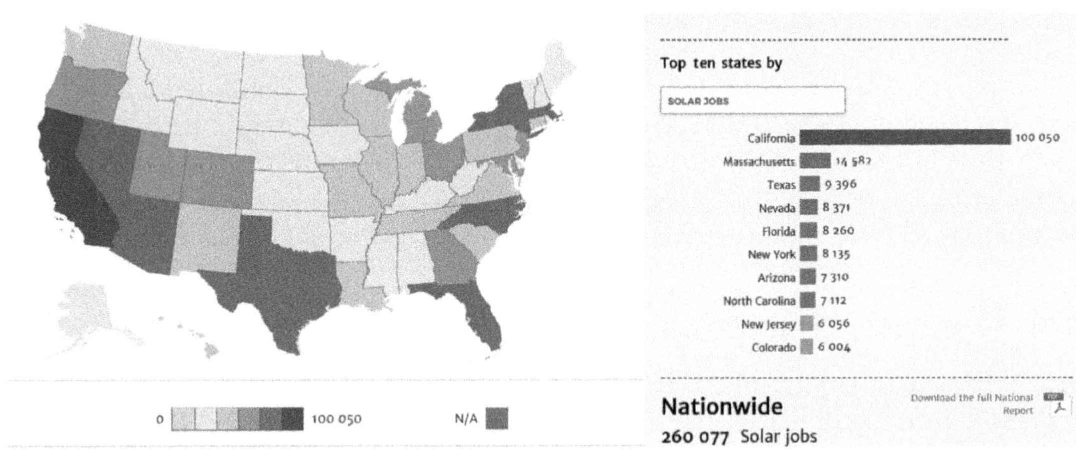

FIGURE 4.23 US solar jobs census 2016 [35].

The impact of the Renewable Energy Academies has been substantial. Since 2011, the consortium trained 284 high school and community college instructors across 41 states. Participants taught 35 000 students, 28 000 of whom received instruction directly from the content covered in the academies [36].

Another important industry initiative to fill the skills gap and raise standards in the RE sector is the North American Board of Certified Energy Practitioners (NABCEP). NABCEP has assembled "Job Task Areas" that categorize the knowledge professionals' need as installers, inspectors, and salespeople. As an American National Standards Institute, ANSI, accredited body, NABCEP credentialed professionals are trusted in the industry and provide a boost to those seeking employment [37].

4.4.2 Technological and Financial Innovations

With the most patents for solar manufacturing in the world, the US remains a leading driver of innovation in the RE sector. The Second installment of the quadrennial energy review (QER) identified the need for system flexibility, metrics for valuing solar, data modeling and analysis for improved understanding of the electrical system, and transmission system upgrades. This section will highlight these limitations and solutions emerging from US R&D and praxis.

Valuing Solar: From a societal perspective, solar PV offers significant advantages to our resource-constrained economy. Integrating utility-scale and distributed generation PV with the grid, however, poses several technical and financial complications, requiring a careful evaluation of solar. It is necessary to establish equitable compensation for the value of solar to the grid and determine which party bears the responsibility for the investments required to upgrade the transmission system, the customer or the utility.

Unlike conventional fuels, which have environmental and public health repercussions, solar PV does not emit GHG or use water in generation. These "externalities" can be quantified and provide a more compelling argument for PV than simple financial calculations. Recent assessments show that the environmental and public health benefits of PV use in the US represent 3.5 cents $(\text{kW h})^{-1}$ [38]. Studies by Columbia University's Center for Life Cycle Analysis show that when fully accounting for the costs of energy, solar is below grid parity in most states [39,40].

PV represents substantial water savings with cumulative impacts from 2015 to 2050 under the DOE SunShot Vision amounting to 174×10^{12} L (46 trillion US gallons) of water withdrawals, or 4% of total power sector withdrawals, and 18.93×10^{12} L (5 trillion gallons) of avoided consumption, or 9% of total power sector consumption [38]. Withdrawals are defined as the amount of water removed or diverted from a water source for use, while consumption is the amount of water evaporated, transpired, incorporated into products or crops, or otherwise removed from the immediate water environment.

This is of particular importance to the water-strapped states of California and the Southwest. A decrease in GHG emissions by 10% from 2015 to 2020 represents savings of \$238–\$252 $\times 10^9$ to the economy [38].

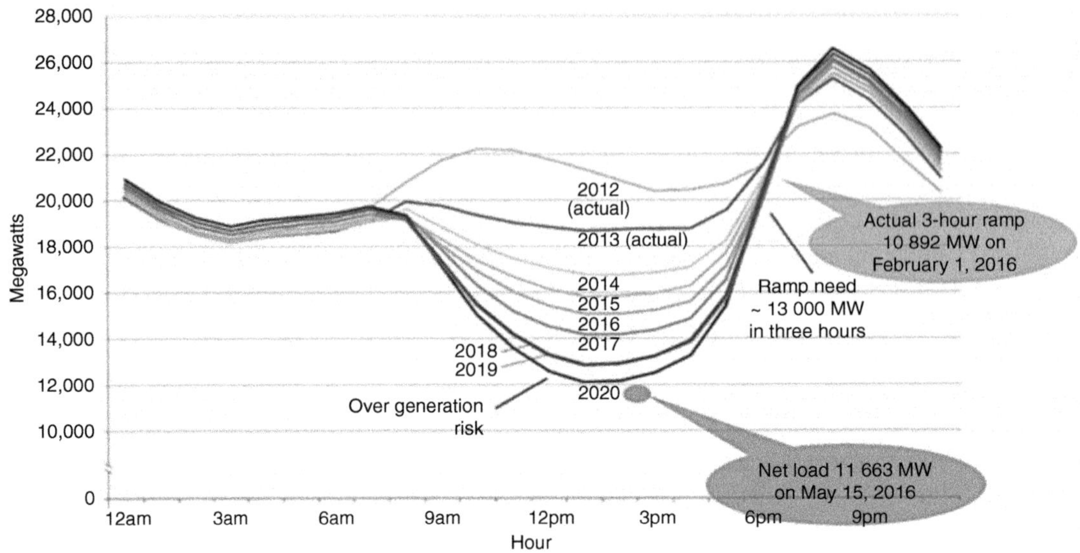

FIGURE 4.24 The duck curve during a typical spring day [41].

In tandem with demand response, solar can provide a benefit to the utility that is not currently accounted for in most states. Utilities often assess commercial and industrial facilities a demand charge based on their peak kilowatt use in a given month. Often the utility applies a "ratchet clause," charging for monthly demand based on a percentage of the peak kilowatt in a year, even if demand is far less during the rest of the year. Therefore, consumers can substantially lower their demand charge by implementing demand response measures in tandem with utility's peak demand. Along with energy conservation, consumers can in fact provide a service to utilities during peak demand by maximizing PV system production and export during such times (i.e., installing west-facing arrays to maximize afternoon and evening production). These services, however, must still be quantified and properly valued with respect to their benefit to the grid.

In contrast, the integration of distributed generation (DG) requires system upgrades to enable bidirectional power flow and energy storage. The issue of intermittency is paramount. The "duck curve," developed by the California Independent System Operator (CAISO), demonstrates the challenge of matching demand and supply with increased RE penetration. The curve (Fig. 4.24) shows that nonsolar supply must be high in the morning, steeply decline during the "solar window," the three hours before and after 12 PM noon, and surge in the afternoon as the sun begins to set [41].

CAISO has taken several steps to better match supply and demand, including time of use rates to deter high use in the evening, and flexible ramping. Flexible ramping is the ability of a generator to start and stop on command. This is useful since during high noon when solar resources are the most productive, conventional "baseload" generators are still producing, often causing oversupply. Cycling of power plants causes large thermal losses, a slight increase in emissions, and the potential for mechanical failures, reducing plant life. In October 2016, the Federal Energy Regulatory Commission approved CAISO's "flexible ramping product" that increases the ramp rate, or speed, of power plants to start and stop production. In this market mechanism, the ISO pays generators to remain off during high periods of solar production to compensate for lost revenue. As such, CAISO shifts costs to those necessitating flexibility [42].

Spotlight on Data-Driven Valuation: In addition to NREL's suite of modeling software, the valuation of solar and grid modeling requires sophisticated analytical tools. One example of such tools is Greenlink's ATHENIA, which can generate hourly and daily generation scenarios of solar in regional service territories, quantify existing generation and assess the potential for future growth. The software is designed for informing state and city policymaking through determining the social, health, and environmental benefits of distributed solar energy generation [43].

Other emerging modeling and forecasting techniques incorporate the use of stochastic modeling and Unit Commitment and Economic Dispatch models based on General Algebraic Modeling System (GAMS) [44]. These methodologies are able to incorporate numerous variables to anticipate, among other issues, the financial value of solar to the grid and system upgrade requirements. Ongoing R&D in this area involves system-specific analyses to identify the optimal combination of technology, grid infrastructure, and operational challenges beyond 30% variable RE, the point at which significantly greater investment will be required.

4.4.3 System Upgrades

As PV comprises an increasing share of US generation, the physical infrastructure of the grid will need to be upgraded. A partnership between the DOE and 13 National Laboratories, the Grid Modernization Laboratory Consortium is working to equip the grid with the technology needed to meet the increase in clean energy and distributed generation [45].

Concerns abound that increased renewable penetration will result in system interruptions. The US, however, would benefit from examining the experience of Denmark and Germany, both of which experienced less power outages per recent years than the US and have four and two times the percentage of renewable energy on their grid (Fig. 4.25) [46].

Nonetheless the increased penetration of renewable energy will require transmission system upgrades. Large "utility-friendly" power plant technologies are being introduced [47] but constraints in small systems remain; these include the need for bidirectional power flow, and inverter electronics that better synchronize with the grid. Bidirectional power flow is required for grid-tie PV systems that export excess energy to the grid. Inverter electronics refer to PV inverter's short circuit current capacity and ability to harmonize to the frequency and voltage of the grid.

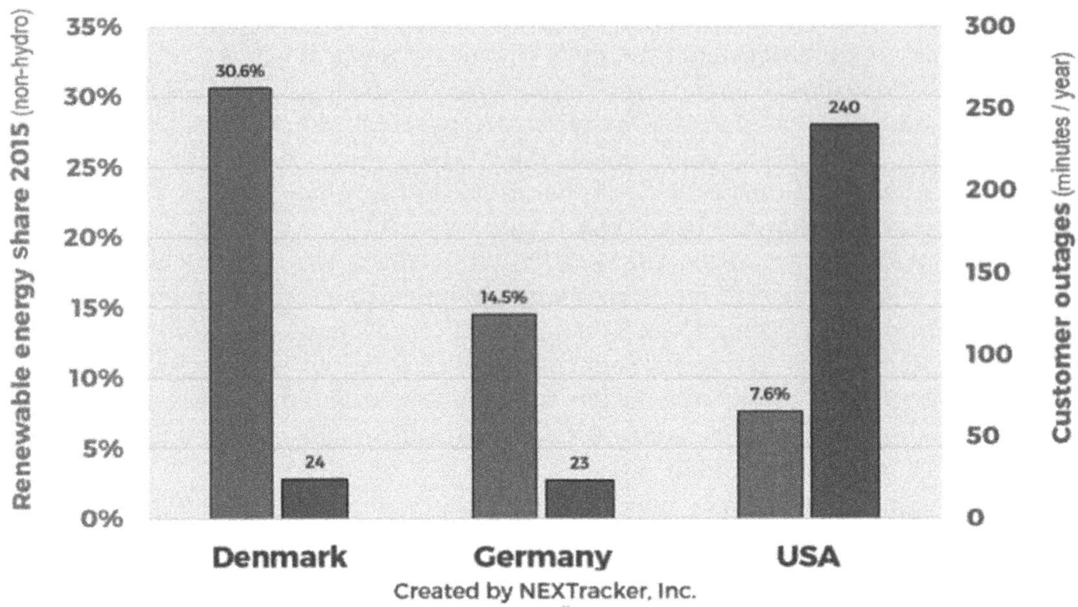

FIGURE 4.25 Renewable penetration versus annual power outages [46].

With the decline in hard costs, module level power electronics (MLPE), including microinverters and power optimizers, have seen a rise in popularity. Recent data has found that the price premium PV systems with MLPEs over conventional string systems was negligible, and even less due to soft cost efficiencies realized in their use (i.e., ease of installation) [2].

A Schneider Electric White Paper on "How New Microgrid Technologies Enable Optimal Cooperation Among Distributed Energy Resources" identified several limitations and potential solutions to increased penetration of RE [48]. Primary requirements include protection system design and power quality control.

Protection System Design: Mixing inverter-based renewable generation with traditional rotating generators presents several problems. During a fault, or short circuit, condition, PV inverter current is limited to roughly 120% of inverter-rated current, to protect the inverter itself. Compared to rotating machines, the magnitude of inverter short circuit current permitted is low and would cause frequent system interruptions [48].

Power Quality Control: Most inverters today harmonize frequency and voltage with other generators. Unlike traditional alternators, inverters are not "grid-forming" in that they cannot rely on themselves to form grid voltage and do not have the "natural inertial link between rotational speed and grid frequency" [48].

An Emerging Solution—Robust Inverter-Based Control Loop Design: New inverter-based generators (IBGs) "embed advanced voltage and frequency control loops" that enable them to be grid-forming, and incorporate power storage that smoothes the variability of renewables by decoupling instantaneous generation from demand [48].

4.4.4 A Vision for the Future of the US Grid—The Internet of Interoperable Microgrids

In response to increased solar and RE penetration on the grid, local communities in the US are creating customer-centric models for power generation, distribution, and purchasing that diverge from the traditional investor-owned utility (IOU).

These new models, which are highly responsive to individual and community preferences, are and will increasingly be enabled by the Internet of Things. Just as one can now monitor sleep cycles, track personal fitness, and control room temperature through smartphone apps, individuals and communities will soon be able to monitor energy consumption and production from local sources, and use this data to make efficient and effective demand response and purchasing decisions.

The centralized IOU brought electricity to every corner of the US. In today's world of individual choice, however, central generation can be viewed as a "one-size fits all" system in which consumers have little to no control over the energy source, how power generation aligns with demand, and pricing.

Net metering, time of use rates, and flexible ramping may soon be seen as a "band-aid," rather than a long-term solution for the needs of the modern US grid. New conceptions of the electrical system are needed. An emerging alternative to the traditional grid is the microgrid. The microgrid integrates increased distributed energy resources as part of the optimal operation and management of an interconnected system. Microgrids may be the next evolution in grid operation as they enable increased renewable energy integration, improved system resilience, and local economic revitalization. In addition, microgrids enable new means of realizing the financial value of renewable energy and demand response.

It is time for a reboot. The advent of Internet of Things (IoT) network connectivity enables the rise of data-driven, community-tailored microgrids, and fleets of microgrids with local distributed energy resources (DERs) that can be responsive to community preferences, serve as a tool for local economic development and work in parallel with the grid. In addition, these microgrids can internally balance DER variability to provide export consistency and overcome other technical issues through innovations such as the Consortium for Electric Reliability Technology Solutions (CERTS) microgrid architecture [49].

Emerging modeling software and graphical user interfaces (GUIs) will need to be employed to advance the grid of the future—an interoperable system that unifies distribution system management with transactive energy purchasing.

References

[1] E (n.d.). Electricity in the United States. U.S. Energy Information Administration. Available from: https://www.eia.gov/energyexplained/index.cfm?page=electricity_in_the_united_states.

[2] Quandrennial Energy review: Second Installment (2017, January 06). US Department of Energy. Available from: https://www.energy.gov/epsa/downloads/quadrennial-energy-review-second-installment.

[3] U.S. Energy Information Administration—EIA—Independent Statistics and Analysis. (n.d.). Available from: https://www.eia.gov/todayinenergy/detail.php?id=30652.

[4] 2017 Sustainable Energy in America Factbook. Bloomberg New Energy Finance. Available from: http://www.bcse.org/sustainableenergyfactbook/#.

[5] Walton, R. (2017, February 09). BNEF: US economy decoupled from energy demand as renewables rise, emissions fall. Available from: http://www.utilitydive.com/news/bnef-us-economy-decoupled-from-energy-demand-as-renewables-rise-emissio/435825/.

[6] Mehos, M, Turchi, C, Jorgenson, J, Denholm, P, Ho, C, Armijo, K. On the Path to SunShot: Advancing Concentrating Solar Power Technology, Performance, and Dispatchability. Golden, CO: National Renewable Energy Laboratory. NREL/TP-5500-65688; 2016. Available from: http://www.nrel.gov/docs/fy16osti/65688.pdf.

[7] On the Path to SunShot. U.S. Department of Energy. Available from: https://energy.gov/eere/sunshot/path-sunshot.

[8] Barbose, G, Darghouth, N. (2016, August). Tracking the Sun IX: The Installed Price of Residential and Non-Residential Photovoltaic Systems in the United States. SunShot U.S. Department of Energy. Lawrence Berkeley National Laboratory. Available from: https://emp.lbl.gov/sites/default/files/tracking_the_sun_ix_report.pdf.

[9] Perea, A, Honeyman, C, Kann, S, Mond, A, Shiao, M, Jones, J, Jones, J, Smith, C, Gallagher, B, Rumery, S, Holm, A, O'Brien, K, Baca, J. Executive Summary of Solar Market Insight Report 2017 Q2. GTM & SEIA Research Team. Available from: http://www.seia.org/research-resources/solar-market-insight-report-2017-q2.

[10] Bolinger, M, Seel, J. (2016, August). Utility-Scale Solar 2015. SunShot U.S. Department of Energy. Lawrence Berkeley National Laboratory. Available from: https://emp.lbl.gov/sites/default/files/lbnl-1006037_report.pdf.

[11] Brooks, B. (2012, July). Expedited Permit Process for PV Systems Revision 2. Solar America Board for Codes and Standards. Available from: http://www.solarabcs.org/about/publications/reports/expedited-permit/pdfs/Expermitprocess.pdf.

[12] CUNY, S. (2016, December). NYSolar Smart Final Report. US DOE SunShot Initiative. Available from: http://www.cuny.edu/about/resources/sustainability/nyssolar/FinalCUNYRSCIIProgressReport-Linked.pdf.

[13] Ginsberg, M. (n.d.). Helioscope. Shading Report for Sean White Residence. Available from: https://docs.wixstatic.com/ugd/c0937a_296500b035314c5c94f1afbd3a016e9a.pdf.

[14] Shwartz, M. (2014, February 26). Stanford scientist unveils 50-state plan to transform U.S. to renewable energy. Available from: http://news.stanford.edu/news/2014/february/fifty-states-renewables-022414.html.

[15] Fthenakis V, Mason J, Zweibel K: The technical, geographical and economic feasibility for solar energy to supply the energy needs of the United States, *Energy Policy* 37:387–399, 2009.

[16] Grandoni, D. (2017, July 18). Analysis | The Energy 202 Perry's electric grid draft draws some blanks. Available from: https://www.washingtonpost.com/news/powerpost/paloma/the-energy-202/2017/07/18/the-energy-202-perry-s-electric-grid-draft-draws-some-blanks/596d1413e9b69b7071abcb61/?utm_term=.73d10238b39b.

[17] President Trump's First 100 Days. (2017, May 01). Available from: https://www.whitehouse.gov/100-days.

[18] Fthenakis, V, Zweibel, K. (2017, June 22). A Shiny Border Wall That Pays for Itself. WSJ. Available from: https://www.wsj.com/articles/a-shiny-border-wall-that-pays-for-itself-1488931053.

[19] U.S. may put emergency tariffs on solar imports. (2017, May 29). Reuters. Available from: http://www.reuters.com/article/us-usa-solar-wto-idUSKBN18P1JL.

[20] Egan, M. (2017, June 8). Trump's renewable energy cuts alarm former officials. CNN. Available from: http://money.cnn.com/2017/06/08/investing/renewable-energy-trump-budget-cuts/index.html.

[21] Wiser, R, Mai, T, Carpenter, A, Keyser, D, Mills, A, A retrospective analysis of the benefits and impacts of U.S. renewable portfolio standards (Lawrence Berkeley National Laboratory and National Renewable Energy Laboratory, 2016), TP-6A20-65005. Available from: http://www.nrel.gov/docs/fy16osti/65005.pdf.

[22] Pickerel, K. (2017, June 02). Solar development still depends on renewable portfolio standards. Solar Power World Online. Available from: https://www.solarpowerworldonline.com/2017/06/solar-development-still-depends-renewable-portfolio-standards/.

[23] Commission, C.E. (n.d.). Total System Electric Generation. Available from: http://www.energy.ca.gov/almanac/electricity_data/total_system_power.html.

[24] Photovoltaic Solar Resource of the United States. (2012, September 19). NREL. Available from: https://www.nrel.gov/gis/images/eere_pv/national_photovoltaic_2012-01.jpg.

[25] California plan for 100% renewable energy by 2045 clears key hurdle. (n.d.). LA Times. Available from: http://www.latimes.com/politics/essential/la-pol-ca-essential-politics-updates-california-plan-for-100-renewable-1496258464-htmlstory.html.

[26] California Electricity Rates. (n.d.). Available from: http://www.electricitylocal.com/states/california/.

[27] Go Solar CA. (n.d.). Available from: http://www.gosolarcalifornia.ca.gov/about/index.php.

[28] Mai, T, Cole, W, Lantz, E, Marcy, C, Sigrin, B. (February 2016). Impacts of federal tax credit extensions on renewable deployment and power sector emissions. NREL. Available from: http://www.nrel.gov/docs/fy16osti/65571.pdf.

[29] Renewable Electricity Futures Study (Entire Report) National Renewable Energy Laboratory. (2012). Renewable Electricity Futures Study. Hand, MM, Baldwin, S, DeMeo, E, Reilly, JM, Mai, T, Arent, D, Porro, G, Meshek, M, Sandor, D, editors. 4 vols. NREL/TP-6A20-52409. Golden, CO: National Renewable Energy Laboratory. Available from: http://www.nrel.gov/analysis/re_futures/.

[30] Nikolakakis T, Fthenakis V: The optimum mix of electricity from wind- and solar-sources in conventional power systems: evaluating the case for New York State, *Energy Policy* 39(11):6972–6980, 2011.

[31] Cochran, J, Denholm, P, Speer, B, Miller, M (August 2015). Grid integration and the carrying capacity of the U.S. grid to incorporate variable renewable energy. NREL. Available from: http://www.nrel.gov/docs/fy15osti/62607.pdf.

[32] Loan Guarantee Program. (n.d.). Available from: http://www.seia.org/policy/finance-tax/loan-guarantee-program.

[33] Reicher, D. $41 billion in guaranteed infrastructure? DOE's loan guarantee program is ready. Brookings Institution. March 2, 2017. Available from: https://www.brookings.edu/blog/the-avenue/2017/03/02/doe-loan-guarantee-program-is-ready.

[34] Rowe, M. Profoundly disconnected: a true confession. 2014; MRW Holdings.

[35] Solar Foundation. (2016). National Solar Jobs Census, 2015. Washington DC, United States: The Solar Foundation. Available from: www.tsfcensus.org.

[36] Walz, KA, Shoemaker, JB (2017). Preparing the Future Sustainable Energy Workforce and The Center for Renewable Energy Advanced Technological Education. *Journal of Sustainability Education*, 13. Available from: http://www.jsedimensions.org/wordpress/content/preparing-the-future-sustainable-energy-workforce-and-the-center-for-renewable-energy-advanced-technological-education_2017_03/.

[37] NABCEP: Raising standards. Promoting confidence. (n.d.). Retrieved July 01, 2017. Available from: http://www.nabcep.org/.

[38] Wiser, R, Mai, T, Millstein, D, Macknick, J, Carpenter, A, Cohen, S, Heath, G. (2016, May). On the path to sunshot: the environmental and public health benefits of achieving high penetrations of solar energy in the United States. NREL. Available from: http://www.nrel.gov/docs/fy16osti/65628.pdf.

[39] Sener C, Fthenakis V: Energy policy and financing options to achieve solar energy grid penetration targets: accounting for external costs, *Renew Sust Energy Rev* 32:854–868, 2014.

[40] Fthenakis V: Considering the total cost of electricity from sunlight and the alternatives, *Proc. IEEE* 103(3):283–286, 2015.

[41] What the duck curve tells us about managing a green grid. (2016). CAISO. Available from: https://www.caiso.com/Documents/FlexibleResourcesHelpRenewables_FastFacts.pdf.

[42] Fast ramping: Without it, the grid's a sitting duck. (n.d.). Enbala Power Networks. Available from: http://cdn2.hubspot.net/hubfs/1537427/NeedforSpeed.pdf?submissionGuid=41eb08c8-555b-49d1-b074-5ded05d82cf9.

[43] Greenlink Group [E-mail interview]. (2016). Available from: http://www.thegreenlinkgroup.com/.

[44] Nikolakakis T, Fthenakis V: The value of compressed air energy storage (CAES) for enhancing variable renewable energy (VRE) integration: a new unit commitment and economic dispatch model for Ireland, *Energy Technol*, 2017doi: 10.1002/ente.201700151.

[45] Grid Modernization Laboratory Consortium. US Department of Energy (August 01, 2017). Retrieved from: https://gridmod.labworks.org/about.

[46] Countries With the Most Wind and Solar Have 10 Times Fewer Outages Than America. (2017, June 19). GTM. Available from: https://www.greentechmedia.com/articles/read/the-countries-with-the-most-wind-and-solar-have-far-fewer-outages.

[47] Morjaria, M, Anichkov, D, Chadliev, V, Soni, S, A grid-friendly plant the role of utility-scale photovoltaic plants in grid stability and reliability, IEEE Power and Energy magazine, May/June 2014.

[48] How Microgrids Enable Optimal Cooperation Among Distributed Energy Resources. (n.d.). Schneider Electric. Available from: http://microgridknowledge.com/white-paper/microgrids-and-distributed-energy-resources/.

[49] CERTS Microgrid. (n.d.). Available from: http://certs.aeptechlab.com/.

Further Reading

[50] Margolis, R, Feldman, D, Boff, D (2017, April 25). Q4 2016/Q1 2017 Solar Industry Update. SunShot DOE. Available from: http://www.nrel.gov/docs/fy17osti/68425.pdf.

5

Sustainable Solar Energy Collection and Storage for Rural Sub-Saharan Africa

Rhys G. Charles*,**, Matthew L. Davies*,**, Peter Douglas†,‡, Ingrid L. Hallin§

*SPECIFIC-IKC, SWANSEA UNIVERSITY, SWANSEA, UNITED KINGDOM; **MATERIALS RESEARCH CENTRE, SWANSEA UNIVERSITY, SWANSEA, UNITED KINGDOM; †CHEMISTRY GROUP, MEDICAL SCHOOL, SWANSEA UNIVERSITY, SWANSEA, UNITED KINGDOM; ‡UNIVERSITY OF KWAZULU-NATAL, DURBAN, SOUTH AFRICA; §FREELANCE RESEARCHER, EDMONTON, AB, CANADA

r.charles@swansea.ac.uk, rhys.charles@hotmail.com; m.l.davies@swansea.ac.uk

5.1 Introduction

We live in a warming world; we burn enormous quantities of fossil fuels; climate change is one of the most pressing issues facing humanity, and yet in 2014, more than 1.2 billion people (16% of the world's population) were without access to electricity, and over half of these were in sub-Saharan Africa (SSA) [1]. How can the ever-growing energy needs of the developing world be met while avoiding the climatic consequences of an accelerating use of fossil fuels? Solar energy is one of the 'environmentally friendly' energy sources, widely touted as at least part of the answer to this question. There are a number of solar energy conversion systems available and/or under development [2]. Of these, photovoltaic energy generation is the most developed, with a well-established manufacturing and commercial base and rapidly expanding deployment. In 2016, the global photovoltaics (PV) market was 77.3 GW, with 320 GW installed capacity. Global PV power consumption has reached 333 TW h, accounting for 1.3% of electricity generated [3]. When considering Africa, installed PV has surged from 500 MW in 2013 to 2100 MW at the end of 2015. About 65% of installed capacity in the continent is in South Africa (1361 MW), Algeria accounts for 13% (274 MW), and Egypt for 1% (25 MW). Uganda, Namibia, and Kenya also account for around 1% each, with between 20 and 24 MW each. In 2015, South Africa and Algeria installed 710 MW, accounting for 95% of the 751 MW of installed capacity for the continent that year [4].

However, although usually viewed as a 'green technology', PV electricity generation and storage have major environmental impacts associated with production, use, and disposal. And, as we discuss here, truly sustainable PV solar energy generation and storage will remain an elusive goal until it is fully integrated into a circular economy [5].

A Comprehensive Guide to Solar Energy Systems. http://dx.doi.org/10.1016/B978-0-12-811479-7.00005-1

In countries with a widespread electricity grid, such as the United Kingdom, PV is used primarily for in-grid, domestic, and small unit generation. Because in-grid systems do not require localized energy storage, in-grid PV can be considered separately from energy storage. However, one of the great advantages of PV solar energy generation for those regions of the world without a good electricity grid is the possibility for small-scale localized off-grid generation. For off-grid PV, solar energy generation *and storage* technologies are required, and sustainable energy storage becomes atleast as important as sustainable energy generation. It is with this combined problem, sustainable energy *and storage* that this chapter is concerned.

To explore off-grid solar in the United Kingdom, D.A. Worsley's team has constructed a building monitored to test and validate localized, off-grid, solar energy collection and storage at the SPECIFIC Innovation and Knowledge Centre in Swansea University [6]. This ~200 m^2 building demonstrates the 'buildings as power stations' principle being developed at SPECIFIC using technologies embedded into the building envelope to generate, store, and release energy (Fig. 5.1) [7]. The building has a 17 kWp building integrated photovoltaic roof installation, transpired solar air collectors integrated into the south elevation for heating, and 60 kW h capacity of aqueous hybrid ion batteries (C2C certified) for clean and safe energy storage [8].

In Africa, PV offers the possibility of localized electricity generation for the millions of African homes, schools, and clinics, without grid connection, and one of the authors (MLD) worked with a team from Swansea University on a project to install a small-scale off-grid solar energy structure with integrated photovoltaics for an orphanage in Mutende, Lulamba, Zambia [9]. The total installed capacity of the solar cells was 1.6 kWp. Two 12 V, 102 A h lead-acid batteries were used for electricity storage (Fig. 5.2).

In this chapter, we discuss the problem of sustainable solar energy generation and storage for rural SSA, and how generation and storage technologies can be integrated into a circular economy to improve sustainability. We begin with the climate and geography, both physical and human, of the region; then introduce the ideas of the circular economy, discuss silicon PV in light of these ideas, identify a basic energy requirement for rural households using South Africa as an example, consider the sustainability issues over the lifecycle of a low cost PV and energy storage system capable of meeting this requirement, and finally, discuss briefly current developments in PV technology, and how these might be used in this application.

5.2 Geography

SSA comprises 49 countries, covers an area of 24.3 million km^2 (Fig. 5.3), and spans four time zones [10]. This vast region accounts for more than 15% of the Earth's land surface and has landmass in all four hemispheres. More than half of SSA lies between the Tropics of Cancer and Capricorn, and as such the area is mainly influenced by a tropical climate, although there is high variability across the region [11].

FIGURE 5.1 The SPECIFIC 'Active Classroom' with roof integrated PV and salt water batteries.

SSA can be divided into four main regions: West Africa, stretching from Senegal to Chad; Central Africa, from Cameroon to the Democratic Republic of Congo; East Africa, from Sudan to Tanzania; and Southern Africa, from Angola to South Africa (Fig. 5.3) [10]. Topographically, SSA is, on the whole, a plateau tilting downwards to the west, but each of the four regions has its own topographic features, climate, and vegetation zones, all of which influence local economies.

West Africa is covered by humid equatorial rainforests along the coast and transitions to semi-arid steppe land in the interior [10]. With the Sahara Desert directly to the north, the northern-most lands of West Africa are drought-prone and difficult to develop, but oil reserves running from Côte D'Ivoire to Angola sustain the economy for this region.

FIGURE 5.2 **Photograph of build in Mutende, Lulamba, Zambia with twelve 136 W amorphous Si PV panels with an inverter/charger for two 102 Ah lead acid batteries feeding a mains outlet.** A small DC system powers internal lighting (six LED bulbs) a radio and two USB charging points through a small battery bank.

The humid equatorial climate of Central Africa, which lies along the equator, sustains tropical rainforests throughout the region and provides the excellent growing conditions needed for high value crops such as bananas, coffee, oil palm, and cacao. Mining of diamonds, uranium, and manganese also contributes significantly to the economy. The economy of East Africa is based largely in agriculture; however, the topography is extremely varied; Kilimanjaro, the Serengeti plains, and Lake Victoria are all located here. Southern Africa is a mix of lowlands and mountains along the coasts—the Cape Ford range in the west and the Drakensberg Mountains in the east. Interior plateau stretches between them, incorporating the vast Zambezi River, Victoria Falls, and the Namib and Kalahari deserts.

As of 2015, the population of SSA was 949.2 million [12]. African countries are among the fastest growing in the world: the UN predicts 1.3 billion people will be added to

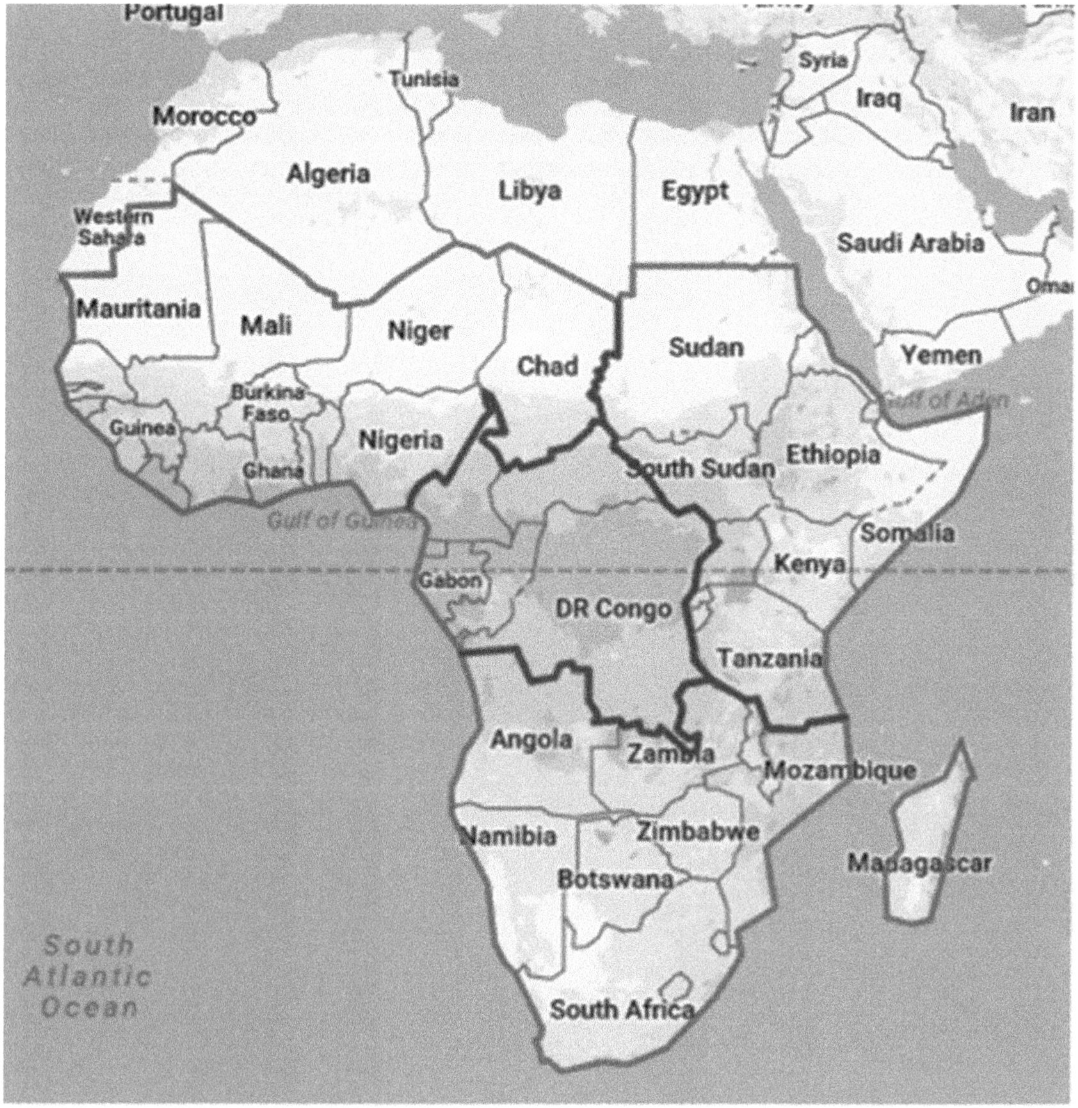

FIGURE 5.3 Regions of sub-Saharan Africa.

the population by 2050 [13]. Decreasing infant mortality rates and increasing fertility rates have resulted in a high under-24 population, and by 2035 the number of working aged Africans (18–64) is expected to exceed that in the rest of the world combined [14]. Currently, the average population density in SSA is approximately 40 people per square kilometer, although the total range across countries is quite high: Namibia has only 2.5 people per square kilometer while Mauritius has 639 people per square kilometer (Fig. 5.4) [14]. Assuming a community with more than 90 000 inhabitants qualifies as 'urban', approximately 37% of SSA is considered to be living in urban areas. SSA has the world's

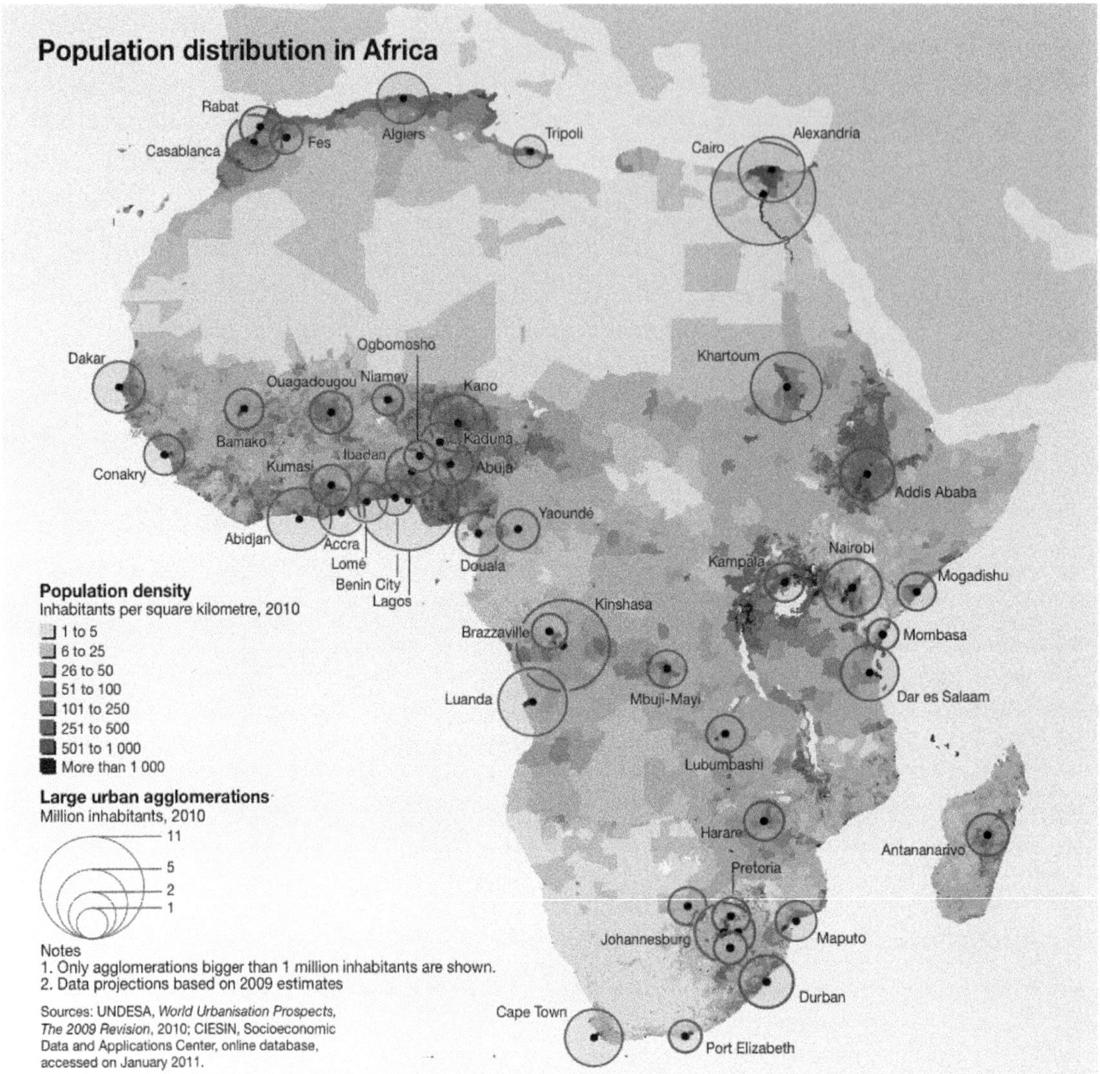

FIGURE 5.4 Population density in Africa [15].

highest rate of urbanization: 4% [14], and by 2050, the UN projects that approximately 55% of the population here will live in urban areas [13]. The growth rate of both total and urban populations is putting more pressure on the land to supply food and energy; more than 80% of SSA still relies on wood for fuel, and without infrastructure in place to reliably supply alternative energy for this growing population, deforestation (and subsequent land degradation) rates will continue to increase [14].

As of 2014, electricity in SSA was provided mainly from coal (45%), hydropower (22%), oil (17%), gas (14%), nuclear (2%) and renewables such as wind and solar PV (<1%) [16]. Diesel-powered generators are frequently used to supplement unreliable electricity supplies in both homes and businesses, and account for approximately 3% of energy usage throughout the region [14,16]. The use of diesel generators causes air pollution through the emission of particulates and NO_x gases, which contribute to climate change, and can cause serious health problems, such as asthma and lung cancer. Furthermore, the cost of electricity generation is approximately twice that of solar PV microgrids over the lifetime of systems [17,18].

Only ~30% of the population of SSA had regular access to electricity in 2014, although that number is rising [16]. Political instability, poorly maintained and inadequate infrastructure, and unaffordable tariffs are major barriers to widespread electricity access. Rural electricity access is very low: in Angola and Chad, less than 5% of the rural population have access to electricity, and electrification has not kept pace with population growth (Fig. 5.5). The International Energy Agency projects that, despite major improvements to electrification across SSA, more than 500 million people living in rural areas will still be without electricity in 2040 [16].

Outside investment in Africa's energy system, particularly from China and USA, has resulted in increased availability of solar energy and other renewable options. The high number of sunny hours makes solar energy an obvious choice to explore for the area, and it is a particularly attractive option for North-eastern and Southern Africa, where annual solar radiation ranges between 2400 and 2800 (kW h) m^{-3} (Fig. 5.6) [16,19]. The solar PV potential in Africa is shown in Fig. 5.7. With this in mind, and because of the Authors' familiarity with the region, further discussion will generally be based on the situation in South Africa. However, it should be noted that the purpose is not to propose South Africa as a model solution for SSA, but to emphasise the importance of considering specific local conditions to identify sustainable PV solutions for SSA, using South Africa as an example.

5.3 The Circular Economy Approach

Fig. 5.8 shows the essential feature of the circular economy approach, an alternative to current 'take-make-use-dispose' linear economic models. Retention of materials within the economy through recovery and regeneration of products at the end of each service life maximizes their economic productivity, offsetting demand for primary resources and decoupling growth from resource consumption. Circular economy is regenerative by design, and replaces the concepts of 'end-of-life' and 'waste' with 'restoration' and 'resources'. Key features include elimination of waste through industrial symbiosis, superior design, appropriate business models, and reverse logistics systems [20]. Cost effective reverse logistics will be a challenge in rural SSA, where technologies are dispersed over vast areas. Circular economy represents a path for sustainable economic development. Resource efficiency, afforded by a circular economy, will yield economic, social, and environmental benefits. The use of renewable energy is also an important feature of circular economy [21]. In light of the rapid increase in population of a working age, the resulting burden on the land and natural resources, and current heavy reliance on fossil fuels in SSA, the

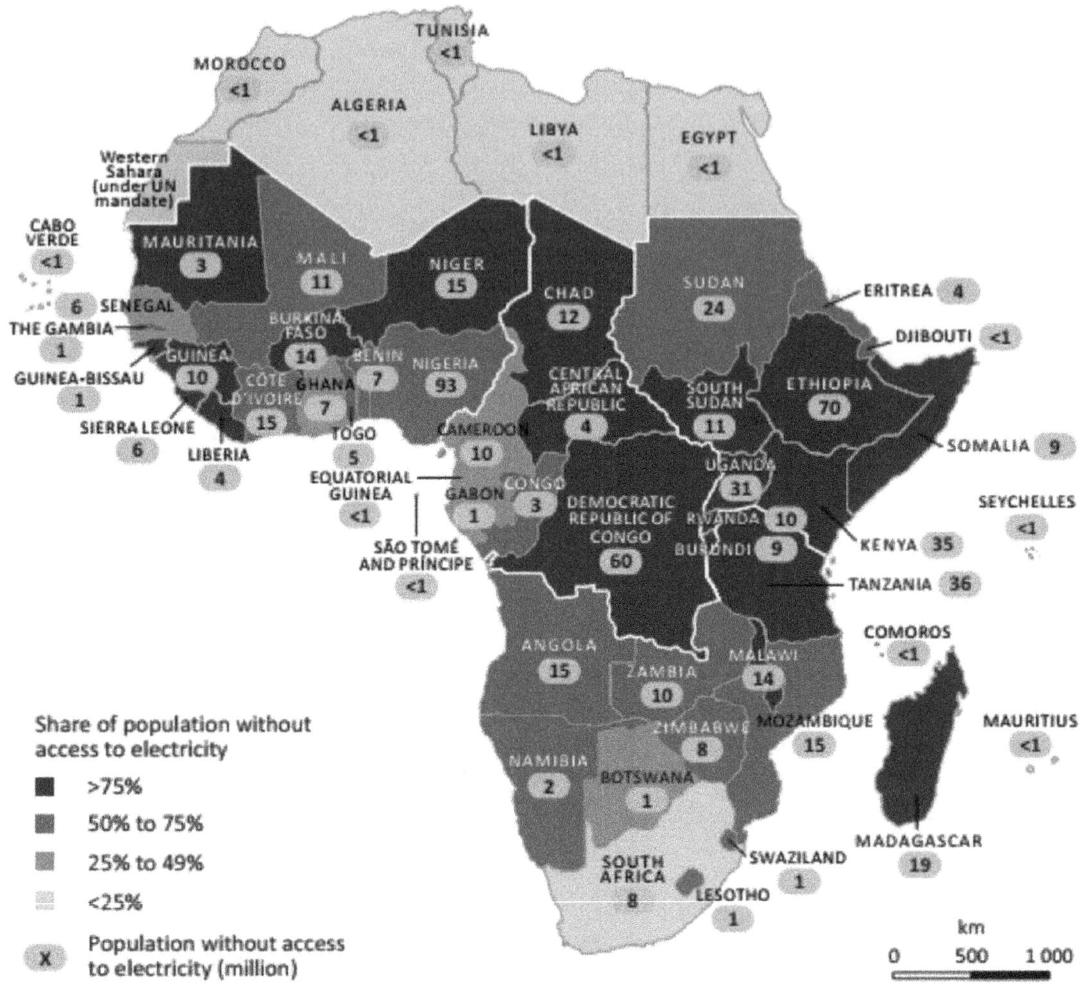

FIGURE 5.5 Electricity access across Africa [16].

requirement for jobs creation through sustainable development and electrification with renewable energy technologies are urgent.

Prioritizing reuse and repair > refurbishment/upgrade > remanufacturing > recycling, results in greatest resource efficiency benefits and larger savings in embedded costs (economic and environmental) of products and components. Whenever costs of reverse logistics and returning products to market are lower than production costs in linear models, circular systems afford greater value than linear alternatives. Benefits are amplified by cycling resources in consecutive product lifecycles and extending the useful life of products. The economic benefits of a circular economy are expected to become more important in the future as the costs of primary raw materials, and safe disposal, rise [22].

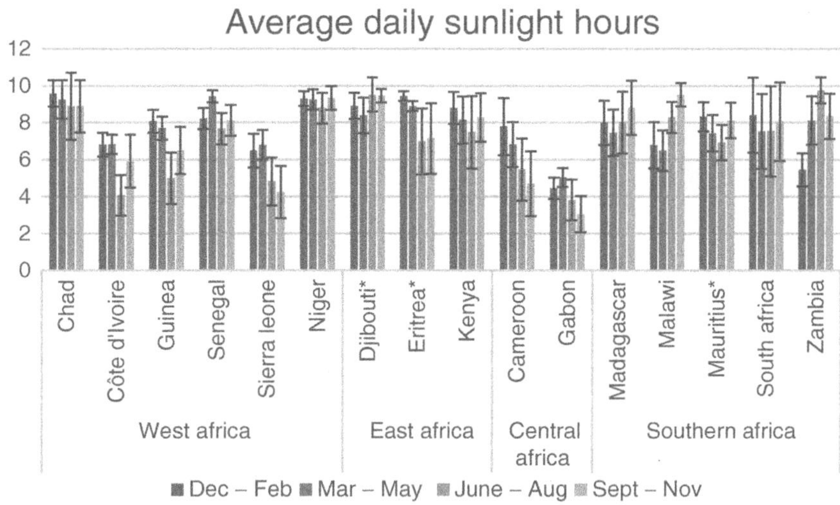

FIGURE 5.6 Average daily hours (±1 SD) of sunlight across sub-Saharan Africa. * Indicates a country with less than 3 monitoring stations available.

Growing recognition of these benefits, rising/volatile resource prices [22], global resource criticality concerns [23], and rising production costs, have made circular economy an attractive prospect, particularly for manufacturers who rely on supplies of critical raw materials. Material criticality issues and environmental impacts associated with the use of toxic materials in devices can also be mitigated with appropriate circular practices. Additionally, intangible company assets such as brand value may be enhanced as consumers become increasingly environmentally aware [22]. To take an example from the PV world, First Solar, the company that produce CdTe PV, have adopted business models that unlock the power of circular economy and generate value through the development of appropriate recycling technologies, long product lifecycles, and linking value chains with other industries and supply chain partners.

The commercial viability of PV is based on the levelized cost of electricity generation, determined by the power conversion efficiency, cost, and lifetime of PV products. Resource efficiency benefits afforded by circular economy can potentially reduce the economic and environmental costs of module production, enhancing commercial viability and increasing competitiveness with alternative renewable energy technologies. In an African context, where cost of systems has been a barrier to adoption of PV, this is of great importance. Energy payback time, emissions associated with electricity generation [CO_2eq $(kW\ h)^{-1}$], and the cost of energy generation (US\$ Wp^{-1}) can all be reduced through adoption of circular practices. Studies have shown that the energy payback time for Si wafer-based PV technologies is reduced by half through use of recycled materials [24]. For CdTe PV, it has been predicted that, as power conversion efficiency improvements are made, and available volumes of end-of-life modules increase, demand for CdTe for PV could be satisfied

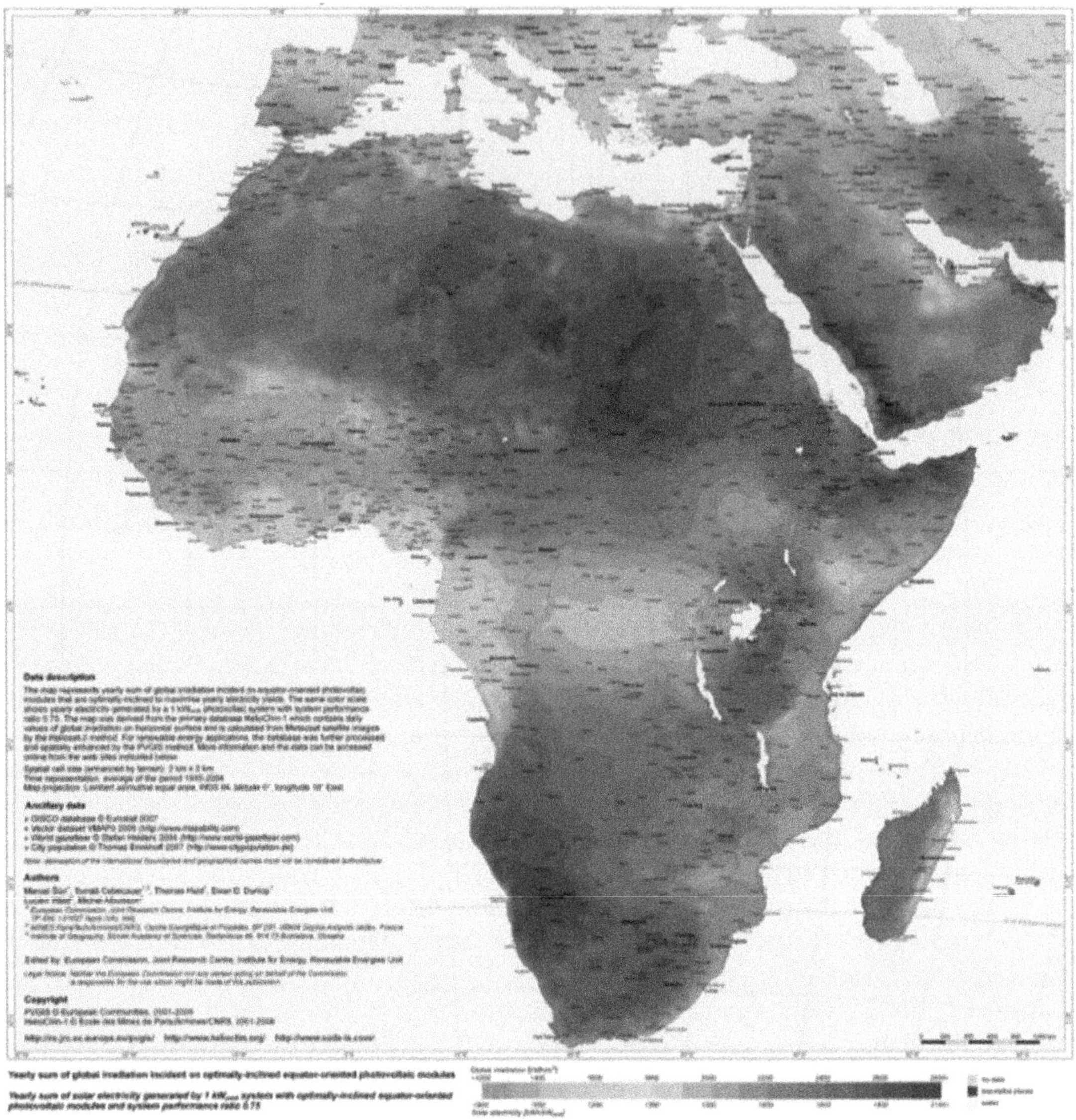

FIGURE 5.7 PV potential across Africa [19].

exclusively with secondary supplies from end-of-life modules [25]. The value of take-back and recovery in this case is enormous. The magnitude of these benefits is determined by the effectiveness of optimizations made to product lifecycles through eco-design, and the greatest benefits are gained when lifecycle optimization for circular economy starts at the earliest possible stage in development of technologies, using a multidisciplinary approach with collaboration between parties involved in all stages of product lifecycles.

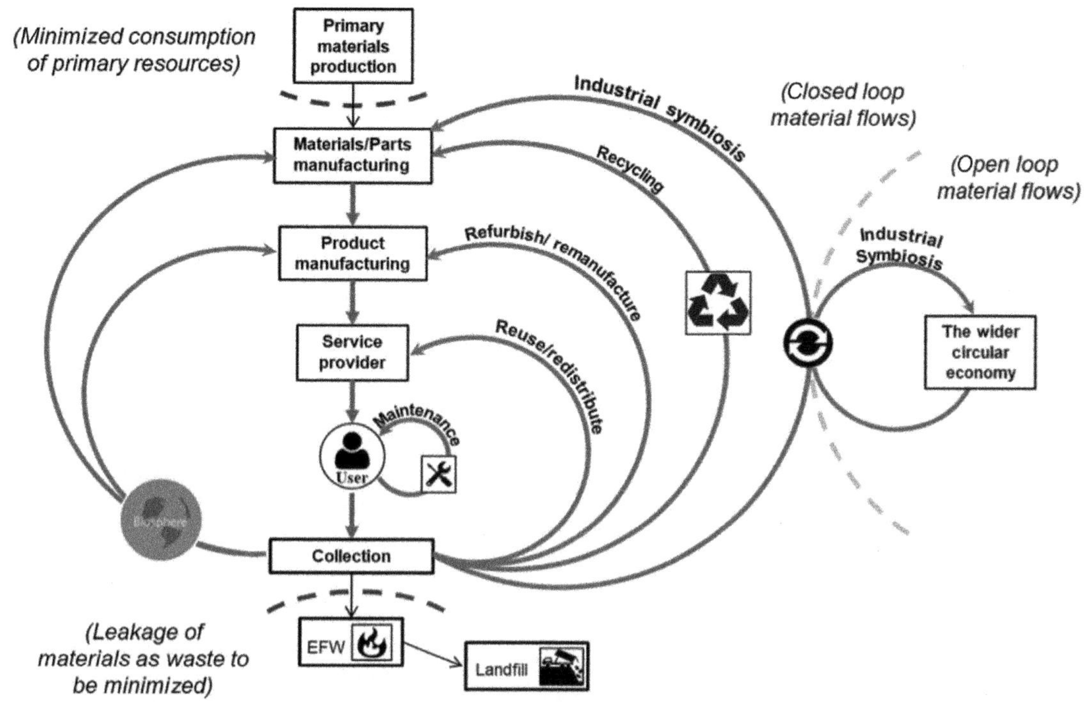

FIGURE 5.8 Material flows within the circular economy. *Adapted from Ellen MacArthur Foundation, Towards a Circular Economy – Economics and business rationale for an accelerated transition, Ellen MAcArthur Foundation, 2013.*

5.4 Photovoltaic Technology

First generation crystalline silicon (c-Si) PV devices are the dominant product on the market today, accounting for ~94% of global PV production [3]. This is the most mature of the PV technologies, and commercial devices exhibit reasonably high power conversion efficiencies and good stability. African governments have set ambitious targets for PV installation. Nigeria aims to install 30 000 MW of PV by 2030, most of this as off-grid systems. Ghana aims to install 30 000 solar home systems by 2020 and invest $230 million into solar energy projects, including mini-grids and stand-alone solar PV systems. Other countries have similarly ambitious targets. The Africa Renewable Energy Initiative has a 30 GW target for installed capacity, and solar PV will be a major component of this [26]. The majority of PV panels installed will be c-Si PV from China, with most of the rest coming from Korea. As production has ramped up in the far East the price of modules has fallen with economies of scale and technological improvements [3].

Panels have a target lifetime of ~25 years, so much of the projected installed capacity will not become waste until this long after their initial installation. However, enormous quantities of PV will require end-of-life management in the nearer future: early PV and low quality panels failing before target lifetimes will already be reaching end-of-life. It is likely that many of these will have arrived in Africa within the large quantities of WEEE that are illegally exported from the developed world, and this quantity is likely to increase significantly into the future. This presents both a potential problem and a potential opportunity.

The silicon used for wafers in panels is a critical raw material which is important for semi-conductor, aluminium alloy, silicone and silane chemical industries, as well as for PV which accounts for 12% of global demand [27]. Manufacturing the panels is a resource and energy intensive process, with production of Si feedstock material and wafer manufacturing accounting for approximately two-thirds of the embodied energy of modules [28]. Wafer production results in a loss of 40% of solar grade silicon as a slurry when wafers are cut from silicon ingots. This slurry is notoriously problematic to recycle [29]. Recovery of wafers is therefore important from an environmental and materials criticality point of view, with reuse of wafers in new modules reducing the carbon footprint by two-thirds, compared to manufacturing from virgin raw materials. When modules reach end-of-life, many of the wafers remain functional. From an economic point of view, wafer recovery is also important. Wafers are the most valuable component of modules, and it is projected that by 2050, PV waste could total 78 million tonnes. The value generated by recovering these materials for injection back into the economy could exceed US\$50 $\times 10^9$ (\$50 billion) [30]. Inverters and other components will also become WEEE at their end-of-life and will require appropriate management to avoid environmental impacts and to generate economic and social value.

An economic opportunity exists for Africa if systems can be established which enable exploitation of this inherent value in c-Si PV waste through reuse, remanufacturing and recycling of domestically generated and imported PV waste. c-Si PV is readily recycled around the world, with PV Cycle in Europe recently achieving record recycling efficiencies for Si PV modules of 96% of module mass [31]. However, current design of c-Si panels embeds wafers within ethyl-vinyl-acetate (EVA), a nonmelting plastic, which makes wafer recovery prohibitively expensive. Consequently, the standard practice is to crush modules to recover lower value materials rather than isolate whole wafers; this recovers <2% of a panel's value [32]. Processes which enable wafer isolation exist, although these are not wide-spread and many remain at the research stage [29].

To the best of the authors' knowledge, no PV recycling currently occurs in Africa, so end-of-life panels would have to be exported for recycling, resulting in a missed opportunity in terms of economic, social, and environmental value for Africa. However, many organizations have established module assembly plants across the continent, aware of the growing opportunities for off-grid electrification in light of the lack of grid infrastructure, and several inverter manufacturers have also established in South Africa [4]. This, atleast, presents opportunities to valorize modules and system components at end-of-life through remanufacturing, given that the necessary skills, knowledge, training, and plants are now available within the continent. It is feasible that the low labor costs, large markets for off-grid PV systems in Africa, and the criticality of Si, might attract organizations to set up PV recycling and remanufacturing in Africa in the future. Achieving this, however, will require significant progress toward the establishment of appropriate reverse logistics systems for the collection of end-of-life modules, and the engagement of companies involved in all stages of PV product lifecycles to establish reverse supply chains. Appropriate legislation, incorporating extended producer responsibility to finance end-of-life costs of PV systems may spark progress in eco-design of modules for easy disassembly for recycling. Detachable frames and glass allow panels to be disassembled for reuse, and using thermoplastics to seal panels in place of EVA plastic allows wafers to be recovered by melting plastics rather than crushing modules [32].

African nations now have a window of opportunity to put in place measures to enable circular economy around the large volumes of c-Si PV and associated system components which will be deployed across the continent in the near future. Organizations such as Mobisol, who offer affordable solar energy solutions in East Africa, have recognized this challenge and opportunity and established a project team to explore recycling of solar components, and already offer a battery recycling service.

5.5 Energy, and Energy Storage, Needs of Households in Rural Africa

To give a basis for discussion, we first need to define the energy needs to be met. In recent years, in South Africa, energy suppliers introduced a scheme to provide 50 kW h month^{-1} 'free basic electricity' to grid-connected households, with a plan to develop off-grid solar powered systems providing 50 kW h month^{-1} to rural households [33]. We have used this 50 kW h month^{-1} as an initial target to achieve with a simple off-grid system comprising of PV panels, a battery, and a charge controller. 50 kW h month^{-1} is ~1.67 kW h day^{-1}, significantly lower than the average daily consumption of South African homes with grid electricity access (~8 kW h day^{-1}), but sufficient for basic commodities such as lighting, TV, radio, cell phone charging, washing, and possibly refrigeration. Table 5.1 provides typical daily energy consumption values (DC appliances are considered to avoid the additional cost of an inverter).

If 90% of this energy was required overnight, then ~1.5 kW h energy storage is needed. Allowing for an annual load growth of 2% year^{-1} over the 20 year lifetime of the PV system requires ~2.1 kW h of battery storage. (We note that increase in electricity demand is such that this year, 2017, the proposed on-grid 'free electricity' allocation in Durban is

Table 5.1 Example Daily Energy Consumption for Small Domestic Appliances

Item	Usage/h	Energy Consumption/kW h day^{-1}
TV (32 in. LCD)	5	0.35
Small DC fridge (50 W)	24	1.2
Compact fluorescent light (one 14 W bulb)	5	0.07
Cell phone charger (5 W)	3	0.01
Washing machine (500 W)	1	0.5
Total energy consumption		2.13

FIGURE 5.9 Photographs of the build in Mutende, Lulamba, Zambia 12 months after installation, demonstrating the problem dust can cause on PV systems in challenging climates. Dust accumulation on PV modules (left) and batteries and inverters, even in enclosed spaces (right), is a major problem.

65 kW h month^{-1}). Battery technology is developing rapidly; here we will concentrate on commercially available technologies which could be put in place immediately.

Operation conditions have a strong influence on battery performance and lifetime. South Africa, and SSA as a whole, is a less than ideal environment for batteries due to high ambient temperature variations, both day-to-day and across the year, and dust (Fig. 5.9). To mitigate this, we propose that batteries be buried at a depth of ~1.5 m, where we expect the insulation and thermal lag from soil cover to limit temperature variations to an acceptable ~12–25°C, thus removing the need for a cooling fan, which will almost certainly be required for most land-surface mounted batteries [34]. We also advocate periodic cleaning of PV panels to ensure power conversion efficiencies are maintained.

To give some idea of what is required in an off-grid solar home system for rural Africa, we have modeled these battery and PV requirements using the Photovoltaic Geographical Information System [35] with Durban as an example location, and estimated the cost of a system capable of generating a minimum of 50 kW h month^{-1}. A 600 Wp photovoltaic system (optimized in terms of orientation and angle) coupled with a 12 V, 250 A h (250 amp-hour) lead-acid battery, with a minimum state of charge cut-off limit of 40% and a daily power consumption of 2 kW h, would yield an average of 56.5 kW h month^{-1} and 678 kW h year^{-1}. We have used a value of R1.3928 (kW h)$^{-1}$ [€0.096 (kW h)$^{-1}$] as the consumer cost of electricity in Durban, giving a yearly value of €65.09 for the 678 kW h. To provide an indicative cost of this system, we have used a price of €0.34 W^{-1} for a multi-Si module [36], totalling €205.32 for 600 Wp, €345.15 for a 250 A h lead-acid battery (based on commercially available batteries in RSA), and €65 for a charge controller and cables. This gives a total initial system cost of €615. Over a period of 20 years, given the typical number of cycles from a lead-acid battery, the battery would need to be replaced roughly once every 4 years (5 batteries in total), bringing the 20 year cost of the system to €1996. Fig. 5.10 shows the cost of the system over time, including these battery changes and the value of the electricity produced, assuming all costs remain constant. What may be surprising to those working in PV is that most of the system cost is associated with the batteries, that is, with the storage of energy rather than its generation. The need to reduce system costs to create cost-effective and affordable PV solutions for rural Africa should be noted.

5.6 Energy Storage—Battery Choices

Four main battery technologies dominate stationary energy storage applications (Table 5.2) [38]. Lithium ion (Li-ion) batteries represent the majority of installed storage capacity, and are commonly used in domestic photovoltaic systems. The merits of lead-acid batteries for applications in rural Africa have already been mentioned in describing their use in the Swansea University Zambia project. Vanadium redox flow batteries (VRFB) require pumps for electrolyte flow and additional energy and storage capacity to support this. This, along with the additional mechanical complexity of VRFB systems, makes them unsuitable for this small-scale application. High temperature NaNiCl batteries are also unsuitable because of the hazards associated with molten metal electrodes.

Cost is of paramount importance for this application, and Table 5.3 compares costs for three candidate technologies for the proposed system: valve regulated lead-acid (VRLA), Li-ion, and Aquion saltwater batteries. Lead-acid batteries are significantly cheaper than lithium-iron-yttrium-phosphate (LFYP), or Aquion, batteries (~1/10th and 1/5th the cost respectively). The easy availability and low capital investment costs of lead-acid batteries are very attractive, but lead-acid has relatively low cycle lives compared with Li-ion and Aquion batteries, and a high sensitivity to deep discharge. Significant oversizing of capacity is therefore required. The relatively short lifetimes for lead-acid batteries mean that these must be replaced 4 times over the lifetime of the proposed system, resulting in a total cost of €1726, which is still only one-quarter the price of the longer lifetime LFYP (€7104) or one-third that of Aquion (€5040) batteries.

Of course, economies of scale and direct bulk purchase from manufacturers may result in lower battery prices. The use of a circular economy approach can also give cost savings by: using remanufactured, refurbished, or repurposed batteries; purchasing batteries manufactured from recovered materials; and valorising end-of-life batteries to recoup costs by diverting them to refurbishment, remanufacturing, and recycling. In addition, opportunities which enhance the longevity of batteries should be explored, to reduce the number of necessary replacements over the system lifetime. Such opportunities for batteries in South Africa are discussed later.

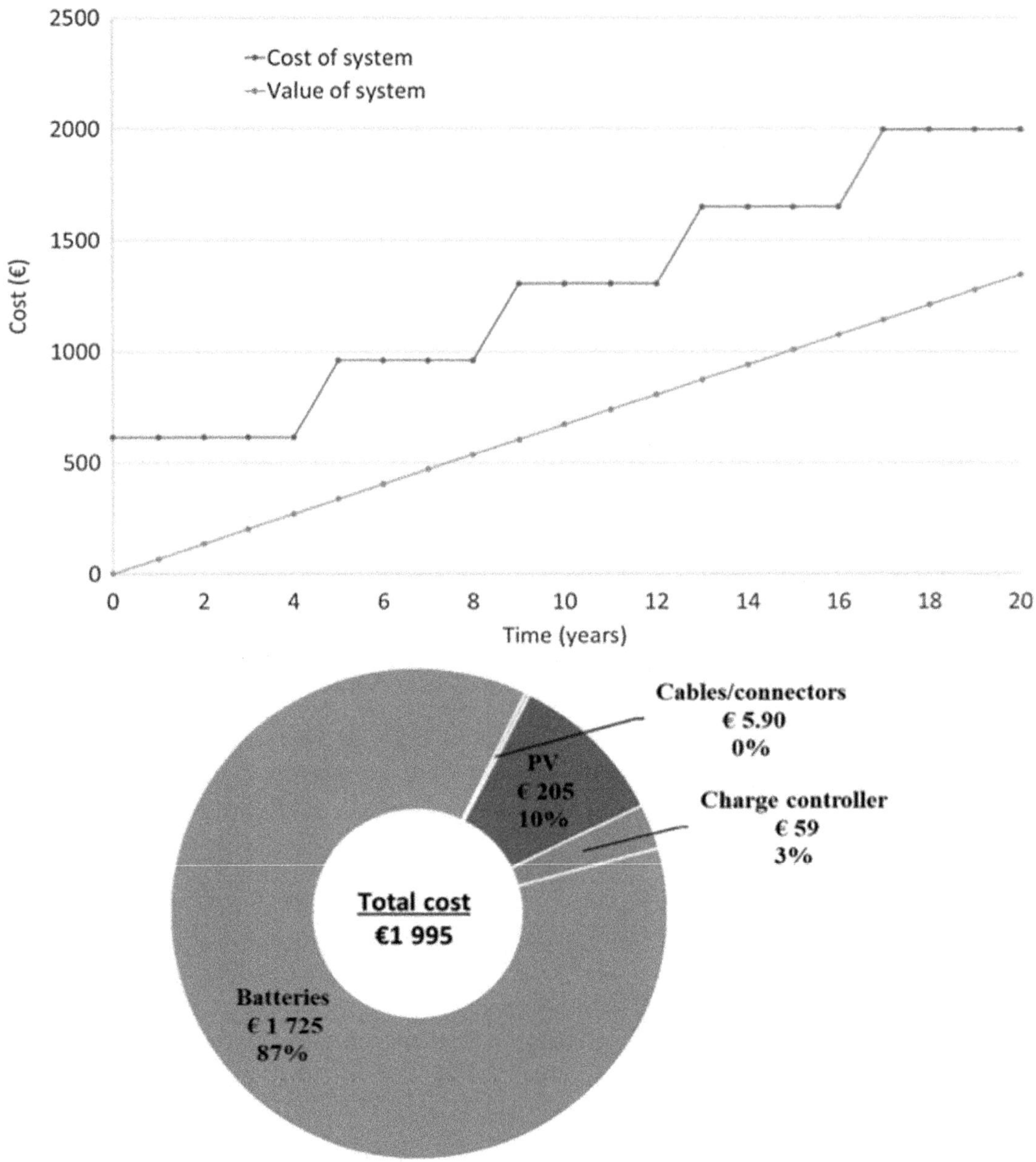

FIGURE 5.10 (Top) The estimated cost of an off-grid DC PV system in Durban, South Africa, consisting of a 600 Wp silicon module, a 12 V, 250 Ah lead-acid battery (and scheduled replacement batteries every 4 years), a charge controller, and cables (top). (Bottom) Breakdown of the estimated cost of the system (€) over 20 years; batteries are the dominant factor in the overall cost.[37].

Table 5.2 Globally Installed Stationary Energy Storage Capacity by Battery Type, 2016

	Installed Capacity	
Battery Technology	**(/MW)**	**(/GW h)**
Li-ion	~1300	1.27
High temperature NaNiCl	171	1.01
Valve regulated lead-acid (VRLA)	196	0.173
Vanadium redox-flow batteries (VRFB)	114	412

Table 5.3 Comparative Costs of Commercially Available Batteries in South Africa Over 20-year Lifetime of System

Battery	Cost/€(A h)$^{-1}$	No. Replacements	Total Cost/€
VRLA	1.38	4	1726
LFYP	14.21	1	7104
Aquion	6.72	2	5040

LFYP, Lithium-iron-yttrium-phosphate battery; *VRLA*, valve regulated lead acid battery.

5.7 Carbon Footprint and Lifecycle Impact Considerations

A comparison of carbon footprints of lead-acid and different types of Li-ion battery production for the system is given in Fig. 5.11 (equivalent data for Aquion batteries is unavailable at the time of writing). When the need for replacement batteries is considered, lead-acid batteries account for a greater contribution to the carbon footprint of the system than Li-ion alternatives, other than lithium titanate batteries (LTO). Lithium-nickel-cobalt-aluminium batteries (NCA) would contribute least to the carbon footprint of the system over its lifetime. Using recovered materials in battery manufacturing could significantly reduce emissions associated with production.

Carbon footprint of batteries production is useful for a comparison of global warming potential. However, this is a limited picture of the environmental impacts of batteries. Further consideration of emissions during production from primary resources should be made; for example, the production of Li-ion batteries from primary raw materials results in considerable SO_2 emissions and water contamination.

Consideration of the hazardous nature of materials within batteries and their potential impacts if improperly managed during use and end-of-life is also important. Issues relating to end-of-life of Li-ion batteries arise primarily from their metal content: Cobalt is present in cathodic materials and electrodes are made with the extremely reactive alkali metal Li. In addition, F, As, and sulfonated compounds are present in electrolytes. Improper treatment of lead-acid batteries at end-of-life results in the release of lead and sulfuric acid to the environment. These materials can directly impact human health through contamination of water and soil, and accumulate in food chains when batteries are landfilled or recycled improperly [39]. This is of particular concern for rural SSA, where batteries are installed in isolated underdeveloped areas and recycling practices lack basic precautions to prevent emissions of these substances to the workplace and environment. Those recycling facilities that are licensed face serious competition from the informal sector, which formerly consisted of small-scale backyard recycling operations which are now increasingly replaced by industrial scale informal smelters [40,41]. Lead poisoning of workers is common and fatal in some cases. The issue of hazardous waste arising from increased deployment of batteries for solar home systems in Africa is significant. In 2016, 1.232 million tonnes of lead-acid batteries were shipped to Africa containing >800 000 tonnes of Pb (equivalent to 10% of global production) [40]. To meet the Nigerian target of 30 GW of installed solar capacity using lead-acid batteries will require an initial installation of over 40 million batteries, with 280 million batteries installed, recovered, and recycled over the lifetime of these systems [26].

The discussion so far indicates that lead-acid batteries are cheapest by a significant margin, and therefore will continue to be widely used for PV systems in place of alternative technologies across Africa. However, their use results in greater global warming potential than most Li-ion alternatives over the 20 year lifetime of the system.

Optimum battery use requires some knowledge of the technology, as does proper handling of waste batteries [40]. Thus any system installation also requires: (1) an additional basic education and training package on the benefits of solar energy, as well as proper operation, maintenance and replacement of components; and (2) full system performance monitoring and analysis for problem/fault prediction/finding.

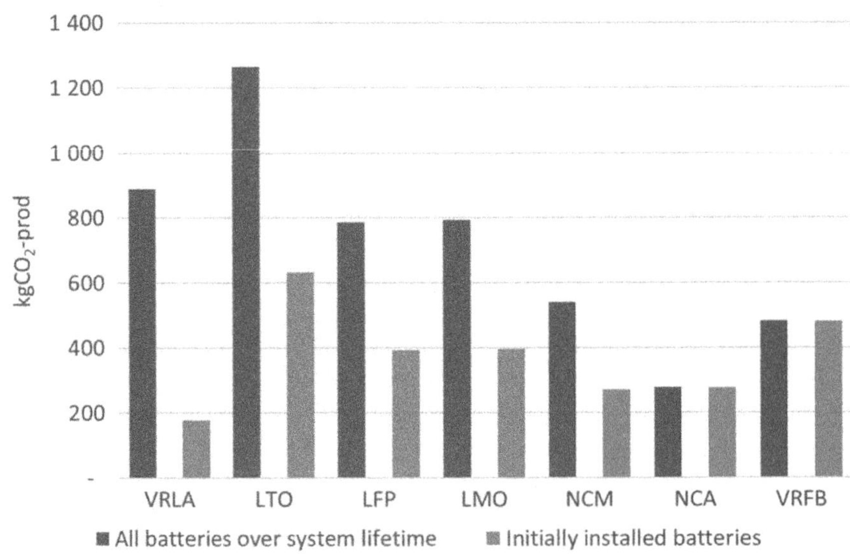

FIGURE 5.11 Carbon footprint of batteries for the system including replacements. *VRLA*, Valve regulated lead-acid; *LTO*, lithium-iron-phosphate with lithium titanate anode; *LFP*, lithium-iron-phosphate with carbon anode; *LMO*, lithium-manganese-oxide; *NCM*, lithium-nickel-cobalt-manganese; *NCA*, lithium nickel-cobalt-aluminium-oxide. *Data used from Baumann M, Peters JF, Weil M, Grunwald A, CO_2 footprint and life-cycle costs of electrochemical energy storage for stationary grid applications, Energy Technol 2017; 7:1071–83.*

5.8 Resource-Efficiency and Circular Economy

5.8.1 Critical Materials

Critical materials used in batteries are shown in Table 5.4, along with their current supply risk index from the British Geological Survey. The high supply risk associated with vanadium may present future resource security issues for VRFBs, further justifying its elimination as a suitable technology for this application. Li-ion batteries face resource security issues due to Li, Co, graphite, and rare earths in the case of lithium-iron-yttrium-phosphate batteries (LFYP), as do Aquion cells which contain graphite. In the interests of global resource security, it is questionable whether technologies containing critical materials should be utilized without further consideration of available infrastructure to support closed-loop recycling, refurbishment and remanufacturing. Lead-acid batteries contain no critical materials.

5.8.2 End-of-life Prospects and Compatibility With Circular Economy

Closed-loop recycling of lead-acid batteries is well established in South Africa. First National Battery operates a network of collection points across South Africa which divert lead-acid batteries to their smelting facility in Benoni for recycling. Recovered Pb and plastics are used to manufacture new batteries with optimized design for disassembly [43]. This suggests end-of-life costs will be low in comparison to other batteries which cannot be recycled domestically, and that lead-acid batteries are an appropriate choice for circular economy in South Africa, with environmental, economic, and social benefits afforded through closed-loop retention of the materials within the South African Economy. Materials cost savings resulting from use of recovered components/materials

Table 5.4 Supply Risks of Materials in Batteries, Critical Materials are Those Highlighted in Bold

Element	Relative Supply Risk Index [42]	Relevant Battery Technology
REEs	**9.5**	LIBs (LFYP)
V	**8.6**	VRFB
Co	**8.1**	LIBs (NMC and NCA)
Li	**7.6**	All LIBs
Graphite	**7.4**	LIBs (LFP, LMO, NMC, and NCA),Aquion
Mn	5.7	LIBs (LMO NMC), Aquion
Ni	5.7	LIBs (NMC and NCA)
Pb	5.5	VRLA
Fe	5.2	LIBs
Ti	4.8	LIBs (LTO)
Al	4.8	LIB

The supply risk index runs from 1 (very low risk) to 10 (very high risk); *LFYP*, Lithium-iron-yttrium-phosphate; *LIBs*, Li-ion batteries; *REEs*, rare earth elements; highlighted elements are included in the EU20 critical list.

should rise with volumes of lead-acid batteries recycled in the future, potentially resulting in cheaper batteries for photovoltaic systems. To maximize this advantage, South African energy companies could engage with First National Battery to optimize circular material flows for lead-acid batteries within South Africa. Several businesses within South Africa operate a lead-acid battery reconditioning service which reverses the sulfation process that limits their working life. This presents opportunities to extend the working lives of lead-acid batteries for the proposed photovoltaic system, improving its economic viability through reduction of battery replacement costs over the system lifetime, and ultimately providing lower cost electrification for rural regions. With this infrastructure in place, opportunities should be explored for utilization of refurbished automotive lead-acid batteries in a second life for the proposed system; these are much cheaper than new batteries, and associated environmental, social, and economic benefits will result from the initiation of new industry around battery refurbishment.

Li-ion batteries are undergoing rapid development and may, in future, challenge lead-acid batteries for this role, but to the best of the Authors' knowledge no Li-ion battery recycling exists in Africa. In South Africa, Li-ion batteries are collected and shipped to Europe for recycling at considerable economic and environmental cost. South Africa also has no Li-ion battery manufacturing through which to valorize recovered materials in closed-loop material flows [44]. This indicates Li-ion end-of-life costs in South Africa will be comparatively high, with significant logistics costs incurred, and little of the social and economic value inherent in Li-ion batteries exploited within South Africa. High costs increase the likelihood of improper end-of-life management, and the resulting potential for impacts on health and the environment is high. However, the South African government has funded research seeking to develop domestic Li-ion battery recycling [44]. Were such an industry to emerge, the derivable economic, environmental, and social benefits from Li-ion battery recycling within South Africa could be improved significantly. No future prospects for Li-ion refurbishment exist at this time. It may be possible to source used automotive batteries for reuse at reduced costs, with associated environmental benefits to the proposed system [45]. As an emerging technology yet to be deployed in Africa, Aquion batteries have few prospects for end-of-life treatment within the continent in the near future.

5.9 Future Solar Cell Technologies

Although they are the dominant product on the market, c-Si PV devices are relatively fragile, expensive, quite heavy, and have relatively high embodied energy compared to following generations of PV [46]. The second generation of thin-film technologies, which includes amorphous silicon (a-Si), cadmium telluride (CdTe), copper-indium-selenide (CIS), and copper-indium-gallium-diselinide (CIGS), have begun gaining market share, accounting for ~7% of global PV production in 2015, with some projections showing an increase to 50% by 2030 [47]. Although giving lower efficiency than c-Si PV, second generation

PV require less materials and energy for manufacturing, and offer lower cost electricity generation, short energy payback time, and reduced emissions associated with electricity generation [48]. In addition, flexible devices can be created. However, manufacturing involves costly vacuum processes, and devices contain toxic materials (such as Cd) and 'critical raw materials' (e.g., In, Ga, Te), the use of which may limit widespread deployment of these technologies [25]. The only PV manufacturing from raw materials to product within Africa to date is carried out by PTiP Innovations in Stellenbosch, South Africa. PTiP is producing thin-film copper-indium-gallium-selenium-sulphite (CIGSSe) modules on glass [49–52].

In light of these issues, we are now witnessing the emergence of printable PV (PPV), which are thin-film devices based on molecular photoactive layers, potentially manufactured from earth abundant materials using cheap roll-to-roll production. Early versions of dye-sensitized solar-cells (DSSC) and organic photovoltaic (OPV) devices for niche applications are now commercially available, and research into new materials, improved device performance, and superior manufacturing processes is ongoing. Flexible OPV products have emerged on the market including Heliatek's Heliafilms for use in building incorporated PV applications, and in the automotive sector for integration with car roofs [53]. Solar phone chargers, solar adhesive tapes, and flexible solar foils are commercially available [54]. Such products are suitable for retrofitting of buildings, windows, and consumer electronics. Perovskite solar cells [55], the newest of the PPV technologies, are yet to emerge on the market as issues with device stability have yet to be fully addressed. However, power conversion efficiencies of lab-based perovskite devices have already reached 22.1% [56], which is comparable to record cell efficiencies for competing thin-film technologies. PPV can be applied vertically to walls and windows, for example, the building incorporated PV DSSCs in the façades of the SwissTech Conference Centre. Additional possibilities arise from the transparent nature of PPV, allowing combinations with existing PV technologies in tandem devices for higher efficiency. The earliest perovskite products will probably be 'tandem cells', in which a perovskite device is combined in tandem with existing PV technologies [51]. This is the goal of Oxford PV who are developing and commercialising thin-film perovskite solar cells for printing directly onto Si or CIGS modules.

PPV is cheap because it can be made using roll-to-roll production on flexible substrates [57] using solution deposition of materials. In such processes, rolls of substrate are run through a series of sequential deposition techniques in which each of the layers of solar cells are deposited as thin films (10 nm–10 μm), with the final coated product recoiled at the end of the line (Fig. 5.12). The result is rapid production at relatively low cost. Substrates include metals such as steel for functionalized building envelopes, or ITO on polyethylene terephthalate (PET) for transparent devices.

The low cost of PPV in comparison to other technologies will make it an interesting option for the African market. Its lightweight nature will also be useful for retrofitting rural homes. However, lifecycle challenges must be addressed for these technologies before they

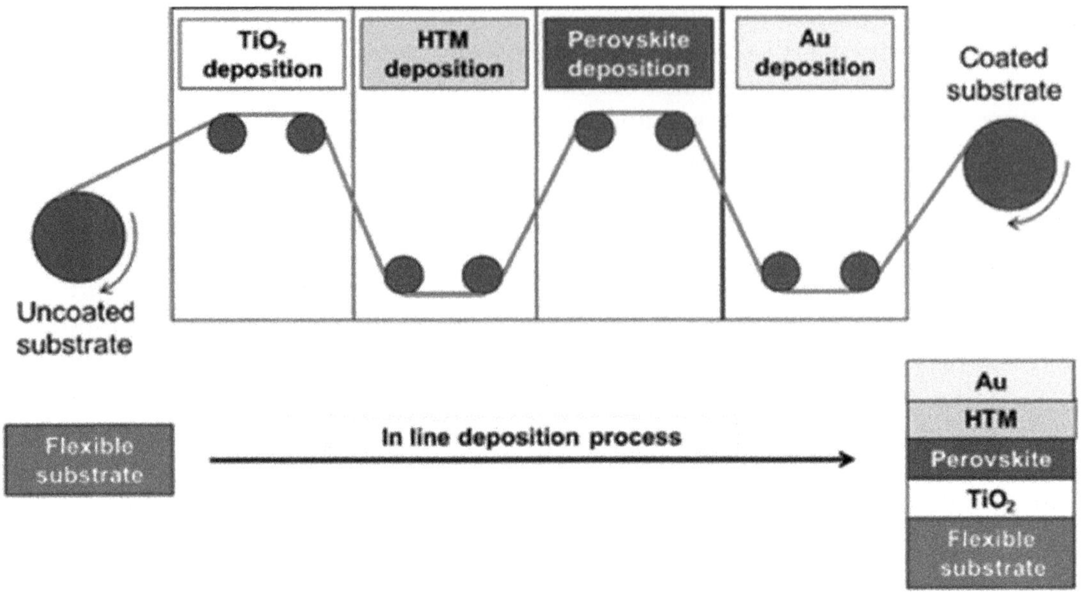

FIGURE 5.12 Principle of roll-to-roll (R2R) production of planar p-i-n perovskite solar cells.

can be considered truly sustainable, or be incorporated effectively into a viable circular economy in South Africa, or elsewhere in SSA.

Numerous lifecycle analysis studies have shown that substrates represent a large proportion of the embodied environmental and economic costs of PV cells [58], and that these costs are lower for PET substrates than glass [59]. Analysis has also shown that laminate materials such as EVA, represent a significant proportion of the embodied costs of thin-film modules [28], making its recovery desirable. So there is a need to develop new delamination methods, new laminate materials, and/or alternative flexible transparent substrates which are compatible with delamination processes.

PPV technologies utilize numerous critical raw materials. The traditional transparent conducting oxide used has been ITO. However, due to the rising price and global criticality of indium, resulting from supply bottlenecks and demand for ITO for flat screens, this has been replaced by FTO on glass in solar cell applications. Replacement with FTO on PET has been problematic due to the high temperature PVD process used for FTO deposition, and degradation of PET in the process. Mitigation strategies enabling ITO substitution may result from research into new low temperature methods for deposition of transparent conducting oxides such as RF magnetron sputtering [60–62]. Substitution with graphene-coated PET may be a solution, however, cost is currently prohibitive for commercial applications. Carbon nanotubes, which can be printed on substrates, have also been explored as conducting layers [63]. Another potential critical raw materials issue arises from the use of Ru in the dyes for DSSCs, and there is much work on fully organic dyes as replacements [64]. Their use mitigates the criticality issue associated with Ru, and they are compatible with current dyeing processes.

An alternative strategy to substitution for mitigating resource criticality issues is to decouple supply from primary production by developing secondary supplies from within the circular economy, including supplies from end-of-life devices and cascaded materials derived from wastes available within the circular economy, that is, industrial symbiosis. Examples of lab-scale processes for production of PPV materials from waste include the production of perovskites from lead-acid car batteries [65], the production of carbon-based counter electrodes from batteries [66], the use of conductive glass from TFT-LCD screens as DSSC counter electrodes [67,68], and the generation of platinised counter electrodes for DSSCs from waste thermocouples [69].

Plastic substrates are derived from crude oil, so biologically derived alternatives are an environmentally attractive prospect. Transparent flexible substrates composed of cellulose nanocrystals can have high transparency and low surface roughness, and these have been used in OPV where they offer easy recyclability due to the solubility of the substrates in water [70]. Where plant-derived materials are used, there is also a carbon sequestration benefit.

Priority research areas to enable full lifecycle optimization include: methods of module lamination/delamination which do not degrade material components of cells and modules; substitution of critical raw materials; processes for generation of secondary resources from 'wastes' available within the circular economy; development of biologically derived components such as cellulose based substrates; and methods which enhance resource and energy efficiency of roll-to-roll manufacturing, such as solvent capture, and recovery of production scrap.

5.10 Conclusions

The convergence of energy demand, population growth, sunlight, and low cost solar cells, suggests that SSA will see, over the next few decades, a huge and rapid expansion in use of off-grid PV solar systems, particularly for rural locations. This will require an enormous volume of solar cells and batteries. The installation, use, and end-of-life management of these systems may cause serious environmental damage if done in an unregulated, irresponsible way. But, if built on circular economy principles, management of PV systems and their waste components could provide a large, widespread, regional industry, with major economic, environmental and social benefits for the region.

Developing and adapting PV technologies for local conditions, materials availability, available logistic chains, and end-of-life infrastructure, education and local skills, will be essential if the off-grid electrification of rural Africa is to be a success, technologically, economically, and environmentally. It is not only a great challenge, but also a great opportunity, for the scientists, industries, businesses, technologists, and regulators of the region. It is vital that the informal sector of SSA be involved in any solution, and universities of the region are ideally placed to coordinate the great number of stakeholders who must be involved in any viable solution.

We have considered the solar cell and battery requirements for a small-scale rural solar energy collection and storage system capable of generating a minimum of 50 kW h month^{-1}. The major economic and environmental costs of such a system are the batteries. The current best choice batteries for SSA are lead-acid batteries, despite lower efficiencies and shorter lifetimes than Li-ion and Aquion batteries. This is justified by the ready availability of lead-acid batteries, the domestic manufacturing capabilities in South Africa and elsewhere in SSA, relatively low costs, and existing infrastructure for refurbishment and closed-loop recycling. Maximum circular economy benefits would be achieved if energy companies engaged with key organizations involved at all stages of lead-acid battery lifecycles, and considered appropriate business models to maximize return of batteries at end-of-life such as 'lease and take back' schemes, or deposit schemes for batteries.

Future developments for Li-ion batteries initiated by the South African government may in time enhance the benefits of Li-ion for this application, however, high costs, critical materials issues, and poor prospects for refurbishment and remanufacturing cast doubt over the suitability of this technology for the proposed system. Even with end-of-life infrastructure in place for Li-ion batteries, the ability to refurbish lead-acid batteries, thereby extending their working life, presents greater opportunities for valorization than could be achieved with recycling, and so their use holds greater potential value for SSA economies, and therefore greater opportunities for sustainable development and employment creation.

Whichever battery technology is used, optimum battery use, maintenance, and disposal will require knowledge of the technology. Thus any system installation also requires an additional basic education and training package on the benefits of solar energy, the proper operation, maintenance, and replacement of components, and full system performance monitoring and analysis for problem/fault prediction/finding.

References

[1] World Energy Outlook, WEO 2016 Electricity Access Database; International Energy Agency (IEA), 2016. Available from: http://www.worldenergyoutlook.org/resources/energydevelopment/energyaccessdatabase/.

[2] Kalogirou S: *Solar energy engineering*, 2nd ed., Boston, 2009, Academic Press.

[3] Fraunhofer ISE and PSE AG, Photovoltaics Report; Fraunhofer Institute for Solar Energy Systems; Freiburg, 2017.

[4] IRENA: *Solar PV in Africa: costs and markets*, International Renewable Energy Agency, 2016.

[5] Charles RG, Davies ML, Douglas P, Third generation photovoltaics; early intervention for circular economy and a sustainable future. In: Electronics Goes Green 2016+ (EGG) 2016; 1–8.

[6] SPECIFIC, Buildings as Power Stations; 2016. Available from: http://specific.eu.com/.

[7] SPECIFIC, Active Classroom; 2016. Available from: http://www.specific.eu.com/assets/downloads/casestudy/Active_Classroom_Web_Case_Study.pdf.

[8] blue.sky energy gmbh, Clean Energy Requires Clean Batteries – Aquion: the world's safest and cleanest energy storage for yoru home; 2014. Available from: http://www.bluesky-energy.eu/aquion-the-saltwater-battery/?gclid=CKmor_-O8tMCFe6_7QoduU4CRg.

[9] Swansea University, Zambia Summer Expeditions; 2016. Available from: http://www.swansea.ac.uk/engineering/zambia/.

[10] Attoh SE: *Geography of Sub-Saharan Africa*, 3rd ed., Upple Saddle River, NJ, 2010, Pearson Education Inc.

[11] NEO, Land Surface Temperature [DAY] – 1 Month (TERRA/MODIS); 2017. Available from: https://neo.sci.gsfc.nasa.gov/view.php?datasetId=MOD11C1_M_LSTDA.

[12] UN, Population, consumption and the environment 2015; Department of economic and social affairs/population division, United Nations, New York; 2015.

[13] UN, World population prospects: 2015 revision; Department of economic and social affairs, United Nations, New York, 2015.

[14] *Frame I, Africa South of the Sahara 2017*, 46th ed., Routledge, New York, 2016, Europa Publications.

[15] Riccardo Pravettoni, Green Hills, Blue Cities - an Ecosystems Approach to Water Resources Management for African Cities, www.grida.no/resources/8200, 2011.

[16] Birol F: *Africa energy outlook*, Paris, France, 2014, IEA.

[17] Knippertz P, Evans MJ, Field PR, Fink AH, Liousse C, Marsham JH: The possible role of local air pollution in climate change in West Africa, *Nat Clim Change* 5:815–822, 2015.

[18] Akinyele DO, Rayudu RK: Techno-economic and life cycle environmental performance analyses of a solar photovoltaic microgrid system for developing countries, *Energy* 109:160–179, 2016.

[19] Huld T, Müller R, Gambardella A: A new solar radiation database for estimating PV performance in Europe and Africa, *Solar Energy* 86:1803–1815, 2012.

[20] Ellen MacArthur Foundation, Towards a Circular Economy – Economics and business rationale for an accelerated transition, Ellen MAcArthur Foundation, 2013.

[21] Bunting G, Huxtable L, Clement M, Network – Integration and communication: a clear route to sustainability? In: 13th International conference of the greening of industry, Cardiff, UK, 2006.

[22] Lacy P, Keeble J, McNamara R, Rutqvist J, Circular advantage: innovative business models and technologies to create value in a world without limits to growth, Accenture; 2014.

[23] EC, Report on critical raw materials for the EU – report of the Ad hoc Working Group on defining critical raw materials, European Commission; 2014.

[24] Müller A, Wambach K, Alsema E, Life cycle analysis of solar module recycling process. In: Proceedings of MRS fall 2005 meeting, Boston, MS; 2005.

[25] Marwede M, Reller A: Future recycling flows of tellurium from cadmium telluride photovoltaic waste, *Resour Conserv Recycl* 69:35–49, 2012.

[26] Schroeder P, Will solar PV create a wave of toxic battery waste in rural Africa?, Institute of Development studies; 2016. Available from: http://www.ids.ac.uk/opinion/will-solar-pv-create-a-wave-of-toxic-battery-waste-in-rural-africa.

[27] EC, Study on the review of the list of critical raw materials: critical raw materials factsheets, Publications Office of the European Union, Brussels; 2016.

[28] De Wild-Scholten MJ: Energy payback time and carbon footprint of commercial photovoltaic systems, *Solar Energy Mater Solar Cells* 119:296–305, 2013.

[29] Tao J, Yu S: Review on feasible recycling pathways and technologies of solar photovoltaic modules, *Solar Energy Mater Solar Cells* 141:108–124, 2015.

[30] Aspire Africa, Solar PV recycling offers significant untapped business opportunity, Clack Global Media; 2016. Available from: http://aspire-africa.com/2016/06/20/solar-pv-recycling-offers-significant-untapped-business-opportunity/.

[31] PV Cycle, Breakthrough in PV module recycling, PC Cycle; 2016. Available from: http://www.pvcycle.org.uk/press/breakthrough-in-pv-module-recycling/.

[32] Coats E, Benton D: *Better products by design: ensuring high standards for UK consumers*, London, 2016, Green Alliance.

[33] Jamal N: Options for the supply of electricity to rural homes in South Africa, *J Energy South Afr* 26:58–65, 2015.

[34] Hillel D: *Introduction to soil physics*, Academic Press Inc, Elsevier Inc, 1982.

[35] European Commission Joint Research Centre (JRC), Photovoltaic geographical information system – interactive maps; 2017. Available from: http://re.jrc.ec.europa.eu/pvgis/apps4/pvest.php?map=africa&lang=en.

[36] European Commission Joint Research Centre (JRC), EnergyTrend PV – PV spot price; 2017. Available from: http://pv.energytrend.com/pricequotes.html.

[37] Charles RG, Davies ML, Douglas P, Atiemo SM, Bates M, Clews A, Mabbett I, Martincigh BS, Mombeshora ET, Morgan JR, Nyamori VO, Worsley DA: Sustainable Solar energy Storage for Rural Africa, 10th International Conference on Sustainable Energy and Environmnetal Protection, 27–30 June 2017, Bled, Slovenia, Energy Storage: Conference Proceedings, pp 177–186.

[38] Baumann M, Peters JF, Weil M, Grunwald A: CO_2 footprint and life-cycle costs of electrochemical energy storage for stationary grid applications, *Energy Technol* 7:1071–1083, 2017.

[39] Cesaro A, Marra A, Belgiorno V, Guida M: Effectiveness of WEEE mechanical treatment: separation yields and recovered material toxicity, *J Clean Prod* 142:2656–2662, 2017.

[40] Manhart A, Amera T, Kuepouo G, Mathai D, Mng'anya S, Schleicher T: *The deadly business – findings from the lead recycling africa project*, Öko-Institut e.V, 2016.

[41] Factor, PV/battery waste management in the context of rural electrification support on PV/battery waste management for a rural electrification program, Clean Energy Solutions Center, NREL, 2016.

[42] British Geological Survey, Risk List 2015; 2015. Available from: https://www.bgs.ac.uk/mineralsuk/statistics/riskList.html.

[43] ESI Africa, Recycling of Batteries in South Africa, ESI Africa – Africa's Power Journal; 2013. Available from: https://www.esi-africa.com/news/recycling-of-batteries-in-south-africa/.

[44] Knights BJD, Saloojee F, Lithium battery recycling – keeping the future fully charged, Department of Environmental Affairs, Greenfund, Development Bank of South Africa; 2015.

[45] Ahmadi L, Young SB, Fowler M, Fraser RA, Achachlouei MA: A cascaded life cycle: reuse of electric vehicle lithium-ion battery packs in energy storage systems, *Int J Life Cycle Assess* 22:111–124, 2017.

[46] Peng J, Lu L, Yang H: Review on life cycle assessment of energy payback and greenhouse gas emission of solar photovoltaic systems, *Renewable Sustainable Energy Rev* 19:255–274, 2013.

[47] Elshkaki A, Graedel TE: Dynamic analysis of the global metals flows and stocks in electricity generation technologies, *J Clean Prod* 59:260–273, 2013.

[48] Baharwani V, Meena N, Dubey A, Sharma D, Brighu U, Mathur J, Dynamic analysis of the global metals flows and stocks in electricity generation technologies, 2014 Power and Energy Systems Conference: Towards Sustainable Energy (PESTSE 2014), 13–15 March 2014, Bangalore, India.

[49] PTiP Innovations, Thin Film Photovoltaic Technology; 2014. Available from: http://www.ptip.co.za/index.php/technology-overview/thin-film-photovoltaic-technology.

[50] Alberts V: A comparison of the material and device properties of homogeneous and compositional-graded Cu(In,Ga)(Se,S) 2 chalcopyrite thin films, *Semicond Sci Technol* 22:585, 2007.

[51] Alberts V: Band gap optimization in $Cu(In_{1-x}Ga_x)(Se_{1-y}S_y)_2$ by controlled Ga and S incorporation during reaction of Cu-(In,Ga) intermetallics in H_2Se and H_2S, *Thin Solid Films* 517:2115–2120, 2009.

[52] Alberts V: Band gap engineering in polycrystalline Cu(In,Ga)$(Se,S)_2$ chalcopyrite thin films, *Mater Sci Eng: B* 107:139–147, 2004.

[53] Heliatek Gmbh, Heliatek the future is light, Available from: http://www.heliatek.com/en/.

[54] infinityPV ApS, Infinity PV; 2017. Available from: https://www.infinitypv.com/.

[55] Snaith HJ: Perovskites: the emergence of a new era for low-cost, high-efficiency solar cells, *J Phys Chem Lett* 4:3623–3630, 2013.

[56] NREL, Best Research Cell Efficiencies; 2017. Available from: http://www.nrel.gov/ncpv/images/efficiency_chart.jpg.

[57] Shah A, Torres P, Tscharner R, Wyrsch N, Keppner H: Photovoltaic technology: the case for thin-film solar cells, *Science* 285:692–698, 1999.

[58] Lizin S, Van Passel S, De Schepper E, Maes W, Lutsen L, Manca J, et al: Life cycle analyses of organic photovoltaics: a review, *Energy Environ Sci* 6:3136–3149, 2013.

[59] Parisi ML, Maranghi S, Basosi R: The evolution of the dye sensitized solar cells from Grätzel prototype to up-scaled solar applications: a life cycle assessment approach, *Renewable Sustainable Energy Rev* 39:124–138, 2014.

60] Hudaya C, Park JH, Lee JK: Effects of process parameters on sheet resistance uniformity of fluorine-doped tin oxide thin films, *Nanoscale Res Lett* 7:17, 2012.

[61] Xu Q, Shen W, Huang Q, Yang Y, Tan R, Zhu K, et al: Flexible transparent conductive films on PET substrates with an AZO/AgNW/AZO sandwich structure, *J Mater Chem C* 2:3750–3755, 2014.

[62] Park SU, Koh JH: Low temperature rf-sputtered in and Al co-doped ZnO thin films deposited on flexible PET substrate, *Ceram Int* 40:10021–10025, 2014.

[63] Brennan LJ, Byrne MT, Bari M, Gun'ko YK: Carbon nanomaterials for dye-sensitized solar cell applications: a bright future, *Adv Energy Mater* 1:472–485, 2011.

[64] Mishra A, Fischer MKR, Bäuerle P: Metal-free organic dyes for dye-sensitized solar cells: from structure:property relationships to design rules, *Angew Chem Int Ed* 48:2474–2499, 2009.

[65] Chen P -Y, Qi J, Klug MT, Dang X, Hammond PT, Belcher A: Environmentally-responsible fabrication of efficient perovskite solar cells from recycled car batteries, *Energy Environ Sci* 7:3659–3665, 2014.

[66] Nair G, Shafawi M, Irwanto M, Yusoff MI, Fitra M, Mariun N: Performance improvement of dye sensitized solar cell by using recycle material for counter electrode, *Appl Mech Mater* 446–447:823–826, 2014.

[67] Ayaz M, Kasi JK, Kasi AK, Samiullah, Ali M: Toward eco green energy: fabrication of DSSC from recycled phone screen, *Int J Resistive Econ* 4:17–22, 2016.

[68] Chen CC, Chang FC, Peng CY, Wang HP: Conducting glasses recovered from thin film transistor liquid crystal display wastes for dye-sensitized solar cell cathodes, *Environ Technol (United Kingdom)* 36:3008–3012, 2015.

[69] Charles R, From E-waste to green energy: waste as a critical material source for photovoltaic technologies: a case study for industrial symbiosis, In: the asian conference on sustainability, energy and the environment 2015, Kobe, Japan, Official Conference Proceedings 2015; p. 427–439.

[70] Zhou Y, Khan TM, Liu J -C, Fuentes-Hernandez C, Shim JW, Najafabadi E, et al: Efficient recyclable organic solar cells on cellulose nanocrystal substrates with a conducting polymer top electrode deposited by film-transfer lamination, Org Electron 15:661–666, 2014.

Further Reading

[71] WMO: *UNdata: sunshine*, Geneva, Switzerland, 2010, World Meteorological Organization.

PART
3

Thermal Solar Energy Technology

6 Solar Water Heaters

Zhangyuan Wang*, Zicong Huang*, Siming Zheng*, Xudong Zhao**
*GUANGDONG UNIVERSITY OF TECHNOLOGY, GUANGZHOU, P.R. CHINA; **UNIVERSITY OF HULL, HULL, UNITED KINGDOM
zwang@gdut.edu.cn

6.1 Introduction

6.1.1 The Marketing Situation of Solar Water Heaters

Over the last two decades, the global solar market has increased significantly. By the end of 2014, the total capacity of the solar thermal collectors in operation in the world was 410.2 GW, of which 70.6% (289.5 GW) was in China. An overview of the different types of collectors installed in the different economic regions in 2014 is shown in Fig. 6.1 [1]. China is the world leader in respect to solar thermal capacity, focusing on evacuated tube collectors for the purposes of preparing hot water and providing space heating. Europe (47.5 GW) is ranked second, largely due to the contribution of the Germany market and the United States is ranked third due to the large number of installation of unglazed collectors used for swimming pool heating.

With the total market share of 93.3%, the global market of glazed solar collectors grew significantly between 2000 and 2011, as shown in Fig. 6.2 [1]. However, in 2014 the market declined by 15.6% due to the shrinking of the Chinese and Australian markets [1]. The dominance of China is driven by its large population and the interest shown in the solar heating sector. The second largest market for glazed collectors, the European market, peaked at 4.4 GW installed capacity in 2008 and has since decreased to 3.4 GW in 2014. In the remaining markets worldwide, an upward trend was observed between 2002 and 2012, which leveled out in 2013 and 2014.

The energy production of all water-based solar thermal systems in 2014 was 335 TWh, corresponding to an energy savings of 36.1×10^6 t (36.1 million tonnes) of oil equivalent (Mtoe) and 116.4×10^6 t of CO_2 [1]. Provisional numbers for 2015 suggested that in 2015, 357 TWh energy was produced, which included 24.8 GW newly installed solar thermal collectors.

The European Solar Thermal Industry Federation [2] has forecasted that by 2030, as a result of research and development activities and policy scenarios, the European installed capacity will reach 1019 GW, and this will contribute about 15% of the low temperature heat demand. Furthermore, they predict that by 2050, the total capacity could reach 2716 GW thus providing about 129 Mtoe equivalent of solar heat, which is roughly 47% of the overall heat demand in the EU.

A Comprehensive Guide to Solar Energy Systems. http://dx.doi.org/10.1016/B978-0-12-811479-7.00006-3

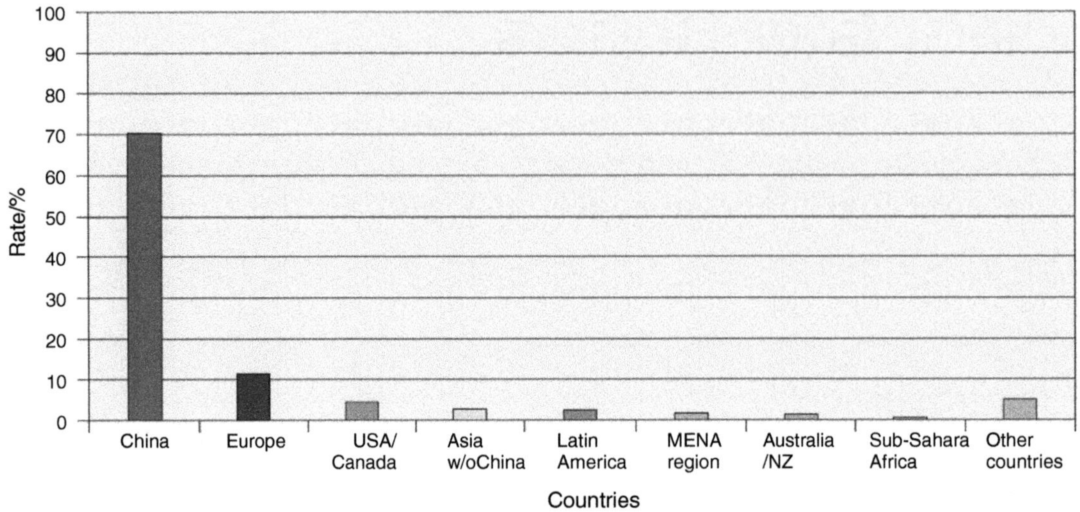

FIGURE 6.1 Total installed capacity of glazed and unglazed water and air collectors shared by economic regions [1].

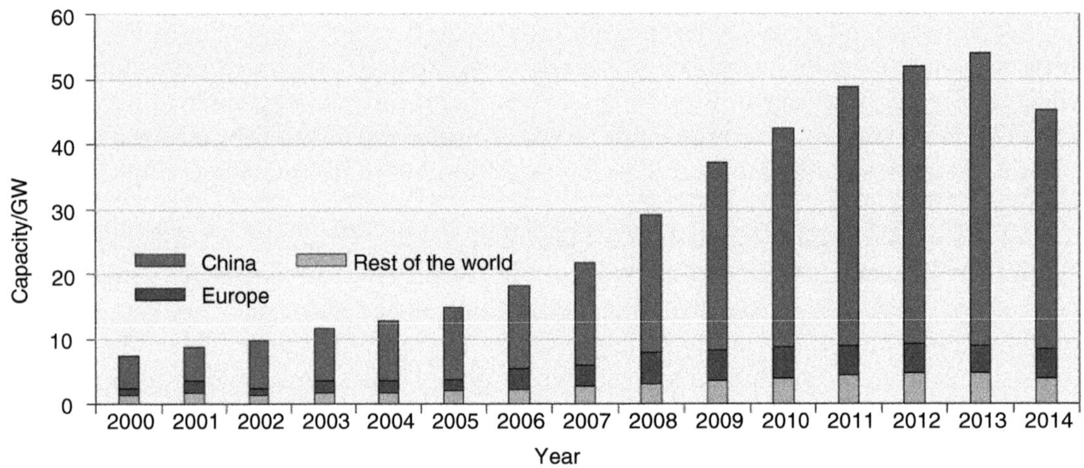

FIGURE 6.2 Annual installed capacity of glazed water collectors from 2000 to 2014 [1].

6.1.2 Driving Forces for the Expansion of the Global Solar Thermal Market

The global solar thermal market is driven by several factors, for example, low costs, financial incentives, regulatory instruments, education, and environmental and local visual impacts.

Solar heating costs vary considerably as a result of weather conditions, the complexity of solar thermal installation, the application of solar thermal system, and the costs of labor and materials [2–4]. Compared to the energy price for heat supplied by conventional energy resources, solar thermal heating can be cost-effective under certain conditions. The energy cost for SWHs range between £7 and £200 per GJ [3].

Financial incentives are used to encourage energy customers to utilize renewable energy sources to meet heat demands and aim to fill the cost gap between the renewable energy sources and conventional energy technologies for use in heating. The incentive schemes can be categorized into financial and fiscal [3,5] schemes. Financial incentives involve direct financial support funded from government budgets, and these include capital grants (subsidy), operating grants, and soft loans. Fiscal incentives include tax credits, reductions, and accelerated depreciation, based on investment costs or energy production.

Governments can intervene in the market by means of regulation, which can include building regulations and standards, and forcing the deployment of renewable energy heating by directly requiring the implementation of SWH technologies [3,6].

Education, to promote renewable energy heating, aims at raising public awareness through information campaigns and training programs. It may take the forms of technical assistance, financial advice, labeling of appliances, and information distribution [3].

The concerns around saving energy and reducing carbon emission could also drive the fast-growing solar thermal market. The manufacturing and decommissioning of solar thermal systems requires relatively small amounts of fossil fuel, thus producing little environmental pollution, and furthermore the SWH systems involve no fossil fuel during its operation [3].

Solar thermal markets could also be driven by local visual impact by placing SWH systems on roofs and building envelopes.

6.1.3 Existing Barriers to the Diffusion of Global Solar Thermal Market

Although the future of solar thermal systems appear to be very positive, there are still barriers to its widespread rollout; these include technical, economic, legal, educational, and behavioral barriers.

Over the past few decades of SWH development, technical problems have focused mainly on the availability of space for the installation of solar thermal collectors and the accompanying; high energy density thermal storage, the availability of appropriate materials for mass productions of collectors, the integration of solar equipment as part of the building's fabric, and the protection of solar collectors from freezing in cold weather [2,7]. Another barrier relating to the utilization of solar energy is that the solar radiation reaching the earth is intermittent, weather-dependent, highly dispersed, and unequally distributed over the surface of the earth in that most of the sun's energy is experienced between 30°N and 30°S [7,8].

Several economic barriers have impacted on the desire for increased utilization of solar thermal in developing countries. These include a lack of public awareness, a lack of energy policies, low levels of income, a lack of subsidies, short-term investing syndrome, and a lack of institutional support. For developed countries, the principal economic barriers are capital cost, poor regulation of promotion, and poor public perceptions [7,9].

The largest barrier is that property developers and building owners have little incentive to invest in energy-saving equipment in new constructions and for rental market. This is because the returns on investments flow directly to the occupants rather than to the building owners or the developers [7]. Another barrier existing in collective dwellings or multistory buildings is that the installation of a single device may become technically complex and would require permits from a majority of co-owners [7]. The diversity of local requirements is another barrier as solar systems need to be considered in terms of their compatibility with existing community aesthetic standards and architectural requirements [7,10]. Finally, some barriers relate to human behavior. These include the reluctance to manage a complex system, the intermittence solar energy for water heating, leading to low comfort levels, and a change in habits [7].

6.2 Working Principle of SWH Systems

The conventional SWH system, as shown in Fig. 6.3 [11], consists of a discrete collector, which is designed to maximize solar absorption and reduce heat losses. The solar collector could be either a black-painted flat-plate absorber bonded to copper piping and covered with a transparent glass (flat-plate collector) or copper tubing surrounded by evacuated and selectively coated glass tubes (evacuated-tube collector). When solar radiation passes through the transparent glass or evacuated tubes and impinges on the collector with its high absorptive surface, a large part of the energy is absorbed by the collector and is then transferred to the fluid to be transported in the pipes. The heat transfer fluid, usually a mixture of water and antifreeze fluid, is either pumped (active system) or driven by natural

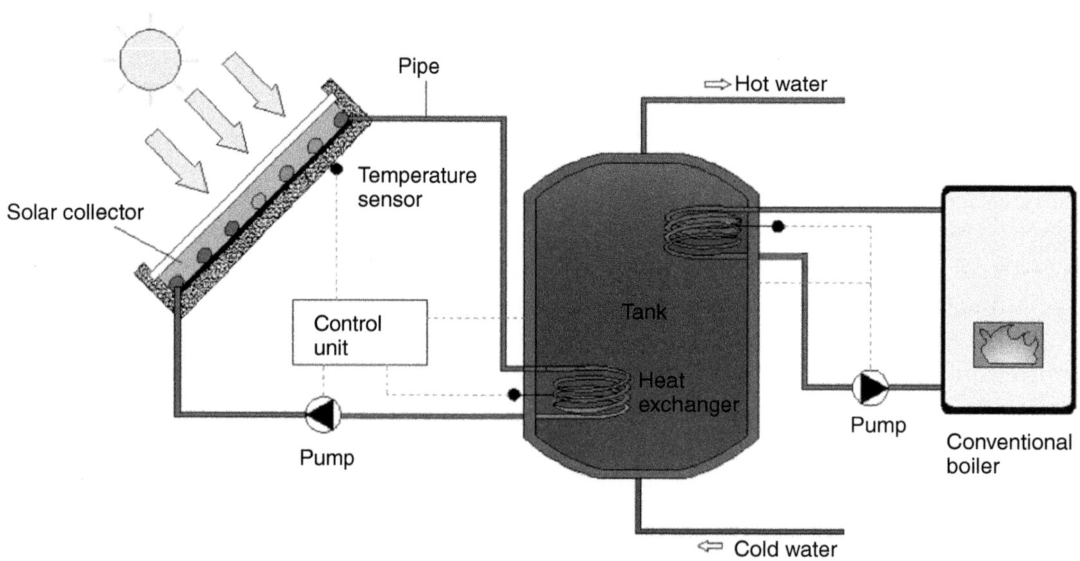

FIGURE 6.3 Schematic of a conventional SWH system [11].

convection (passive system) through the collector to a coil heat exchanger at the bottom of a cylinder tank (indirect system), where the heat is further transferred to a storage tank or is used directly. The tank is usually insulated and may contain an auxiliary heater, for example, electric immersion heater or conventional boiler for winter use [12].

6.3 The Classification of SWH Systems

6.3.1 Passive and Active Systems

Based on whether the SWH system requires pumps or not to function, SWHs are categorized into two basic configurations: passive or active systems, as shown in Fig. 6.4 [13]. Passive systems transfer heat from the collector to the tank located above the collector by natural circulation, which could supply hot water at a temperature of the order of 60°C, and are the most commonly used solar water heaters for domestic applications [14]. Active systems use an electric pump to circulate water through the collector. A check valve may be required to prevent reverse water circulation [13].

The efficiency of an active SWH system is usually between 35% and 80% [15] while that of the passive system is in the range of 30%–50% [14]. An advantage of the active system is that the collector does not need to be close to the tank and hence can be used in multistory buildings. However, the drawbacks of the active system include: its dependent on electricity and the fact that it is more complicated in nature and requires the need of experienced personnel to ensure optimal operation. This leads to the active system, having much higher running costs than the passive system.

6.3.2 Direct and Indirect Systems

SWH systems, which do not include a heat exchanger, are called direct systems (shown in Fig. 6.4) while the SWHs, which are fitted with heat exchangers, are called indirect systems

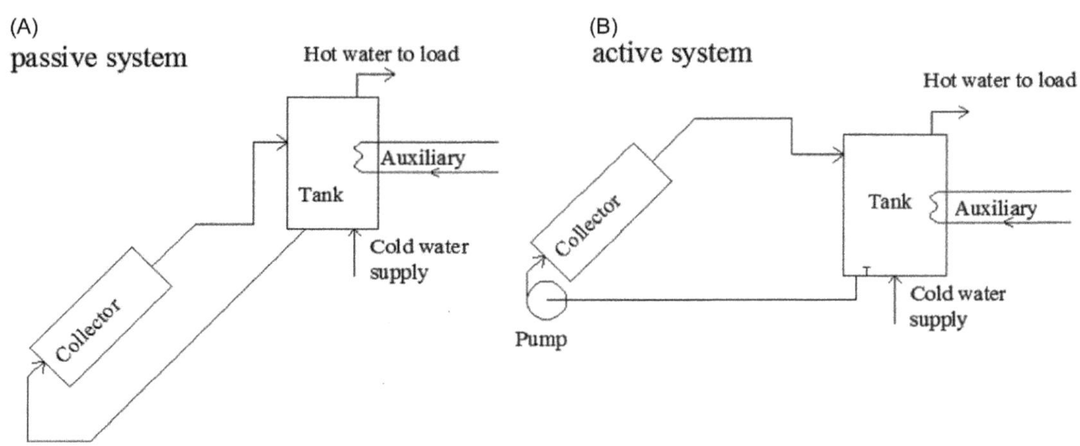

FIGURE 6.4 Schematic of the passive and active SWH systems [13].

(shown in Fig. 6.5 [16]). In a direct system, the service water is directly circulated between the water tank and the collector, while in an indirect system, a heat transfer fluid, usually antifreeze, distilled water, or an organic fluid, is circulated through the solar collector. A heat exchanger is employed to affect the heat transfer from the collector to the service water in the tank. The heat exchanger could be inside or outside the hot water tank as shown in Fig. 6.5A and B.

The indirect system, in most situations, performs better than the direct one. The indirect system is less climate-selective and more suitable for use in regions experiencing cooler temperatures.

6.3.3 SWH Systems in Different Solar Collector Configurations

Solar collectors can have many variations according to their operating temperatures as shown in Fig. 6.6 [17]. Unglazed panels and flat-plate water and air collectors are categorized as low temperature solar collectors. Evacuated-tube, line-focus, and point-focus collectors are classified as high temperature solar collectors.

6.3.3.1 Low Temperature Solar Collectors

Unglazed panels are most suitable for swimming pool heating as it is only necessary for the water temperature to rise by a few degrees above ambient air temperature.

Flat-plate water collectors are the mainstay of domestic solar water heating (SWH) worldwide. They are usually single glazing but may have an extra second glazing. The absorber plate should have a high thermal conductivity to transfer the collected energy to the water with a minimum temperature drop. It usually has a black surface or a selective coating that has both high optical absorption and low emission, to cut heat losses.

Flat-plate air collectors are not as popular as water collectors and flat-plate air collectors are usually used for space heating. One application of this type of collector involves the linking up with a photovoltaic panel to produce both heat and electricity.

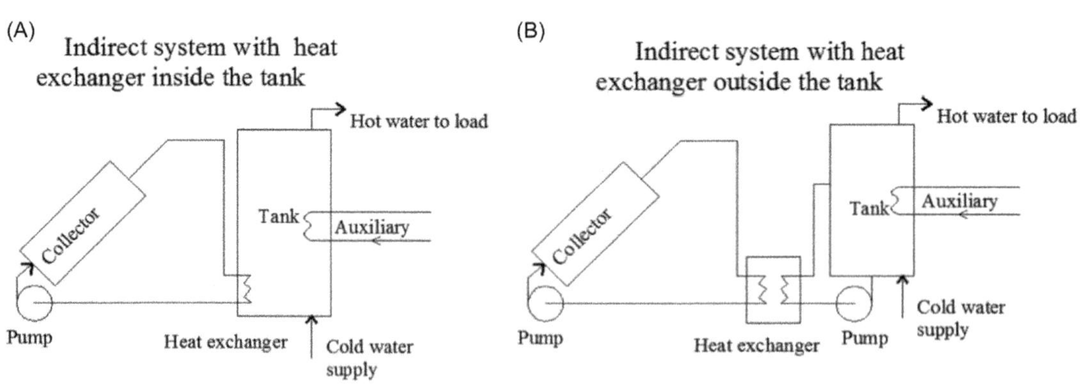

FIGURE 6.5 Schematic of the indirect SWH systems [16].

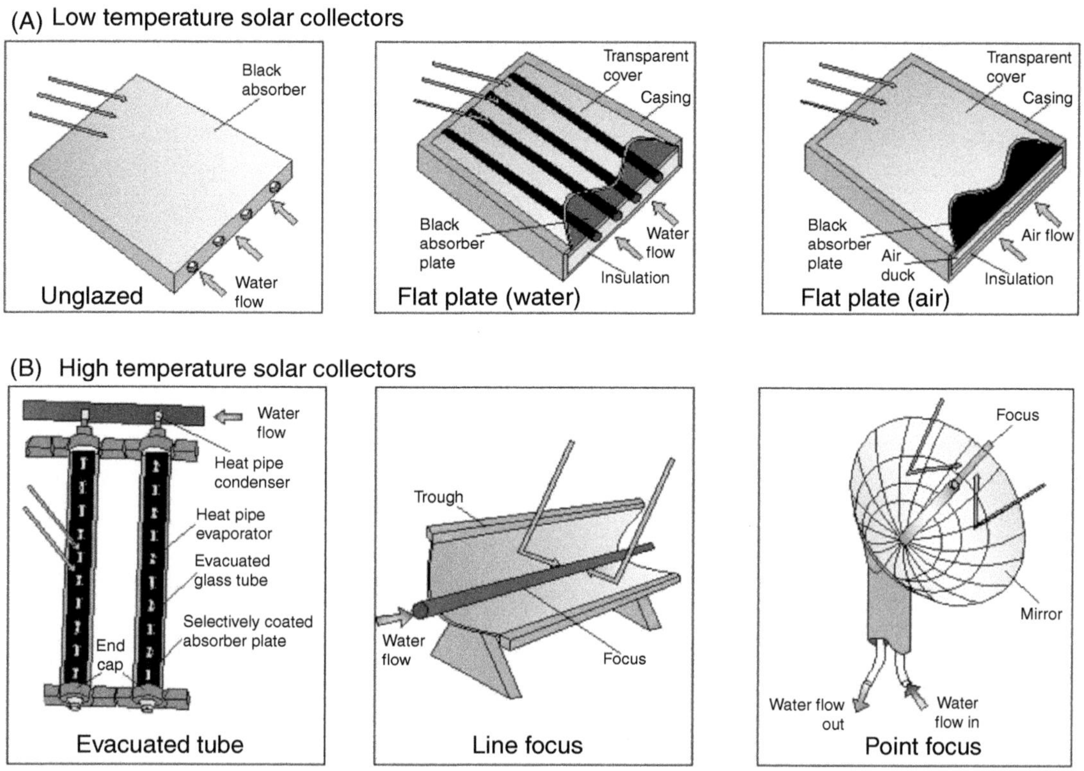

FIGURE 6.6 Classification of solar collectors [17].

6.3.3.2 High Temperature Solar Collectors

An evacuated-tube collector consists of a set of modular tubes, where convective heat losses are minimized by virtue of the vacuum in the tubes. The absorber plate is a metal strip in the center of each tube, and a heat pipe is used to carry the collected energy to the water, which circulates along a header at the top of the pipe array.

Line-focus collectors can concentrate sunlight onto a pipe running down the center of a trough that could be pivoted to track the sun up and down or east to west. It is mainly used for generating steam for electricity plants. A line-focus collector can be orientated with its axis in either a horizontal or a vertical plane.

Point-focus collectors are also used for steam generation or driving a Stirling engine, but needs to track the sun in two dimensions.

6.3.3.3 Comparison of the Evacuated-Tube and Flat-Plate Collectors

Flat-plate SWH systems have been used worldwide owing to their structural simplicity and low cost. However, evacuated-tube collectors with heat pipe arrays are growing in popularity as they have many advantages over the flat-plate collectors. The main features of the evacuated-tube and flat-plate collectors are compared in Table 6.1 [13].

Table 6.1 Comparison of the Evacuated-Tube and Flat-Plate Collectors [13]

	Evacuated-Tube Collector	Flat-Plate Collector
Heat production	Rapid	Slow
Heat losses during daytime	Negligible	High
Influence of the sunrays' incidence angle	Maximum solar absorption through the day	Maximum solar absorption at noon
Cold weather operation	Satisfactory performance	Limiting effect, freezing risk
Maximum operating temperature range	Above 95°C	Up to 80°C
Cost-effective	Advanced technology at competitive price	Old technology at higher price
Hot water availability	Greater number of days throughout the year	Lesser number of days throughout the year
Position of the collector	Assembled onto the roof surface	Preassembled

6.4 Most Advanced Technologies of SWHs

Three kinds of the most advanced technologies of SWH systems: (1) SWHs with PCMs, (2) SWHs with LHPs, and (3) SWHs with MCHPs.

6.4.1 SWHs With Phase Change Materials

PCM, also called latent heat storage material, has a high capability of storing and releasing large amount of heat within a constant or a narrow temperature range [1,18,19]. Two properties that make PCMs attractive in SWH systems are: their compactness and also their small volume change during a phase change [20–22]. The SWH systems, which include PCMs, can be divided into two types: those where the PCM is directly linked to the solar collector and those where the storage unit is filled with the PCM.

A schematic diagram of a flat-plate solar collector involving a PCM is given in Fig. 6.7. The impure PCM surrounds the solar collector tubes and is covered with black absorber [23]. Typically the system can maintain an operating temperature of the collector of under 40°C for 80 min with a constant solar radiation of 1000 Wm^{-2} [22]. Such systems have been shown to have efficiencies of between 42% and 55% higher than that of conventional SWH systems [24].

A refined version of this type of SWH system was investigated by Chen et al. [25] (Fig. 6.8) and has the tubes of the flat-plate solar collector embedded within a high porous aluminum foam incorporating paraffin. They found that the performance of the system was improved significantly compared with the paraffin system without aluminum foam as can be seen in Figs. 6.9 and 6.10 [25].

SWH systems involving PCMs in the storage unit have been investigated by Tarhan et al. [26]. The PCMs they used were lauric acid and myristic acid. The results showed that the lauric acid storage could retain a stable water temperature, and the myristic acid storage could reduce the heat losses during night.

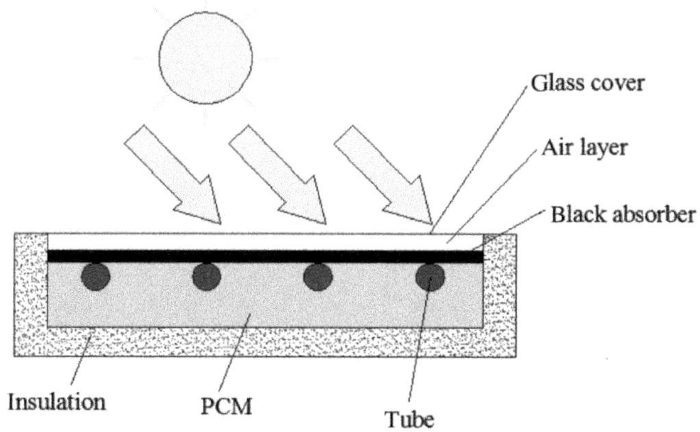

FIGURE 6.7 Schematic of the flat-plate solar collector with PCM [23]. *PCM*, phase change material.

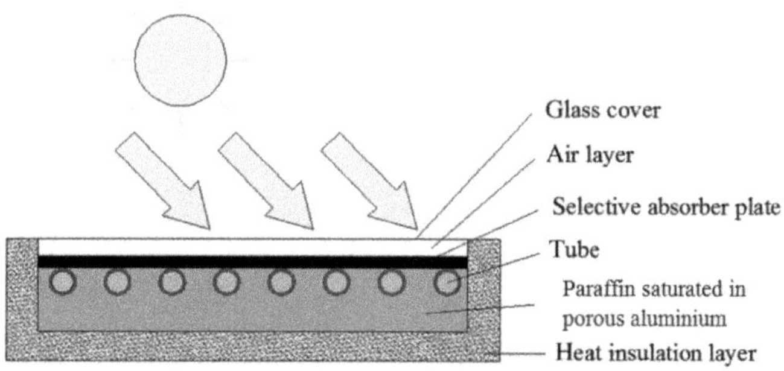

FIGURE 6.8 Schematic of the solar collector with high porous aluminium foam incorporating with paraffin [25].

Fazilati and Alemrajabi [21] investigated the performance of a solar water heater using paraffin wax in spherical capsules as the storage medium in the jacketed shell type tank. It was evaluated that 25% of service time of the SWH system could be prolonged by using PCMs.

A SWH system with a galvanized steel storage tank containing paraffin was investigated by Al-Hinti et al. [27] (Fig. 6.11). The total volume of the storage tank was at 107.4 L with paraffin occupying 49.4 L and the water encapsulated making up the remaining volume (58 L). This water-paraffin PCM storage system reached a temperature of 45°C higher than the ambient temperature after 24 h of operation.

6.4.2 SWHs With Loop Heat Pipe

A LHP [28–31], as shown in Fig. 6.12, is a two-phase (liquid/vapor) heat transfer device allowing a high thermal flux to be transported over a distance of up to several tens of meters in a horizontal or vertical position owing to its capillary or gravitational structure. Such LHP systems consist of an evaporator, condenser, compensation chamber, liquid line, and vapor line and are particularly suitable for applications linked to SWHs.

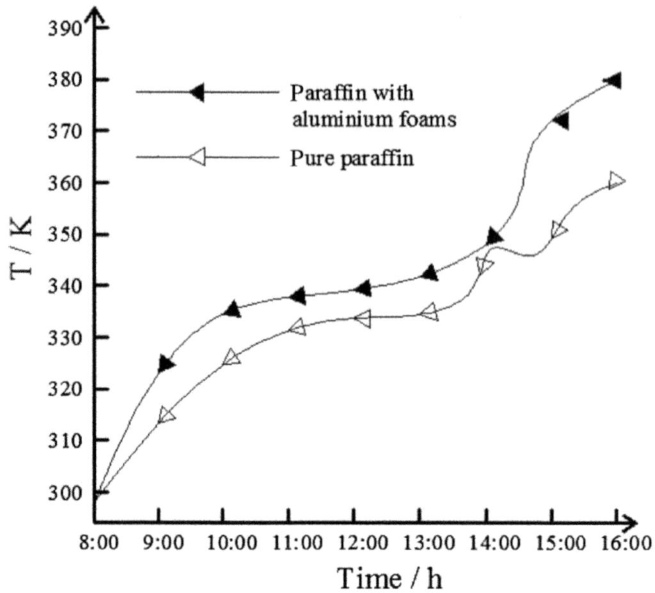

FIGURE 6.9 Variations of the temperatures of the paraffin and aluminium foams in solar collector [25].

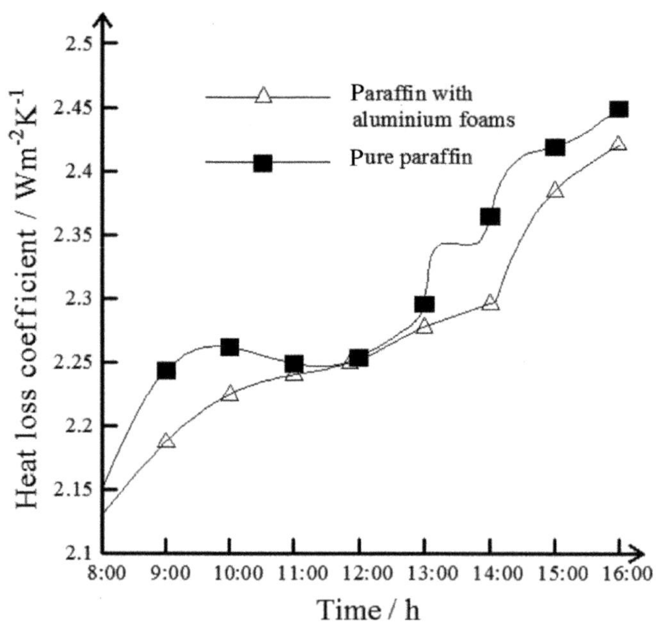

FIGURE 6.10 Variations of the heat loss coefficient of the collector [25].

Wang and Zhao [32] designed a novel LHP-based SWH system (shown in Fig. 6.13) and established theoretical models related to the following six important heat transport capacity properties: capillary action, entrainment, viscosity, boiling, sonic, and filled liquid

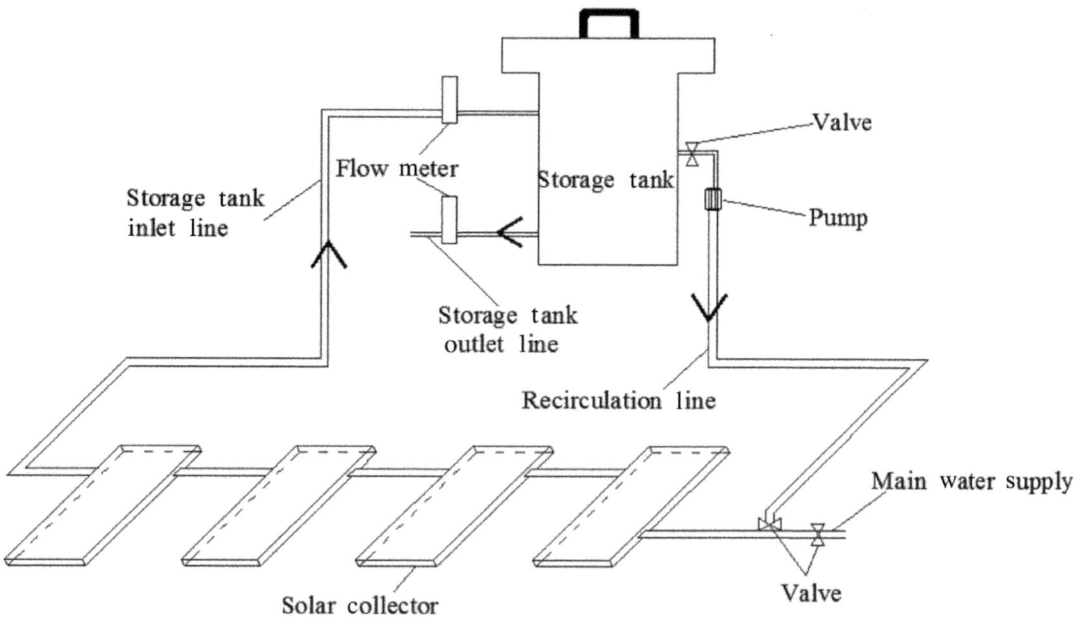

FIGURE 6.11 Schematic of the SWH system with water-PCM storage [27]. *PCM*, phase change material.

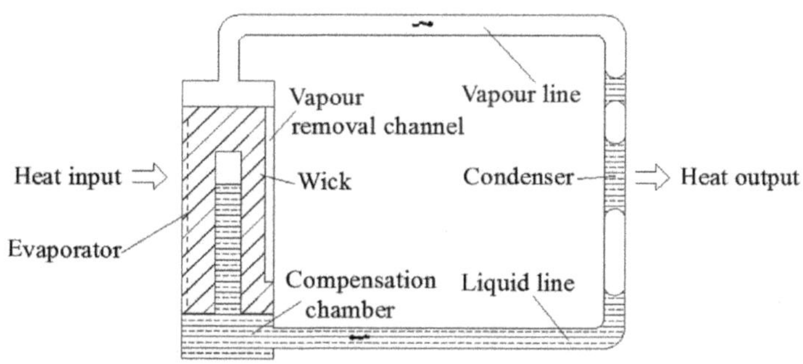

FIGURE 6.12 Schematic of a conventional LHP device [28]. *LHP*, loop heat pipe.

mass. The extremely high vapor speed, for example, as high as sonic or supersonic level, will limit heat pipe's further heat take-up because the heat pipe would be choked. The results show that the capillary was usually the governing limit of the system heat transfer, while the entrainment may be the alternative when the pipe diameter was below 5.6 mm.

He et al. [33] and Zhang et al. [34] proposed a novel heat pump assisted solar facade LHP water heating system (Fig. 6.14). The average thermal efficiency of the LHP module was much better than that of the non-heat-pump-assisted LHP module under the same solar radiation and ambient temperature condition. The water temperature increased throughout the heat pump turn-on period and could reach the maximum temperature of around 58°C.

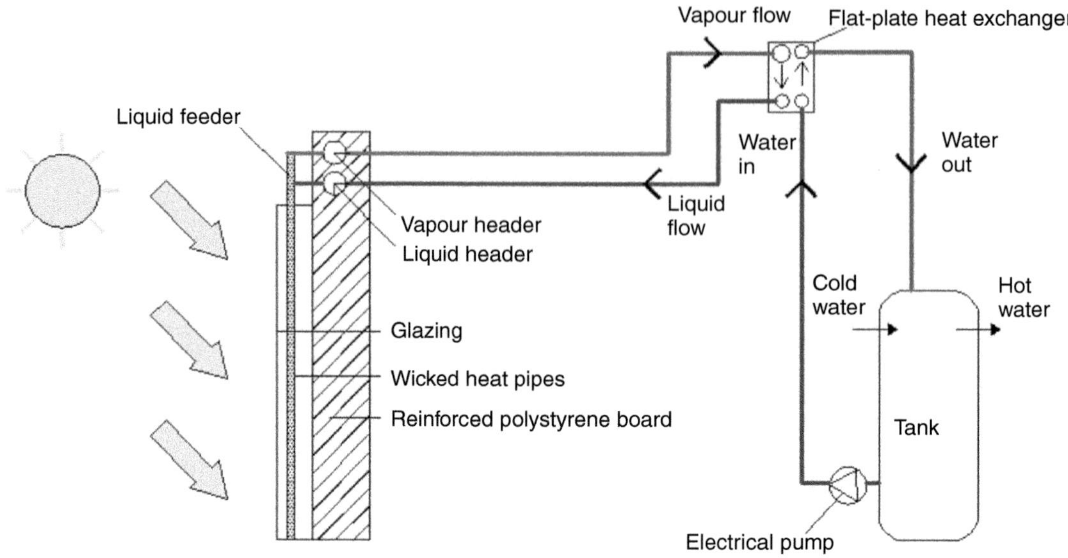

FIGURE 6.13 Schematic of the novel LHP system [32]. *LHP*, loop heat pipe.

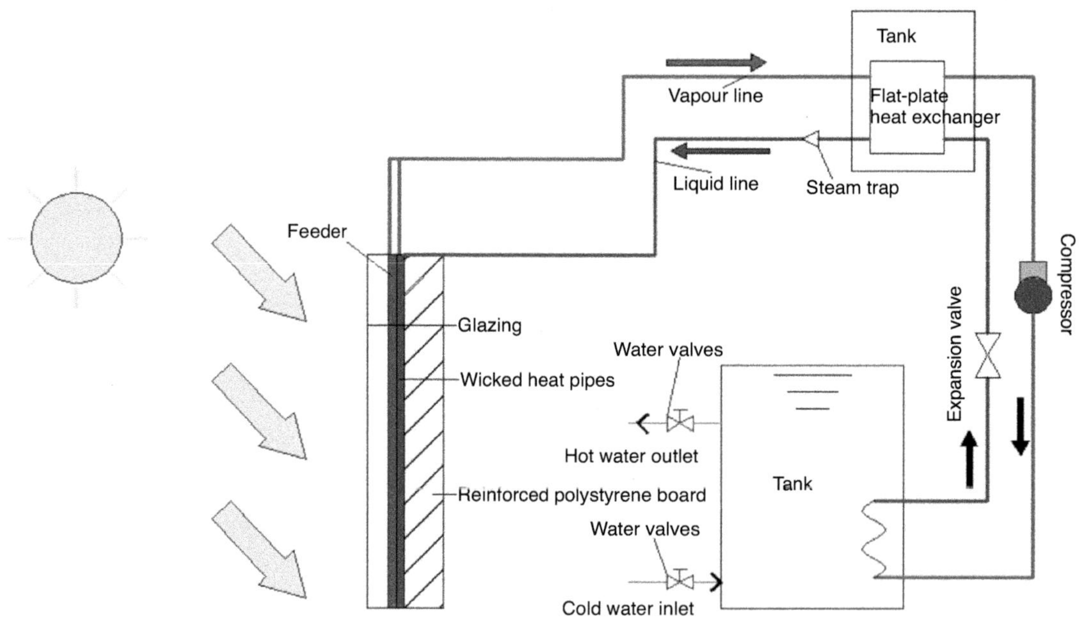

FIGURE 6.14 Schematic of the heat pump assisted solar LHP heating system [33]. *LHP*, loop heat pipe.

6.4.3 SWHs With Microchannel Heat Pipe Array

The MCHP uses the capillary-driven property to transport large amount of heat during the phase change of the working liquid [35,36], which contained evaporator and condenser sections made of inner micro-grooves (or microfins) to enhance the heat transfer [35]. The thermal performance of the MCHP was governed by the strength of evaporation and the circulation effectiveness of condensate from the condenser to the evaporator [36]. With the advantages of high heat transfer performance, high reliability, low cost, and small contact thermal resistance, MCHPs makes an excellent addition to a SWH system.

A novel flat-plate solar collector with a MCHP installed at the angle of local latitude to receive maximum annual solar irradiance was investigated by Deng et al. [37] (Fig. 6.15). The evaporator of the MCHP was located below the condenser of the MCHP to enhance the ability of condensate returning to the liquid pool in the evaporator. It was found that the temperature difference between the evaporator and condenser was less than 1°C, leading to excellent isothermal ability and quick thermal respond speed. The surface temperature along the length of the MCHP was stable within 2 min of operation.

Zhu et al. [38] investigated the thermal performance of a vacuum tube solar collector based on a MCHP as shown in Fig. 6.16. The heat was collected in the vacuum tube and released at the end of the MCHP, which was fitted with fins. The result showed that the efficiency of the collector increases with a decrease of the inlet temperature due to lower heat losses to the environment.

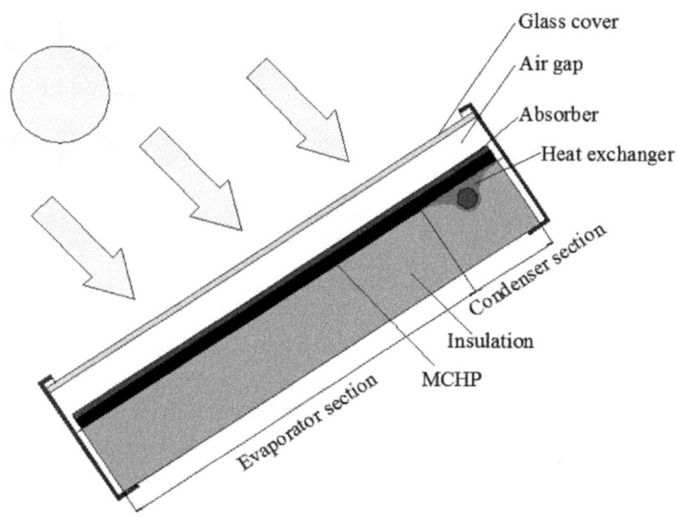

FIGURE 6.15 Schematic of the novel flat-plate solar collector with a MCHP [37]. *MCHP*, microchannel heat pipe array.

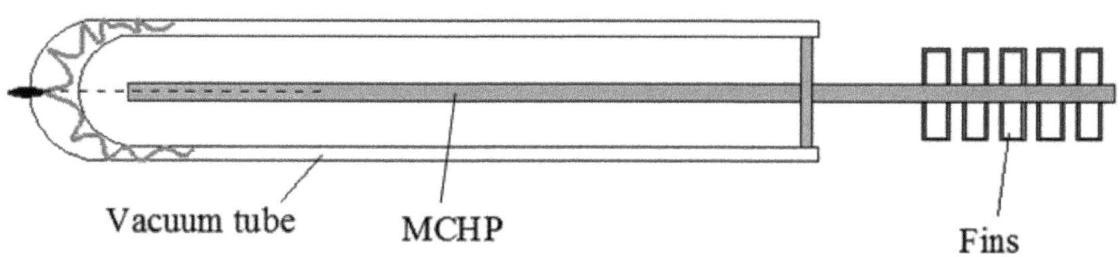

FIGURE 6.16 Schematic of the MCHP-based vacuum tube [38]. *MCHP*, microchannel heat pipe array.

References

[1] Mauthner F, Weiss W, Spork-Dur M: *Solar heat worldwide: markets and contribution to the energy supply 2014*, 2016 ed., UK, 2016, IEA Solar Heating & Cooling Programme.

[2] Weiss W, Biermayr P: *Potential of solar thermal in Europe*, EU, 2009, ESTIF.

[3] IEA: *Renewables for heating and cooling*, OECD/IEA, 2007.

[4] IEA: *World energy outlook 2008*, OECD/IEA, 2008.

[5] IEA: *Renewable energy- market and policy trends in IEA countries*, UK, 2004, OECD/IEA.

[6] Renewable energy policy network for the 21st century (REN21). Renewables 2010. Global Status Report; 2010.

[7] Philibert C. Barriers to technology diffusion: the case of solar thermal technologies. Organisation for Economic Cooperation and Development/Inter-national Energy Agency. UK: OECD/IEA; 2006.

[8] Sun in action II – a solar thermal strategy for Europe. Market overview, perspectives and strategy for growth, vol. 1. UK: ESTIF; 2003.

[9] IEA: *Renewables information 2005*, OECD/IEA, 2005.

[10] Karen F, Neumann K: *Removing barriers to solar energy use in Santa Barbara county*, Santa Barbara, CA, 2004, Community Environmental Council.

[11] Jamar A, Majid ZAA, Azmi WH, Norhafana M, Razak AA: A review of water heating system for solar energy applications, *Int Commun Heat Mass Transf* 76:178–187, 2016.

[12] Hastings S, Wall M. Sustainable solar housing, vol. 2. Exemplary buildings and technology. UK: Earthscan; 2007.

[13] Lee DW, Sharma A: Thermal performances of the active and passive water heating systems based on annual operation, *Sol Energy* 81:207–215, 2007.

[14] Nahar N: Capital cost and economic viability of thermosyphonic solar water heaters manufactured from alternate materials in India, *Renew Energy* 26:623–635, 2002.

[15] Khalifa A: Forced versus natural circulation solar water heaters: a comparative performance study, *Renew Energy* 14:77–82, 1998.

[16] Wang Z, Guo P, Zhang H, Yang W, Mei S: Comprehensive review on the development of SAHP for domestic hot water, *Renew Sustain Energy Rev* 72:871–881, 2017.

[17] Boyle G: *Renewable energy: power for a sustainable future*, 2nd ed., UK, 2004, Oxford University Press.

[18] Weiss W, Mauthner F: *Solar heat worldwide: markets and contribution to the energy supply 2009*, 2011 ed., UK, 2011, IEA Solar Heating & Cooling Programme.

[19] Huang X, Alva G, Jia Y, Fang G: Morphological characterization and applications of phase change materials in thermal energy storage: a review, *Renew Sustain Energy Rev* 72:128–145, 2017.

[20] Mahfuz MH, Anisur MR, Kibria MA, Saidur R, Metselaar IHSC: Performance investigation of thermal energy storage system with phase change material (PCM) for solar water heating application, *Int Commun Heat Mass* 57:132–139, 2014.

[21] Fazilati MA, Alemrajabi AA: Phase change material for enhancing solar water heater, an experimental approach, *Energy Convers Manag* 71:138–145, 2013.

[22] Biwole PH, Eclache P, Kuznik F: Phase-change materials to improve solar panel's performance, *Energy Build* 62:59–67, 2013.

[23] Wang Z, Qiu F, Yang W, Zhao X: Applications of solar water heating system with phase change material, *Renew Sustain Energy Rev* 52:645–652, 2015.

[24] Khalifa AJN, Abdul Jabbar RA: Conventional versus storage domestic solar hot water systems: a comparative performance study, *Energy Convers Manag* 51:265–270, 2010.

[25] Chen Z, Gu M, Peng D: Heat transfer performance analysis of a solar flat-plate collector with an integrated metal foam porous structure filled with paraffin, *Appl Therm Eng* 30:1967–1973, 2010.

[26] Tarhan S, Sari A, Yardim MH: Temperature distributions in trapezoidal built in storage solar water heaters with/without phase change materials, *Energy Convers Manag* 47:2143–2154, 2006.

[27] Al-Hinti I, Al-Ghandoor A, Maaly A, Abu Naqeera I, Al-Khateeb Z, Al-Sheikh O: Experimental investigation on the use of water-phase change material storage in conventional solar water heating system, *Energy Convers Manag* 51:1735–1740, 2010.

[28] Maydanik YF: Loop heat pipes, *Appl Therm Eng* 25:635–657, 2005.

[29] Reay D, Kew P: *Heat Pipe*, 5th ed., UK, 2006, Elsevier.

[30] Xu X, Wang S, Wang J, Xiao F: Active pipe-embedded structures in buildings for utilizing low-grade energy sources: a review, *Energy Build* 42:1567–1581, 2010.

[31] Butler D, Ku F, Swanson T: Loop heat pipes and capillary pumped loops – an applications perspective, *Space Technol Appl Int Forum* 608:49–56, 2002.

[32] Wang Z, Zhao X: Analytical study of the heat transfer limits of a novel loop heat pipe system, *Int J Energy Res* 35:404–414, 2011.

[33] He W, Hong X, Zhao X, Zhang X, Shen J, Ji J: Operational performance of a novel heat pump assisted solar facade loop-heat-pipe water heating system, *Appl Energy* 146:371–382, 2015.

[34] Zhang X, Zhao X, Shen J, Hu X, Liu X, Xu J: Design, fabrication and experimental study of a solar photovoltaic/loop-heat-pipe based heat pump system, *Sol Energy* 97:551–568, 2013.

[35] Ling L, Zhang Q, Yu Y, Wu Y, Liao S: Study on thermal performance of micro-channel separate heat pipe for telecommunication stations: Experiment and simulation, *Int J Refrig* 59:198–209, 2015.

[36] Chang FL, Hung YM: Dielectric liquid pumping flow in optimally operated micro heat pipes, *Int J Heat Mass Tranfer* 108:257–270, 2017.

[37] Deng Y, Zhao Y, Wang W, Zhenhua Q, Lincheng W, Dan Y: Experimental investigation of performance for the novel flat plate solar collector with micro-channel heat pipe array (MHPA-FPC), *Appl Therm Eng* 54:440–449, 2013.

[38] Zhu T-T, Zhao Y-H, Diao Y-H, Li F-F, Quan Z-H: Experimental investigation and performance evaluation of a vacuum tube solar air collector based on micro heat pipe arrays, J Clean Prod 142:1–10, 2017.

7 Concentrating Solar Thermal Power

Eduardo Zarza-Moya
CIEMAT-PSA, ALMERÍA, SPAIN
Eduardo.zarza@psa.es

7.1 Introduction

There are many technologies that convert solar radiation into electricity. Although the most popular systems are the photovoltaic (PV) systems, there exist other technologies to convert solar radiation into electricity. Concentrating solar thermal power (CSTP) systems convert direct solar radiation into thermal energy at a medium or high-temperature (from 125°C to even above 1000°C) and the thermal energy is then converted into electricity using a thermodynamic cycle. The more common cycles used nowadays are: Brayton, Rankine, or Stirling cycles.

Historically only concentrating solar power (CSP) was universally referred to and was used in place of solar thermal electricity (STE). Only in recent years has the term STE become widespread and some organizations have moved the CSP definition to a higher level to include both STE and concentrating photovoltaics (CPV). However, some organizations still use CSP to refer to and in place of STE, and in these cases CSP does not include CPV. Therefore, the meaning of CSP varies among organizations; it is without a clear definition and is an ambiguous term, and is not used in this chapter. The term concentrating solar thermal (CST) is used globally to refer to the technologies used to concentrate and convert solar radiation into thermal energy (i.e., CST technology or technologies). In this chapter concentrated solar thermal power refers to electricity generated using CST technologies.

Fig. 7.1 shows the simplified scheme of a typical CSTP system. Starting from the primary energy source (i.e., the direct solar radiation), the first component is the *optical concentrator*, which increases the flux density of the direct solar radiation, so that the solar flux density on the surface of the *receiver* may be of up to several thousand times higher than the flux density of the direct solar radiation reaching the Earth's surface. The latter is usually of the order of, or lower than, 1000 W m^{-2}. This concentration is usually achieved by reflecting the collected direct solar radiation onto a receiver with a surface smaller than that of the *concentrator*. The concentrated solar radiation reaching the *receiver's* outer surface is converted into thermal energy by increasing the enthalpy of a working fluid as it circulates inside the receiver. Several working fluids are used nowadays (e.g., water, thermal oils, molten salts, or air).

A Comprehensive Guide to Solar Energy Systems. http://dx.doi.org/10.1016/B978-0-12-811479-7.00007-5

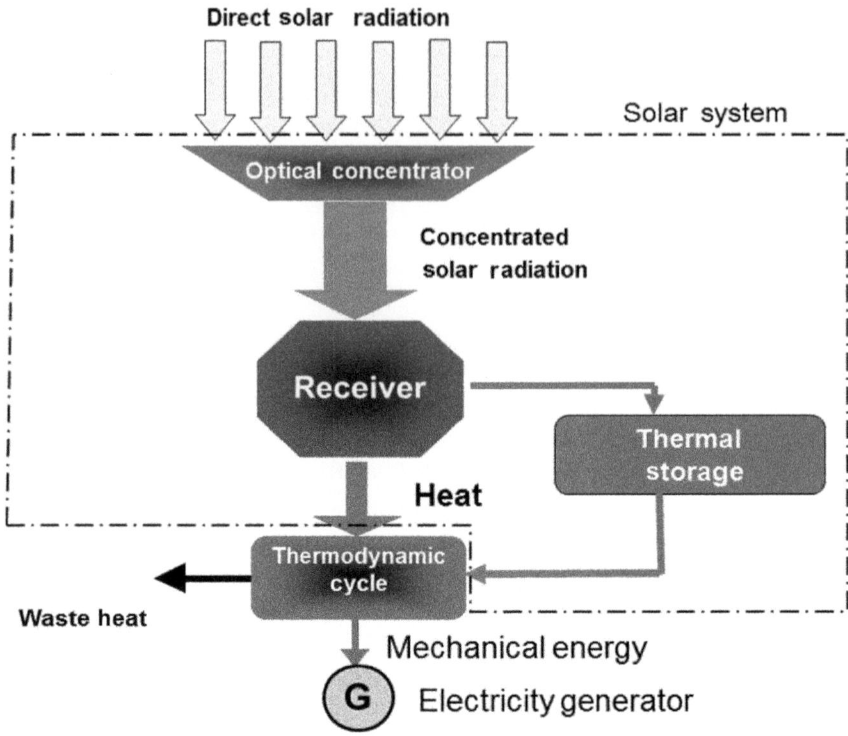

FIGURE 7.1 Simplified scheme and main components of a Concentrating Solar Thermal Power System.

Solar concentrators are basically optical devices that are designed according to the optical law that the incidence and reflection angles of a ray on a reflecting surface are equal in value; the solar radiation impinging on a concentrator must arrive from a proper direction to assure that it will go to the receiver after being reflected. Hence a very important limitation of CST systems is that they can only use direct solar radiation; diffuse radiation is not effective as it propagates in all directions and can't be concentrated in an efficient manner.

The thermal energy gained by the working fluid at the *receiver*, in the form of sensible and/or latent heat, is then either converted into electricity by means of a *thermodynamic cycle* producing the mechanical energy required to drive the *electricity generator* or sent to a *thermal storage system* to be used later to feed the *thermodynamic cycle* when direct solar radiation is not available (i.e., overnight or during cloudy periods). Some waste heat from the *thermodynamic cycle* is usually available as a by-product of electricity generation.

The set of elements where the thermal energy delivered by the receiver is converted into electricity is called the *Power Block*, which includes not only the *thermodynamic cycle* and the *electricity generator*, but also the auxiliary components required for the thermodynamic cycle to work (e.g., fluid circulation pumps, heat exchangers, and steam condensers).

Although CSTP plants are still at an early stage of technical development, they are already considered a promising candidate to play a significant role in a future carbon-free energy market.

The electricity generated by a CSTP plant is usually known as "Solar Thermal Electricity (STE)." The International Energy Agency (IEA) has predicted in its STE Roadmap published in 2014 [1] that STE plants will probably be the dominant technology in the future for Middle East (with a 40% of market share) and African countries (26% of market share) and they will play a significant role in other regions (e.g., 18% of market share in the United States) by 2050. This forecast is based not only on the huge potential of solar radiation but also on the many socioeconomic benefits that CSTP plants have for local economies [2,3].

A significant commercial deployment of CSTP plants has taken place since 2000, achieving a total installed capacity of about 5 MW_e in mid-2017. A database with information of all the CSTP plants existing worldwide can be found in Ref. [4].

Spain has played a significant role in this commercial deployment, with 50 commercial plants in routine operation and a total installed capacity of 2.3 GW_e. Detailed information about the CSTP plants located in Spain is given in Ref. [5]. However, the very low cost already achieved by large PV plants (about \$0.03 $(kW\ h)^{-1}$ in sunny countries) is at present a significant barrier to continue with this commercial deployment, because the levelized cost of electricity produced by CSTP plants is in the order of about \$0.1 $(kW\ h)^{-1}$ for similar solar conditions (www.estelasolar.org). However, CSTP plants should not be considered as competitors to PV plants, but a complement to deliver electricity not only during sunlight hours but also at night when solar radiation is not available. This ability to store thermal energy is the main advantage of CSTP plants [6]. This high degree of dispatchability of CSTP plants makes them an excellent complement for PV plants which can't store high amount of electricity at an affordable price. So, the joint implementation of PV and CSTP plants can meet the electricity demand not only during sunlight hours, but also overnight. The average electricity cost of a hybrid PV and CSTP plant is not as low as PV electricity, but it is affordable at about \$0.07 $(kW\ h)^{-1}$ and is competitive with fossil fuel plants. This cost is even more competitive when the main benefits of the CSTP plants are taken into account [7].

On the basis of the latest IEA estimates [1], between €39×10^9 and €57×10^9 (39–57 billion) will be invested in STE developments on average every year between 2015 and 2030, creating 275 000–520 000 jobs worldwide. Up to 150 000 qualified jobs will be available in Europe during these 15 years [6] covering a wide spectrum of direct activities related to: (a) engineering, development, and financing, (b) manufacturing of components: reflectors, receivers, etc., (c) construction, civil, installation, and commissioning works, and (d) operation and maintenance (O&M). In addition to direct activities, the European STE industry will in this case also create numerous indirect jobs: research, training, transport, information and communication activities, general maintenance services, etc. The benefits that have been identified for Europe can also be extrapolated to other continents because the high local content of CSTP plants is a major benefit. The Spanish experience has shown that most of the investment required to implement CSTP plants is spent at regional or national level, thus empowering the local economy and reducing currency flight to foreign economies.

At present, four different technologies are available for CSTP systems:

1. Parabolic-trough collectors
2. Central receiver systems
3. Compact linear Fresnel concentrators
4. Stirling dishes

The main differences among these four technologies are: the way they concentrate the direct solar radiation onto the receiver; the geometry of the receiver, and the concentration ratio (i.e., the quotient between the solar flux density at the surface of the receiver and the flux density of the primary direct solar radiation). The higher the concentration ratio, the higher the optimum temperature that can be achieved by the working fluid in the receiver. The optimum temperature is that achieving the highest efficiency of the conversion from thermal energy to electricity in the power block.

If a simplified model of a CSTP plant is used to assess the overall plant efficiency versus the concentration ratio, the curves given in Fig. 7.2 are obtained. These curves show that the optimum working temperature and the associated maximum efficiency increases with the concentration ratio, *C*. Hence the importance of achieving high concentration ratios to increase the temperatures and efficiencies.

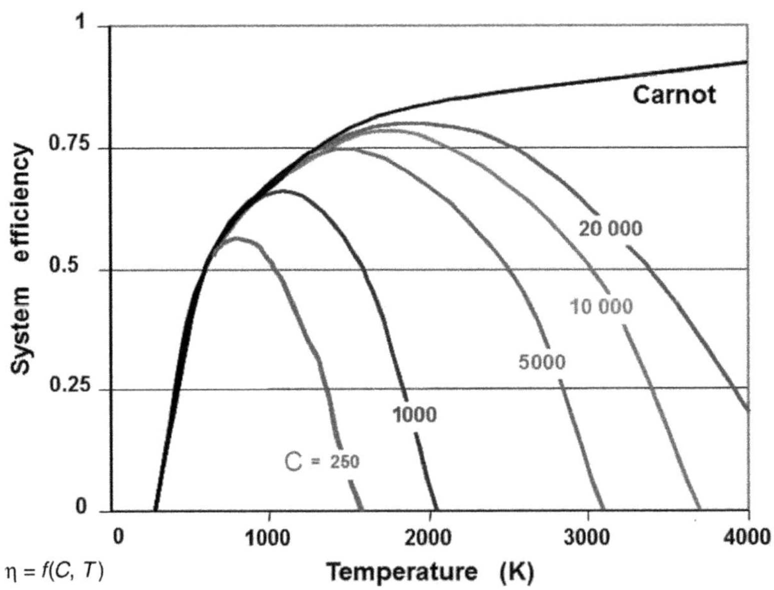

FIGURE 7.2 Theoretical efficiency and associated working temperature versus the concentration ratio, *C*. [8].

7.2 Parabolic-Trough Collectors

The deployment of parabolic-trough collectors (PTC) has experienced a surge since 2000, with about 4 GW_e of installed power worldwide at the end of 2016. This section gives an overview of the state-of-the-art of this technology, describing the main components, working fluids, and commercial applications of PTC. Further information on the design, energy balance, operation, and maintenance of PTC can be found in Ref. [9].

7.2.1 Main Components

Fig. 7.3 shows a typical PTC with its three main components: (1) the parabolic-trough concentrator, (2) the receiver pipe, and (3) the supporting structure providing the required rigidity and mechanical stiffness. The parabolic-trough concentrator is used to concentrate the direct solar radiation onto a receiver pipe placed at its focal line. This concentrated solar radiation increases the enthalpy of a working fluid as it circulates inside the receiver tube from the collector inlet to its outlet. The maximum temperature of the working fluid in current commercial applications is 395°C. However, one of the current research and development (R + D) topics is the increase of this temperature up to 500°C or even 550°C to increase the overall plant efficiency.

Back-silvered thick-glass reflectors are the most widely used type of reflector for parabolic-trough concentrators. This type of reflector is similar to conventional mirrors apart from the curvature and the type of glass used. As standard glass has some iron content that reduces its transmissivity for solar radiation, borosilicate glass (glass with much lower iron content than standard glass) is commonly used for PTC, and for glass components used in CST systems in general. Although there are other types of reflectors available (e.g., polymeric-film reflectors, aluminum-sheet reflectors, and thin-glass reflectors),

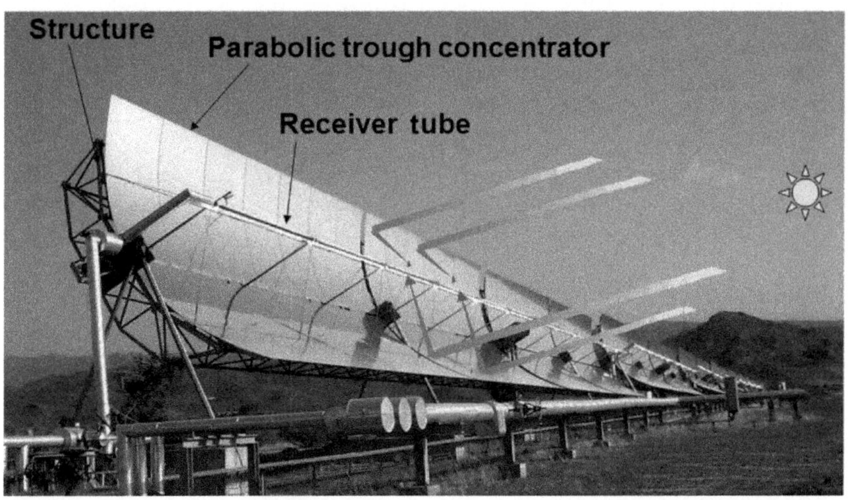

FIGURE 7.3 A Typical parabolic-trough collector (PTC).

back-silvered thick-glass reflectors are the preferred ones for CST plants due to their better solar reflectance, good durability, and affordable cost. Sandwich-glass reflectors (i.e., reflectors made of two pieces of glass with a silver layer deposited in between) have good optical properties, but their higher cost limits their use.

The receiver pipe of a PTC is composed of several receiver tubes connected in series and placed at the focal line of the parabolic-trough concentrator. The place where two adjacent tubes are connected is used to install the brackets and metallic supports required to hold the receiver tubes at the focal line. These metallic supports can be seen in Fig. 7.3. Receiver tubes of PTC used in CSTP plants are composed of two pipes: (1) an inner steel pipe inside which the working fluid circulates, and (2) an outer glass envelope surrounding the inner steel tube with an annular gap in between. Both ends of the cylindrical glass envelope are connected to the inner steel tube by means of stainless-steel flexible bellows, which compensates for the different thermal expansion of the steel and glass tubes. One end of the flexible bellows is welded to the steel tube, while the other end is connected to the glass envelope using a glass-to-metal special weld. A hermetic annular gap is thus created between the steel tube and the glass envelope. The air can be extracted from this annulus creating a high-vacuum condition ($P < 10^{-2}$ Pa), thus avoiding thermal losses by convection and increasing the efficiency of the system. This efficiency is further increased by depositing a selective coating on the outer surface of the steel tube. Such selective coating has a high solar absorptance (≥ 0.94) and a low emittance (≤ 0.1 at 400°C) in the infrared region. Additionally, both sides of the glass envelope are provided with an antireflective coating that increases the solar transmittance to 0.96. These receiver tubes are sophisticated from a technical standpoint to efficiently convert the concentrated solar radiation reflected by the parabolic-trough concentrator into heat.

Another key element of a PTC is the sun tracking system, which keeps the parabolic concentrator correctly aimed at the sun during operation to reflect the concentrated solar radiation toward the receiver tube. All concentrating solar systems must be provided with a sun tracking system to ensure that the incoming direct solar radiation is impinging on the concentrator with the correct incidence angle. The sun tracking system of a PTC must have a high accuracy (equal or better than 0.1 degree) to ensure that most of the reflected radiation goes to the focal line where the receiver tube is located. An error of 0.5 degree would be enough to lose a significant fraction (>25%) of the concentrated radiation.

7.2.2 Working Fluids for PTC

The working fluid currently used in CSTP plants with PTC is a thermal oil composed of an eutectic mixture of two very stable compounds: (1) biphenyl ($C_{12}H_{10}$) and (2) diphenyl oxide ($C_{12}H_{10}O$). This thermal oil has a maximum working temperature of 398°C and is stable with time so long as this maximum temperature is not surpassed and a suitable treatment system (the so-called "oil ullage system") is installed in the plant. Under these conditions, this thermal oil can be used for more than 30 years. Another advantage of this thermal oil is its low vapor pressure (1.06 MPa at 398°C), which reduces the pressure required in the solar field piping to keep the oil in a liquid phase when it is at its maximum working temperature. These properties, together with its affordable price, are the reasons why no other thermal oil is nowadays used in CSTP plants. However, the main problem with this oil is that it is an environmental hazard (if leaks occur) and can be harmful to workers. The thermal limit of 398°C is another constraint of this thermal oil, because the overall CSTP plant efficiency depends on the temperature of the superheated steam delivered to the power block, and such temperature is limited by the temperature of the oil used to generate it.

The thermal limit of this oil and its environmental hazards are the reasons why alternative working fluids are being investigated. The three main alternative fluids are: (1) liquid water/steam (direct steam generation), (2) molten salt mixtures, and (3) compressed gases. The advantages and disadvantages of these working fluids are analyzed in Ref. [10].

As silicone-based oils have higher thermal limits and are less harmful to the environment and to humans, they are also being considered an interesting option to replace the current thermal oils in CSTP plants. However, long-term experience with silicone oils working under real solar conditions is lacking. Trials are ongoing and if successful, the first commercial CSTP plants using silicone oil could be in operation in China in 2018–19.

7.2.3 Main Applications of PTC

PTC has two main applications:

(a) Electricity generation, and
(b) Industrial process heat (IPH) applications

Electricity generation is the most widely used application nowadays, and such systems using PTC are called CSTP or STE plants.

Fig. 7.4 shows the simplified scheme of a typical CSTP plant with PTC and thermal oil as working fluid. The plant is composed of four main subsystems: (1) *solar field*, (2) *thermal energy storage* (TES), (3) *steam generator*, and (4) *power block* (also called *power conversion system*, PCS). The thermal oil is heated in the solar field from 295°C up to about 395°C. Once heated, the hot oil is sent to the *steam generator* or/and to the TES, depending on the decision taken by the plant operator. The hot oil sent to the steam generator is used to produce the superheated steam required to drive the steam turbine of the PCS, which is mechanically coupled to the electricity generator. The PCS of this type of CSTP plant is based on a Rankine cycle with steam reheating because the steam leaving the high-pressure stage of the turbine is reheated before entering the low-pressure stage. A small fraction of the hot oil delivered by the solar field is used for this steam reheating. Because of the thermal stability limitation of current thermal oils (398°C), the maximum steam temperature that can be achieved in the steam generator is about 385°C, thus limiting the PCS efficiency, and therefore the overall plant efficiency.

The most common TES system used nowadays in CSTP plants with PTC is based on molten salts (Fig. 7.4), so that the thermal energy is stored as sensible heat (i.e., temperature increase) of a binary mixture of molten salts, the so-called "Solar Salt" (i.e., 40% of potassium

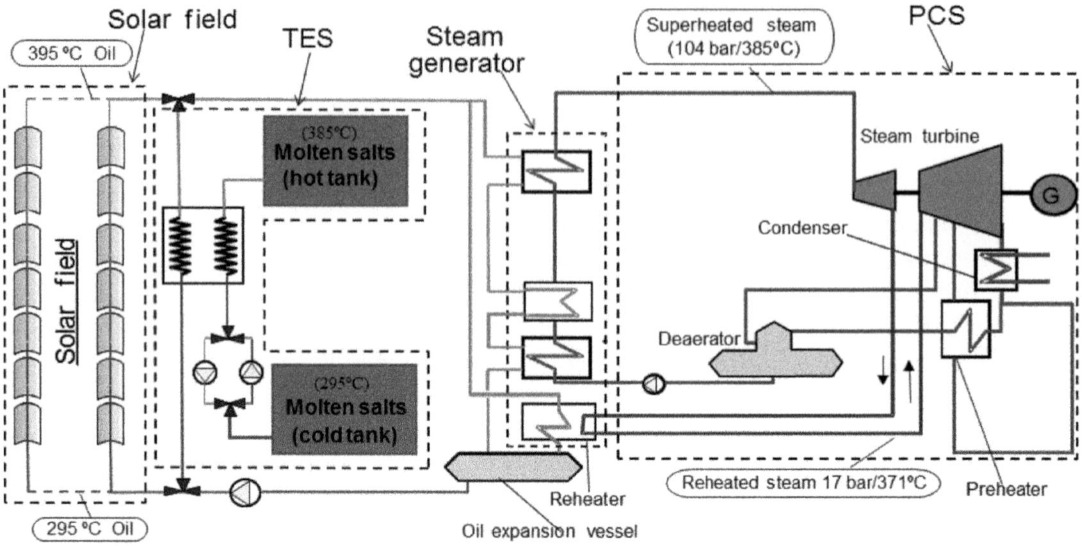

FIGURE 7.4 Simplified scheme of a typical CSTP plant with parabolic-trough collectors.

nitrate + 60% of sodium nitrate). This type of TES system had been implemented in 22 STE plants with PTC (as of the end of 2016). With a specific investment cost of about €35–40 $(kW\,h)^{-1}$ of capacity, TES systems using molten salts are cost-effective as they can meet the electricity demand not only during sun-light hours but also overnight.

When the hot oil is sent to the TES (TES charging process), it circulates through a heat exchanger where its thermal energy is transferred to the molten salt at 290°C that is pumped from the "cold tank" to the "hot tank," where it is stored at 385°C. Once cooled down in the TES heat exchanger, the oil is used to feed the solar field again.

When the plant operator wants to recover the thermal energy stored in the TES, cold oil is circulated through the TES heat exchanger while the hot salt at 385°C is pumped from the hot tank to the cold tank through the same heat exchanger, thus transferring thermal energy from the salt to the oil, which then goes to the steam generator to produce superheated steam for the turbine, which keeps the electricity generator running. This hot oil is coming from the TES is cooled down at 295°C in the steam generator and it is then sent to the TES heat exchanger again to continue with the discharging process, which can continue until all the molten salt is in the cold tank and the hot tank is empty.

The first CSTP plants with PTC installed in the world were designed by the company LUZ International and built in the Mohave dessert (California, USA) between 1985 and 1991. These plants were named SEGS (solar electricity generating system) and numbered from I to IX [11]. The SEGS plants were the precursor of the modern CSTP plants installed worldwide a few decades later with PTC.

A special design of a CSTP plant with PTC is called an *Integrated Solar Combined Cycle System* (*ISCCS*) *plant*. This type of plant is basically a natural gas combined-cycle plant with a small PTC solar field coupled to the Rankine cycle, so that the thermal energy

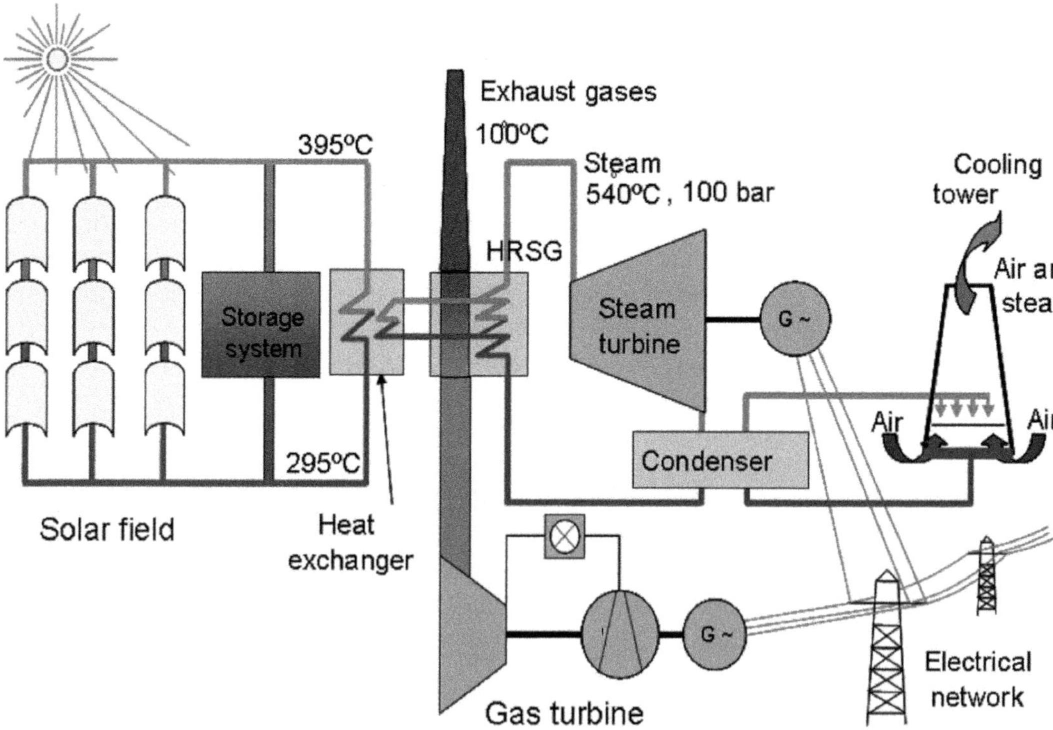

FIGURE 7.5 Simplified schema of an Integrated Solar Combined Cycle System (ISCCS).

delivered by the solar field complements the waste heat provided by the Brayton cycle to increase the electricity production of the steam turbine. An ISCCS plant can produce electricity at any time without a TES, because the main fuel is natural gas and the thermal energy delivered by the PTC solar field is only used to increase the electricity production associated with the Rankine cycle, which works at 80% load when the solar field is not in operation. This type of solar plant is a very interesting option for those countries willing to employ CST systems (PTC technology) without facing a high risk, because the plant will produce electricity whenever natural gas is available to feed the Brayton cycle. The contribution of the solar system to the total electricity production is small (about 11% of the total yearly production). Fig. 7.5 shows the simplified scheme of an ISCCS plant.

Solar fields using PTC for industrial process heat (IPH) applications are designed to deliver thermal energy to an endothermic process (e.g., food processing, dairy, and biodiesel production). These solar fields usually have some differences from those of CSTP plants. One of the differences is the size of the PTC used in the solar field. Usually small-sized collectors are used for IPH applications. The collecting area of a small PTC is usually smaller than 250 m^2, while typical sizes of big PTC are usually in the range from 500 to 1000 m^2, with a unit length of between 100 and 150 m. There is even a PTC model with a collecting area higher than 2000 m^2 (the Ultimate Trough design [12]). By comparison,

CSTP plants involve solar fields greater than 250 000 m^2, while solar fields for IPH applications are usually smaller than 50 000 m^2.

A solar field with PTC is composed of parallel rows, with several PTC connected in series within each row. The total length of each row depends on how much the working fluid must increase its temperature as it circulates from the row inlet to outlet: the higher the temperature increase, the more PTC connected in series are required in each row. The number of parallel rows depends on the power of the plant; the higher the power, the more parallel rows are installed in the solar field. The number of parallel rows for IPH applications is usually small, while solar fields with more than 100 parallel rows are common in CSTP plants. The length of each row is usually smaller in IPH applications, while the length of the rows in a CSTP plant is usually 600 m (each row is composed of four 150 m collectors or six 100 m collectors connected in series).

The maximum working temperature is also lower in IPH applications, with maximum temperatures usually below 300°C. These lower operation temperatures make the use of pressurized liquid water a feasible option as working fluid instead of thermal oil. Several companies are commercializing PTC solar fields using pressurized water for IPH applications. The Mexican company Inventive Power (http://inventivepower.com.mx) and the Turkish company Lucida Solar (https://lucidasolar.com/) are two companies offering PTC solar fields to provide hot liquid water at temperatures of about 200°C for IPH applications.

Another difference that is usually found in PTC used for IPH applications is the type of sun tracking system, because the systems used for small PTC are usually based on an electric motor with a speed-reducing gear box, while hydraulic units are preferred for big PTC because they can provide higher torque at lower cost.

Although electricity generation is the commercial application most widely used for PTC so far, IPH applications are of interest because of the huge commercial potential. According to Ref. [13], the total energy consumption of the industry in 2014 was 115 EJ, and 16% out of these 115 EJ was thermal energy consumption within the temperature range from 150°C to 400°C, which is very suitable for PTC systems. This means that each year 18.7 EJ could be delivered by solar systems using PTC. This is the reason why the number of companies interested in IPH applications for PTC is increasing and a high number of IPH commercial projects are expected to be developed in the coming years.

7.3 Central Receiver Systems

In a central receiver system, the receiver is placed at the top of a tower and many reflecting elements called *heliostats* are placed around the tower reflecting and focusing the direct solar radiation onto the receiver (Fig. 7.6). Though commercial systems using this technology are working at temperatures below 600°C, temperatures of 1000°C and even higher can be achieved with this technology because the solar flux incident on the receiver is significantly higher than in a PTC (about 1 MW m^{-2}).

FIGURE 7.6 Aerial view of the plant SOLAR-II, a concentrating solar thermal power (CSTP) plant with central receiver.

For the time being, unlike CST plants with PTC, central receiver systems have been commercially used for electricity production only. IPH applications are not considered as convenient for central receiver systems as they are for PTC systems. The technical complexity of central receiver systems makes their coupling to industrial processes more difficult. However, this type of commercial application for central receiver systems can't be completely excluded in a long-term.

In Spain, after initial experimental projects in the 1980s and 1990s, the first commercial CSTP plant was put in operation in 2007. It was the plant PS-10, promoted by the Spanish company ABENGOA [14]. Since then, the PS-10 plant has been in routine operation with a nominal output power of 10 MW_e. Other commercial plants with central receiver technology were implemented later on in Spain (PS-20 and GEMASOLAR plants) and other countries for electricity generation (such as the Ivanpah and Crescent Dunes plants in USA and the Khi Solar One plant in South Africa).

Although commercial deployment of central receiver systems started more slowly than PTC plants, an increasing number of commercial CSTP plants with central receivers are being considered for implementation over the next few years. The main reason for this is because their overall efficiency (i.e., from solar to electricity) is higher (~17.5%) than that of plants with PTC (16%). The reason why most of the initial projects for commercial CSTP plants in the present century were designed with PTC instead of central receiver was the higher confidence of investors in PTC technology due to the operational experience accumulated by the SEGS plants installed in California (USA) between 1985 and 1991 [11]. The lack of previous operational experience with central receiver plants was a significant barrier to their deployment and financing. Once this barrier had been overcome with the implementation of the first commercial plants (i.e., PS-10, PS-20, and GEMASOLAR), the funding required for central receiver plants has been much easier to obtain.

7.3.1 Main Components

The main four elements in a central receiver plants are (Fig. 7.6):

1. the heliostat field
2. the receiver
3. the power conversion system (PCS), and
4. the thermal storage system (TES)

The heliostat field of a plant with a central receiver is usually composed of thousands of heliostats, which are reflecting surfaces supported on a vertical pylon. The reflecting surface has two degrees of freedom (up/down, and right/left movements) to reflect the collected direct solar radiation onto the receiver. Each heliostat is provided with a local control in continuous communication with the central control room to receive the commands related to the position to be adopted by the reflecting surface to assure that the collected radiation is reflected in the right direction toward the receiver. The local control manages the drive unit of the heliostat according to the commands received from the central computer located in the control room. The reflecting surface of each heliostat is composed of mirror segments, called "facets," which are attached to a metallic structure and placed altogether in a predefined way to have the small curvature required to project onto the receiver a reflected image with a total area smaller than the area of the reflecting surface itself. This positioning of each facet during the heliostat assembly process is performed with a very high accuracy. As the distance between the central receiver and the outer heliostats of the solar field may be of 1 km and even more in large plants, a very high degree of accuracy in the positioning of the facets within the total reflecting surface of the heliostats is required (Fig. 7.7).

When the nominal power of the CSTP plant is small (>20 MW_e) the heliostats are placed at one side of the tower only (at the North side in plants located in the North Hemisphere and at the South side in plants located in the South Hemisphere) because a circular receiver would be less efficient due to higher optical losses.

There are different types of receivers for these solar systems. The first commercial plant with a central receiver was the PS-10 plant built in 2007 by the company ABENGOA in the village of Sanlucar La Mayor (Seville, Spain) [14]. This plant has a saturated-steam receiver placed on a tower of 110 m, where the liquid water entering the receiver is converted into saturated steam at 250°C and 40×10^5 Pa (40 bar) by the solar radiation reflected by 627 heliostats with a total reflecting surface of 75 216 m^2. The saturated steam feeds a 10 MW_e (net power) turbo-generator.

Because of the lower efficiency of steam turbines using saturated steam, other options have been commercially implemented. For example, the company BrightSource Energy (http://www.brightsourceenergy.com/) built the superheated steam plant, IVANPAH, in Ivanpah Dry Lake (California, USA) , composed of three units of central receiver systems with a total net power of 377 MW_e. Since its start-up this is the largest central receiver plant in operation to date. It is composed of 173 500 heliostats with a total reflecting surface of 2.6×10^6 m^2. The receiver installed in each of the three 140 m high towers, convert the feed

FIGURE 7.7 Front view of a typical heliostat. The 32 facets composing its reflecting surface are clearly shown.

water at 249°C into superheated steam at 565°C and 160×10^5 Pa. Because of the high thermodynamic quality of this superheated steam the annual solar-to-electricity efficiency of this plant is 28.7%, which is significantly higher than the 14.4% efficiency of the plant PS-10.

The size of the heliostats used in a CSTP plant with central receiver does not depend on the nominal power of the plant. As there is no consensus about the most cost-effective strategy to achieve the required total reflective surface there are companies using heliostats with a large reflective surface per unit (>150 m^2), while other companies are using small heliostats (<20 m^2). PS-10 and IVANPAH plants are good examples of these opposite approaches: the PS-10 plant uses 120 m^2 heliostats for a total nominal power of 10 MW_e (net), while IVANPAH uses 15 m^2 heliostats for a total nominal power of 377 MW_e (net). A common approach among all the manufacturers of heliostats is the use of silvered glass mirrors for the facets, with two variants: back-silvered thick-glass mirrors or sandwich-type mirrors (a silver layer deposited between two glass pieces). Borosilicate glass is preferred whenever possible due to its high solar transmittance.

Central receivers for saturated or superheated steam are composed of walls made of many parallel steel tubes inside which the feed water is converted into steam as it circulates from the receiver inlet to the outlet. The outer wall of the steel tubes composing the receiver are painted with a black thermal paint providing a high solar absorptance and suitable for high-temperatures. Because of the high solar flux onto the receiver it appears as bright as a fluorescent lamp when the receiver is in operation, despite the external black color provided by the thermal paint (Fig. 7.6).

Both, the saturated-steam and superheated-steam central receiver plants have a common problem: the difficulty to store thermal energy during sunlight hours to produce electricity when solar radiation is not available. Only small-capacity thermal storage systems are cost-effective for this type of plant with the result that the stored energy can only keep the turbine running at nominal power for minutes rather than hours. The reason is that the saturated water and steam must be stored in very large and expensive thick-wall steel vessels to provide a stable operation of the turbine during solar radiation transients. Higher storage capacity would require many high-pressure vessels connected in parallel, which are very expensive and would jeopardize the profitability of the plant.

As the possibility of producing electricity at night is one of the main benefits of CSTP plants, other central receiver technologies more suitable for thermal storage during sunlight hours have been developed for commercial plants using central receivers. The most popular one involves molten-salt receivers. A molten-salt receiver looks similar to a saturated or superheated-steam receiver, because it is made of many parallel steel tubes coated with a black thermal paint. The main difference is the raw material used for the steel tubes and the wall thickness of the tubes, because the working pressure with molten-salt is much lower (just a few bars at the receiver outlet). The molten-salt used in these central receiver plants is the same as is used for thermal storage in plants with PTC: a binary mixture of potassium nitrate (40%) and sodium nitrate (60%). The affordable price, low viscosity and good thermal stability (up to 600°C) of this binary salt make it a very good option for use as both the thermal storage medium and the working fluid in the central receiver. The first commercial CSTP plant with a central receiver using molten salt in both the receiver and the thermal storage system was the GEMASOLAR plant, which was developed by the company Torresol Energy in 2011 in the province of Seville (Spain) (http://www.power-technology.com/projects/gemasolar-concentrated-solar-power/).

This plant has a nominal net output of 19.9 MW_e and is fitted with a molten salt thermal storage system able to keep the plant running at full load for 15 h after sunset. Fig. 7.8 shows the scheme of this plant. Molten salt at 290°C is pumped to the receiver from the cold storage tank to increase its temperature to 565°C as it circulates through the receiver. Once heated, the molten salt is sent to the hot storage tank where it remains until used in the steam generator to produce the superheated steam required to drive the turbo-generator and thus produce the electricity exported to the grid. After the steam generator, the molten salt is sent to the cold storage tank. As the amount of hot salt delivered by the receiver during sunlight hours is more than that required to run the turbo-generator at full load, the level of the molten salt inside the hot storage tank increases, thus charging the storage system to produce electricity after sunset.

The good dispatchability and overall efficiency of this type of CSTP plant are boosting the number of commercial projects being developed with this technology in many countries such as United States of America, Morocco, South Africa, and China.

The PCS, also called Power Block, so far used in commercial CSTP plants with central receivers is composed of a water/steam Rankine cycle with superheated steam, similar

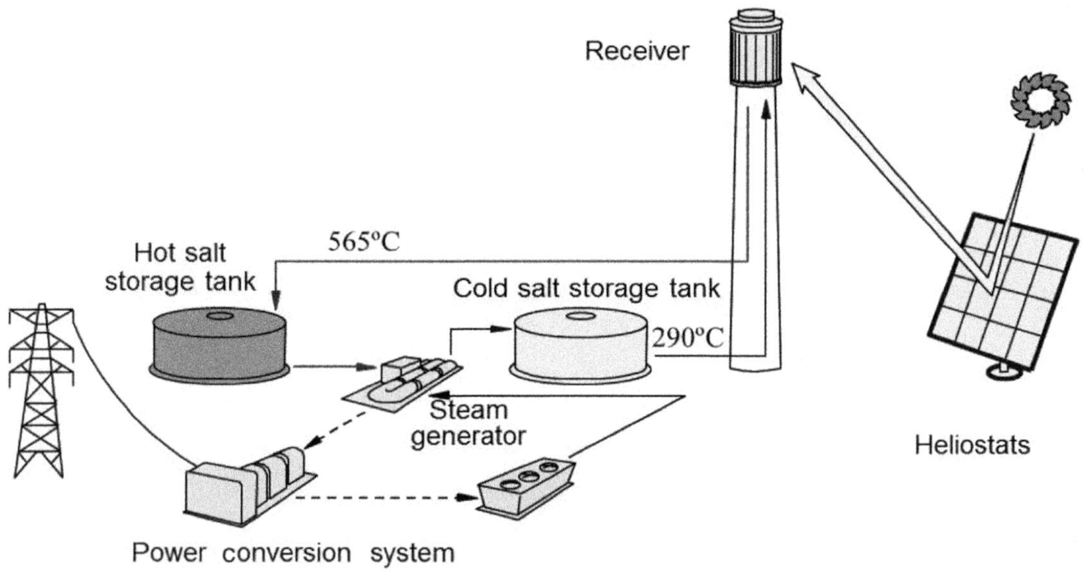

FIGURE 7.8 Scheme of the Plant GEMASOLAR with a molten-salt central receiver and thermal storage system.
Courtesy of SENER Ingeniería.

to the PCS used in plants with PTC. The main difference is the steam temperature delivered to the steam turbine. Although the steam produced in the PS-10 and PS-20 plants is saturated steam at 240°C and 40 $\times 10^5$ Pa due to lack of suppliers at that time for a central receiver with higher temperature, the later use of molten salt receivers increased the superheated steam temperature up to 550°C, thus significantly increasing the PCS efficiency, and therefore, the overall plant efficiency. The state-of-the-art of CSTP plants with central receiver is based on the use of molten salt as both working fluid in the receiver and storage media in the thermal storage system, as depicted in Fig. 7.8.

A very important environmental parameter that must be evaluated before building a CSTP plant with a central receiver is the local atmospheric attenuation, because at least 15% of the solar radiation reflected by the heliostats located at the outer part of the solar field can be absorbed by the air before reaching the receiver. This property is especially important in desert and windy areas because dust particles in the air absorb a significant amount of the radiation reflected by the heliostats that are placed far away from the receiver.

7.4 Compact Linear Fresnel Concentrators

Fig. 7.9 shows a typical linear Fresnel concentrator (LFC) with its long rectangular reflectors that can rotate independently from each other to reflect the incoming direct solar radiation toward the receiver tube located above the reflectors. The solar radiation reflected by each rectangular mirror overlap onto the receiver tube and the flux density is thus increased. The working fluid circulating through the receiver tube is heated as it circulates

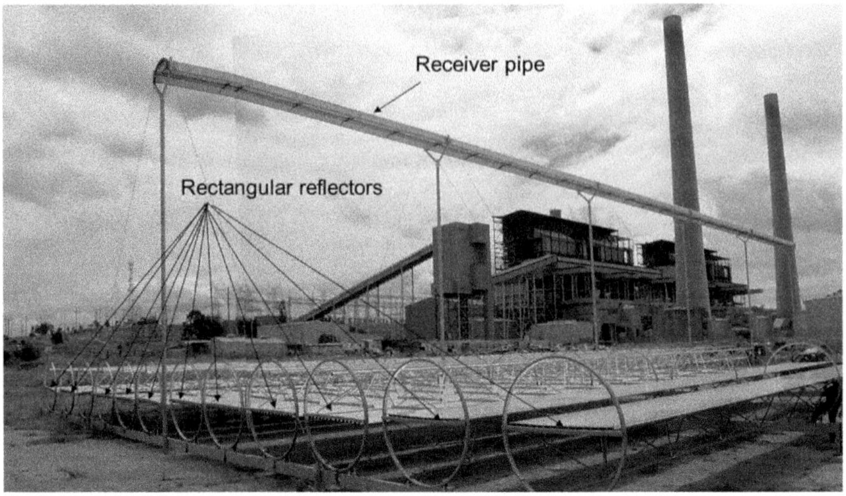

FIGURE 7.9 A typical linear Fresnel concentrator.

along the tube from the collector inlet to the outlet. The receiver tube does not change its position during the operation.

The concentration ratio that can be achieved in a LFC is similar to that of a PTC, and therefore, the working temperatures are similar too. The main advantages of a LFC when compared to a PTC are its lower investment cost and better land coverage factor, while the main disadvantage is a lower efficiency due to optical and geometrical constraints. However, LFC systems are not having a great success for electricity generation because the final cost of the electricity generated is higher than for PTC plants. The overall efficiency from solar to electricity of a LFC power plant is about 40% lower than that of a plant with PTC, while the investment cost of the LFC plant is only 30% lower [15]. This is the reason why current research and development efforts related to LFC are devoted to either reducing the investment cost or increasing the efficiency to become more competitive with PTC systems.

Only two commercial CSTP plants with LFC were in routine operation at the end of 2016: the Puerto Herrado I and II plants located in Southern Spain with unit powers of 1.4 and 30 MW_e. One LFC plant with a nominal power of 125 MW_e erected in Rajasthan (India) has been facing operational problems since its start-up in 2014. In spite of the small commercial success of this CST technology, the design of several LFC plants of 50 MW_e was underway in China in 2017.

Further information about the past and current state of LFR technology is given in Ref. [16].

7.5 Parabolic Dishes

The shape of a parabolic dish concentrator is that of a paraboloid and the receiver is placed at the focal point of the concentrator (Fig. 7.10). The most usual application of this type of concentrator is for electricity generation using a Stirling engine

FIGURE 7.10 One of the parabolic dishes installed at the Plataforma Solar de Almería (www.psa.es).

(http://www.robertstirlingengine.com/principles.php). The absorber of the Stirling engine is placed at the focal point of the solar concentrator and the thermal energy transferred to the working fluid (usually hydrogen or helium) at the absorber is converted into mechanical energy by the Stirling engine, which drives a small electricity generator (<50 kW_e). These systems are very modular (the dish diameter is usually smaller than 10 m) and can be used to supply electricity to small consumers.

The Stirling dish CST systems have the highest solar to electric efficiency with peak efficiency values greater than 30% having been reported. However, no commercial CSTP plant using this technology is currently in operation because of the reliability problems showed by the Stirling engines when working at the high-pressures (15–20 MPa) and temperatures (>600°C) required to achieve high efficiencies and become competitive with the other CST technologies. It seems that the commercial deployment of this technology will not take place until the problems with the Stirling engine are solved. Many companies (Stirling Energy Systems, INFINIA, SOLO, Tessera Solar, etc.) have unsuccessfully devoted great efforts to solve these problems. Meanwhile, a few experimental units of Stirling dishes are being tested worldwide, such as the units installed at the Plataforma Solar de Almería (www.psa.es). This research establishment is the largest public research and development (R + D) centre devoted to CST technologies in the world.

7.6 Technology Trends

The main challenge for CSTP plants at present is the achievement of a significant cost reduction in the short term so as to become more cost-effective and competitive with other renewable energy options for electricity production (i.e., wind farms and photovoltaic plants). The significant cost reduction recently experienced by PV plants is a major barrier to the deployment of CSTP plants. Nevertheless, the proven ability of CSTP plants to store energy is something that PV plants can't offer.

Most of the current R + D efforts related to CST technologies are guided by this need to reduce costs and also the need to improvement overall efficiency.

Concerning PTC technology, the two main R + D technology trends being investigated are: new working fluids for the solar field; and new collector designs. The replacement of the thermal oil currently used as working fluid in the solar field by silicone-based oils would not only allow higher steam temperatures for the PCS (and therefore, higher plant efficiencies), but also less operation and maintenance costs due to the lower environmental hazard of silicone oils. The use of molten salt, water/steam and compressed gases as working fluids for PTC is also under investigation and the advantages and disadvantages of these three innovative working fluids have been analysed by Zarza [10].

The development of new PTC designs with higher concentration ratios is also being researched, as it would create a temperature increase in the solar field without increasing the thermal losses. Achievement of higher concentration ratio requires mirrors with better optical properties as well as sun tracking systems with a higher accuracy.

Concerning central receiver technology, the main technology trends are related to new working fluids to achieve higher temperatures and therefore higher efficiencies. In some cases, the new working fluids investigated demand a re-design of the main plant components (i.e., central receiver, thermal storage system, and PCS). New working fluids under investigation are:

- molten salts suitable for temperatures higher than 600°C
- falling particles
- supercritical CO_2, and
- atmospheric or compressed air

Molten salt currently used in CSTP plants is composed of a binary mixture of potassium nitrate (40%) and sodium nitrate (60%). This mixture is cheap (about €1 kg^{-1}) and it has good thermal stability up to 600°C. New salt compositions are required to increase the thermal stability up to 700°C or more, while keeping a moderate melting point and low viscosity in the liquid phase.

The use of small solid particles, with a diameter of 1–2 mm, carried by an air stream is also being investigated as an option to increase the working temperature to temperatures above 600°C and hence to increase the efficiency of CSTP plants [17]. The solid particles heated in the receiver can then be stored in a thermally insulated vessel to be used later on when solar radiation is not available. The main challenges when using a falling particle receiver are the erosion produced by the friction of the particles inside the piping and vessels, and the stability of the particles after many heating-cooling cycles. Transport of the particles from one component to another is also a technical constraint because conventional blowers or pumps can't be used for this purpose.

The feasibility of supercritical CO_2, (s-CO_2), is also being investigated to increase the efficiency of CSTP plants with central receiver technology because it would allow higher working temperatures in the receiver and the Rankine cycle of the PCS could be replaced by a Brayton cycle resulting in higher efficiencies [18]. Because s-CO_2 has good thermo-physical properties as a heat carrier medium, a significant effort is being devoted to using it by R + D centres, located in countries such as the United States of America and Australia. The density of s-CO_2 is similar to that of liquid water and allows for the pumping power needed in a compressor to be significantly reduced, thus increasing the thermal-to-electric energy conversion efficiency of the Brayton cycle. This is the main reason why the use of s-CO_2 is one of the key R + D topics included in the American SunShot initiative (https://energy.gov/eere/sunshot/sunshot-initiative), which is a national effort supported by the US Department of Energy to drive down the cost of solar electricity and support the use of solar energy to replace fossil fuels. The target of SunShot for CSTP plants is to lower the cost of STE to \$0.06 $(kW\,h)^{-1}$. At present, the main short-term challenge in this technology program is the implementation of a small ($\leq$10 MW_e) experimental plant using a Brayton cycle with s-CO_2. Such a plant requires the design and manufacture of special equipment that is not yet available in the market (this includes s-CO_2 heat exchangers and turbine). A complete set of technical documents concerning power cycles and equipment for s-CO_2, as well as the main problems associated with this working fluid (i.e., corrosion and erosion in the associated equipment) are available at: http://energy.sandia.gov/energy/renewable-energy/supercritical-co2.

Another interesting research topic related to central receiver technology is the use of air, either at atmospheric pressure or under pressure. As air is freely available, the development of a central receiver technology using air as working fluid is a very compelling option. Central receivers using air at atmospheric pressure are made of ceramic or metallic porous elements that heat the air flow circulating through them. The ceramic or metallic porous elements composing the receivers are heated by the concentrated solar radiation reflected by the heliostats, and the heat is then transferred to the air. The hot air leaving the receiver is then sent to a thermal storage system and/or to a steam generator (air/water heat exchanger) where the superheated steam needed for the Rankine cycle of the PCS is produced. The thermal storage vessel and medium suitable for this type of central receiver system must withstand high-temperatures (>800°C) and heavy thermal cycling without degradation. So far, pellets of alumina (AL_2O_3) have been often used as high-temperature storage medium in experimental facilities. Fig. 7.11 shows the scheme of the pilot plant installed at the Plataforma Solar de Almería research facility (PSA, www.psa.es).

Central receiver plants using atmospheric air have a lower efficiency than those with molten salt receivers (Fig. 7.8). However, the simplicity and ease of operation of volumetric receivers with atmospheric air could compensate for their lower efficiency. It must be kept in mind that overall plant efficiency is not the only key aspect of a CSTP plant, because the investment and operation and maintenance costs also play a significant role in the

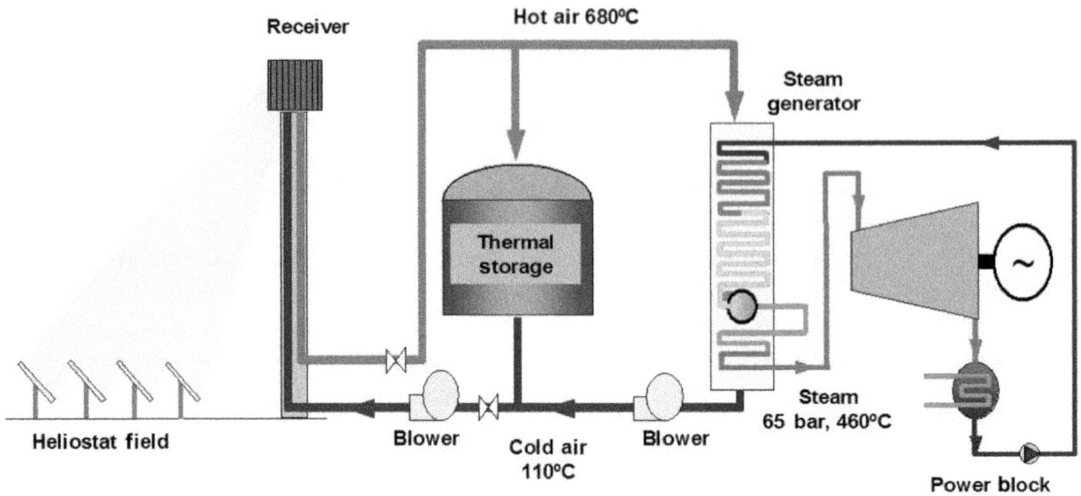

FIGURE 7.11 Scheme of a central receiver plant using atmospheric air.

final cost of the electricity produced by a plant. As these costs are significantly lower for a volumetric receiver than for a molten salt receiver, it is not yet fully clear which option is more profitable and cost-effective. This is the reason why volumetric air receivers are still considered a feasible option for future CSTP plants with central receivers.

Some R + D projects are developing pressurized-air central receivers, so that the hot air delivered by the receiver is then expanded in a Brayton cycle, without any intermediate fluid between the receiver and the PCS. In this technology, there are two options for the central receiver: (1) receiver composed of tubes inside which the pressurized air is heated as it circulates through the tubes, or (2) a chamber closed by a quartz window to allow the concentrated solar radiation reflected by the heliostats to enter into the chamber and thus heat the pressurized air flowing inside. The thermal energy delivered to the air flow in the receiver replaces the energy delivered by the fossil fuel used in conventional Brayton cycles.

Concerning the LFC technology, current R + D efforts are mainly aimed at developing innovative designs for the receiver tube and the reflectors to reduce thermal, optical, and geometrical losses [19,20]. Also, the use of molten salt as working fluid in the receiver tube is one of the high-priority R + D topics for LFC technology [21]. The operational experience to be gathered at the new CSTP plants that will be implemented in China in 2018 and 2019 will show to what extent this technology can compete with PTC and central receiver plants. Meanwhile, it seems that LFC are more suitable for IPH applications than for electricity generation.

The situation with parabolic dishes is somewhat discouraging because all the efforts so far has been unsuccessfully devoted to solve the durability and reliability problems of the Stirling engines when operating at high-temperature/pressure.

Many other R + D topics have been identified by the CSTP sector and they all are discussed in Ref. [22].

References

[1] IEA (International Energy Agency): *Solar thermal electricity technology roadmap*, OECD/IEA, 2014, Available from: https://www.iea.org/publications/freepublications/publication/TechnologyRoadmapSolarThermalElectricity_2014edition.pdf.

[2] Ernst & Young: *Etude des retombées économiques potentielles de la filière solaire thermodynamique française*, France, 2013, Syndicat des Energies Renouvelables.

[3] Deloitte: Macroeconomic impact of the solar thermal electricity industry in Spain, edited by Protermosolar. 2010.

[4] NREL (National Renewable Energy Laboratory): *Solar thermal power plants data base*, 2017. Available from: https://www.nrel.gov/csp/solarpaces/index.cfm.

[5] PROTERMOSOLAR (Asociación Española para la Promoción de la Industria Termosolar): *Data base of CSTP plants in Spain*, 2017. Available from: http://www.protermosolar.com/proyectos-termosolares/mapa-de-proyectos-en-espana/.

[6] ESTELA: *The value of thermal storage. The challenge of integrating intermittent renewable generation into power system operation*. Technical report published by the European Solar Thermal Electricity Association, 2016. Available from: www.estelasolar.org.

[7] CSPA (Concentrating Solar Power Alliance): *The economic and reliability benefits of CSP with thermal energy storage: literature review and research needs*. Technical report, 2014. Available from: www.estelasolar.org.

[8] Romero M, Zarza E: Concentrating solar thermal power. In Goswami Y, Kreith F, editors: *Handbook of energy efficiency and renewable energy*, CRC Press, Taylor Francis Group, 2007 [chapter 42; ISBN: 0-8493-1730-4, 978-0-8493-1730-9].

[9] Zarza E: Parabolic trough concentrating solar power (CSP) systems. In Lovegrove K, Stein W, editors: *Handbook on concentrated solar thermal technologies: principles, development and applications*, Woodhead, 2012 [chapter 7, ISBN: 978-1-84569-769-3].

[10] Zarza E: Innovative working fluids for parabolic trough collectors. In Blanco M, Ramirez L, editors: *Advances in concentrating solar thermal research and technology*, Woodhead, 2016 [chapter 5; ISBN: 978-0-08-100516-3 (printed), 978-0-08-100517-0 (online)].

[11] Lotker M: *Barriers to commercialization of large Scale solar electricity. The LUZ experience*, Albuquerque, 1991, Sandia National Laboratories, Technical Report SAND91-7014.

[12] Riffelmann K, Richert T, Nava P, Achweitzer A: Ultimate trough—a significant step towards cost-competitive CSP, *Energy Procedia* 49:1831–1839, 2014.

[13] International Energy Agency (IEA): World Energy Statistics 2016, online tables. Available from: www.iea.org/statistics/.

[14] Osuna R, Olavarría R, Morillo R, Sánchez M, Cantero F, Fernández-Quero V, et al: PS10, construction of a 11 MW solar thermal tower plant in Seville, Spain. In *Proceedings of the 13th International Symposium on Concentrated Solar Power and Chemical Energy Technologies (SolarPACES)*, Seville, Spain, June 20–23, 2006.

[15] Morin G, Dersch J, Eck M, Häberle A, Platzer W: Comparison of linear Fresnel and parabolic trough collector systems—influence of linear Fresnel collector design variations on break even cost. In *Proceedings of the 15th International Symposium on Concentrated Solar Power and Chemical Energy Technologies (SolarPACES)*, Berlin, Germany, September 15–18, 2009.

[16] Zhu G, Wendelin T, Wagner MJ, Kutscher C: History, current state, and future of linear Fresnel concentrating solar collectors, *Solar Energy* 103:639–652, 2014.

[17] Ho C, Christian J, Gill D, Moya A, Jeter S, Abdel-Khalik S, et al: Technology advancements for next generation falling particle receivers, *Energy Procedia* 49:398–407, 2014.

[18] Turchi C: Supercritical CO2 for application in concentrating solar power systems. In *Proceedings of SCCO2 Power Cycle Symposium*, New York, USA, April 29–30, 2009.

[19] Canavarro D, Collares M: New dual asymmetric CEC linear Fresnel concentrator for evacuated tubular receivers. In *Proceedings of the 22nd International Symposium on Concentrated Solar Power and Chemical Energy Technologies (SolarPACES)*, Abu-Dhabi, United Arab Emirates, October 11–14, 2016.

[20] Canavarro D, Collares M, Lopes L: Increasing the efficiency of conventional LFR technologies: a new CLFC "Etendue matched" CSP collector. In *Proceedings of the 17th International Symposium on Concentrated Solar Power and Chemical Energy Technologies (SolarPACES)*, Granada, Spain, September 20–23, 2011.

[21] Morin G, Karl M, Mertins M, Selig M: Molten salt as a heat transfer fluid in a linear Fresnel collector—commercial application backed by demonstration, *Solar Energy* 69:689–698, 2015.

[22] ESTELA: *Solar thermal electricity. Strategic research agenda 2020–2025*. Technical report published by the European Solar Thermal Electricity Association, 2012. Available from: www.estelasolar.org.

PART 4

Photo Voltaic Solar Energy–Generation of Electricity

8

Photovoltaics: The Basics

Vítězslav Benda
CZECH TECHNICAL UNIVERSITY IN PRAGUE, PRAGUE, CZECH REPUBLIC
benda@fel.cvut.cz

8.1 Introduction

The photovoltaic (PV) industry has recently shown an unprecedented rate of growth with the installed global PV power increasing by more than 30-fold over the last 10 years: from 9.1 GW_p in 2007 to over 300 GW_p in 2016. This impressive progress has been due to a number of factors, including a lowering to production costs, an increase in PV panel efficiency, and a greater improvement in PV panel reliability.

The success of the PV industry lies with the development of the PV (solar) cells. PV cells are simply semiconductor diode structures that have been carefully designed and constructed to efficiently absorb and convert solar irradiation energy into electrical energy. For practical purposes, PV cells are interconnected and environmentally protected in PV modules that form the basic elements of PV systems.

This chapter introduces the basic principles and information concerning PV cells, module, characteristics, and design rules. Monographs [1–7] describing particular phenomena in details are used as references.

8.2 Light Absorption in Materials and Excess Carrier Generation

Light of the wavelength λ can be represented by a flux of photons that have energy hc/λ ($h = 6.626 \times 10^{-34}$ J s is the Planck's constant and $c = 2.998 \times 10^{8}$ ms^{-1} is speed of the light in vacuum). When a flux of photons Φ_{in} falls on a material surface, a part is reflected and a part, the flux of photons Φ_0, penetrates the material

$$\Phi_0(\lambda) = \Phi_{\text{in}}(\lambda)[1 - \mathcal{R}(\lambda)], \tag{8.1}$$

where $\mathcal{R}(\lambda)$ is the surface reflectivity that depends on the photon energy.

The photons penetrating into the material interact with material particles. In the case of monochromatic light of wavelength λ, the interaction can be considered as an interaction of a photon with energy $h\nu = hc/\lambda$ and momentum ≈ 0 with the particles of the material (electrons and nucleus). The interaction has to satisfy the conservation of both energy and

A Comprehensive Guide to Solar Energy Systems. http://dx.doi.org/10.1016/B978-0-12-811479-7.00008-7

momentum. If there are no possible interactions satisfying the conservation laws , the energy of photons cannot be absorbed by the material and the material is transparent. If there are possible interactions by which the material particles can increase their energy (the original energy plus photon energy is one of the possible energy states of the particle in the material), the photon can be absorbed. Suppose that $\Phi(x)$ is the flux of photons in the distance x from the surface, the change of the photon flux $d\Phi$ in the path dx due to absorption can be expressed as

$$d\Phi = -\alpha\Phi dx, \tag{8.2}$$

where α is a proportionality constant called the absorption coefficient that depends on the photon energy, that is, $\alpha = \alpha(\lambda)$.

If we integrate Eq. (8.1) for a constant photon energy (illumination with monochromatic light), the transmitted photon flux decreases exponentially with the coordinate, x,

$$\Phi(x;\lambda) = \Phi_0(\lambda)\exp[-\alpha(\lambda)x] = \Phi_0 \exp\left(-\frac{x}{x_L(\lambda)}\right), \tag{8.3}$$

where x_L is the so-called absorption length. The absorption length gives the distance from the surface, at which the photon flux (light intensity) decreases to 36.8% of the flux Φ_0 entering (semiconductor) at the material surface. On the path from the surface to the absorption length, 63.2% of incident photon flux is absorbed. Practically all photons are absorbed in a layer of thickness $3x_L$.

The interactions of the photons with the material particles may be

- interactions with the lattice,
- interactions with free electrons, or
- interactions with bonded electrons.

Interactions with the lattice (nucleus) are possible for low energy photons (deep infrared), and interactions with free charge carriers (electrons) are important when the carrier concentration is high. Both interactions result in increasing the kinetic energy of particles and, consequently, in an increase of the materials temperature.

Interactions with bonded electrons are the most important. If the photon energy is higher than the energy of the bond, the electron can "liberalize" and move in the material, the unoccupied place behaves like a positive charged "hole". In this way, the incident light generates some excess carriers (electron/hole pairs).

In crystalline semiconductors, all possible energies of bonded valence electrons create the valence band, and all possible energies of free electrons create the conduction band. The dependence of energy on momentum is called the structure of the band. The difference between the maximum energy in the valence band W_v and the minimum energy in the conduction band W_c, $W_g = W_c - W_v$, is the so-called bandgap. At the temperature $T > 0$ some of bonds may be incomplete due to interactions with the lattice or impurities or defects. In thermal equilibrium, free electron concentration n_0 and hole concentration p_0 are connected by

$$n_0 p_0 = n_i^2 = BT^3 \exp\left(\frac{-W_g}{kT}\right), \tag{8.4}$$

where n_i is the so-called *intrinsic concentration* and B is a characteristic material constant. In real semiconductors, there exist some impurities or defects, from which electron may be liberated (donors) or which may generate free holes (acceptors) by overcoming a lower energy barrier than W_g. At room temperature, free carrier concentration is controlled by those impurity concentrations, donor concentration N_D and acceptor concentration N_A. If $N_D > N_A$ (N-type semiconductor), electron concentration $n_0 = N_D - N_A \gg p = n_i^2 / n_0$. If $N_A > N_D$ (P-type semiconductor), hole concentration $p_0 = N_A - N_D \gg n = n_i^2 / p_0$. The control of free carrier concentration in thermal equilibrium by doping with suitable impurities is one of fundamental features of semiconductor technology [1,2]. At thermal equilibrium, free electrons occupy states with energy close to the bottom of conductive band; the holes occupy states close to the top of the valence band (in dependence on the band structure).

8.2.1 Carrier Generation

Let us consider interaction of irradiation with a semiconductor. For photon energy $h\nu < W_g$, the photon can be absorbed by lattice or free carriers only, and absorption coefficient is low for wavelengths longer than hc/W_g.

If the photon energy $h\nu > W_g$, the band-to-band (interband) absorption takes place and the absorption coefficient α increases rapidly with photon energy. The electron–hole pair generation is schematically shown in Fig. 8.1.

The absorption coefficient depends on the photon energy and the band structures. Both electron and hole energy have a minimum at some momentum value, in the surrounding of this minimum the energy increases as a quadratic function of the momentum. Two principal possibilities of the band structure are shown in Fig. 8.2. Fig. 8.2A demonstrates the so-called direct bandgap structure, where both energy maximum in the

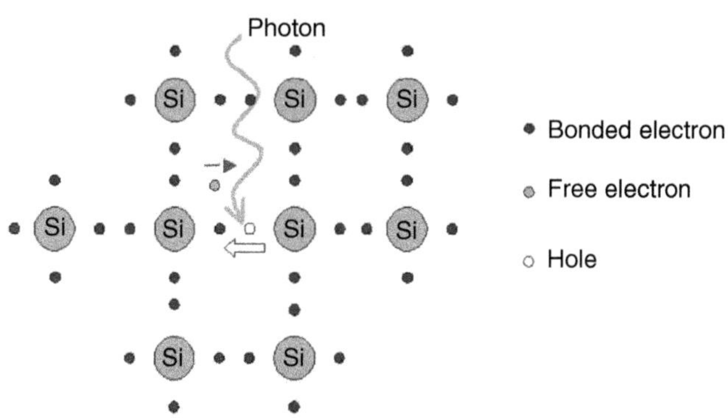

FIGURE 8.1 Optical generation of electron–hole pairs.

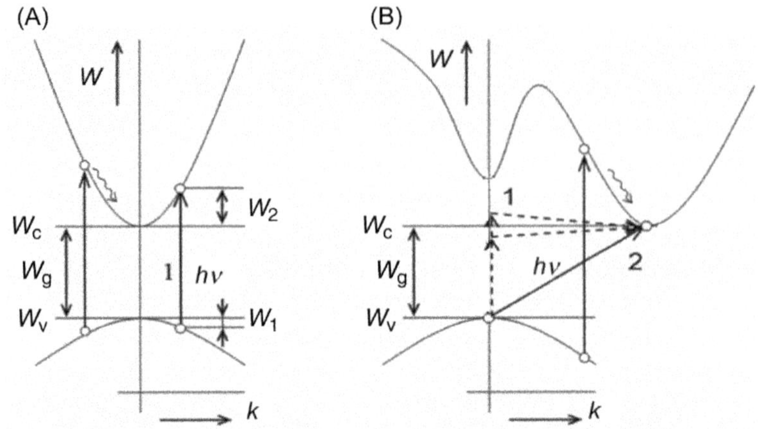

FIGURE 8.2 Carrier generation by photon absorption in the energy band scheme. (A) Direct band structure. (B) Indirect band structure.

valence band and the energy minimum in the conductive band have the same momentum. Examples of "direct" semiconductors include GaAs, CdTe, and $CuInSe_2$. Fig. 8.2A also illustrates the transition of an electron from the valence band to the conductive band following the absorption of a photon with energy $hv > W_g$. As the photon momentum $h/\lambda \approx 0$, the generated electron and hole have practically the same momentum. Increasing photon energy also increases the kinetic energy of the electrons and the holes generated. To reach thermal equilibrium, the excess energy is lost to the lattice vibration as heat as the electrons and holes are scattered from lattice vibrations (phonons), as indicated in Fig. 8.2A. This process is called *thermalization*. The excess energy dissipation is fast; it takes in the order of 10^{-12} s. In the case of the direct transitions, for $hv > W_g$ the absorption coefficient $\alpha(\lambda)$ shows a steep rise with the photon energy up to levels in order 10^4 cm^{-1} (penetration depth x_L in order 1 μm).

The so-called indirect band structure is shown in Fig. 8.2B. The minimum of electron energy in the conductive band and the maximum of energy in the valence band have a different momentum value. Examples of "indirect" semiconductors are Si, Ge, and GaP. In this case, the transition between the maximum of valence band to the minimum of the conductive band is not possible with only the absorption of photon with energy hv close to the bandgap W_g. As the photon momentum $h/\lambda \approx 0$, the transitions can be realized only with the absorption of the photon and simultaneous absorption or emission of the phonon (interaction with the lattice vibration). The requirement of simultaneous electron–photon–phonon interaction in the case of "indirect transitions" results in a relatively small absorption coefficient α in comparison with the case of "direct" transitions. The absorption coefficient increases relatively slowly with the photon energy. For sufficiently high photon energy, "direct" transitions can also be realized in the "indirect" band structure, as indicated in Fig. 8.2B. This results in a steeper increase of the absorption coefficient at higher photon energies. The effect of generated carrier thermalization occurs in indirect semiconductors, too.

A comparison of the absorption coefficient on photon energy of CdTe (direct band structure) and Si (indirect band structure) is shown in Fig. 8.3.

As shown earlier, electron–hole pairs are generated when photon energy $h\nu > W_g$. In this way, the carrier concentration n is higher than the concentration n_0 corresponding to the thermal equilibrium. The difference $\Delta n = n - n_0$ is called excess carrier concentration ($\Delta n = \Delta p$ because electron–hole pairs are generated).

Let suppose that each absorbed photon can generate β electron–hole pairs. For a photon with energy $h\nu < W_g$, no electron–hole pairs are generated and $\beta = 0$. For absorbed photons with energy $h\nu > W_g$, $\beta = 1$ is generally expected. Note that high energy photons may generate more than one electron–hole pair if the kinetic energy given to a generated free electron or hole is higher than $2W_g$. Then, following an impact with a neutral atom, it may generate another electron–hole pair. (This effect may be considerable only for semiconductors with a narrow bandgap.) The excess carrier generation rate G depends on the band structure, the wavelength λ (the photon energy), and the flux of the photons $\Phi(\lambda)$. At a distance x from the surface, it can be expressed by

$$G(x;\lambda) = \left(\frac{d\Delta n}{dt}\right)_{\text{gen}} = \alpha(\lambda)\beta(\lambda)\Phi(x;\lambda) = \alpha(\lambda)\beta(\lambda)\Phi_0(\lambda)\exp[-\alpha(\lambda)x]. \tag{8.5}$$

From Eq. (8.5) it follows that the generation rate strongly depends on the wavelength and intensity of the incident light (and also on the semiconductor material). Under conditions of solar radiation, the total generation rate in a distance x from the surface can be expressed by

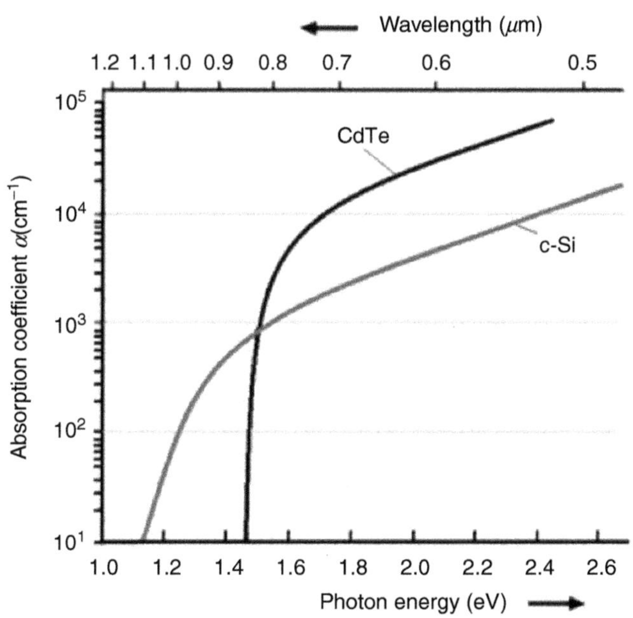

FIGURE 8.3 A comparison of absorption coefficients of silicon and CdTe (after Ref. [7]).

$$G_{\text{tot}}(x) = \int_0^\infty G(\lambda;x)d\lambda = \int_0^\infty \alpha(\lambda)\beta(\lambda)\Phi(\lambda;x)d\lambda. \quad (8.6)$$

The number of carriers generated in solar energy conversion depends strongly on the type of semiconductor (bandgap, band structure).

Photons with too low energy are not absorbed and their energy cannot be transferred in excess carrier generation. Surplus energy of photons with energy higher than the bandgap energy is mostly transformed into heat. Therefore, only a part of incident solar energy can be converted in free charge generation and, consequently, into electric power. The part of the solar spectrum converted in carrier generation in the case of crystalline silicon is shown in Fig. 8.4.

In some materials, after photon absorption an exciton can be generated, that is, an excited electron/hole pair that is still in a bound state due to the Coulomb forces between the particles. Such exciton can diffuse in material and it can dissociate in an electron–hole pair after obtaining an additional energy higher than its bounding energy, or recombine. Exciton generation is an important phenomenon in organic materials and at organic–inorganic material junctions. Further details will be discussed in Chapters 11 and 12.

8.2.2 Carrier Recombination

As discussed in the previous section, excess carrier pairs are created by photogeneration and the carrier concentration is higher than it is at thermodynamic equilibrium. The system tends to reach equilibrium and free electrons and holes recombine (electron takes

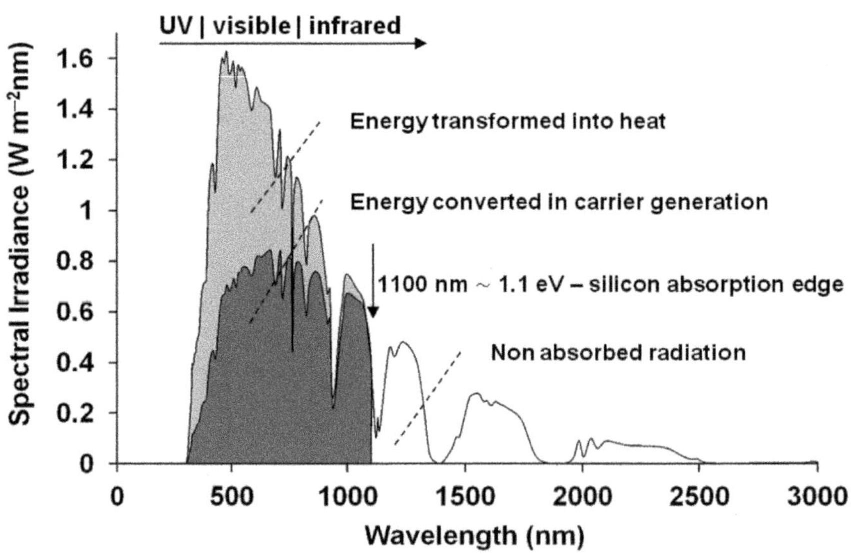

FIGURE 8.4 Energy converted in carrier generation in the case of crystalline silicon.

back a position in an atomic bond, which results in the annihilation of an electron–hole pair). The recombination rate is described by

$$\left(\frac{d\Delta n}{dt}\right)_{\text{rec}} = -\frac{\Delta n}{\tau}, \tag{8.7}$$

where τ is the so-called carrier lifetime that characterizes the excess carrier recombination rate.

In the recombination process, the conservation of both energy and momentum has to be satisfied. The excess energy released as either photons (irradiative recombination) or it is transformed into heat (nonirradiative recombination). The following are the three most important recombination mechanisms.

Radiative recombination. This involves a conduction band electron falling from an allowed conduction band state into a vacant valence band state (a hole). It is the inverse process to optical generation, the excess energy being released mainly as a photon with energy close to that of the bandgap. The irradiative recombination rate depends on the concentration of free electrons, n, and free holes, p, that is, on doping concentration (donor or acceptor) and also on excess carrier concentration Δn. For low injection level ($\Delta n \ll n_0 + p_0$), the irradiative lifetime τ_r can be expressed by

$$\tau_r = \frac{1}{C_r N}, \tag{8.8}$$

where C_r is a constant characteristic for the material and N is the doping concentration (donor concentration N_D in N-type or acceptor concentration N_A in P-type). The irradiative recombination coefficient C_r depends on the band structure. For an "indirect" semiconductor, for example, silicon, C_r is very low (for Si, $C_r \approx 2 \times 10^{-13}$ cm^{-3}) and irradiative recombination is not an important recombination mechanism. In the case of InGaAs and some other semiconductors with so-called direct band structure, C_r is much higher.

Auger recombination. This can be considered as a three-particle interaction where a conduction band electron and a valence band hole recombine, with the excess energy being transferred to a third particle (free electron or hole) as kinetic energy and transferred to heat by the thermalization process. The Auger recombination may be very important in a highly doped semiconductor, as the carrier lifetime strongly depends on free carrier concentration. For a low injection level (i.e., $\Delta n \ll n_0 + p_0$), the carrier lifetime due to the Auger recombination in an N-type semiconductor is given by

$$\tau_A = \frac{1}{C_{An} N_D^2}, \tag{8.9a}$$

and in P-type semiconductor by

$$\tau_A = \frac{1}{C_{Ap} N_A^2}, \tag{8.9b}$$

where C_{An} is the coefficient of Auger recombination in N-type semiconductor and C_{Ap} is the coefficient of Auger recombination in P-type. In silicon, $C_{An} \approx 10^{-31}$ cm^{-3} and Auger recombination may be the dominant recombination mechanism in layers with doping concentration higher than 10^{19} cm^{-3}.

Recombination through local centers. The presence of defects within a semiconductor crystal (from impurities or crystallographic imperfections such as dislocations) produces discrete energy levels within the bandgap. Some of energy levels lie deep in the middle of the bandgap. These defect levels, also known as traps, greatly facilitate recombination through a two-step process. In the first step, a free electron from the conduction band relaxes to the defect level and then (the second step) relaxes to the valence band where it annihilates a hole. The recombination rate is proportional to center concentration, N_t. In an approximation, carrier lifetime can be expressed by

$$\tau_t = \frac{1}{C_t N_t}, \tag{8.10}$$

where C_t depends on center capture cross sections for electrons σ_n and holes σ_p, the center energy level W_t, and it also slightly depends on carrier concentration. By controlling the recombination center concentration N_t, it is possible to control the carrier lifetime.

In general, the recombination processes can be considered to occur independently and the resulting recombination rate is simply the sum of the individual rates. From that it follows that the resulting carrier lifetime, τ, is given by

$$\frac{1}{\tau} = \frac{1}{\tau_r} + \frac{1}{\tau_A} + \frac{1}{\tau_t}. \tag{8.11}$$

At the surface, a higher recombination rate can occur due to surface local states. More details about recombination processes can be found in Refs. [1,3,4].

8.2.3 Excess Carrier Concentration

As shown earlier, excess carrier generation is not uniform and an excess carrier concentration gradient is caused by the decreasing generation rate with depth, below the illuminated surface. A flow of carriers in the direction from higher to lower concentration is connected with the concentration gradient. The flow of charged particles is an electric current. This way, diffusion current density of electrons of the charge $-q(q = 1.602 \times 10^{-19}$ C) is given by

$$J_{ndif} = qD_n \frac{dn}{dx}. \tag{8.12a}$$

The hole diffusion current density (hole charge is q) is given by

$$J_{pdif} = -qD_p \frac{dp}{dx}, \tag{8.12b}$$

where D_n and D_p are diffusion coefficients of electrons and holes, respectively. Diffusion coefficients are connected with electron and hole mobility by $D_n = kT\mu_n/q$, $D_p = kT\mu_p/q$ ($k = 1.381 \times 10^{-23}$ J K^{-1} is the Boltzmann constant). Details describing influence of temperature and doping concentration on the carrier mobility in crystalline silicon can be found in [4].

Considering carrier generation, recombination, and diffusion, the change of excess carrier concentration can be described by the so-called continuity equations

$$\frac{\partial n}{\partial t} = G_n - \frac{\Delta n}{\tau_n} + \frac{1}{q} divJ_n, \tag{8.13a}$$

$$\frac{\partial p}{\partial t} = G_p - \frac{\Delta p}{\tau_p} - \frac{1}{q} divJ_p. \tag{8.13b}$$

Under steady-state conditions ($\partial n / \partial t = 0$ describes a dynamic equilibrium), excess carrier distribution generated by light, in the one-dimensional case for electrons in P-type semiconductor, is given by

$$\frac{d^2 \Delta n}{dx^2} = \frac{\Delta n}{L_n^2} - \frac{G(\lambda; x)}{D_n}, \tag{8.14a}$$

and for holes in N-type semiconductor by

$$\frac{d^2 \Delta p}{dx^2} = \frac{\Delta p}{L_p^2} - \frac{G(\lambda; x)}{D_p}. \tag{8.14b}$$

$L_n = \sqrt{D_n \tau}$ and $L_p = \sqrt{D_p \tau}$ are the so-called diffusion lengths of electrons and holes, respectively, and as usually $\tau_n = \tau_p = \tau$. Using proper boundary conditions, solution of Eq. (8.14) gives information about excess carrier distribution in the irradiated area.

8.3 Photovoltaic Effect and Basic Solar Cell Parameters

8.3.1 Photovoltaic Effect

In homogeneous material, incident light induces excess carrier generation, but through recombination processes practically all the energy is converted into heat (some of the input energy may be reradiated in the form of recombination irradiation). For obtaining a potential difference that may be used as a source of electrical energy, an inhomogeneous structure with an internal electric field is necessary. Suitable structures may be

- PN junction,
- heterojunction (contact of different materials), and
- PIN structure.

A simple structure of a cell is shown in Fig. 8.5A, and in Fig. 8.5B a PN junction energy band diagram is shown. The energy band diagram of a structure with one type of

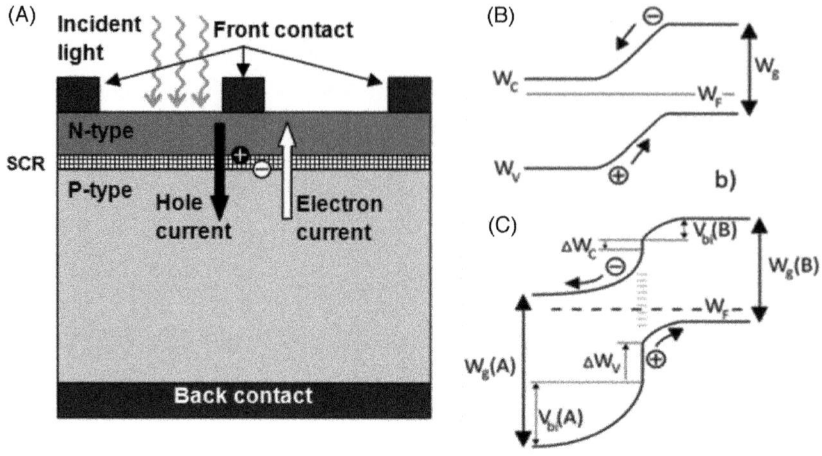

FIGURE 8.5 (A) A schematic solar cell structure. (B) PN junction energy band diagram. (C) PN heterojunction energy band diagram.

heterojunction is shown in Fig. 8.5C. The most common is the PN junction and the principle for this case will be explained later.

Incident light enters the structure and generates excess electron–hole pairs. The generated carriers diffuse to the space charge region at the junction (PN junction or heterojunction). All holes generated in the N-type region reaching the space charge region boundary, drift into the P-type region by a strong built-in electric field and all electrons generated in the P-type region reaching the space charge region drift into the N-type region, as indicated in Fig. 8.5. This way, the N-type region is charged negatively, the P-type region is charged positively and a potential difference between the regions is created. This voltage is capable of driving a current through an external circuit and thereby producing useful work in a load.

8.3.2 *I–V* Characteristics and Basic Parameters of Photovoltaic Cells

The light-induced voltage biases the PN junction in the forward direction. This way, the maximum voltage is limited by the forward *I–V* characteristic of the junction. Then, the solar cell characteristics can be modeled by a superposition of light-induced current I_{PV} and the *I–V* characteristic of the junction. This can be modeled as a current source in parallel with a diode, connected to an external load, as indicated in Fig. 8.6A.

In a real device, there can be some imperfections in the PN junction that result in a parallel resistance R_p across the junction. The current of carriers collected by the PN junction flows to the output contacts through material with a finite resistance. It means that any solar cell has a series resistance R_s between the PN junction and output contacts. The equivalent circuit of a real solar cell is shown in Fig. 8.6B. This equivalent circuit can be used for modeling *I–V* characteristics of PV cells.

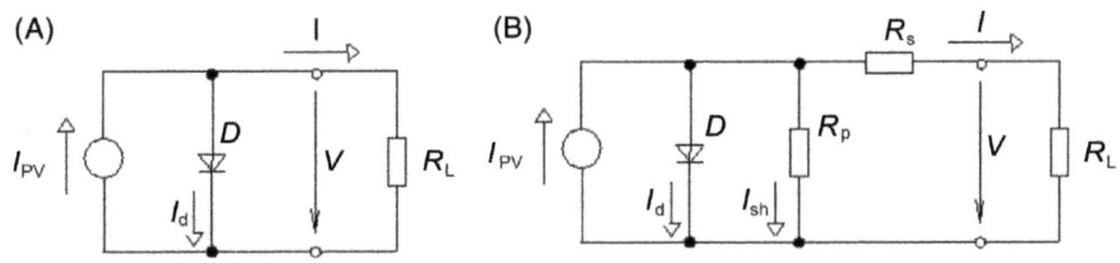

FIGURE 8.6 The equivalent circuit. (A) An ideal PV cell. (B) A real PV cell.

If the surface of the PV cell is irradiated with a constant irradiance, then J_{PV} is the density of the current generated in the volume of the cell structure. In the irradiated area A_{ill} of the cell structure (a part of the total cell area A may be shaded by metallic contacts), the current $I_{PV} = A_{ill}J_{PV}$ is generated. The voltage V_j across the junction is influenced by the load resistance R_L. Then, a part of photo-generated current I_d flows through the diode and a part of generated current I_{sh} flows through the parallel resistance R_p, $I_{sh} = V_j/R_p$. The output current I can be expressed by

$$I = I_{PV} - I_d - I_{sh}. \tag{8.15}$$

The analytical solution is relatively easy for solar cells with PN junction. From PN junction theory, it follows that the voltage V_j across the junction and the junction current density J are connected by the equation

$$J = J_{01}\left[\exp\left(\frac{qV_j}{\zeta_1 kT}\right) - 1\right] + J_{02}\left[\exp\left(\frac{qV_j}{\zeta_2 kT}\right) - 1\right], \tag{8.16}$$

where ζ_1 and ζ_2 are so-called diode factors ($1 \le \zeta_1 \le 2$, $\zeta_2 \ge 2$); $J_{01} = n_i^2 q\left(\frac{D_n}{L_n}\frac{1}{p_{p0}} + \frac{D_p}{L_p}\frac{1}{n_{n0}}\right)$ and $J_{02} = \frac{qn_i d}{\tau_{sc}}$ represent diffusion and generation-recombination components of the PN junction reverse current density. In Eq. (8.16), the full form of the PN junction *I–V* characteristics including generation-recombination processes in the PN junction space charge region has been used. Usually, the simplified form of *I–V* characteristics is used, which is a relatively good approximation (e.g., in the case of crystalline silicon) for $V_j > 0.4$ V,

$$J = J_{01}\left[\exp\left(\frac{qV_j}{\zeta_1 kT}\right) - 1\right]. \tag{8.17}$$

If V is the output voltage and I is the output current and R_s is the series resistance, there is a voltage $V + IR_s$ across the PN junction forward biasing the diode. Therefore, if A is the total solar cell area, the output current, Eq. (8.15), is given by

$$I = A_{ill}J_{PV} - I_{01}\left[\exp\left(q\frac{V + R_s I}{\zeta_1 kT}\right) - 1\right] - I_{02}\left[\exp\left(q\frac{V + R_s I}{\zeta_2 kT}\right) - 1\right] - \frac{V + R_s I}{R_p}, \tag{8.18}$$

where J_{PV} is density of current generated in the volume of the cell structure, $A_{ill} \le A$ is the illuminated area of the cell, $I_{01} = AJ_{01}$ and $I_{02} = AJ_{02}$.

Considering solar cell as an energy source, the characteristics are used in the form shown in Fig. 8.7. The output power depends on the load, as indicated in Fig. 8.8.

The most important points on the *I–V* characteristic are

- open-circuit voltage V_{OC},
- short-circuit current I_{SC},
- maximum output power characterized by the voltage V_{mp} and current I_{mp}. For characteristic evaluation, the fill factor (*FF*) of the solar cell has been introduced,

$$FF = \frac{V_{mp} I_{mp}}{V_{OC} I_{SC}}. \tag{8.19}$$

The cell efficiency η is the next very important parameter that is defined as the ratio of the maximum output power (the photovoltaically generated electrical output of the cell in a case of the optimum load) to the radiation power falling on the cell area:

$$\eta = \frac{V_{mp} I_{mp}}{P_{in}} = \frac{V_{OC} I_{SC} FF}{P_{in}}. \tag{8.20}$$

All these parameters depend on both input radiation power and spectral distribution, which must be precisely defined. Usually, parameters P_m, V_{OC}, I_{SC}, *FF*, and η are given for so-called *standard testing conditions* (STCs), that is, the radiation power 1000 W m^{-2} of spectrum AM 1.5, and cell temperature 25°C. (The AM 1.5 spectrum is defined in the International Standard IEC 60904/3 and it is based on the spectral distribution of solar irradiance tilted at 42 degree to the horizontal.)

Important features of the *I–V* characteristic can be derived from Eq. (8.18). The short-circuit current I_{SC} can be obtained from Eq. (8.18) by putting $V = 0$. For a very low series resistance R_s, the short-circuit current $I_{SC} \approx A_{ill} J_{PV}$; it depends only very weakly on temperature (a very slight increase due to decrease of the recombination rate). I_{SC} decreases with increasing series resistance R_s. In an ideal case, $R_s = 0$ and the $FF \approx 0.89$. With increasing R_s, both maximum power $I_{mp} V_{mp}$ and *FF* decrease and, consequently, efficiency decreases. The influence of series resistance R_s on cell characteristics is schematically shown in Fig. 8.9A.

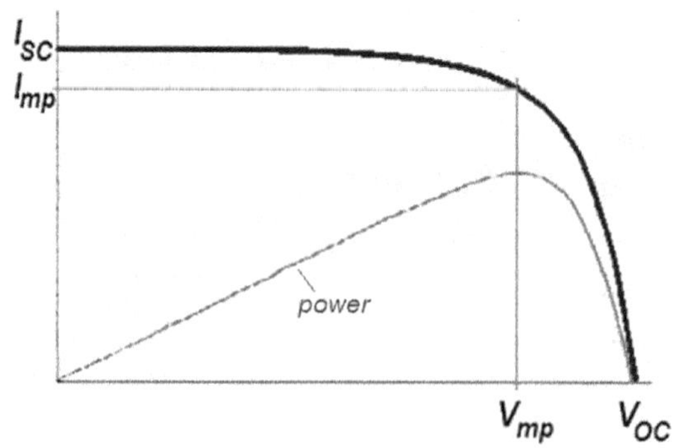

FIGURE 8.7 ***I–V*** **characteristic of a PV cell.**

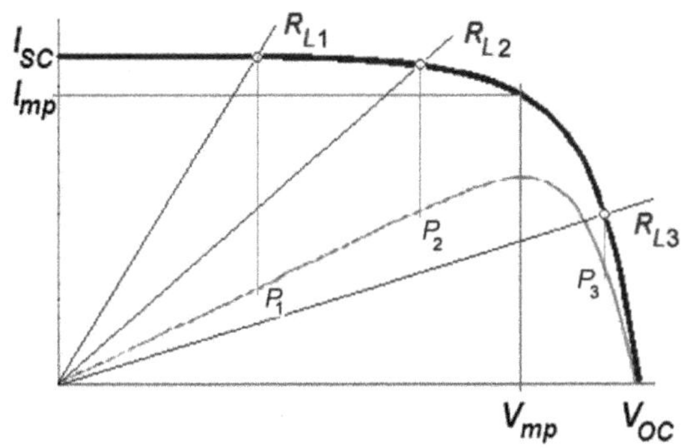

FIGURE 8.8 A load influence on PV cell output power.

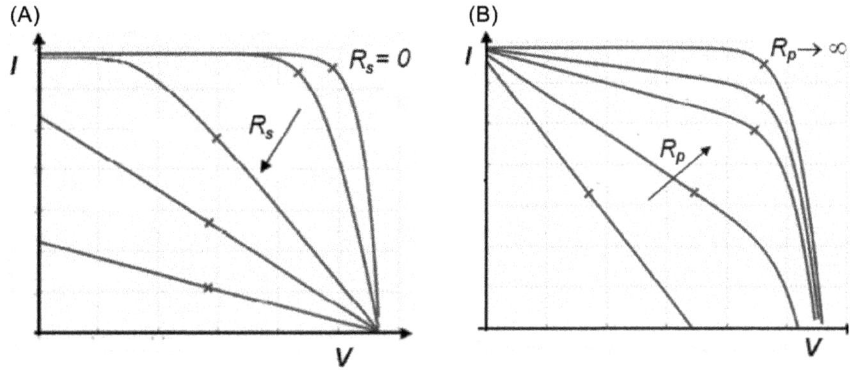

FIGURE 8.9 The effect of parasitic resistances on PV cell *I–V* characteristics. (A) Series resistance R_s. (B) Parallel resistance R_p.

The open-circuit voltage V_{OC} can be found from Eq. (8.18) putting $I = 0$. For quality cells with a high parallel resistance R_p, open-circuit voltage can be expressed as

$$V_{OC} \approx \frac{\zeta_1 kT}{q} \ln \frac{I_{PV}}{I_{01}}. \tag{8.21}$$

The *I–V* characteristics are influenced also by the parallel resistance R_p. A decrease of R_p results in decreasing V_{OC}, maximum power $I_{mp}V_{mp}$, and *FF.* The influence of series resistance R_p on cell characteristics is schematically shown in Fig. 8.9B.

As follows from Eq. (8.21), the open-circuit voltage V_{OC} depends on the junction quality that can be described by the PN junction quality factor ζ and the reverse current density J_0, as demonstrated in Fig. 8.10.

The *I–V* characteristics of PV cells depend on irradiance, as shown in Fig. 8.11A. The PV current I_{PV} is directly proportional to the irradiative power. As follows from Eq. (8.18), in an ideal PV cell with $R_s = 0$ and infinite R_p, both I_{SC} and V_{OC} increase with irradiance and, consequently, both the maximum power and efficiency increase with irradiance, as shown in Fig. 8.11B. In the case of real cells, the *I–V* characteristics are influenced by the series

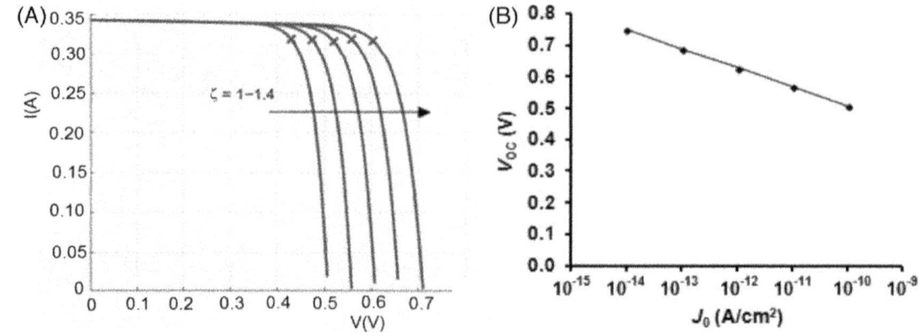

FIGURE 8.10 (A) The effect of the PN junction quality factor ζ on a PV cell *I–V* characteristics. (B) The effect of the reverse current density J_0 on a PV cell *I–V* characteristics.

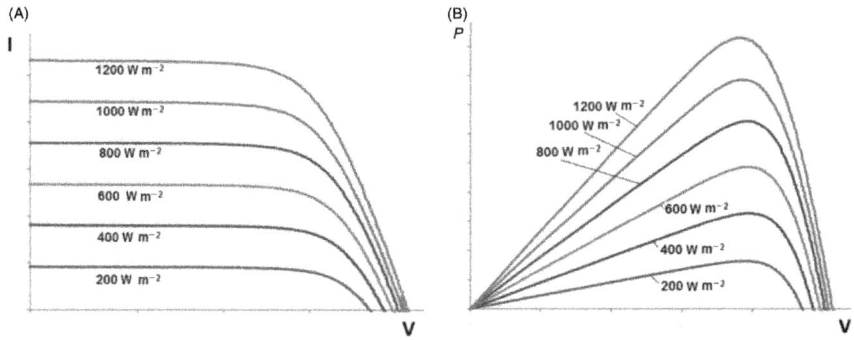

FIGURE 8.11 The influence of irradiance on (A) PV cell *I–V* characteristics and (B) generated power characteristics.

resistance R_s. With increasing irradiance the efficiency at the beginning increases, but with increasing voltage drop across the series resistance R_s, the efficiency increases in the current through diode I_d, and with high irradiances, the efficiency decreases with increasing irradiance. A calculated effect of series resistance R_s on a silicon PV cell efficiency as a function of irradiance is shown in Fig. 8.12.

The *I–V* characteristics of PV cells depend also on temperature. Open-circuit voltage V_{OC} is given in Eq. (8.21) and it is proportional to $\ln(I_{PV}/I_0)$. Although the photo-generated current I_{PV} is nearly independent on temperature, the diode reverse current I_0 is proportional to n_i^2 that strongly increases with temperature with respect to Eq. (8.4). Consequently, as follows from Eq. (8.21), V_{OC} decreases with temperature. Short-circuit current I_{SC} only very slowly increases with temperature and the maximum power current I_{mp} is nearly independent on temperature. With increasing temperature there is also an increase in the series resistance R_s and, consequently, a decrease of the *FF*. Therefore, the decrease of maximum power with increasing temperature is higher than the decrease of the open-circuit voltage V_{OC}. An example of an *I–V* characteristic change with temperature is shown in Fig. 8.13A. The corresponding change of maximum power $P_m = I_m V_m$ with temperature is shown in Fig. 8.13B.

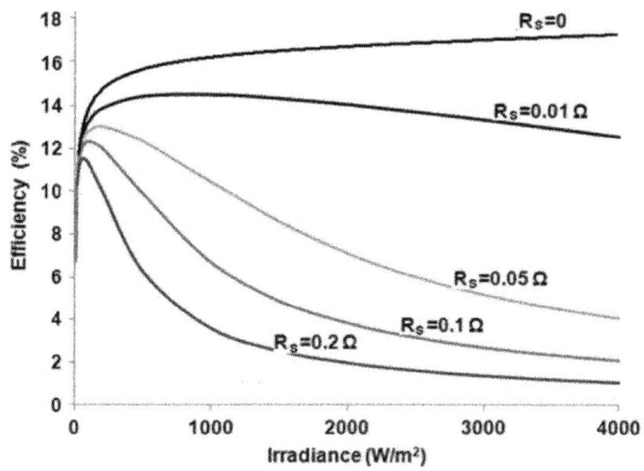

FIGURE 8.12 The influence of series resistance R_s on PV cell efficiency in dependence on irradiance.

Figs. 8.12 and 8.13 demonstrate that operating temperature is a very important parameter for the efficiency of PV energy conversion. In a linear approximation, the temperature dependence of cell efficiency can be described by

$$\eta(T) = \eta_{STC} + K_{T\eta}(T_{STC} - T), \tag{8.22}$$

where η_{STC} is the PV cell efficiency at STC, $K_{T\eta}$ is the thermal coefficient of efficiency, and T_{STC} = 25°C. The thermal coefficient $K_{T\eta}$ depends on the material bandgap, as shown in Fig. 8.14.

The PV cell equivalent circuit and the *I–V* characteristics have been discussed for the case of a constant or slowly varying irradiance. For fast changes (e.g., using a flash simulator), it is also necessary to take into account the PN junction capacitance that can be (under conditions of a forward biased junction) relatively large and that can influence the measured *I–V* characteristic shape.

8.3.3 In-Series and In-Parallel Connection of PV Cells

The open-circuit voltage V_{OC} of individual cells is usually lower than 1.5 V and the photogenerated current density is in order of tens of milliampere per square centimeter. Solar cells can be connected in series to obtain a higher output voltage or in parallel to obtain higher output currents. The case of a series connected cell is demonstrated in Fig. 8.15A. In this case, the same current flows through all solar cells. If the current generated in individual cells differs, the short-circuit current I_{SC} of the series connected cells will be the minimum I_{SC} of all in-series connected cells and the voltage for currents $I \leq I_{SC}$ will be the sum of voltages at individual cells at the current I, as indicated in Fig. 8.15B. The maximum power current I_{mp} of a series connected cell will be limited by a cell with the minimum I_{SC} and, consequently, the output power of the series connected cells will be lower than the sum of output power of individual cells. Only in the case when the current I_{mp} of a series connected cells is the same, will the output power be the sum of output power of individual cells, as indicated in Fig. 8.15C.

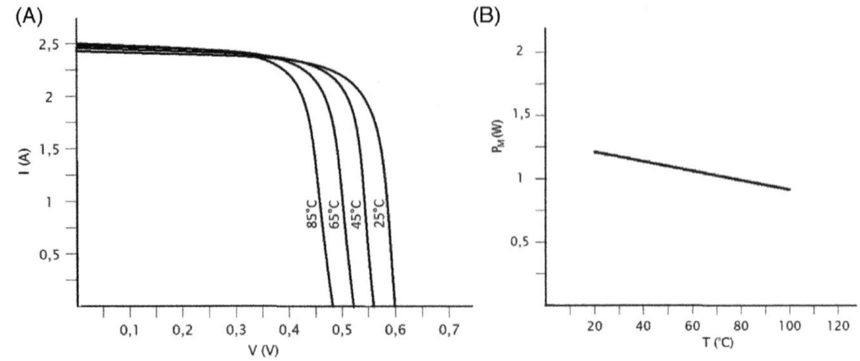

FIGURE 8.13 An example of changes of solar cell *I–V* characteristic with temperature (A) and an example of the measured dependence of maximum power P_m on temperature (B)

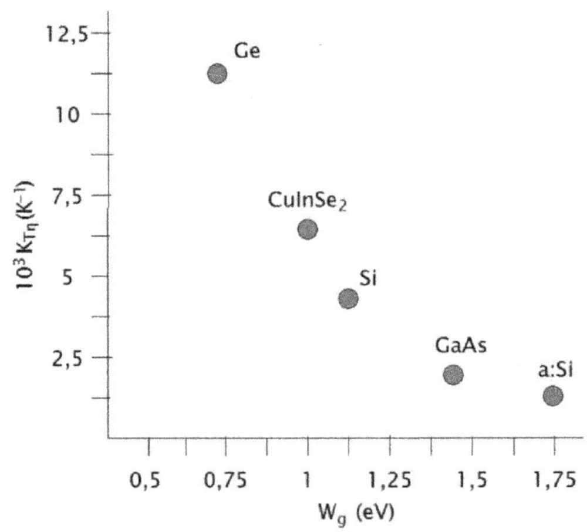

FIGURE 8.14 The dependence of thermal coefficient of efficiency on the material bandgap (after Ref. [3]).

In the case of parallel connected cells, there is the same voltage across all cells, as follows from Fig. 8.16A. Therefore, across a set of parallel connected cells there will be a minimum voltage for all the cells and through individual cells will flow the current corresponding to this voltage. The resulting current will be the sum of the currents flowing through individual cells, as indicated in Fig. 8.16B. In the case when the voltage V_{mp} of parallel connected cells is the same, the output power will be the sum of output power of individual cells, as indicated in Fig. 8.16C. Large area cells with inhomogeneous distribution of both recombination centers and junction defects can also be considered as a number of parallel connected cells with different parameters. Mismatch in cell parameters decreases the output power and, therefore, homogeneity of large area cells is very important in reaching high efficiencies.

Note that even the characteristics of connected cells are the same at STC, differences in cell characteristics can be initiated by external conditions. Nonuniform irradiance

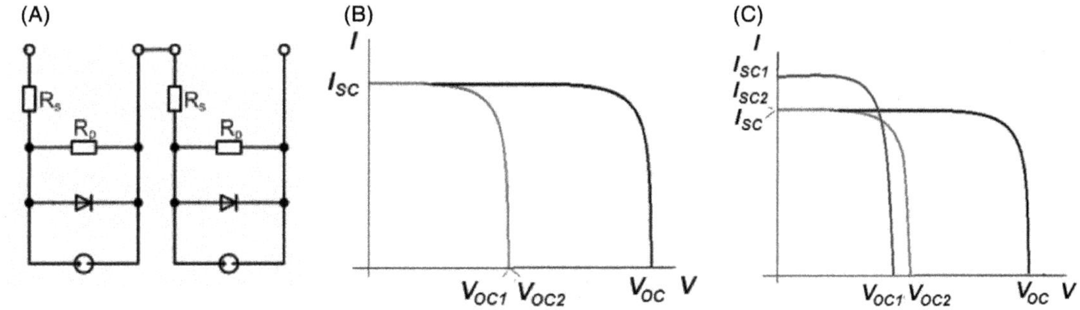

FIGURE 8.15 (A) In-series connection of PV cells. (B) Cells of identical *I–V* characteristics. (C) Cells with different *I–V* characteristics.

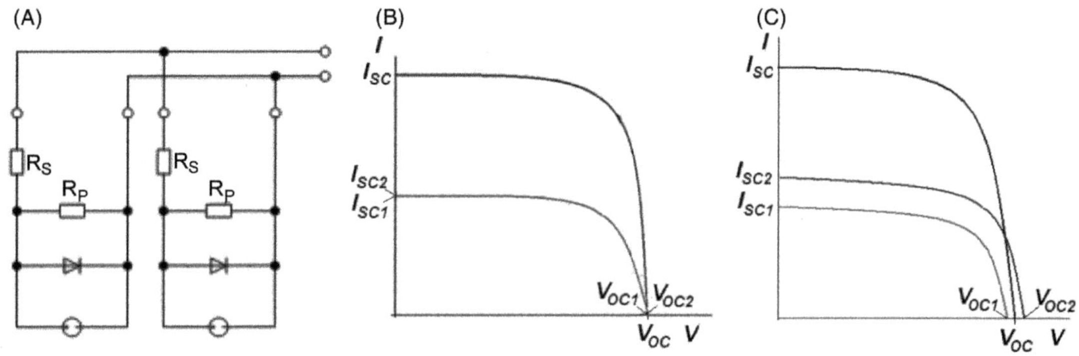

FIGURE 8.16 (A) In-parallel connection of PV cells. (B) Cells of identical *I–V* characteristics. (C) Cells with different *I–V* characteristics.

(e.g., local shading) can cause differences both in the photo-generated current and in the voltage. Furthermore, nonuniform temperatures can cause differences in the cell output voltage and negatively influence the characteristics of cells.

8.4 Principles of Solar Cell Construction

From Eq. (8.18) it follows that for obtaining a high efficient solar cell, it is necessary to maximize the photo-generated current I_{PV} and minimize losses due to parasitic resistances R_p and R_s. At first, we concentrate on optimizing cell construction with the aim of obtaining a maximum photo-generated current.

The structure of a solar cell can be approximated by the diagram, as shown in Fig. 8.17. In the illuminated area excess carriers are generated, which diffuse toward the PN junction. The current density J_{FV} is the current density of carriers collected by the junction space charge region and it can be expressed by

$$J_{PV}(\lambda) = J_{PVN}(\lambda) + J_{PVP}(\lambda) + J_{OPN}(\lambda), \tag{8.23}$$

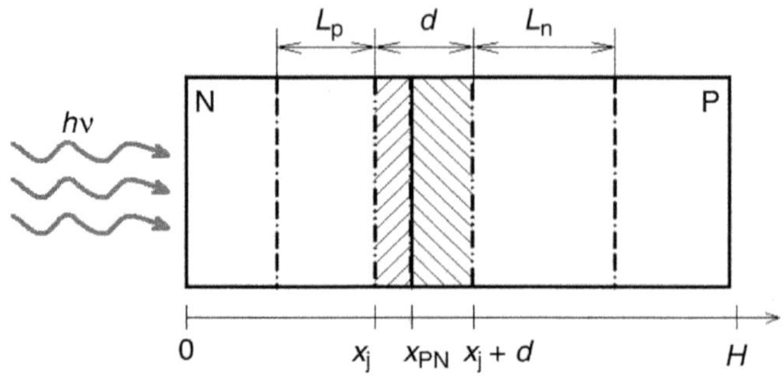

FIGURE 8.17 PN junction solar cell structure.

where J_{PVN} is the current density generated in the N-type region, J_{PVP} is the current density generated in the P-type region, and J_{OPN} is the current density generated in the space charge region of the PN junction. Integrating Eq. (8.13b) over the N-type region, we can obtain for the hole current density generated by light of wavelength λ in the N-type region [2]

$$J_{PVN}(\lambda) = q\int_0^{x_j} G(x;\lambda)dx - q\int_0^{x_j} \frac{\Delta n}{\tau}dx - J_{sr}(0), \tag{8.24}$$

where $J_{sr}(0)$ represents the surface recombination at $x = 0$. Integrating Eq. (8.13a) over the P-type region, we can obtain the electron current density generated by light of wavelength λ in the P-type region

$$J_{PVP}(\lambda) = q\int_{x_j+d}^{H} G(x;\lambda)dx - q\int_{x_j+d}^{H} \frac{\Delta n}{\tau}dx - J_{sr}(H), \tag{8.25}$$

where $J_{sr}(H)$ represents the surface recombination at $x = H$. The third component of the generated current density originates in the PN junction space charge region

$$J_{OPN}(\lambda) = q\int_{x_j}^{x_j+d_j} G(x;\lambda)dx - q\int_{x_j}^{x_j+d_j} \frac{\Delta n}{\tau}dx. \tag{8.26}$$

The generation rate $G(x;\lambda)$ is given by Eq. (8.5) and depends on the cell material and the incident light wavelength λ. The resulting photo-generated current density J_{PV} can be obtained by integrating Eq. (8.23) over the incident light spectrum. The exact analytical solution for the components of current density J_{PV} gives relatively complicated expressions and it can be obtained by solving Eqs. (8.5), (8.14a), and (8.14b) with proper boundary conditions [1,2,5].

Nevertheless, the principles for PV cell construction may be derived from Eqs. (8.24)–(8.26). In all these equations, the photo-generated current density is the difference between carrier generation and recombination in the region (N-type region, P-type region, space charge region, and surface). It follows that it is necessary to maximize carrier generation, and minimize losses by both bulk and surface recombination. From the solar spectral distribution shown in Fig. 8.18, it can be seen that maximum irradiative power is in the visible part of the spectrum, but even energy in the infrared part of the spectrum cannot be neglected.

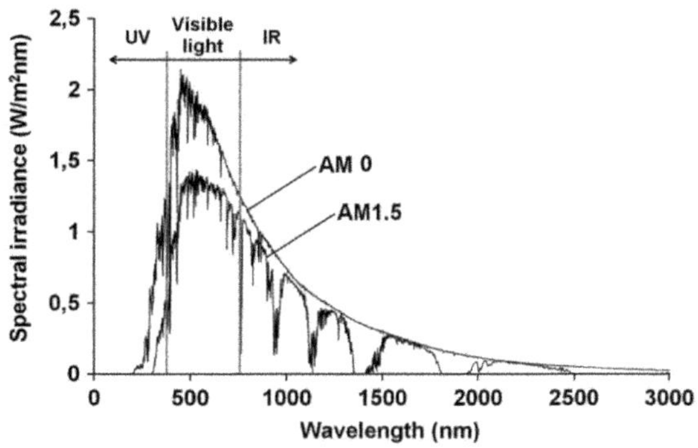

FIGURE 8.18 Spectral distribution of solar radiation power [3].

8.4.1 PV Cell Efficiency Limit

For PV cells, the efficiency of conversion of irradiative energy to electrical energy is very important. As discussed in Section 8.2.1, the number of carrier generated electrons depends strongly on the type of semiconductor (bandgap, band structure). Only photons with energy $hc/\lambda > W_g$ can generate electron–hole pairs. Due to the thermalization process, the part $hc/\lambda - W_g$ of the absorbed energy is transformed into heat. If all the irradiation power $\int_{\lambda_{min}}^{\infty} \frac{hc}{\lambda}\Phi_0(\lambda)d\lambda$ entering the cell body is absorbed, the irradiation power $\int_{\lambda_{min}}^{hc/W_g} W_g\Phi_0(\lambda)d\lambda$ is transferred into electron–hole pair generation. Assuming an ideal PV cell with only radiative recombination and ideal junction characteristic without parasitic resistances, it is possible to state that the ultimate limit for the maximum conversion efficiency η_u is dependent on the absorbing material bandgap and the spectral distribution of the incident radiation power (Quasier–Shockley criteria) as

$$\frac{\int_{\lambda_{min}}^{hC/W_R} W_g\Phi_0(\lambda)d\lambda}{\int_{\lambda_{min}}^{\infty} \frac{hC}{\lambda}\Phi_0(\lambda)d\lambda} = \eta_u. \tag{8.27}$$

The dependence of the maximum efficiency of conversion on the bandgap energy for solar radiation spectra AM 1.5 [3,6,7] is shown in Fig. 8.19. In this case, the ultimate efficiency limit for one absorbing material is 33% for material with a bandgap of $W_g = 1.36$ eV. This value can change if another type of radiation is used (the maximum efficiency can be obtained using monochromatic light with photon energy only a little higher than W_g).

8.4.1.1 Tandem Structures

One possible way of overcoming the Quasier–Shockley limit for PV cell efficiency is by using so-called tandem structures, that is, stacks of cell structures of materials with different

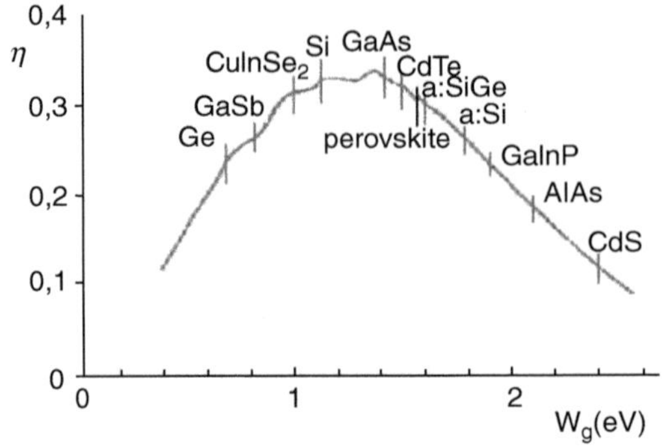

FIGURE 8.19 Maximum conversion efficiency in dependence on bandgap (spectrum AM 1.5).

bandgaps. Consider that the light enters the first cell of material with W_{g1}, where a part of the spectrum (shorter wavelengths) is absorbed. The unabsorbed light then penetrates into the second cell of material with W_{g2} ($W_{g2} < W_{g1}$), where it is absorbed. If the irradiation power $\int_{\lambda_{min}}^{hc/W_{g1}} W_{g1}\Phi_0(\lambda)d\lambda$ is transferred in electron–hole pair generation in the first cell and the irradiation power $\int_{hc/W_{g1}}^{hc/W_{g2}} W_{g2}\Phi_0(\lambda)d\lambda$ is transferred in electron–hole pair generation in the second cell, the ultimate efficiency of the tandem structure η_{ut} is given by

$$\frac{\int_{\lambda_{min}}^{hC/W_{e1}} W_{g1}\Phi_0(\lambda)d\lambda + \int_{hC/W_{e1}}^{hC/W_{e1}} W_{g2}\Phi_0(\lambda)d\lambda}{\int_{\lambda_{min}}^{\infty} \frac{hC}{\lambda}\Phi_0(\lambda)d\lambda} = n_{ut}. \tag{8.28}$$

Comparing Eqs. (8.27) and (8.28) it is clear that the ultimate efficiency of the tandem structure consisting of two cells from materials of different bandgap can overcome the Quasier–Shockley limit and $\eta_{ut} > \eta_u$. This is demonstrated in Fig. 8.20 for a tandem of GaInP and of Ge.

Adding more layers can bring a further increase in the ultimate efficiency of multimaterial structures with suitably selected bandgaps [3, 7], as demonstrated in Table 8.1. For infinite number of layer, the maximum efficiency is 68% for terrestrial applications.

Note that in multimaterial structures the individual cells are connected in series. Therefore, the current in-series connected cells are limited by the cell that delivers the lowest current generated—that makes a special demand on material selection.

Joining of different materials may have a limit in a lattice structure mismatch and there are also some more technological difficulties. However, multilayer cells with efficiency over 45% have already been realized. More details about construction and technology of these structures can be found in literature(e.g. [3, 7]). There are also some other

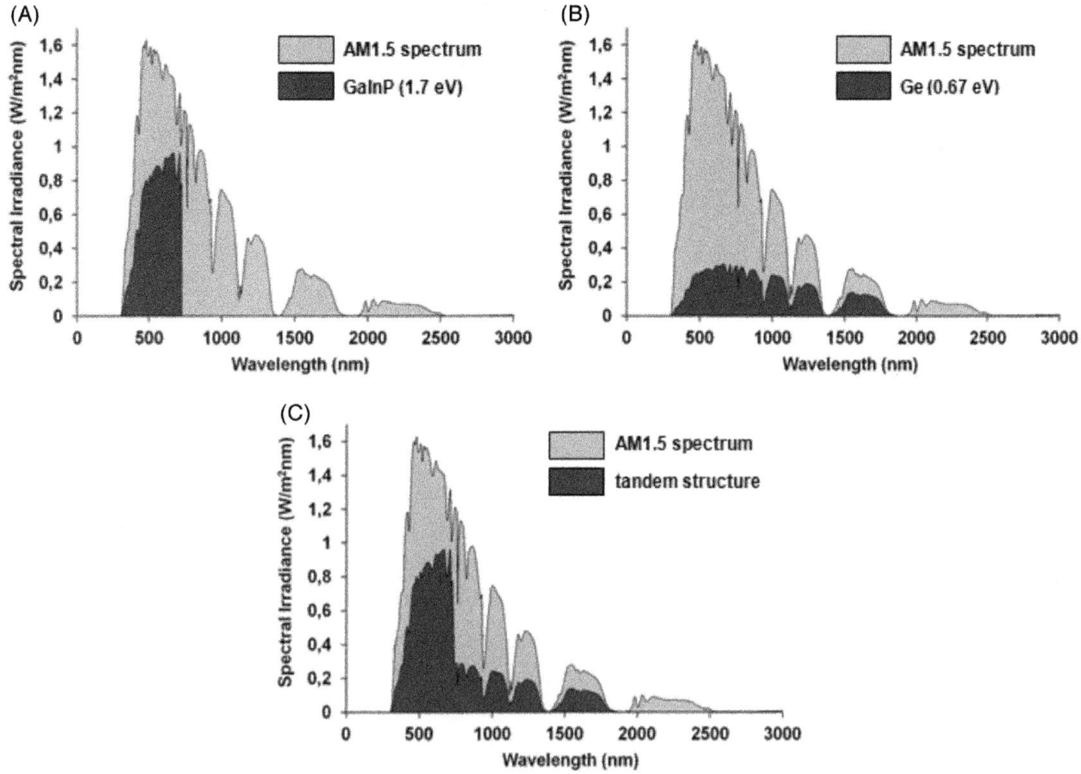

FIGURE 8.20 Irradiative power converted in carrier generation. (A) InGaP (W_g = 1.7 eV). (B) Germanium (W_g = 0.67 eV). (C) Tandem of InGaP and Ge.

Table 8.1 Efficiency Limits of Multimaterial Structures

Number of Cells	Efficiency (%)	Bandgap (eV)
1	33	1.36
2	45	1.63; 0.74
3	51	2.02; 1.21; 0.59
4	55	2.31; 1.55; 0.99; 0.5

ways for overcoming the Quasier–Shockley limit using tools of material engineering that will be discussed in more details in Chapter 13.

8.4.2 Wafer-Based and Thin Film Construction

As follows from Eqs. (8.5), (8.24)–(8.26), the cell construction with respect to individual layer thicknesses strongly depends on the absorption coefficient with photon energy in the solar cell material (Fig. 8.3 and Fig. 8.21).

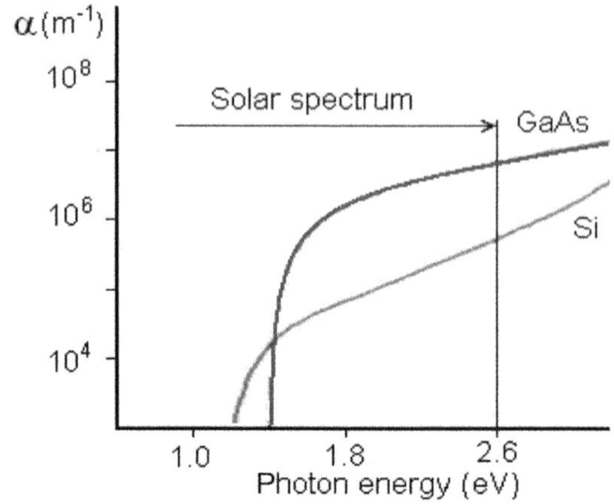

FIGURE 8.21 A comparison of absorption coefficients of silicon and GaAs (after Ref. [1,7]).

Semiconductors with "indirect" energy band structure (e.g., crystalline silicon) have a relatively low absorption coefficient for wavelengths in the infrared part of the spectrum. Therefore, the thickness of the base material should be 150 μm or more for a high efficiency solar cell and have a form of a wafer. In semiconductors with "direct" energy band structure (e.g., GaAs, CdTe, $CuInSe_2$, amorphous silicon), the absorption coefficient increases very quickly with photon energy, as shown in Fig. 8.21, which compares the absorption coefficient of GaAs with that of silicon. In this case, the whole useable part of the solar spectrum is absorbed in a layer of thickness of about 3 μm. In such a case, the solar cell can be realized in a thin film structure, as indicated in Fig. 8.22. Thin film cells can be realized from amorphous silicon (or in combination with microcrystalline silicon), CdTe/CdS, $CuInSe_2$ (CIS), or $Cu(In_xGa_{1-x})Se_2$ (CIGS), deposited on a transparent substrate using thin film techniques. Details about construction and technology of cells from the most important materials will be given in Chapters 9,10 and 18.

8.4.3 Losses in Real PV Cell Structures

Choosing the proper material, the cell structure can be optimized to collect the maximum possible numbers of carriers generated. On the contrary, there may be losses in the solar cell structure, as shown schematically in Fig. 8.23. To obtain a high efficiency in solar cells, the construction and fabricated technology should be prepared in the way to minimize all kind of losses in the structure.

8.4.3.1 Optical Losses

As follows from Eq. (8.1), a part of incident radiation is reflected from the surface. Excess carrier generation rate is proportional to the irradiative power entering the material. To maximize the carrier generation rate, it is necessary to minimize surface reflection. This may be achieved in two basic ways—by thin film antireflection coating or by surface texturing.

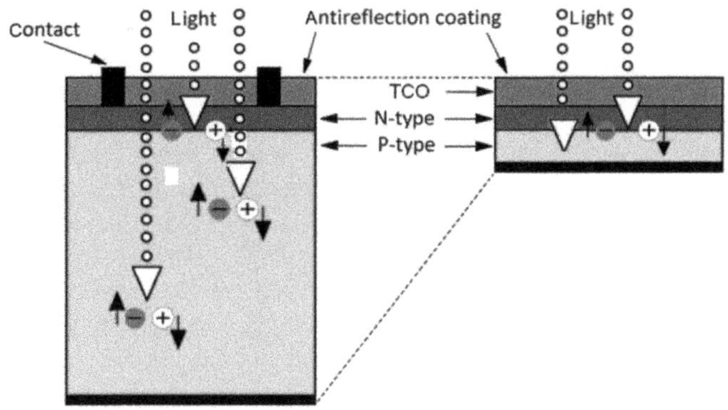

FIGURE 8.22 A comparison of wafer-based and thin film solar cell construction.

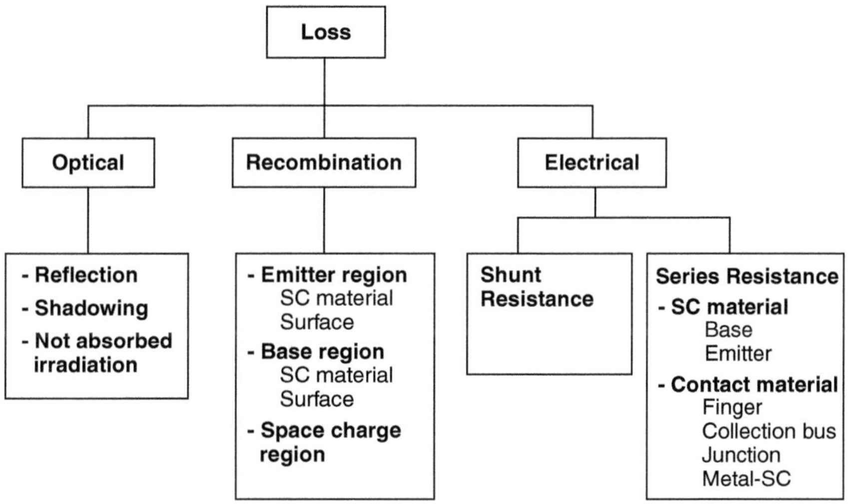

FIGURE 8.23 Possible losses in solar cell structures (after Ref. [5]).

The principle of the thin film antireflection coating is as follows. The incident light beam penetrates the medium of refractive index n_0, through a thin film of transparent material of thickness d_a and refractive index n_1 deposited on the surface of the semiconductor, then into the semiconductor of refractive index n_2. For a monochromatic light, the minimum reflectivity occurs when the layer thickness is $d_a = \lambda / 4n_1$.

Zero reflection occurs when the refractive index of the thin layer is $n_1^2 = n_0 n_2$. The antireflection coating is usually designed to present the minimum of reflectivity at $\lambda \approx 550$ nm, where the flux of photons is a maximum in the solar spectrum. The effect of the antireflection coating could be enhanced by the use of a multilayer antireflection coating, but this involves an increase in the cost of the fabrication.

The total reflection can be also considerably decreased, if the surface is textured, that is, it is covered with microscopic protrusions, produced possibly by anisotropic etching. Reflection on the textured surface of a pyramidal structure is demonstrated in Fig. 8.24. Reflection is at the same angle, so the reflected part of the light beam meets the opposite surface and again a part penetrates the crystal. In this way it is possible to decrease reflection by about one-third of that on a plane surface.

Both principles can be combined to obtain a higher effect in minimizing optical losses by surface reflection. For demonstration, reflectance of both plane and textured surfaces with antireflection coating (optimized for $\lambda = 550$ nm) are compared with Fig. 8.25 for a case of crystalline silicon.

The texturing of the back surface of the cell is also a useful tool; this can increase the reflection on the back surface allowing the light to penetrate the cell from the rear end of the cell. The light reflected back into the cell bulk can be absorbed and can contribute to the enhancing of the generated carrier concentration. This principle, demonstrated in Fig. 8.26, involves thin film cell technology (light trapping).

Another kind of optical loss is due to the shadowing of a part of the cell area by a metallic contact at the front surface of the cell, thus decreasing the illuminated area A_{ill}. The influence of the contact pattern must be taken into account in the cell construction.

8.4.3.2 Recombination Losses

Recombination decreases the photo-generated current density in all parts of the PV cell structure, as follows from Eqs. (8.24)–(8.26). The recombination rate in the case of radiative and Auger recombination is influenced by doping concentration (see Eqs. 8.8, 8.9a, and 8.9b). To keep radiative and Auger recombination low, it is necessary to avoid using highly doped layers in active cell areas. To decrease recombination via local centers (Eq. 8.10), it is necessary to decrease the concentration of both impurities and the defects, thus creating recombination centers to a minimum, using pure starting materials, keeping processes clean during PV cell fabrication. There is also the surface recombination that may have a role at both front and rear surface of the cell structure and on grain boundaries in

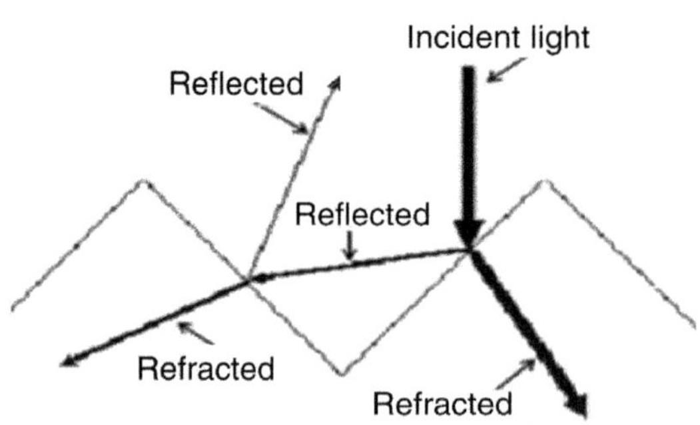

FIGURE 8.24 Reflections on the textured surface [5].

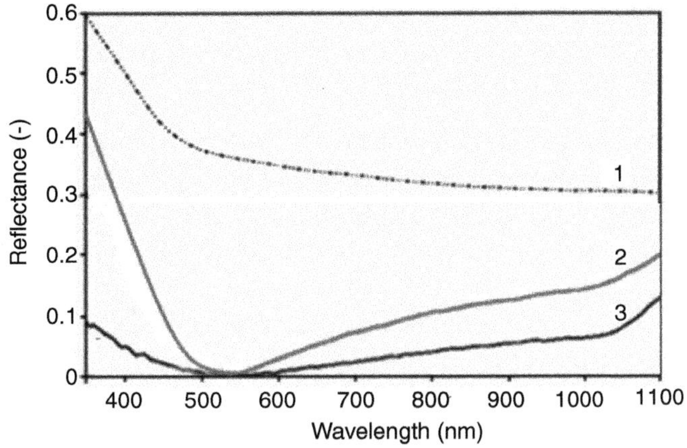

FIGURE 8.25 Reflectance of crystalline silicon surface [5,7]: plane bare surface (1), plane surface with antireflection coating (2), and textured surface with antireflection coating (3).

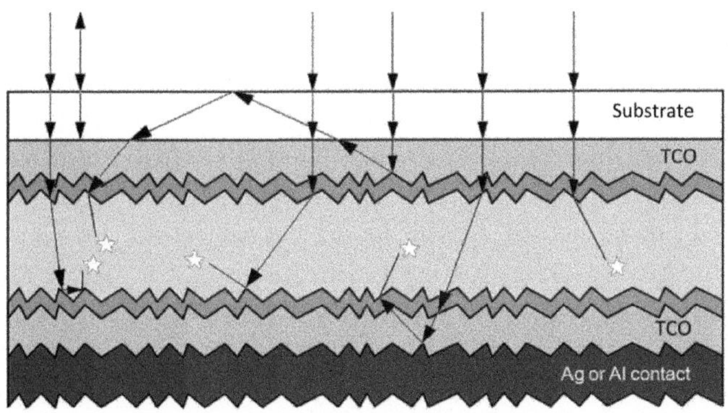

FIGURE 8.26 Light trapping effect in thin film solar cell with textured TCO (transparent conducting oxide).

the case of using a multicrystalline material. Such surface defects should be passivated using a proper manufacturing technology. Methods of decreasing both volume and surface recombination are specific for particular PV cell materials and will be discussed in following sections.

8.4.3.3 Electrical Losses

Optimizing the cell construction with the aim to minimize the series resistance R_s requires different types of contacts for different types of cells such as crystalline silicon cells and thin film cells. In general, the total series resistance R_s consists of following components:

- resistance of contact on the back surface;
- resistance of contact metal-semiconductor on the back surface;
- resistance of base semiconductor material;

- resistance of emitter semiconductor material;
- resistance of contact metal-semiconductor on the front surface; and
- resistance of contact on the front surface.

These individual resistances depend on the PV cell material, contact materials, possible technological tools, and contact pattern, and they should be minimized to decrease electric losses in the cell structure. Therefore, details about minimizing electric losses will be discussed in connection with particular materials and technologies used in Chapters 9–13 and 18.

Some electrical loss is also caused by the parallel resistance R_p. The existence of finite parallel resistance is connected with microshunts in the junction, the fabrication technology imperfections, and with defects created during aging and other degradation processes.

8.5 Photovoltaic Modules—Principles and Construction

As the voltage of a single PV cell is low (usually less than 1 V), several cells must be connected in series to make a practical generator. The PV generator should operate under open air conditions, should be easy and safe to manipulate. And for practical applications, PV cells should be interconnected in a PV module—complete and environmentally protected assembly of interconnected PV cells. The module construction and technology is closely connected with the technology of PV cells. In the following sections, only general features of PV modules will be given, whereas details will be given in later chapters describing PV cells and modules fabricated from particular materials.

8.5.1 PV Modules and Their Characteristics

For applications, a number of cells connected in series are usually encapsulated in modules. As already discussed in Section 8.3.3, the current of series connected cells is limited by the cell that delivers the lowest generated current. Hence, the total current in a string of series connected cells is equal to that of the lowest current. Therefore, to obtain a high efficiency for series connected cells, it is very important that all series connected cells have the same I_{mp}. In this case, the resulting voltage V_{mp} will be the sum of individual cell voltages and resulting power will be the sum of individual cell power.

An example of an analytical description of the module I–V characteristics, an equivalent circuit, shown in Fig. 8.27, can be used.

If m cells of identical parameters are connected in series, the module I–V characteristic can be approximated as

$$I = A_{\mathrm{ill}} J_{\mathrm{PV}} - I_{01}\left[\exp\left(q\frac{V+R_s I}{m\zeta_1 kT}\right)-1\right] - I_{02}\left[\exp\left(q\frac{V+R_s I}{m\zeta_2 kT}\right)-1\right] - \frac{V+R_s I}{R_p}. \qquad (8.29)$$

The parameters of the equivalent circuit suited for Eq. (8.29) can be extracted from the real measured I–V characteristics.

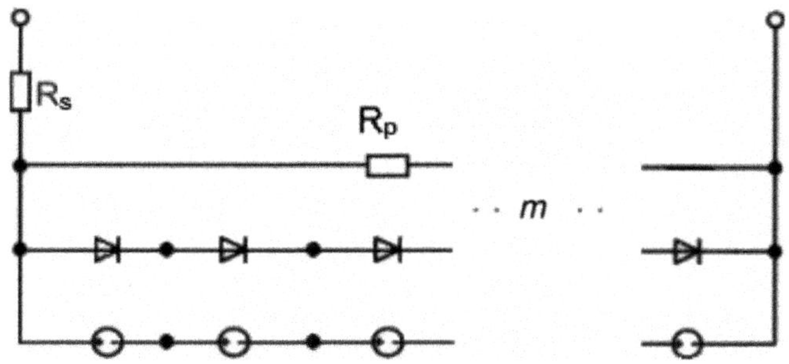

FIGURE 8.27 An equivalent circuit of a module consisting of *m* in-series connected cells.

From the module characteristics, it is possible to apply the same parameters as for PV cells, that is, short-circuit current I_{SC}, open-circuit voltage V_{OC}, maximum power P_m, voltage V_{mp}, and current I_{mp} at the maximum power point, *FF*, and efficiency η. As follows from the rules for connecting cells (Section 8.3.3), in the case of in-series cell connections, the output current of the string will be limited by the minimum current of all in-series connected cells; in the case of in-parallel connection the output voltage will be influenced by the minimum voltage of all in-parallel connected cells.

It is difficult to ensure exactly the same characteristics for cells connected in a module. The problem comes from unavoidable parameter dispersion during cell fabrication and causes so-called mismatch losses, so that the module output power is less than the sum of the powers of the individual cells.

The module is the building unit for the PV generator and it is the PV product on the market. The parameter list contains information measured at the output under STC (the STCs are defined in Section 8.3.2). The following are the parameters usually derived from the module *I–V* characteristic measured using a flash simulator:

- Maximum power P_m
- Maximum power voltage V_{mp}
- Maximum power current I_{mp}
- Open-circuit voltage V_{OC}
- Short-circuit current I_{SC}
- Module efficiency
- Temperature coefficient of I_{SC}
- Temperature coefficient of V_{OC}
- Temperature coefficient of maximum power (or efficiency)
- Maximum system voltage
- NOCT (nominal operating cell temperature)

8.5.2 PV Module Optical, Mechanical, and Thermal Properties

The encapsulation is a very important part of the module. The front should be transparent, allowing the solar radiation to penetrate the PV cell surface with a minimum of optical loss. The output contacts should enable easy and reliable connection to an external circuit. For reliable outdoor operation over a long period (20–30 years), the cells must be properly encapsulated to ensure protection from the weather, including humidity; the withstanding of mechanical loads; protection from impacts; and electrical isolation from supporting constructions. Most modules are constructed in a form of a flat plate creating a solid substrate to support the PV cells. Such flat modules are usually fitted in a frame to enhance the mechanical stability. On the back surface of flat modules a junction box is fitted, in which the electrical connections to PV cells are connected to the output wires which are used to connect the module to the PV system. In thin film technologies, flexible modules can be made for special applications.

Optical losses can occur as a result of refraction at the interface between air and the transparent encapsulating material; absorption in the encapsulating material; and refraction on the interface between the encapsulating material and PV cells (refraction index of the cell antireflection coating should be accommodated to refraction index of the encapsulating material). Optical losses can also occur as a result of the angular distribution of light.

The operating temperature T_c of cell encapsulated in a module is higher than the ambient temperature. The input radiation power absorbed in PV cells is partially converted in the output electrical power and the rest is converted to heat that is dissipated into the surroundings. Considering a relatively low temperature difference between the PV cells and the ambient, convection is assumed as the main mechanism for heat dissipation in terrestrial, flat plate applications. On this basis, a simplifying assumption is made that the cell-ambient temperature increases linearly with irradiance G_r and that the cell temperature T_c in a module of efficiency η can be expressed as

$$T_c = T_a + r_{\text{thca}} G_r (1-\eta), \quad (8.30)$$

where r_{thca} is the thermal resistivity between the cell and ambient.

The thickness and thermal conductivity of individual layers of the module structure are important for reaching a low thermal resistivity r_{thca} of the module that depends also on convection conditions (e.g., speed of wind).

To estimate the average cell temperature in the module, manufacturers provide (in the parameter list) the NOCT parameter, which is defined as the cell temperature under conditions: irradiance 800 W m^{-2}, ambient temperature 25°C, and wind speed 1 m s^{-1}. The NOCT parameter can help evaluate loses connected with enhanced cell temperature under real operating conditions.

Under external conditions, modules are exposed to various external stresses such as temperature changes, mechanical stress (from wind, snow, hail, etc.), moisture (originating from rain, dew, and frost), agents transported via atmosphere (dust, sand, and other agents), and solar radiance (including the highly energetic UV radiation).

8.5.3 Local Shading and Hot Spot Formation

Under real conditions, PV modules can be partially shaded from a neighboring object (a tree, a part of building, etc.) or by a local front surface obstructions (e.g., by leaves fallen on the module surface). Partial shading can significantly influence the PV module output. If there is a shading of one of cells, the current generated in the shaded cells is reduced. As cells are connected in series, the module output current is limited to the current generated by the shaded cell and the output power is decreased. If differences are significant, the nonshaded cells can act like a reverse bias source on the shaded cell. Then, the shaded cell does not generate energy, but it dissipates energy and heats up and that can potentially result in module damage. Similar situation can occur, if part of a PV cell has been damaged by external impact or degradation processes.

To minimize losses due to local shading, a bypass diode is put in parallel, but in opposite polarity, with a cell (or with a group of cells). When the current forced through the shaded substring is such that the reverse bias equals the diode threshold voltage, the bypass diode sinks all necessary current to keep the string at this biasing point thus preventing the increase of the power dissipated in the shaded cell. The bypass diode allows the module to keep delivering the power generated by the unaffected cells. The ability of withstanding local shadings is also part of approval tests of every module type.

It is clear then that the minimum efficiency loss for a shading condition will be if a bypass diode is at each cell, but this has limits in both fabrication technology and cost of processing. Usually, bypass diodes are placed over an internal substring, in some cases over a whole module. The number of bypass diodes and the junction box type together with mechanical characteristics of the module are typically given in information data sheets.

References

[1] Sze SM: *Physics of semiconductor devices*, New York, NY, 1981, John Wiley & Sons.

[2] Fonash SJ: *Solar cell device physics*, New York, 1981, Academic Press.

[3] *Handbook of photovoltaic science and engineering*. In Luque A, Hegedus S, editors: Chichester, 2003, John Wiley & Sons.

[4] Benda V, Gowar J, Grant DA: *Power semiconductor devices—theory and applications*, Chichester, 1999, John Wiley & Sons.

[5] Goetzberger A, Knobloch J, Voss B: *Crystalline silicon solar cells*, Chichester, 1998, John Wiley & Sons.

[6] Würfel P: *Physics of solar cells*, Weinheim, 2005, Wiley-VCH Verlag.

[7] Solar cells—*materials, manufacture and operation*. In Markvart T, Castaner L, editors: Oxford, 2005, Elsevier.

Further reading

[8] *Photovoltaic and photoactive materials—properties, technology and applications (NATO Science Series)*. In Marshall JM, Dimova-Malinovska D, editors: Dortrecht, 2002, Kluwer Academic Publishers.

9

Crystalline Silicon Solar Cell and Module Technology

Vítězslav Benda
CZECH TECHNICAL UNIVERSITY, PRAGUE, CZECH REPUBLIC
benda@fel.cvut.cz

9.1 Introduction

Over the last 40 years, photovoltaic (PV) cells have shown an impressive reduction in price, decreasing by more than 20% every time the cumulative sold volume of PV modules has doubled. This has resulted in a considerable decrease in the overall cost of PV power systems and, consequently, in a decrease of the cost of electrical energy produced by PVs. This cost is now close to the long-term cost of traditionally produced and supplied grid power. Since 1970, crystalline silicon (c-Si) has been the most important material for PV cell and module fabrication and today more than 90% of all PV modules are made from c-Si. Despite 4 decades of research and manufacturing, scientists and engineers are still finding new ways to improve the performance of Si wafer-based PVs and at the same time new ways of reducing the cost. The present state of art and future trends in silicon wafer-based technology will be discussed in this chapter.

9.2 Semiconductor Silicon

Silicon is the most commonly occurring element on the earth. In nature, it never occurs free but in the form of oxides and silicates. In its elementary form, silicon crystallizes in a diamond cubic structure. Some material properties are given in Table 9.1.

c-Si can be prepared in the form of single crystals (monocrystalline silicon, mono c-Si). In this form the lattice parameters and orientation are constant throughout the material. Usually the c-Si is in the form of polycrystalline silicon (poly c-Si) consisting of silicon single-crystal grains separated by grain boundaries. The crystalline grains in poly c-Si can have different sizes and crystalline orientation. The grain size can be sorted into

- multicrystalline silicon (multi c-Si, mc-Si) is composed of many smaller silicon grains of varied crystallographic orientation, typically >1 mm in size (1 mm–10 cm);
- poly c-Si is composed of many silicon grains of varied crystallographic orientation of grain size between 1 μm and 1 mm;

A Comprehensive Guide to Solar Energy Systems. http://dx.doi.org/10.1016/B978-0-12-811479-7.00009-9

Table 9.1 Some Material Properties of Crystalline Silicon at Temperature 300 K

Atomic Density	Density	Lattice Constant	Melting Point	Thermal Conductivity	Thermal Expansion Coefficient
$5 \times 10^{28}\ m^{-3}$	$2328\ kg\ m^{-3}$	0.5431 nm	1415°C	$150\ W\ m^{-1}\ K^{-1}$	$2.6 \times 10^{-6}\ K^{-1}$

Energy Bandgap	Intrinsic Carrier Concentration	Relative Permittivity	Maximum Electron Mobility	Maximum Hole Mobility
1.12 eV	$1 \times 10^{16}\ m^{-3}$	11.9	$0.143\ m^2\ V^{-1}\ s^{-1}$	$0.047\ m^2\ V^{-1}\ s^{-1}$

- microcrystalline silicon (μc-Si) with grain size below 1 μm; and
- nanocrystalline silicon refers to a range of materials around the transition region from microcrystalline to amorphous phase.

The unique properties of Si and SiO_2 enabled the development of integrated circuit technology that has been the basis of present-day microelectronics. Many fabrication tools have been developed and are used in silicon devices technology. Details about particular techniques like diffusion, photolithography, ion implantation, chemical wet and dry processes, and so on can be found in Ref. [1].

The silicon energy bandgap determines the ultimate efficiency of PV cells made from c-Si; this value is 29.4%. As already explained in Section 8.4.2, c-Si solar cells have to be fabricated from wafers of multi-crystalline or mono-crystalline silicon. In the following sections, the technological processes from preparing pure silicon, to silicon wafer fabrication, to cell design and fabrication, and finally to PV module design and fabrication will be discussed.

9.2.1 Semiconductor Silicon Manufacture Technology

In nature, silicon occurs only in the form of oxides and silicates. Silicon is produced by the carbothermic reduction of silica. The so-called metallurgic-grade silicon is produced in a graphite crucible from silicon dioxide (SiO_2) of high-quality lumpy quartz, by reduction with carbon (metallurgical coal) in an arc furnace, as illustrated in Fig. 9.1. The reduction process takes place at approximately 1800°C, according to

$$SiO_2 + 2C \rightarrow Si + 2CO$$

The liquid silicon of purity of approximately 98% is collected by drawing it off at the bottom of the crucible. The reduction process is described in more detail in Refs. [2,3]. To obtain electronic grade silicon, the impurities (Fe, Al, O, Ca, Cu, and others) must be removed. There are several processes that can be used to produce silicon with a purity of greater than 99.9999%. The two most important processes are the Siemens method and the fluidized bed reactor (FBR) method.

9.2.1.1 The Siemens Method

The main features of this technology are described in more details in Refs. [2,3]. In the first stage, pulverized metallurgic silicon is exposed in a FBR (at temperature approximately 350°C) to hydrochloric gas to prepare trichlorsilane $HSiCl_3$:

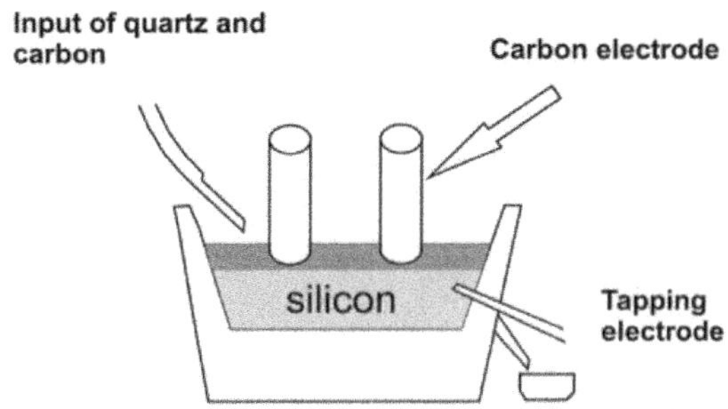

FIGURE 9.1 Arc furnace for metallurgic-grade silicon production.

$$Si + 3HCl \rightarrow HSiCl_3 + H_2$$

Trichlorosilane is chosen because of its high deposition rate, its low boiling point (31.8°C), and its comparatively high volatility. It can easily be separated from hydrogen and other silanes that form during the reaction [the boiling point of other silanes frequently found with trichlorosilane is SiH_4 (−112°C), SiH_2Cl_2 (8.6°C), and $SiCl_4$ (57.6°C)].

In the second stage, remaining impurities are easily be separated from the trichlorosilane by fractional distillation. Trichlorosilane distilled in this way fulfills the requirements for electronic grade silicon with impurity concentrations below the parts per billion level.

Poly c-Si is produced by the reduction of trichlorosilane using hydrogen on a silicon wire (a slim silicon rod of diameter of approximately 4 mm) heated up to 1100°C:

$$HSiCl_3 + H_2 \rightarrow Si + 3HCl$$

The preparing of poly c-Si is schematically shown in Fig. 9.2. Silicon prepared in this way is very clean and fulfills the requirements for electronic grade silicon that can be used in microelectronic technology. A schematic process flow graph for preparing semiconductor silicon from quartz to high-quality semiconductor silicon is shown in Fig. 9.3. Gas flows and electrical power have to be adjusted during the process to obtain optimal deposition rate.

9.2.1.2 The Fluidized Bed Reactor Method

Very pure silicon can be obtained by the decomposition of silane SiH_4. The monosilane SiH_4 can be obtained from trichlorsilane by the reaction:

$$4\,HSiCl_3 \rightarrow SiH_4 + 3\,SiCl_4$$

Tetrachlorsilane $SiCl_4$ can be reduced with hydrogen to trichlorsilane $HSiCl_3$ again. Silane, SiH_4, has the boiling point at −112°C and it can be easily separated in a very pure form.

The silane decomposition occurs on a c-Si surface by reaction:

$$SiH_4 \rightarrow Si + 2H_2$$

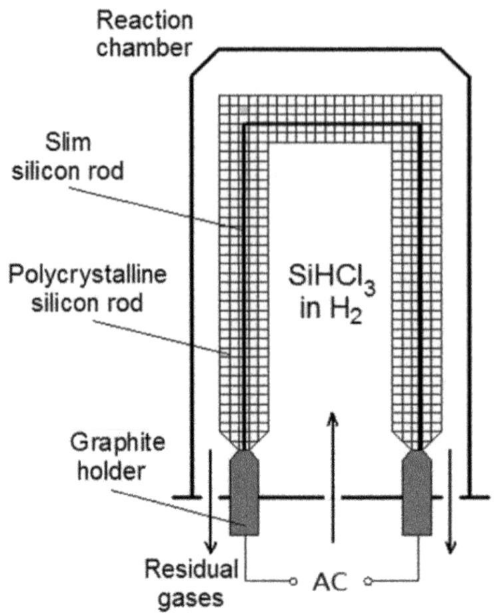

FIGURE 9.2 Schematic of reactor for electronic grade polycrystalline silicon production.

at temperatures of approximately 500°C. The temperature of silane decomposition is much lower and, consequently, less energy consuming than the trichlorsilane decomposition.

The final stage involves using dynamic silicon seed spheres in a fluidized bed sustained by a gas stream of silane and hydrogen. The pure silicon, decomposed from silane, grown on the silicon spheres and the end product are small granules of polysilicon. The reactor is schematically shown in Fig. 9.4.

At present approximately 400×10^6 kg of poly c-Si is produced annually; its price has decreased from hundreds of US dollars per kilogram in 2008 to less than 20 US dollars per kilogram. Today approximately 86% of the silicon production is by the Siemens process, and over 10% is obtained by using the FBR method. It is expected that in the future the share of FBR technology on the silicon production will increase [4].

9.3 Crystalline Silicon Wafer Fabrication

Silicon is a semiconductor with an indirect band structure. As explained in Section 8.1, the absorption coefficient relates to the infrared part of solar spectra; a relatively thick layer (over 250 μm) of c-Si is necessary to absorb all the photons of energy higher than the bandgap. Such thick layers in the form of wafers of a defined rectangular shape (mostly square) and resistivity are used in module fabrication. The wafers are usually cut from a c-Si ingot with a square (or quasi-square) cross section.

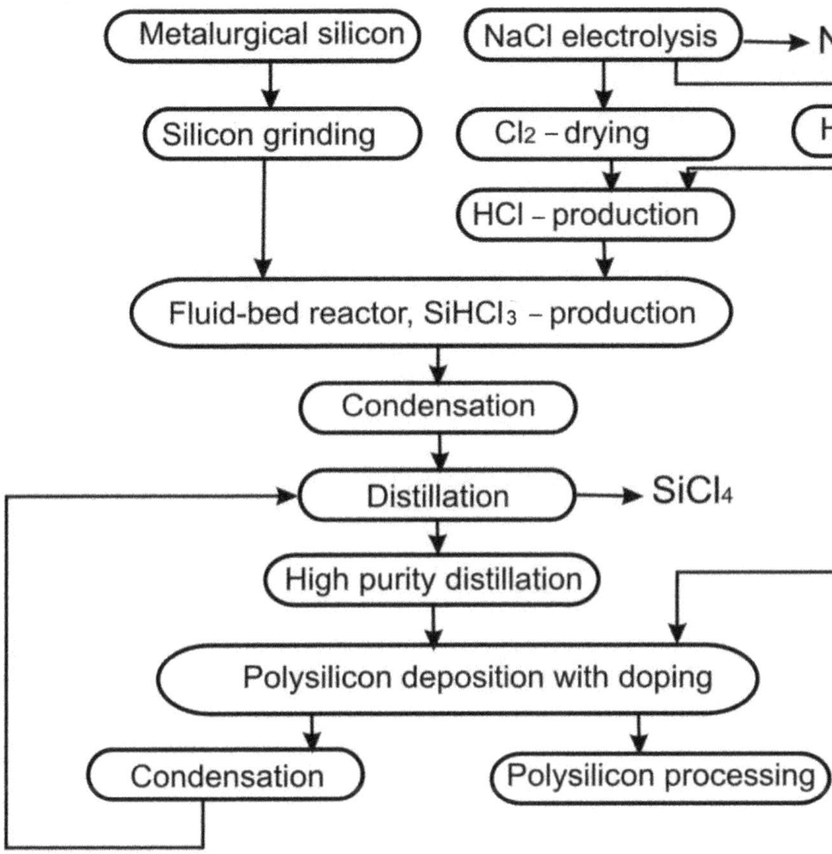

FIGURE 9.3 A schematic process flow chart for preparing semiconductor silicon.

9.3.1 Crystalline Silicon Ingot Fabrication

The c-Si ingot can be a single-crystal or multicrystalline block of well-defined cross section from which can be cut wafers of defined shape and thickness, suitable for solar cell mass production.

9.3.1.1 Silicon Single-Crystal Ingot Fabrication

The Czochralski method is mostly used in the preparation of silicon single crystals. The equipment consists of a chamber in which the feedstock material (poly c-Si pieces or residues from single crystals) is melted in a quartz crucible, doped with the proper concentration of acceptors (to prepare P-type silicon) or donors (to prepare N-type silicon). A seed of a single crystal of defined crystallographic orientation is first dipped into the melt, and then the seed is slowly withdrawn vertically to the melt surface whereby the liquid crystallizes on the seed surface. Then the pulling velocity is raised to the specific value at

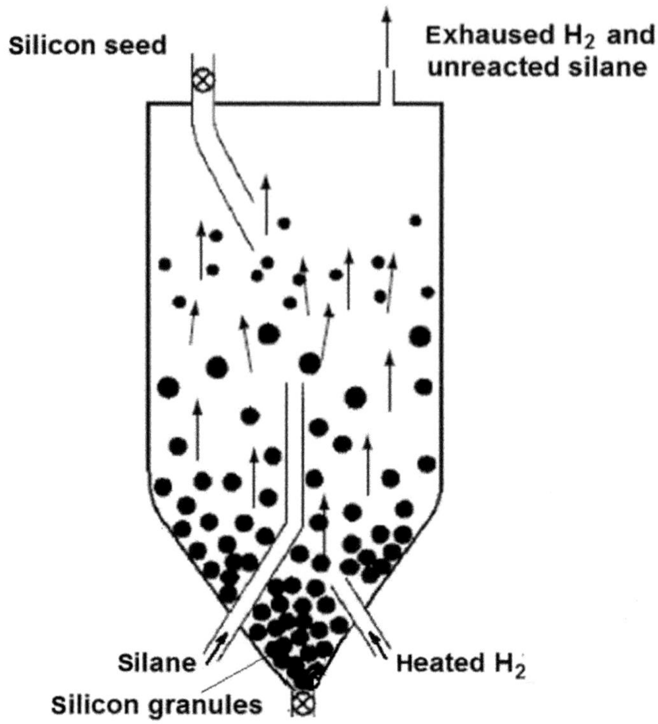

FIGURE 9.4 The fluidized bed reactor.

which the crystal grows to the required diameter. As a result of the seed rotation, the crystal growth is cylindrical shape as indicated in Fig. 9.5. Because of the high reactivity of the molten silicon, the pulling is done under a stream of inert argon gas. The liquid Si reacts with the quartz crucible that supplies considerable amounts of oxygen to the melt.

For preparing P-type silicon, boron is usually used as the doping element. The boron concentration is normally adjusted by adding the equivalent amount of B_2O_3 to the silicon raw material prior to the melting of the silicon. With a segregation coefficient (the relation between the concentration of impurity atoms in the growing crystal and that of the melt) of 0.8, boron doping gives only a small resistivity change throughout the silicon ingot. It is also possible to use gallium, Ga, as a doping element, but due to its small segregation coefficient (0.07) this could result in large variations of doping concentration within the ingot [5].

To prepare N-type silicon, phosphorous is used as the doping element. The phosphorous concentration can be adjusted by adding P_2O_5 to the silicon prior to the melting. Because the segregation coefficient of phosphorous is 0.35, the resistivity distribution over the silicon ingot is less homogeneous. For high-power semiconductor devices, float-zone grown, neutron-doped single crystals are prepared [6], as these devices need very clean and homogeneous material with a long carrier lifetime. The high cost of material is compensated by the improved performance of the final product.

For solar cell technology, P-type (resistivity 0.1–1 Ω cm) single crystals with <100> orientation with a diameter of between 170 and 220 mm and mass of up to 200 kg are mostly used [4]. N-type single crystals are prepared for some types of high efficiency solar cells.

After pulling, the crystal is ground and cut into ingots of an exactly defined shape (normalized). For the solar cell technology, round single-crystal ingots are cut, using a diamond saw, into ingots with a square (or semisquare) cross section, as indicated in Fig. 9.5C.

9.3.1.2 Multicrystalline Block Fabrication

Effective solar cells can be made using mc-Si starting material. Mc-Si offers some advantages over mono c-Si; one being considerably lower manufacturing costs at slightly reduced efficiencies. Another advantage of mc-Si is the rectangular or square wafer shape resulting in a better utilization of the module area in comparison to the mostly pseudo-square monocrystalline wafers. The two main methods of preparing mc-Si ingots are the Bridgman and the block-casting processes [2,3,7].

In the case of the *Bridgman process*, poly c-Si with doping element (B_2O_3) is melted in a quartz crucible coated with silicon nitride (Si_3N_4) of a rectangular cross section. The Si_3N_4 coating serves as an antisticking layer preventing the adhesion of the silicon to the crucible walls. Crystallization is realized by slowly moving the liquid silicon-containing crucible

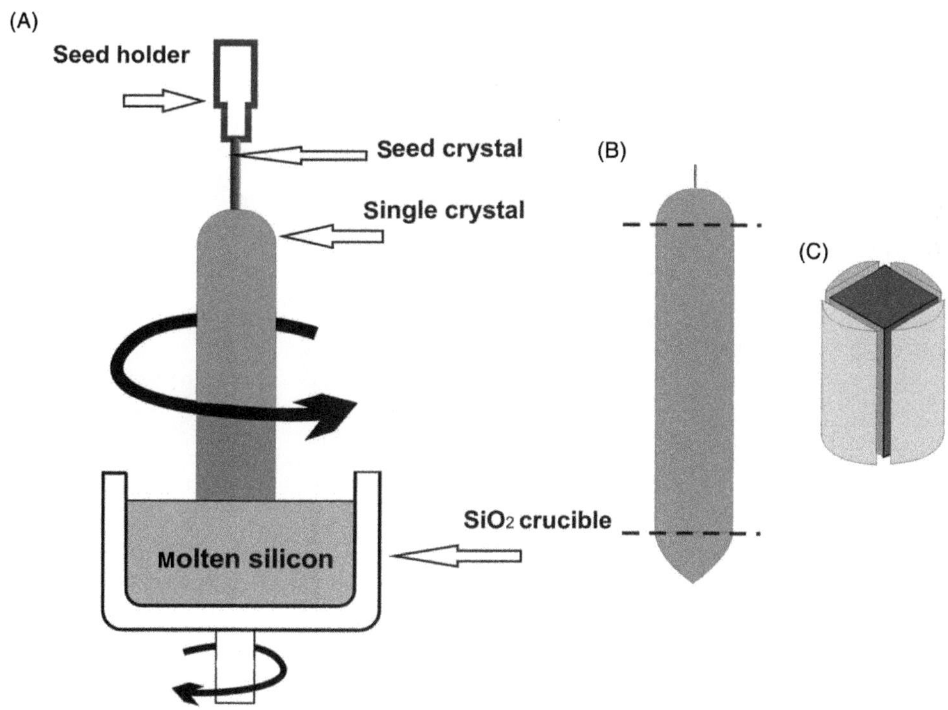

FIGURE 9.5 Preparing silicon single-crystalline rod by the Czochralski method. (A) The crystal pulling equipment. (B) Cylindrical single crystal. (C) Preparing ingots for wafering.

out of the inductively heated hot zone of the process chamber, as indicated in Fig. 9.6A. Crystallization starts at the bottom of the crucible with the temperature below the silicon melting temperature of 1410°C, and the liquid–solid interphase moves in a vertical upward direction through the crystallization crucible. Column-like crystal grains grow spontaneously and need not be of the same crystallographic orientation.

In case of the *block-casting process*, after melting the silicon in a quartz pot, the silicon is poured into a second quartz crucible with a Si_3N_4 coating, as indicated in Fig. 9.6B. Controlled cooling from one side produces a mc-Si block with a large grain structure. In comparison with the Bridgman method, considerably higher crystallization speeds can be achieved owing to the more variable heater system.

A disadvantage of large grains can be a high dislocation density connected with a high recombination rate. For the material quality improvement, the so-called high-performance mc-Si was developed [8]. Using this method, the crystal growth of mc-Si by directional solidification is initiated from uniform small grains having a high fraction of random grain boundaries. The grains developed from such grain structures significantly relax thermal stress and suppress the massive generation and propagation of dislocation clusters. Nowadays, most of commercial mc-Si is grown by this method, which could be implemented by either seeding with silicon crystalline grains or controlled nucleation, for example, using a coating of a nucleation agent.

The mc-Si blocks can be as large as 800 kg [4]; they are cut into smaller rectangular square base ingots of a required area from which wafers are prepared.

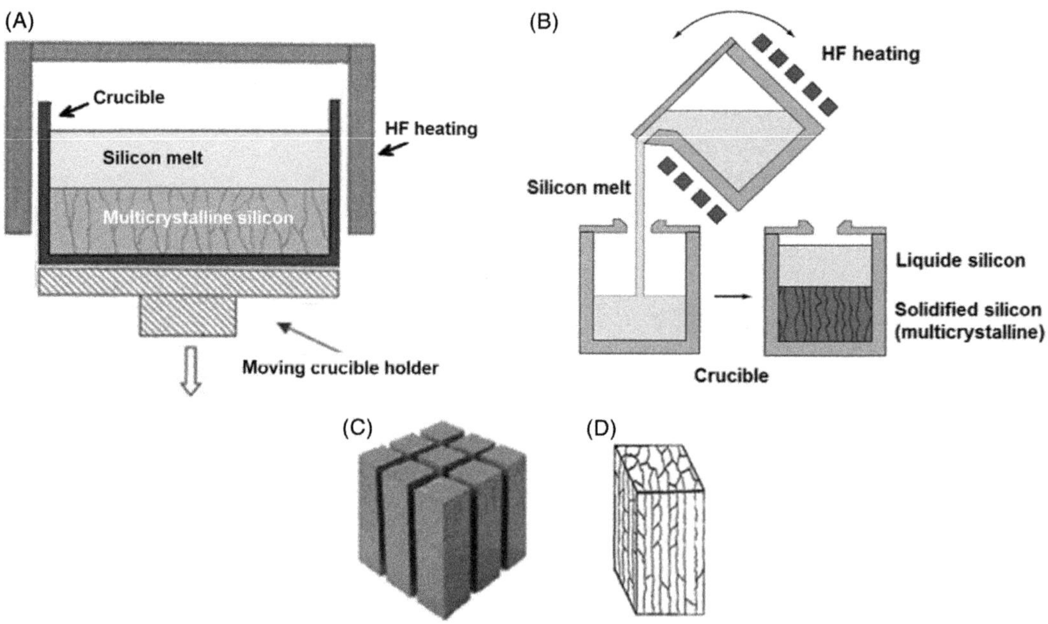

FIGURE 9.6 Multicrystalline silicon ingot fabrication. (A) The Bridgman method. (B) The block casting method. (C) A multicrystalline block divided in ingots. (D) A multicrystalline ingot for wafer cutting.

9.3.2 The Wafering Process

The silicon ingots (both single crystal and multicrystaline) have a cross section that is determined by the final wafer size. At present the standard size is approximately 156 × 156 mm^2, and the largest cross section on the market is 210 × 210 mm^2. In the case of large ingot cross sections, the multiwire cutting technique is generally used [2,3,7]. The principle of the multiwire technology is depicted in Fig. 9.7.

A standard wire saw made of a steel wire is wound around two or four so-called wire-guides, to make a web (500–700) of parallel and regularly spaced wires. The wire is unwound from a spool and is driven into a tension control circuit. The wire length is around 400–800 km, depending on its diameter (generally between 100 and 160 μm, but wire diameters will be reduced in a few next years). The silicon ingot is glued to a substrate holder (usually a glass plate glued to a steel plate) and then it is pushed against the moving wire and sliced into hundreds of wafers at the same time. Cutting is achieved by an abrasive slurry, which is supplied over the wire web and carried by the wire into the sawing channel suspension of hard grinding particles (usually SiC). An alternative is using diamond plated wires [9] in combination with the cooling liquid that allows approximately 3 times higher cutting rate in comparison with slurry-based cutting. Diamond plated wire cutting is used more for mono c-Si [4]. The kerf loss represents approximately 40% of c-Si material.

9.3.3 Ribbon Silicon

The high material losses from cutting silicon ingots induced an effort to prepare the mc-Si starting material in the form of a ribbon. The edge-defined film-fed growth has been used in high-volume production. The process involves placing silicon carbide plates in a crucible-containing molten silicon (doped to the required level). The silicon rises by capillary action between the plates. The poly c-Si sheets prepared using this method are approximately 300 μm thick. The sheets are cut with a laser to create wafers of appropriate

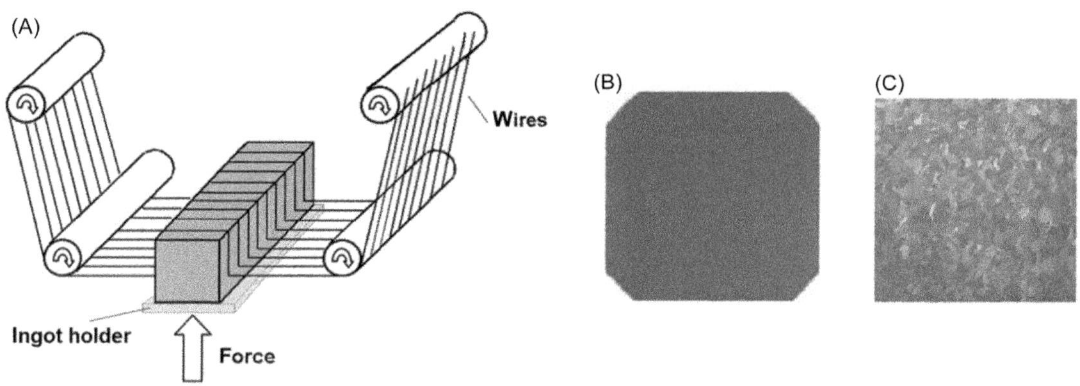

FIGURE 9.7 Preparing wafers by cutting crystalline silicon ingot. (A) The multiwire cutting principle. (B) A monocrystalline wafer (pseudo-square). (C) A multicrystalline wafer.

dimensions. Details about this technology can be found in Ref. [2]. Although ribbon solar cell technology was expected to be very promising, at present it has become marginal and represents less than 1% of world production [4].

9.4 Crystalline Silicon PV Cell Design and Fabrication Technology

As described in this chapter, the cell design and fabrication processes have to maximize power production from incoming irradiance and also minimize all kind of losses (optical, recombination, and electrical) while the processing of materials and the techniques are selected to minimize costs while maintaining a relatively high efficiency.

9.4.1 BSF Solar Cells

After 30 years of development, the majority of c-Si solar cell production is currently based on a very standardized process [2,3,7]. At present, more than 80% of PV module production start from P-type c-Si wafers (both monocrystalline and multicrystalline). These wafers are made with a PN junction over the entire front surface and a full-area aluminum-based metallization with a PP^+ structure on the rear. The design of this cell type is shown in Fig. 9.8. An electric field forms at the PP^+ interface which introduces a barrier to minor carrier flows to the rear surface. The minor carrier concentration is thus maintained at higher levels in the bulk of the device and, consequently, it results in a decrease of surface recombination at the rear contact. This type of structures is known as the BSF (back surface field) technology [10].

On the front side, surface reflectivity is decreased by combination of surface texturing and antireflection coating. The N^+P junction is situated approximately 0.4 μm below the

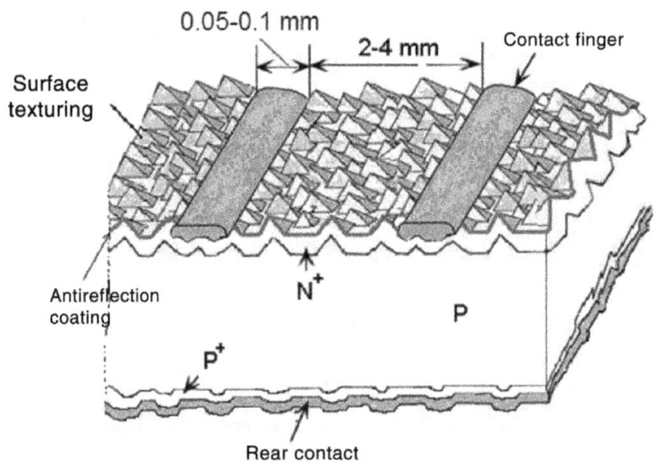

FIGURE 9.8 The BSF cell structure, widely used for mass production.

surface according to dependence of penetration length x_L on the light wavelength, demonstrated in Fig. 9.9A. It is also expected that the junction is below a damaged surface layer to maximize the parallel resistance R_p.

Conventional silicon solar cells are metallized on the front side withthin contact strips, the so-called gridlines (contact fingers), collecting current generated at N^+P junction that flows laterally through the N^+ emitter layer to the contact. Perpendicular to the gridlines wide rectangular-shape strips, the so-called busbars collect the current from the fingers allowing for output connections. Key in conventional The common solar cell design and production is to strike an optimal balance between busbar and finger resistance losses and shading losses. The common solar cell design has an area of 125 × 125 mm^2 and two full-line busbars printed onto the cell. Increasing cell area to 156 × 156 mm^2 involves increasing the length of fingers and consequently the contact grid series resistance also increases. To shorten the fingers, the majority of all PV modules use three busbars at present and there is a trend to increase the number of narrower busbars to four or five or even more [11]. An example of a common contact grid pattern is demonstrated in Fig. 9.10.

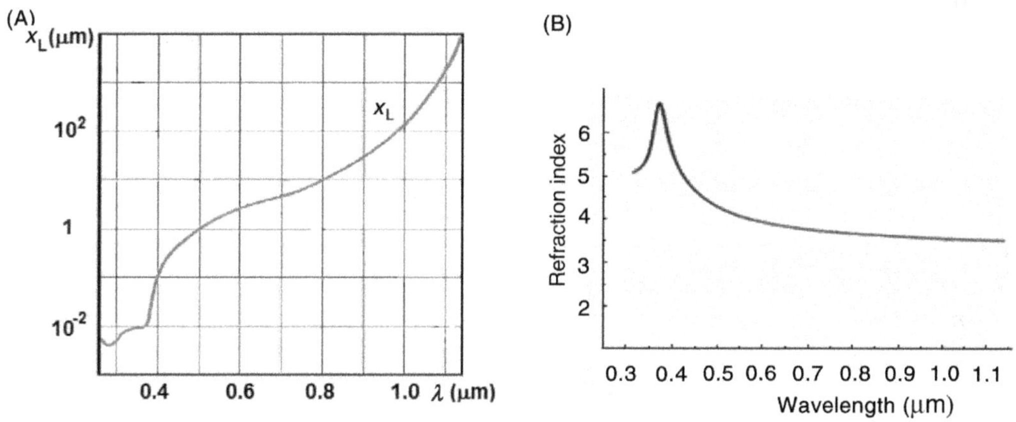

FIGURE 9.9 Absorption depth (A) and refraction index (B) of crystalline silicon in dependence on incident light wavelength [3].

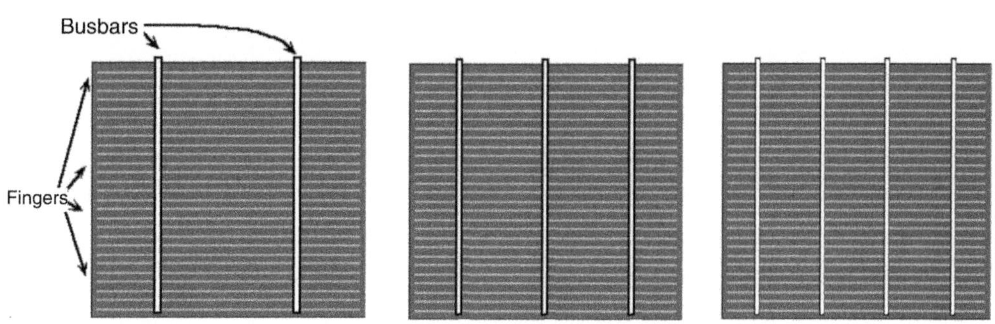

FIGURE 9.10 A comparison of front contact grid pattern for two, three, and four busbars.

The fabrication process for both monocrystalline and multicrystalline is almost the same. At present, the most common thickness of wafers is 180 μm. The process has been developed to avoid relatively high-cost operations like photolithography and vacuum deposition techniques.

Initially, the wafers are etched in an isotropic chemical polishing etchant (e.g., HNO_3:HF:CH_3COOH = 5:1:1) followed by washing in deionized water to remove surface layers (approximately 10 μm) damaged during the cutting process. After removing the damaged layer, wafers are anisotropic-etched to prepare the surface texturing.

In the case of monocrystalline (1 0 0) silicon wafers, an alkaline (NaOH or KOH) anisotropic etching results in a spontaneously generated pyramidal structures on the surface, as shown in Fig. 9.11A.

In the case of multicrystalline wafers, the surface texturing is made using an anisotropic acid etching solution, together with HF, HNO_3, H_2O_2 isotropic stain etching solutions to form a random texture on the surface. An example of the textured surface is shown in Fig. 9.11B.

Recently, a metal-assisted etching technique [12,13] has been introduced to the industrial process. It involves a noble metal (Au, Ag, Pt) nanoparticles used as a catalyst used as a catalyst to enhance the reaction speed locally: it can be used for both monocrystalline and multicrystalline wafers. The resulting surface (usually called "black Si") has very low reflection values with J_{SC} values increasing by approximately 2%–3% together with an increase in the cell efficiency [14].

After texturing, the surface is cleaned and phosphorous is diffused to a depth of approximately 0.4 μm. The diffusion is done at temperature 820°C–900°C using $PClO_3$ in gas containing in a quartz tube in a diffusion furnace or in a continuous annealing furnace using a phosphorous-doped paste screen printed on one side of the wafer. Lower diffusion temperatures are acceptable for multicrystalline starting material because of parameter deterioration when processing temperature exceeds 900°C [15]. Slow cooling after the diffusion process helps to decrease recombination center concentrations by a gettering

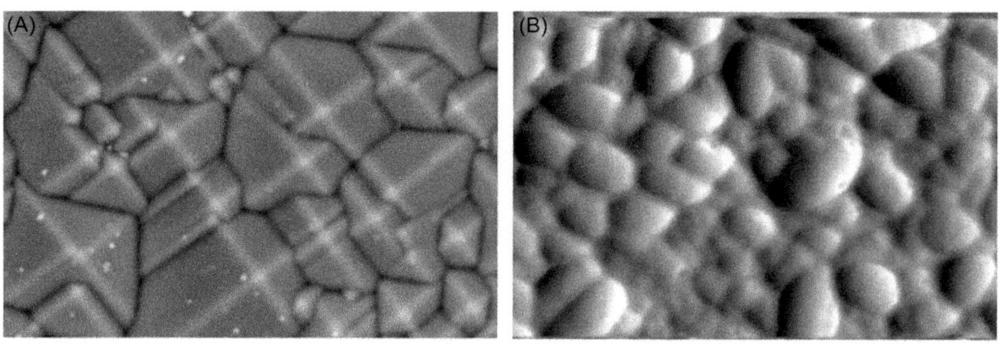

FIGURE 9.11 The surface texturing. (A) Alkaline anisotropy etching of (1; 0; 0) oriented monocrystalline Si wafer. (B) Anisotropic acid etching of a multicrystalline Si wafer.

process. After diffusion, a phosphosilicate glass remains at the surface, but that is usually etched off with diluted HF.

The next step is the surface passivation with a dielectric layer. A negatively charged dielectric (e.g., Al_2O_3) on P-type surface or a positively charged dielectric (SiO_2 or Si_3N_4) on N-type surface is desirable because they create an accumulation layer below the silicon surface and decrease the surface recombination effect. It can be combined with deposition of antireflection coating on the surface after low-level phosphorous diffusion. An effective surface passivation in combination with antireflection layer fabrication can be made by depositing a silicon nitride layer using a PECVD (plasma-enhanced chemical vapor deposition) technique from a mixture of silane and ammonia:

$$3SiH_4 + 4NH_3 \rightarrow Si_3N_4 + 12\,H_2$$

This step is very important for muticrystalline wafers cells. Atomic hydrogen from the plasma interacts with impurities and defects and passivates the grain boundaries. At this fabrication step, an amorphous silicon nitride film is produced with up to 40 atomic percent of hydrogen (the so-called SiN_x layers). The layer composition influences the refraction index and it can be controlled during the layer fabrication. The layer thickness is approximately 75 nm.

Front metallization is a key process in c-Si cell fabrication. The contact grid on the front surface should produce very narrow but thick and highly conductive metal lines with a low contact resistance to Si. This is mostly done by screen printing the phosphorous-doped silver paste on a N-type layer covered with a silicon nitride layer. The paste consists of 60%–80% conducting material (a powder of silver particles of approximately 1 μm), 5%–10% glass frit (a powder of different oxides of lead, bismuth, silicon, etc.) organic solvent, and an organic binder that binds the active powder before thermal activation. The present trend is to decrease the silver consumption by reducing the finger width and decreasing the silver content in the busbars.

On the back surface, a layer of the paste containing aluminum is deposited by screen printing. This paste either contains silver allowing soldering metallic contacts during a module assembly, or strips of silver paste are printed in the second step over the Al containing paste. After each print, wafers are dried at a temperature of approximately 200°C.

The screen-printed paste layers are fired through the nitride layer to make a good electrical contact with the underlying semiconductor at a temperature over 800°C. At the back surface, the Al-Si eutectic is formed and during cooling recrystallized Al-doped P^+ layer creates the back surface field and ensures a good contact. At the front surface, particles of silver (embedded in a compact amorphous glass) come in contact with N^+ silicon layer ensuring a low contact resistance to Si. The composition of the pastes and the thermal profile of cofiring must be adjusted very carefully.

After contact fabrication, the possible shunts on the edge of the wafer must be removed. It is done mostly by laser ablation of the wafer edges. An alternative is plasma etching of wafer edges after the phosphorous diffusion.

The final step is measuring *I–V* characteristics of the cell using a solar simulator and sorting cells into groups according their output (e.g., by cell current at fixed voltage near the maximum power point). It is important that the modules are built from cells with a minimum mismatch of characteristics. The process sequence is demonstrated in Fig. 9.12.

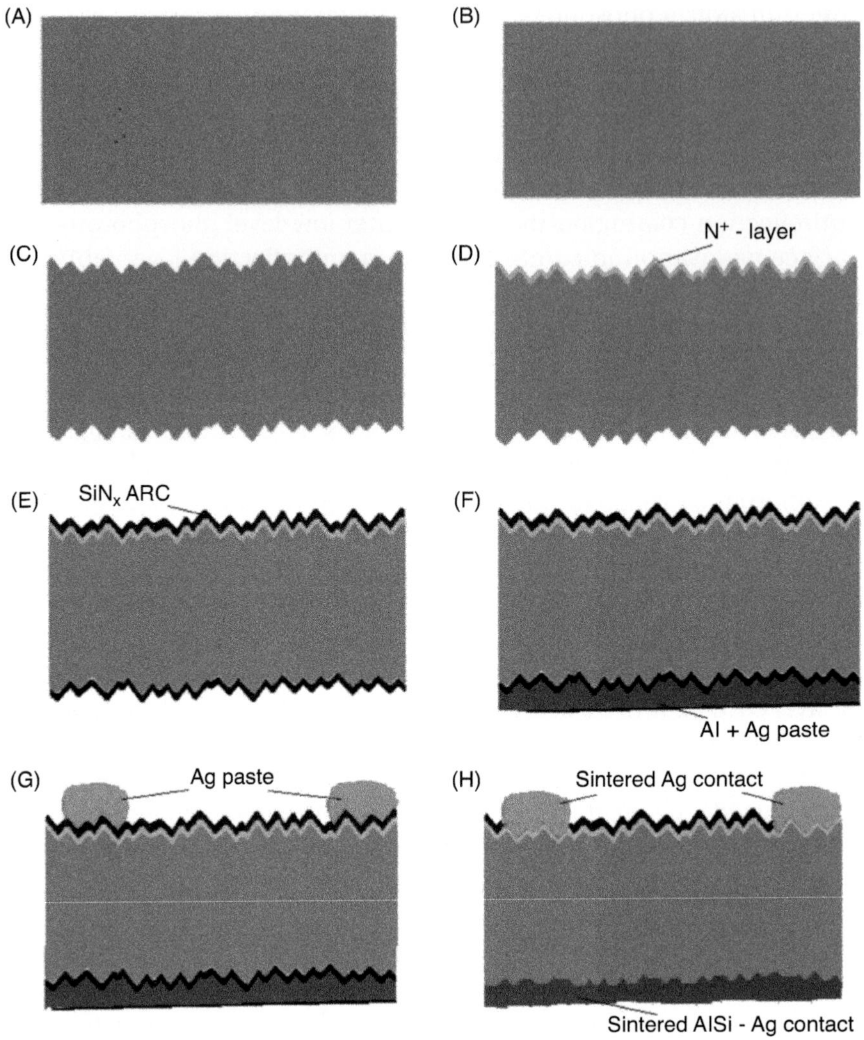

FIGURE 9.12 The standard crystalline silicon solar cell fabrication process sequence. (A) Starting c-Si wafer wire cut without grinding. (B) Silicon wafer after isotropic etchingof damaged surface layer eched-off. (C) Silicon wafer after surface texturing by anisotropic etching. (D) Silicon wafer after phosphorous diffusion. (E) Antireflection coating with silicon nitrid layer (thickness about 75 nm). (F) Print-screenining Al + Ag on the rear of the cell (a layer 20–50 μm thick). (G) Screen printing Ag paste on the front surface of the cell. (H) Contact sintering at temperature about 800°C, edge grinding.

This simple fabrication technology consists of six fabrication steps, resulting in cells of efficiency 15%–19% for monocrystalline and 14 %–18 % for multicrystaline wafers. Besides improving cell efficiency, there is also a trend to use thinner wafers for cell manufacturing. Results showing an increase in efficiency and a decrease in material consumption are indicated in Fig. 9.13.

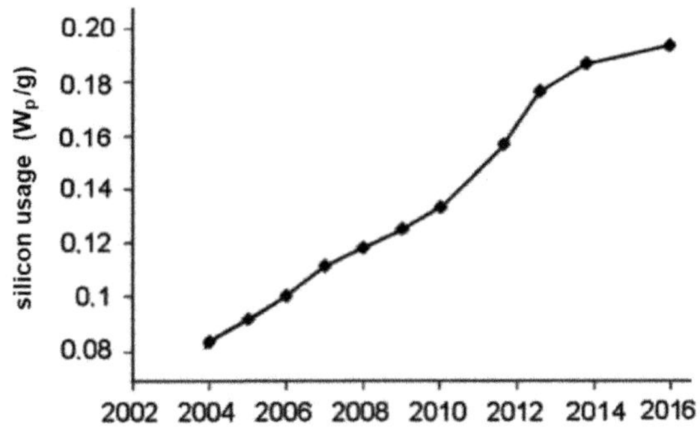

FIGURE 9.13 Development of peak power produced from 1 g of c-Si.

One of tools used for improvements of c-Si PV cells fabricated from starting P-type material involves a selective emitter resulting in a decrease of both Auger recombination losses in the high-doped N^+ emitter layer and in the contact resistance [17]. This construction is based on a heavily doped emitter N^+ region that is narrowly focused at the point of contact between Si and the front side metal, in addition to a less doped N-type emitter region over the entire wafer front surface. The overall expected result when using such a design is a lowered overall value, as well as a slight increase of the J_{SC} (from Eq. 8.24). This follows from the lower recombination in the N-region and results in an increase of the J_{PV} of the cell, which in turn results in a correspondingly higher V_{OC}. Cells of this type can be made in several ways. The most popular is the forming of a pattern of trenches in the silicon wafer with a laser ablation, as demonstrated in Fig. 9.14. The laser grooves are advantageous for either the printed dopant paste or aqueous-based approaches to emitter drive-in, often using the laser-assisted doping. The contact grid may be made using a precisely aligned screen printing of Ag paste or (more advantageously) using electroless plating in the N^+ region of the groove with a thin nickel layer followed by electroplating with a copper alloy [18], as demonstrated in Fig. 9.15. Using an electroplated Cu contact layer has the advantage of reducing Ag consumption for cell fabrication. This technology allows for much narrower grid line widths (of around 30 μm) and reduces losses due to shadowing of light from the front side metal contacts.

Reducing losses due to shadowing from contact busbars on the front side of cells can also be done by connecting the contact grid (consisting from narrow lines) to busbars on the rear of cells by a limited number of holes through the wafer [the so-called metal wrap through (MWT) technology]. The structure is demonstrated in Fig. 9.16. By transferring the busbars to the rear surface, the front shading losses can be reduced and rear busbar contact technology makes for easier module fabrication [19]. In comparison with the standard technology, the MWT technology needs only two additional steps: laser via drilling and rear contact isolation.

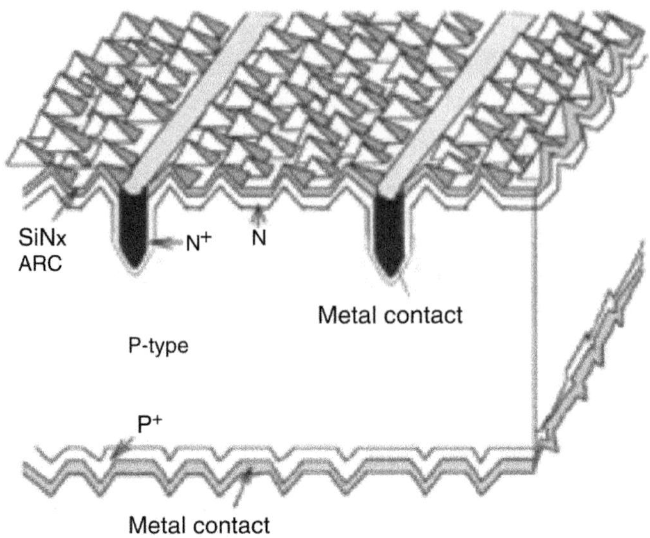

FIGURE 9.14 The selective emitter in trenches prepared by a laser ablation.

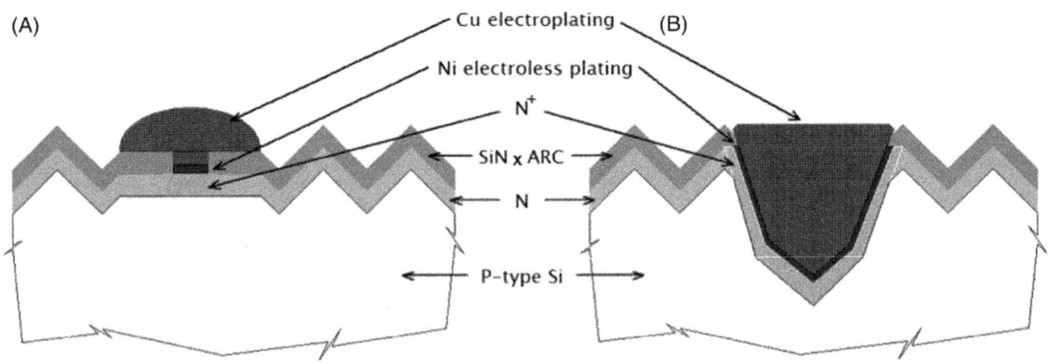

FIGURE 9.15 The plated contact structure. (A) Planar electroplated contact. (B) Electroplated contact in a trenche.

9.4.2 High Efficiency Cells

9.4.2.1 PERC and PERL Cells

Another way to increase cell efficiency is to decrease the surface recombination at the rear of the cell. The passivated emitter and rear contact (PERC) cell reduces recombination at the rear by placing a patterned dielectric layer between the silicon and the aluminum, such that the aluminum only comes into contact with a small fraction of the cell area. The extra dielectric layer of SiO_2 or Al_2O_3 substantially reduces electron surface recombination that results in increased cell efficiency. The PERC structure is shown in Fig. 9.17A.

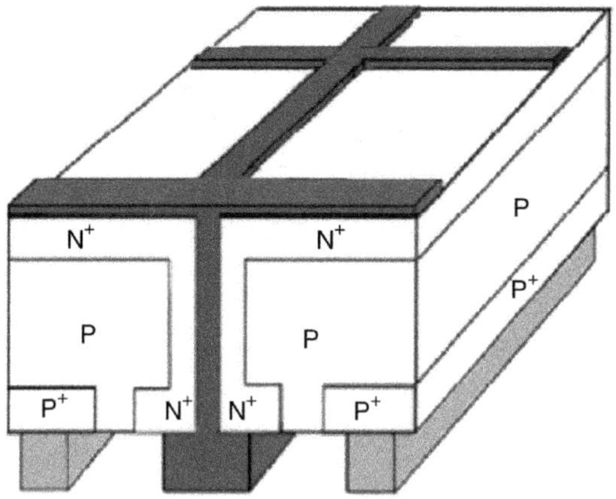

FIGURE 9.16 The MWT solar cells.

The PERC cell fabrication process is as follows. After surface texturing and phosphorous diffusion, the rear side of the wafer is polished followed by phosphor-silicate glass removal and edge isolation. A thin layer Al_2O_3 is deposited on the rear P-type surface (PECVD seems to be the best from the view of the process throughput [20]), and PECVD SiN_x (antireflection coating) is deposited on the both front and rear surfaces. After deposition, a laser is used for the local opening on the rear dielectric layers. After the laser ablation, screen printing of Ag paste on the front and Al paste on the rear side with subsequent cofiring is carried out to form the emitter contact and the local BSF contact on the rear side. PERC cells can be fabricated from both mono c-Si and mc-Si. In comparison with the standard BSF technology, it needs three more fabrication steps; this results in an increase of the cell efficiency of approximately 1% (from 19% to 20%). At present, the PERC technology share is approximately 10%, but is expected to increase significantly in future [4].

A similar cell structure with passivated rear side of the wafer is the passivated emitter, rear locally diffused (PERL) structure. In this case, the rear surface is passivated with

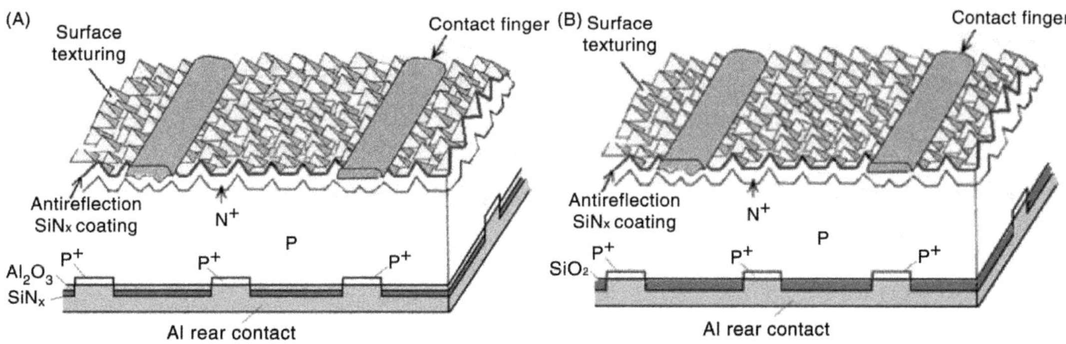

FIGURE 9.17 PV cells with passivated emitter. (A) The PERC cell structure. (B) The PERL structure.

thermally grown SiO_2 layer, in which some windows are opened. Local boron diffusion in these windows creates a local P^+ back surface field. The PERL structure is shown in Fig. 9.17B.

Both monocrystalline and multicrystalline boron-doped silicon contain relatively high concentrations of oxygen. After illumination or carrier injection, B–O complexes create recombination centers that result in severe carrier lifetime degradation, and consequently, in a decrease in efficiency. The oxygen content limits the maximum efficiency that is possible using P-type boron-doped silicon. The monocrystalline P-type silicon prepared by the float zone method does not contain oxygen, and this type of solar cell has the record efficiency of 24%. This material and fabrication technology (PERL [21] prepared using microelectronic technology) is too expensive to be used in mass industrial production.

9.4.2.2 PERT, TOPCon, and Bifacial Cells

Phosphorous-doped N-type silicon wafers retain lifetimes on the order of milliseconds under the same stresses [22] and therefore can be used as a starting material for high-efficient solar cells. The PN junction is formed by boron diffusion [23]. A disadvantage of this technology is that it needs a higher diffusion temperature than the phosphorous diffusion and for P-type surface passivation there must be a layer of thermally grown SiO_2, on which a thin layer of SiN_x:H or TiO_2 is deposited to form an effective antireflection coating. This technology needs a process temperature of over 1000°C; this is also not compatible with multicrystalline material because this material does not tolerate temperatures above 900°C [15]. Therefore, at present only monocrystalline starting N-type material is used in this process for mass production. The multicrystalline N-type material cells technology is still an object of research and development, even though recent research brings very promising results [24].

N-type PERT (passivated emitter rear totally diffused) cells are from the view of the construction similar to PERC cells fabricated from P-type silicon. The structure is shown in Fig. 9.18. After surface texturing and boron diffusion, the rear of the wafer is polished and phosphorous is diffused into the rear to create a N^+N structure on the back surface. The front surface doped with boron needs a thin layer of SiO_2 or Al_2O_3 as effective passivation that can be overlapped with a layer of SiN_x to create the antireflection coat. The phosphorous diffused layer on the rear side requires a layer of SiN_x for passivation. After surface passivation, a laser is used for the local opening of the rear dielectric layers. After the laser ablation, aluminum is evaporated (or sputtered) to create the rear contact. These PERT cells from N-type monocrystalline Si can reach efficiencies of over 22% [25].

The next high-efficient PV cell structure on N-type starting material is the so-called TOPCon (tunnel oxide passivated contact) structure [26], as shown in Fig. 9.19. The front side of the cell is fabricated in the same way as the PERT cells. On the rear side of the N-type bulk materiál, a SiO_x passivation layer of thickness one or two nanometers is deposited which allows the charge carriers to "tunnel" through. Then, a thin N^+ layer is deposited over the entire layer of the ultra-thin tunnel oxide and covered with a metallic layer to create the rear contact. The TOPCon structure has been shown to have an efficiency of

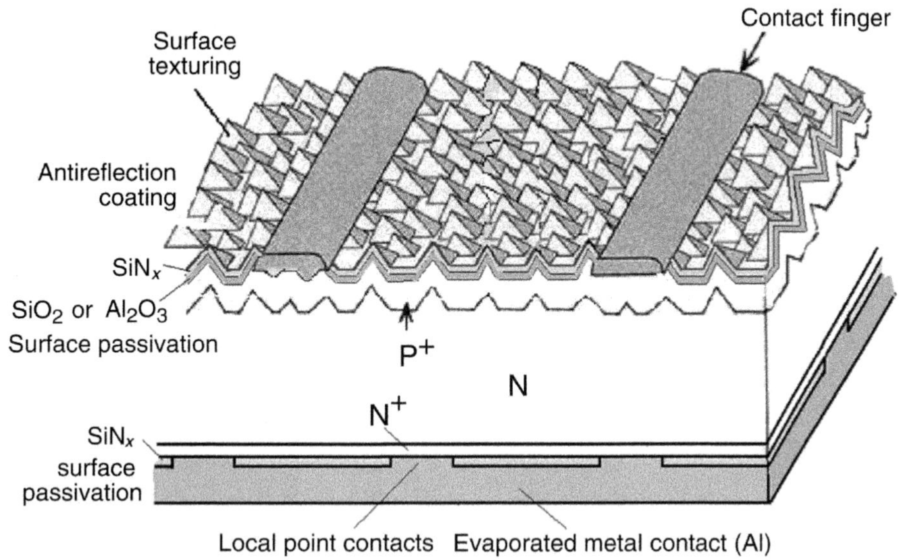

FIGURE 9.18 The PERT cell structure.

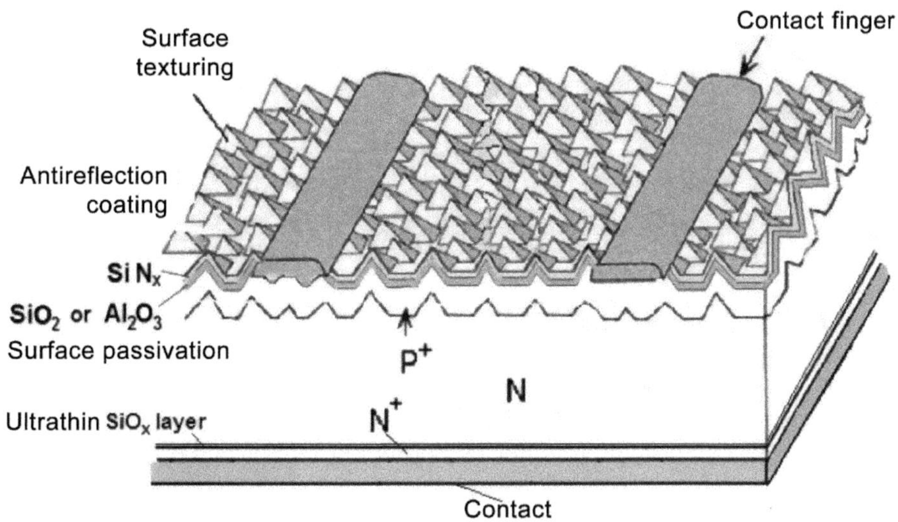

FIGURE 9.19 The TOPCon cell structure.

25.1%. It can be relatively easily produced by mass production because the metal contacts applied to the rear side can be done without patterning.

BIFACIAL SOLAR CELLS

Bifacial solar cells have the capability to generate electricity by capturing light from both sides. This may be useful in some applications [27]. Such cells, for example, are capable of collecting diffused light from the back surface, thus increasing the total efficiency. One variant of a simple bifacial cell structure is shown in Fig. 9.20.

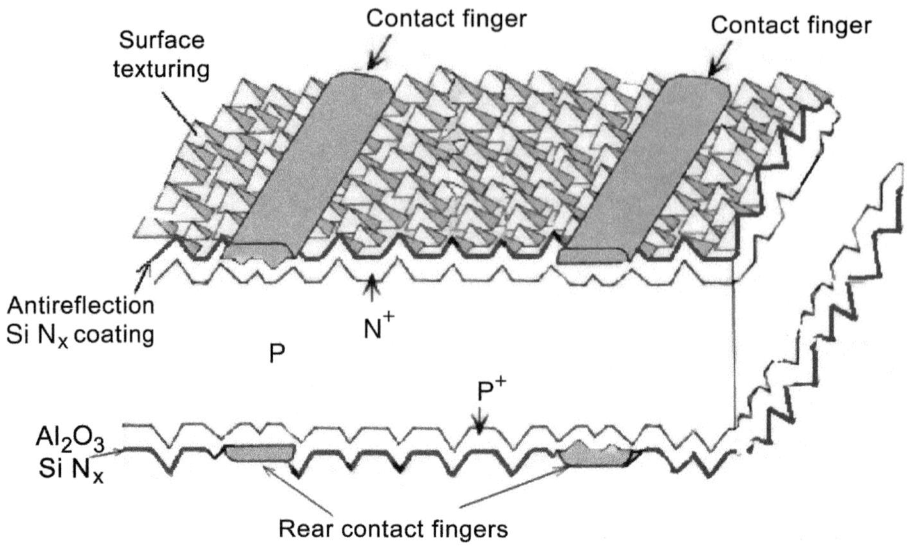

FIGURE 9.20 A bifacial solar cell structure.

The major point of difference between standard and bifacial cells is the structure of the rear surface contact. Rather than covering the entire back surface with a reflective aluminum contact, a "finger" grid is used in its place to allow sunlight (albedo) to come through the rear side.

The fabrication of bifacial cells needs two more operations in comparison to the standard BSF cells. After surface layer etching, boron diffusion is used to create the rear P^+ layer. After the diffusion, a SiO_2 layer is grown as the P^+ surface passivation. Then one side the surface layer is etched off, the surface is texturized, and phosphorous is diffused to create N^+P junction.

Typically, the silicon material used for bifacial solar cells must be of superior quality such that carriers generated near the rear surface can contribute to power production as they diffuse toward the "emitter" on the front surface [28,29].

With respect to higher carrier lifetime in N-type silicon, the bifacial cells are often prepared from N-type starting material. In this case the collecting P^+N junction is made by boron diffusion and the rear N^+N structure is made by phosphorous diffusion.

9.4.2.3 IBC Cells

The high carrier lifetime of N-type starting wafers allows localization of current collecting contacts of both polarities exclusively on the rear side of the PV cell (IBC—interdigitated back contact) [30]. Front contact shaded area typically occupies up to 7 %–10% of a PV cell's available front surface; thus moving metallic contacts to the rear side will increase the PV cell efficiency. The cell structure is demonstrated in Fig. 9.21. The N^+ phosphorous-doped and SiO_2-passivated layers on the front side decrease the front surface recombination. The N^+ and P^+ parallel narrow strips (no overlapping and separated by an efficient

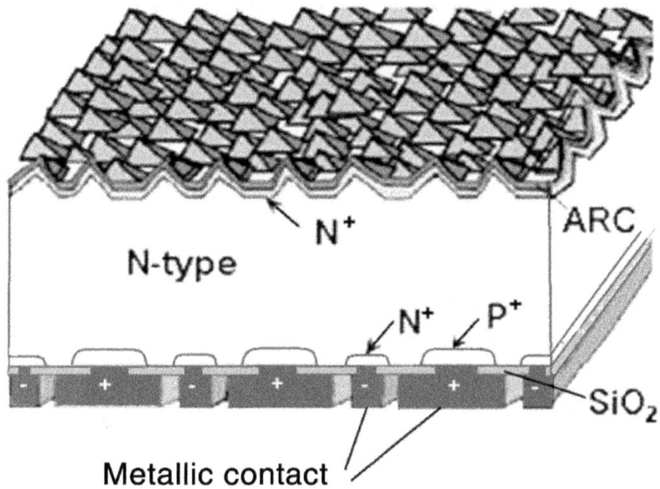

FIGURE 9.21 The IBC cell structure.

electrical insulation) are produced by sequential diffusion processes on the rear side of the IBC cell. The rear side surface passivation is provided by covering with a SiO_2 layer, through which holes for metallic contacting are made. The efficiency of such mass produced cells is now over 23% [26]. On the other hand, the number of operations increases to over 12; this increases the cost of fabrication, resulting in the cost per watt being higher than in the case of standard technology [31].

9.4.2.4 Heterojunction Technology Cells

A further increase of efficiency can be obtained by using a heterostructure with the top layer formed from a semiconductor with a wider bandgap. In such structures fabrication problems occur because the lattice constants of all the structure components need to match to give a good performance. However, amorphous Si has no well-defined lattice constant and surprisingly seems to perform well as a heterostructure barrier that confines excited carriers to the c-Si and away from ohmic contacts. This is the basic principle of heterojunction technology (HJT), also known as HIT (heterojunction with intrinsic thin layer) technology. The basic structure is shown in Fig. 9.22. Cells on the base of the heterojunction, between the amorphous and c-Si, utilize very thin (10–20 nm) a-Si:H layer stacks of N-type wafers to provide surface passivation, emitter formation, and a back surface field [32,33]. The a-Si:H layers are deposited on c-Si by PECVD using a low-temperature process (below 200°C), to avoid carrier lifetime degradation of the bulk material. On both doped layers, transparent conductive oxide layers are formed by sputtering, and metal fingers are screen printed. However, the a-Si:H layers cannot be taken to a temperature above 200°C, and this requirement then excludes the use of the standard screen-printed metal pastes and low-temperature pastes have 3 times higher resistivity than the standard ones. This requirement for low-temperature metallization can be a significant drawback for HJT cells.

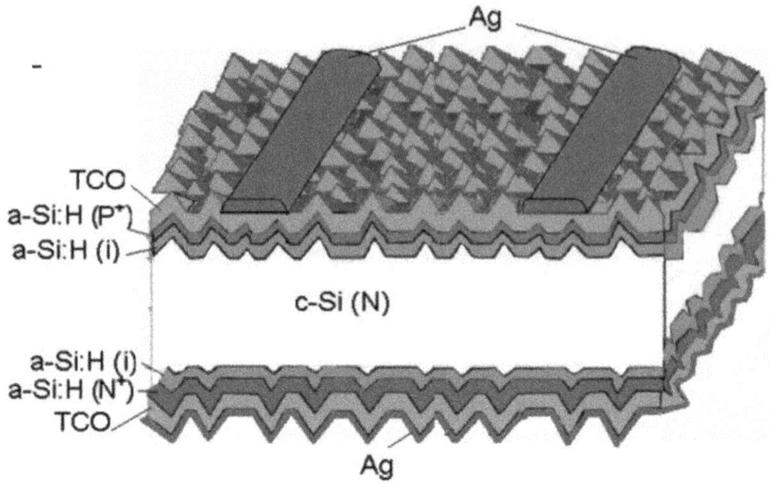

FIGURE 9.22 The HJT (HIT) cell structure.

However, the SmartWire contacting technique (Section 9.5.2.2) can open possibilities of mass production in the near future [34].

The HJT cells have higher V_{OC} values (0.72–0.73 V), an efficiency of over 22%, and offer a lower temperature coefficient (−0.2 to −0.3% K^{-1}) which can be almost half that of a standard c-Si cell [35]; this may be very important for some applications.

The potential of the HJT technology has recently been highlighted with the report that silicon heterojunction solar cells with interdigitated back contacts have been shown to have efficiencies of 26.7% [36].

9.4.3 Si Wafer-Based Multijunction Cells

Further improvements in efficiency may be reached by creating a tandem structure of a high-efficient c-Si cell with a thin film cell [36]. Also, further development involving inexpensive, high-quality Si wafers could provide ideal templates for the overgrowth of thin, crystalline, wider bandgap cells, boosting the efficiency potential of Si wafer-based multijunction cells. However, there are many problems in overcoming lattice mismatch at the interfaces [11]. Amorphous upper cells would remove the need for lattice matching, but it is also necessary to match PV-generated current densities of the c-Si bottom cell and the amorphous top cell. A structure of the c-Si cell overlaid with a thin film cell is shown in Fig. 9.23. There is one limiting factor—current densities over 30 mA cm^{-2} at STC are necessary for a considerable efficiency increase that has not yet been reached by most of thin film structures. With the emergence of perovskite cells, the promising combination of perovskites/c-Si can be considered with the possibility of efficiencies reaching 35% [37]. The perovskite deposition is a relatively low-cost process and it could result in both a decrease of cost per watt and an increase in efficiency. However, there are questions about

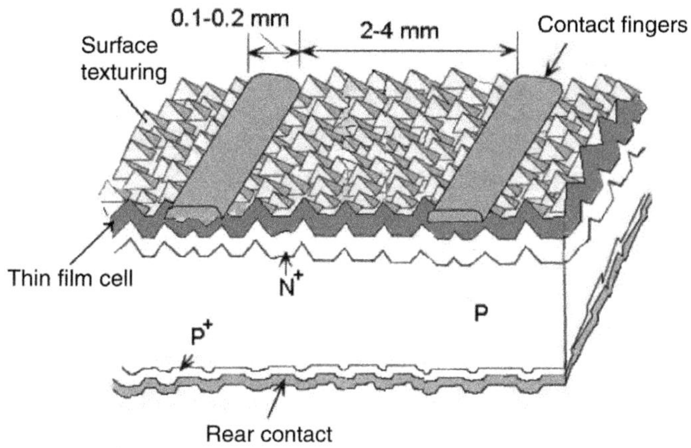

FIGURE 9.23 A structure of the c-Si cell overlaid with a thin film cell.

perovskite cell stability which may be a limiting factor in their development for use in general applications, even though recent research brings very promising results [38].

9.5 Crystalline Si Module Design and Fabrication

For practical applications, PV cells must be linked to form a PV module—complete and environmentally protected assembly of interconnected PV cells. Principles and construction rules of PV modules are explained in Section 8.4. Usually, a number of cells are connected in series and encapsulated in modules to create a suitable voltage.

9.5.1 Standard PV Module Fabrication Technology

c-Si PV cells are prone to mechanical damage unless protected. In addition, the metal contacts on the top surface of the cells may be corroded by water or water vapor. Therefore, the encapsulation has to prevent mechanical damage to the solar cells and to prevent water or water vapor from corroding the electrical contacts. PV modules must have adequate strength and rigidity to allow normal handling during installation, the modules must be able to accommodate some degree of twisting in the mounting structure, as well as to withstand wind-induced vibrations and the loads imposed by high winds, snow, and ice.

The most obvious choice for the front-cover material to cater for the major requirements of providing mechanical stability, high transparency in the spectral response range of the PV cell, protection of the cell and metallization against exterior impacts is toughened high-transmission (low iron) glass. Rear materials are also expected to provide mechanical stability, electrical safety, and protection of the cells and other module components from exterior impacts.

Usually, 2–3 mm thick highly transparent (low iron content) soda lime glass is used as a cover that provides mechanical rigidity and protection to the module while allowing light through [2,3,7,15,16]. Usually the glass has an antireflection coating to decrease the surface reflection [4]. The cell matrix is sandwiched between two layers of the transparent encapsulated material. The strings are interconnected with auxiliary tabs to form the cell matrix, which usually consists of several single strings, as shown in Fig. 9.24. Terminals of the strings are brought outside the module to permit flexible circuit configuration.

Solar cells of very similar parameters (sorted at the end of cell fabrication) are used for the module assembly. In standard technology, tinned copper ribbons (tabs) are soldered to the bus bars at the front of the cell to connect the back surface of the adjacent cell, as shown in Fig. 9.25A and B. The tabs must overlap a long distance along the bus bar length because the conductance of the printed bus bars is relatively low. Tabs provide a flexible link between cells that allow thermal expansion mismatch to be accommodated. The assembly is simplified if both cathode and anode contacts are on the rear of the cells (e.g., MWT technology), as indicated in Fig. 9.25C.

The outer layer on the nonilluminated module side is usually a composite plastic (tedlar) sheet or another glass acting as a barrier against humidity and corrosion that could accelerate degradation reactions by penetrating the PV module through the surface of the polymeric backsheet [38] and by diffusing through the encapsulation polymer until they reach the area between the solar cell and the front glass. Using the glass sheet as the rear covering layer decreases the degradation rate and increases the module durability (up to 40 years); on the other hand, using glass as the rear covering increases the module weight and cost.

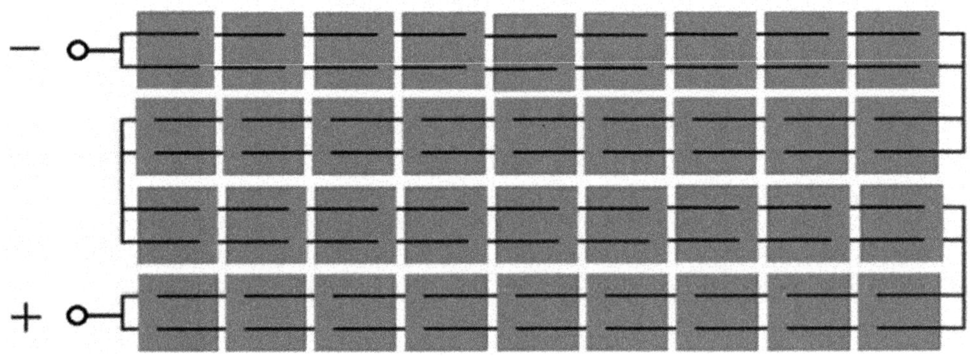

FIGURE 9.24 A common module configuration of 36 series-connected cells.

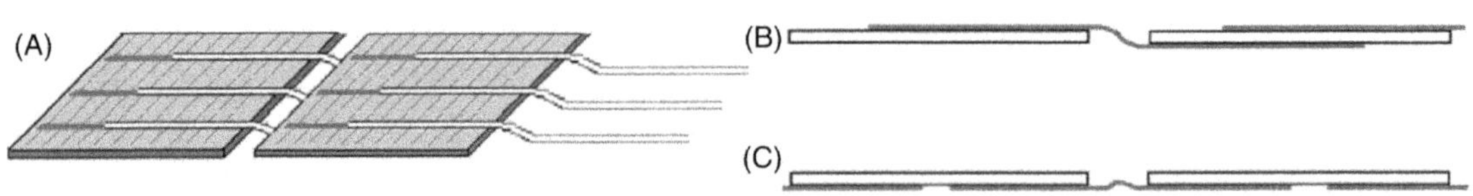

FIGURE 9.25 Cell interconnection in a module. (A) In series connection of two cells using tabs soldered to bus bars. (B) Front to back tab interconnection. (C) Contacts on the rear of cells.

The requirements for properties of encapsulants [39] are as follows:

- A low light absorption and an adapted refractive index to minimize interface reflectance.
- A high thermal conductivity to reduce cell operating temperatures and improve electrical yield.
- A high resistivity to ensure very low leakage currents (in accordance with IEC 61215).

In terms of PV module reliability, the encapsulant properties are critical in respect of UV irradiation, humidity, temperature cycles, extremely low or high ambient temperatures, mechanical loads, electric potential relative to ground, and so on. The encapsulant must preserve strong adhesion to the other module components and protect the cell and metallization from external impacts.

Module manufacturers must consider material costs, processing costs, processing time, shelf life, and quality assurance issues.

The lamination process depends on the material used. The most commonly used encapsulant (more than 90% of present production) is the thermoplastic, ethylene–vinyl–acetate (EVA). It is produced as an extruded film around 0.5 mm thick. Along with the polymer, the film contains curing agents and stabilizers whose role is in improving the resulting properties of the encapsulating layer [40].

The next fabrication steps are lamination and curing [2,3,7]. These steps are carried out in a laminator, a table that can be heated and furnished with a cover containing two chambers separated with a diaphragm. Both chambers can be independently evacuated and filled with air. The module is put in the lower chamber on the table. In the lamination stage, both chambers are evacuated while the temperature is raised above the EVA melting point at around 120°C. Vacuum is important to extract air to prevent voids. The melted EVA embeds the cells and fills the space between glass, solar cells, and the rear sheet (plastic or glass). After a few minutes, the upper chamber is filled with air so that the diaphragm presses the laminate. Then the temperature is increased to 150°C and the curing agents induce cross-linking of the EVA molecule chains; the EVA then acquires rubberlike properties. The curing takes between 10 and 60 min, depending on the curing agent used.

EVA is used mostly for the fabrication of modules with the glass–cells–plastic foil structure, as shown in Fig. 9.26A. For modules with the structure glass–cells–glass (Fig. 9.26B), other materials such as polyvinyl butyral (PVB), which allows for standard lamination processes, are to be used. Silicone has excellent properties, but it is only rarely used owing to its high price and the need for special processing machines and techniques [41].

After lamination, the encapsulant is removed from the edges which are sealed with silicone rubber and the module framed (if required). The module output contacts are then placed in a plastic junction box, which is fixed to the back of the laminate.

To eliminate possible hot spots caused by local shading, the approach followed is to put a diode (bypass diode) in parallel, but in opposite polarity, with a group of cells, as shown in Fig. 9.27. The bypass diodes (usually Schottky diodes) for each substring are connected in the junction box (see Fig. 9.27 for 18 cells in series).

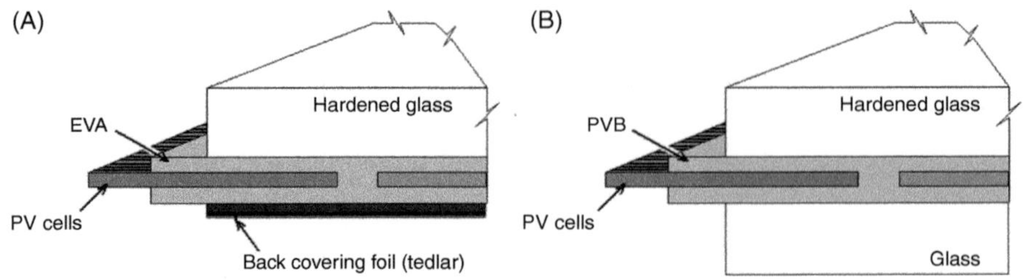

FIGURE 9.26 A common module structures. (A) Front glass and rear plastic foil. (B) Front and rear covering glass.

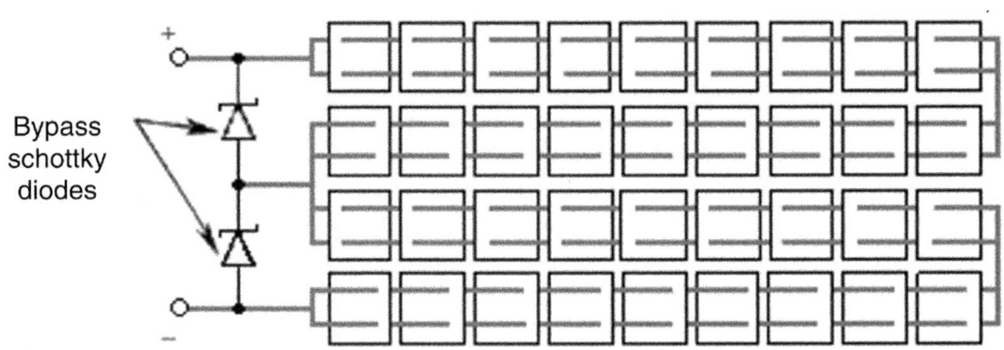

FIGURE 9.27 Bypass diodes in a module.

An embedded power optimizer (a circuit keeping PV module operation in the maximum power point) or a microinverter can also be included the junction box. Such "smart PV modules" nowadays make up approximately 4% of the total module production [4].

The PV module power output is realized by cables connected to the junction box and provided with special connectors to facilitate easy weatherproof connections to the PV array.

At present, most c-Si modules consists of 60 cells, but the shift to 72 cells is expected in future [4]. Another module sizes are also prepared for niche markets.

9.5.2 Emerging Module Technologies

9.5.2.1 Shingled Cell Modules

There is a need for achieving higher module power densities by improving cell interconnections. For reducing the cell-to-module (CTM) losses, a cell shingling design is now being used [42]. These cells are usually rectangular, with the long side having the length of a standard solar wafer, and the short side only a few centimeters in length. These cell strips are cut from a processing device for standard size wafers. The cells have busbars or rows of solder pads along the long edge, one on the front and one on the back (opposite edge). The shingling scheme of cell connection is schematically shown in Fig. 9.28. To create a cell string, an interconnection material is applied to connect the rear busbar of a cell with

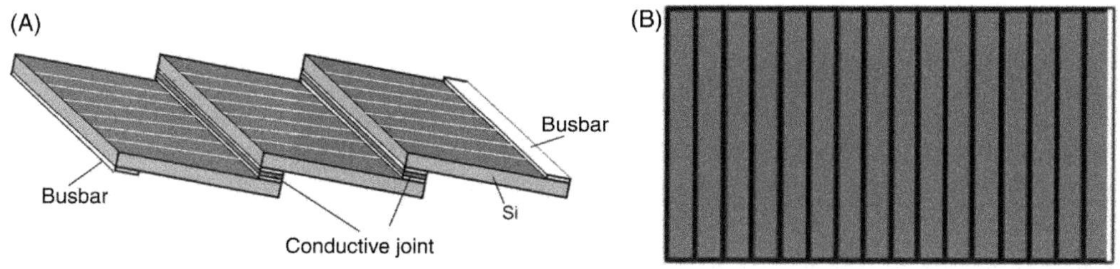

FIGURE 9.28 PV cell shingling design. (A) Cell interconnection. (B) Top view.

the front busbar of the next cell. This structure results into modules with an extremely high ratio of active area to total area, allowing in principle for high module efficiency. This is because there is no spacing between cells as in conventional modules, and also the cell area is not shaded by the front busbar.

The conductive joints between cells are made using conductive adhesives [42]. Strip-like cells are connected in series into a block and blocks are interconnected in a module using combined serial and parallel connections. Shingling requires new solutions for string interconnection, junction boxes, and bypass diodes placement.

9.5.2.2 SmartWires Contact Technology

The SmartWire Contacting Technology [34,43] is an innovative interconnection technology for c-Si solar cells: standard busbars and tabs are replaced by rows of fine copper wires of diameter 0.2–0.3 mm coated with a thin low melting point alloy layer (Bi-Sn or In-Sn solders which have melting points below 150°C).

The coated copper wires run on top of and perpendicular to the thin silver grid fingers, connecting them to neighboring cells, as demonstrated in Fig. 9.29. Typically, between 15 and 38 wires are used on both sides of the solar cell. The wires are embedded in an adhesive and aligned on a plastic film to simplify the fabrication process. The foil with wires is applied directly to the metallized cell. The stack is then laminated together with the soldering done during the lamination process. This contact has lower shading losses and lower contact resistance in comparison with the standard technology. As a result the efficiency is increased and silver consumption is reduced. The low soldering temperature is advantage for the HJT technology and could, in future, be used in Si wafer-based multi-junction cells technology.

9.5.3 Module Reliability and Durability

The performance of a PV module decreases over time due to degradation and aging processes. Degradation may include the effects of solar irradiation, temperature, humidity, mechanical stress, and voltage bias. Other factors affecting degradation include the quality of materials, the manufacturing process, and the assembly and packaging of the cells

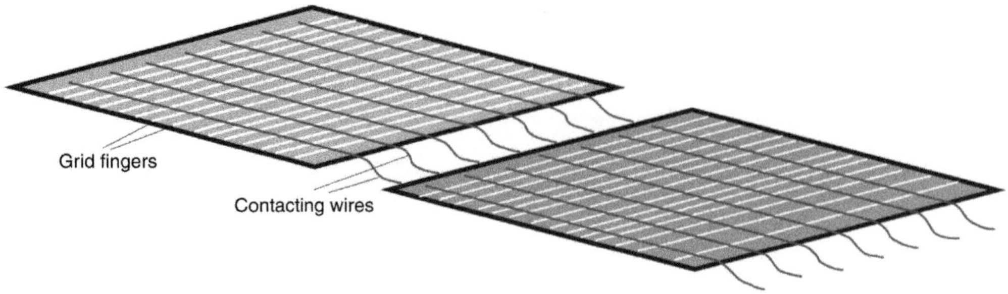

FIGURE 9.29 SmartWire contacting technology.

into modules. The degradation rate of modules depends also on the specific characteristics of the module being used and the local climatic conditions.

For crystalline modules prepared from P-type silicon, the degradation rate is typically higher in the first year upon initial exposure to light (light-induced degradation as discussed in Section 9.4.2) and then stabilizes. Additional degradation for c-Si modules may be caused by

- environment effects on the surface of the module (e.g., pollution),
- discoloration or haze of the encapsulant or glass cover,
- lamination defects,
- mechanical stress and humidity on the contacts,
- cell contact breakdown, and
- wiring degradation.

The long-term reliability of a PV module is highly affected by the degradation behavior of the polymeric components within the module, such as the encapsulant and backsheet [44]. For example, corrosion, a major field failure mode leading to loss of power, is strongly accelerated by acetic acid, a product from the degradation of encapsulant EVA. The cracking and delamination of backsheet due to degradation can lead to the dielectric breakdown of PV systems and safety concerns as well as lower reliability of PV modules. Different types of failures are described in Ref. [45].

Because of these problems standardized methods (Performance standard IEC 61215) are used for quality testing of PV modules. These tests can be divided into the following sections:

- *Diagnostic*: Visual inspection, hot spot.
- *Electrical*: Insulation resistance, wet leakage current.
- *Performance*: P_{max} at STC, temperature coefficients, NOCT, P_{max} at low irradiance.
- *Thermal*: Bypass diode test, hot spot.
- *Irradiance*: Outdoor exposure, UV exposure, light soaking.
- *Environmental*: Temperature cycles, humidity freeze, damp heat.
- *Mechanical*: Mechanical load, robustness of terminations, hail impact.

The PV modules usually may have a long-term power output degradation rate of between 0.3% and 1.0% per annum. For c-Si modules, a generic degradation rate of 0.4% per annum is usually considered acceptable. Power degradation over time is warranted by the manufacturers as a service time, that is, time period between starting operation and decreasing output power below 80% of the nominate value. At present, the service time of c-Si modules with polymer back sheets is 25 years and over 30 years for modules with glass–cell–glass structures.

9.6 Conclusions

The production processes in the solar industry still have great potential for technology optimization. Wafer-based c-Si technologies are the workhorse of the current PV power generation. This has been achieved by the combination of increased cell and module performance in conjunction with significantly reduced manufacturing costs, especially at the cell and module level, by the efficient use of Si and non-Si materials. In 2016, the crystalline Si wafer-based PV technology accounted for approximately 94% of the total production that was nearly 82 GW_p (c-Si: 77 GW_p; thin film: 4.9 GW_p) [46]. Today, mc-Si modules dominate the market; in 2016, its share of global production was nearly 70% (Fig. 9.30). Fig. 9.30 also demonstrates the present dominancy of wafer-based c-Si technology over thin film technologies. It is predicted that in the future there will be a shift toward mono c-Si cells [4]. Present developments in c-Si technology are driven by efficiency improvements, integrated factories scaling up processes providing more efficient production processes, and improvements in supply chain management. Over the past 10 years the efficiency of average commercial wafer-based silicon modules has increased from approximately 12% to 17% and modules based on IBC technology with efficiency 22% are now available on the market.

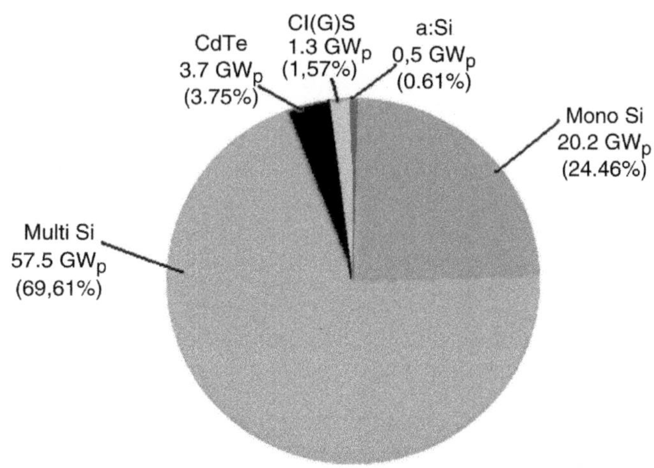

FIGURE 9.30 Technology distribution in 2016.

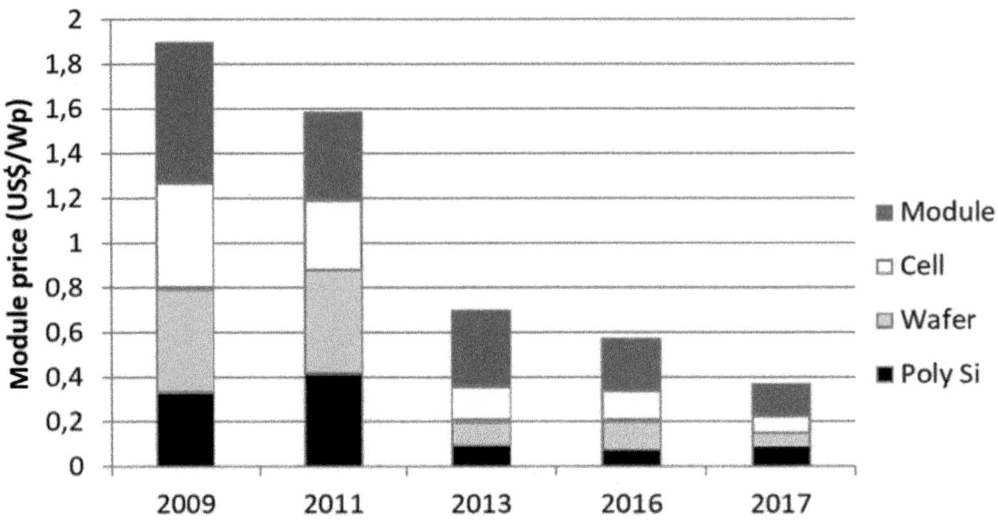

FIGURE 9.31 The PV modules cost development in the period 2009–17.

The cost development in the period 2009–17 is demonstrated in Fig. 9.31. In the future, further progress can be achieved by new processes, new tools based on these processes, new materials, and new designs of products. At present, PV systems operating with c-Si modules seem to be the optimum system for producing low-cost electrical energy and are expected to remain dominant in the field of PVs over the next 5 years, ensuring the long-term competitiveness of PV power generation [4,47].

References

[1] Middleman S, Hochberg AK: *Process engineering analysis in semiconductor device fabrication: silicon production*, New York, 1993, McGrawHill Inc.

[2] Luque A, Hegedus S, editors: *Handbook of photovoltaic science and engineering*, Chichester, 2003, John Wiley & Sons.

[3] Goetzberger A, Knobloch J, Bernhard V: *Crystalline silicon solar cells*, Chichester, 1998, John Wiley & Sons, Inc.

[4] International technology roadmap for photovoltaics—results 2016. 8th ed. www.ITRPV.org; 2017.

[5] Meemongko lkiat V, Nakayashiki K, Rohatgi A, Crabtree G, Nickerson J, Jester TL: Resistivity and lifetime variation along commercially grown Ga- and B-doped Czochralski Si ingots and its effect on light-induced degradation and performance of solar cells, *Prog Photovolt Res Appl* 14:125–134, 2006.

[6] Benda V, Gowar J, Grant DA: *Power semiconductor devices—theory and applications*, Chichester, 1999, John Wiley & Sons.

[7] Markvart T, Castaner L, editors: *Solar cells—materials, manufacture and operation*, Oxford, 2005, Elsevier.

[8] Lan CW, Lan A, Yang CF, Hsu HP, Yang M, Yu A, et al: The emergence of high-performance multicrystalline silicon in photovoltaics, *J Cryst Growth* 468:17–23, 2017.

[9] Chunduri SK: Diamond wire—the die has been cast, *Photon Int* 5:258–263, 2011.

[10] Fossum JG: Physical operations on back-surface field solar cells, *IEEE Trans Electron Devices* 24:322–325, 1977.

[11] Braun S, Hahn G, Nissler R, Pönisch Ch, Habermann D: The multi-busbar design: an overview, *Energy Procedia* 43:86–92, 2013.

[12] Koynov S, Brandt MS, Stutzmann M: Black multi-crystalline silicon solar cells, *Phys Status Solidi (RRL)* 1(2):R53–R55, 2007.

[13] Firat ES. Novel metal assisted etching technique for enhanced light management in black crystalline Si solar cells. Thesis, Middle East Technical University; 2006.

[14] Joos W. Overview and latest technologies for texturing of multicrystalline silicon. Available at: www.itrpv.net/.cm4all/iproc.php/04-2017_06_01_ITRPV%20Roadmapmeeting%20-%20Intersolar%20Europe_final_RCT.pdf?cdp=a

[15] Fraas L, Partainning L, editors: *Solar cells and their applications*, 2nd ed., Singapore, 2010, John Wiley & Sons.

[16] Kaplanis S, Kaplani E, editors: *Renewable energy systems: theory, innovations, and intelligent applications*, New York, 2013, Nova Science Publishers, Inc.

[17] Hahn G. Status of selective emitter technology. In: Proceedings of 25th European photovoltaic solar energy conference and exhibition, Valencia, Spain; 2010. p. 1091–6.

[18] Lennon A, Yao Y, Wenham S: Evolution of metal plating for silicon solar cell metallisation, *Prog Photovolt Res Appl* 21:1454–1468, 2013.

[19] Clement F, Thaidigsmann B, Hönig R, Fellmeth T, Spribille A, Lohmüller E, et al. Pilot-line processing oh highly-efficient MWT silicon solar cells. In: Proceedings of 25th European photovoltaic solar energy conference and exhibition, Valencia, Spain; 2010. p. 1097–1101.

[20] Wang X, Wu J: The road to industrializing PERC solar cells, *Photovolt Int* 29:48–52, 2015.

[21] Zhao J, Wang A, Green MA. 24% Efficient perl structure silicon solar cells. In: Proceedings of IEEE photovoltaic specialists conference, vol. 1; 1990. p. 333–5.

[22] Cotter JE, Guo JH, Cousins PJ, Abbott MD, Chen FW, Fisher KC: P-type versus n-type silicon wafers: prospects for high-efficiency commercial silicon solar cells, *IEEE Trans Electron Devices* 53:1893–1901, 2006.

[23] Burgers AR, Naber RCG, Carr AJ, Barton PC, Geerligs LJ, Jingfenget X, et al. 19% Efficient N-type Si solar cells made in pilot production. In: Proceedings of 25th European photovoltaic solar energy conference and exhibition, Valencia, Spain; 2010. p. 1106–1109.

[24] Schindler F, Michl B, Krenckel P, Riepe S, Feldmann F, Benick J, et al: Efficiency potential of p- and n-type high performance multicrystalline silicon, *Energy Procedia* 77:633–638, 2015.

[25] Cai W, Yuan S, Sheng Y, Duan W, Wang Z, Chen Y, et al: 22.2% Efficiency n-type PERT solar cell, *Energy Procedia* 92:399–403, 2016.

[26] Glunz SW, Feldmann F, Richter A, Bivour M, Reichel C, Steinkemper H, et al. The irresistible charm of a simple current flow pattern—25% with a solar cell featuring a full-area back contact. In: Proceedings of the 31st European photovoltaic solar energy conference and exhibition, Hamburg; 2015. p. 259–63.

[27] Cuevas A, Luege A, Eguren J, Del Alamo J: 50% more output power from an albedo collecting flat panel using bifacial solar cells, *Sol Energy* 29:419–420, 1982.

[28] Dullweber T, Kranz C, Peibst R, Baumann U, Hannebauer H, Fülle A, et al. The PERC+ Cell: a 21% efficient industrial bifacial PERC solar cell. In: Proceedings of 31st European photovoltaic solar energy conference, Hamburg, Germany; 2015. p. 341–50.

[29] Wöhrle N, Lohmüller E, Mittag M, Moldovan A, Baliozian P, Fellmeth T, et al: Solar cell demand for bifacial and singulated-cell module architectures, *Photovolt Int* 36:48–62, 2017.

[30] Schwartz RJ, Lammert MD. Silicon solar cells for high concentration applications. In: Proceedings of the IEEE international electron devices meeting, Washington, DC; 1975. p. 350–2.

[31] Goodrich A, Hacke P, Wang Q, Sopori B, Margolis R, James TL, et al: A wafer-based monocrystalline silicon photovoltaics road map: utilizing known technology improvement opportunities for further reductions in manufacturing cost, *Sol Energy Mater Sol Cells* 114:110–135, 2013.

[32] Tanaka M, Taguchi M, Matsuyama T, Sawada T, Tsuda S, Nakano S, et al: Development of new a-Si/c-Si heterojunction solar cells: ACJ-HIT (artificially constructed junction/heterojunction with intrinsic thin-layer), *Jpn J Appl Phys* 31:3518–3522, 1992.

[33] Sakata H, Tsunomura Y, Inoue H, Taira S, Baba T, Kanno H, et al. R&D progress of next-generation very thin HITTM solar cells. In: Proceedings of 25th European photovoltaic solar energy conference, Valencia, Spain; 2010. p. 1102–4.

[34] Faes A, Badel N, Kiaee M, Despeisse M, Levrat J, Champliaud J, et al. SmartWire Solar Cell Interconnection Technology. Available at: http://www.metallizationworkshop.info/fileadmin/metallization-workshop/Faes_2014MetallizationWorkshop_SmartWire_Faes-CSEM-v2.pdf

[35] Taguchi M, Yano A, Tohoda S, Matsuyama K, Nakamura Y, Nishiwaki T, et al: 24.7% record efficiency HIT solar cell on thin silicon wafer, *IEEE J Photovolt* 4(1):96–99, 2014.

[36] Yoshikawa K, Kawasaki H, Yoshida W, Irie T, Konishi K, Nakano K, Uzu H, et al: Silicon heterojunction solar cell with interdigitated back contacts for a photoconversion efficiency over 26%, *Nat Energy* 2:17032, 2017.

[37] Loper P, Niesen B, Moon SJ, Martin de Nicolas S, Holovský J, Remeš Z, et al: Organic-inorganic halide perovskites: perspectives for silicon-based tandem solar cells, *IEEE J Photovolt* 4:1545–1551, 2014.

[38] Miyano K, Yanagida M, Tripathi N, Shirai Y: Simple characterization of electronic processes in perovskite photovoltaic cells, *Appl Phys Lett* 106:093903, 2015.

[39] Peike C, Hädrich I, Weiß KA, Dürr I: Overview of PV module encapsulation materials, *Photovolt Int* 19:85–92, 2013.

[40] Gong H, Wang G: Reliability and durability impact of high UV transmission EVA for PV modules, *Photovolt Int* 29:101–106, 2015.

[41] Geretschläger KJ, Wallner GM, Fischer J: Structure and basic properties of photovoltaic module back-sheet films, *Sol Energy Mater Sol Cells* 144:451–456, 2016.

[42] Forniés E, Silva JP: Cell-to-module losses in standard crystalline PV modules—an industrial approach, *Photovolt Int* 29:91–100, 2015.

[43] Söderström T, Papet P, Yao Y, Ufheil J. SmartWire Connection Technology. Available at: https://www.meyerburger.com/user_upload/dashboard_news_bundle/376409e022f7d2ae6f6e29318f8055410774c7fd.pdf

[44] Peike C, Hülsmann P, Bluml M, Schmid P, Weiß K-A, Kohl M: Impact of permeation properties and backsheet-encapsulant interactions on the reliability of PV modules, *ISRN Renew Energy*, 2012.doi: 10.5402/2012/459731, Article ID 459731.

[45] Köntges M, Kurtz S, Packard C, Jahn U, Berger KA, Kato K, et al. Review of failures of photovoltaic modules. External final report IEA-PVPS. Available at: http://www.iea-pvps.org/fileadmin/dam/intranet/ExCo/IEA-PVPS_T13-01_2014_Review_of_Failures_of_Photovoltaic_Modules_Final.pdf; 2014.

[46] Burger B, Kieferet K, Kost C, Nold S, Philipps S, Preu R, et al. Photovoltaics report. Available at: www.ise.fraunhofer.de; 2017.

[47] Benda V: Photovoltaics towards terawatts—progress in photovoltaic cells and modules, *IET Power Electron* 8:2343–2351, 2015.

Further Readings

[48] Green MA, Hao X, Bremner S, Conibeer GJ, Al Mansouri I, Song N, et al. Silicon wafer-based tandem cells: the ultimate photovoltaic solution? In: Proceedings of the 28th European photovoltaic solar energy conference; 2013. p. 7–10.

[49] Beaucarne G: Materials challenge for shingled cells interconnection, *Energy Procedia* 98:115–124, 2016.

10

CdTe Solar Cells

Tom Baines, Thomas P. Shalvey, Jonathan D. Major
STEPHENSON INSTITUTE FOR RENEWABLE ENERGY, THE UNIVERSITY OF LIVERPOOL, LIVERPOOL, UNITED KINGDOM
jon.major@liverpool.ac.uk

10.1 Introduction

Cadmium telluride (CdTe) solar cells have quietly established themselves as a mass market PV technology. Despite the market remaining dominated by silicon, CdTe now accounts for around a 7% market share [1] and is the first of the second generation thin film technologies to effectively make the leap to truly mass deployment. Blessed with a direct 1.5 eV bandgap, good optical absorption $\sim 1 \times 10^3\ cm^{-1}$ [2], and a simple binary phase chemistry, CdTe has been shown to be an eminently scalable technology. Device efficiencies on the lab scale have now exceed 22% [3] and modules have reached 18.6%, in excess of multi-crystalline silicon modules (Fig. 10.1). It is also now purported to be the lowest cost per watt technology, have the shortest energy payback time and be the least carbon intensive in production. CdTe solar cells clearly have a lot going for them but there remains a number of key technological challenges that, if overcome, could push the conversion efficiencies closer to the theoretical maximum of >30%. This chapter will discuss the current state of play for this technology, reviewing each of the key cell1s in turn before briefly looking at areas that are a focus for future development.

10.2 The CdTe Solar Cell: History, Layers, and Processes

Although early research focused on CdTe homojunction cells, that is *n*-CdTe/*p*-CdTe, in the mold of silicon equivalents, this structure was quickly abandoned owing to the high rates of surface recombination and strong unwanted optical absorption in the *n*-CdTe layer. What is now thought of as the "standard" CdTe "superstrate" structure *n*-CdS/*p*-CdTe, Fig. 10.2A, first emerged in the 1970s from the work of Bonnet and Rabenhorst [4]. As with many of the PV technologies, it has progressed in the intervening decades via a series of empirically arrived at process improvements such as the chloride treatment, improved back-contacting, and improved window layers. By 2001 device efficiencies

A Comprehensive Guide to Solar Energy Systems. http://dx.doi.org/10.1016/B978-0-12-811479-7.00010-5

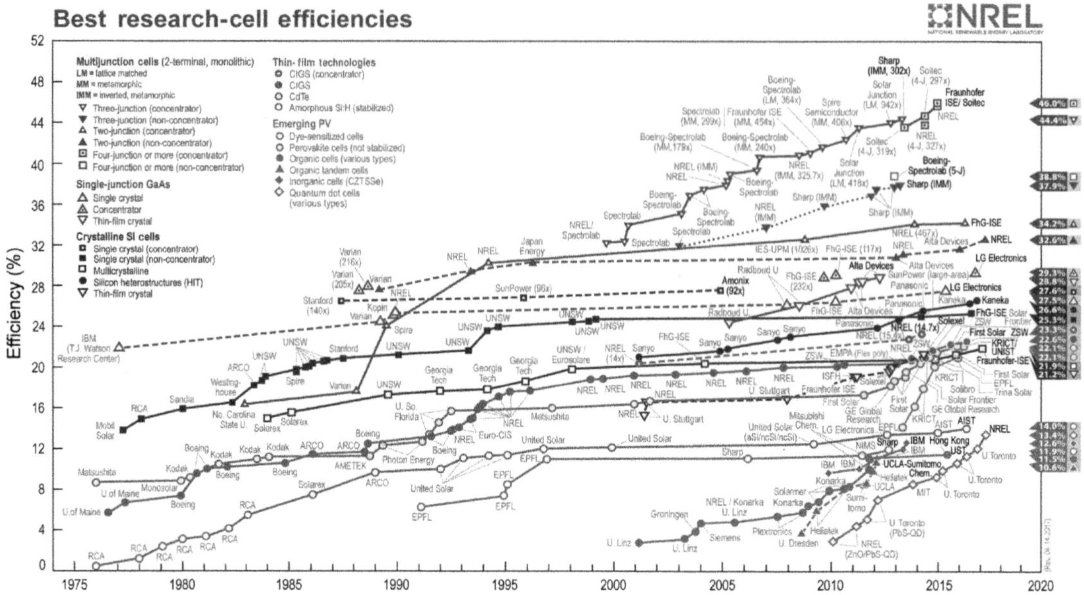

FIGURE 10.1 The NREL solar cell efficiency chart. This plot is courtesy of the National Renewable Energy Laboratory, Golden, CO, USA.

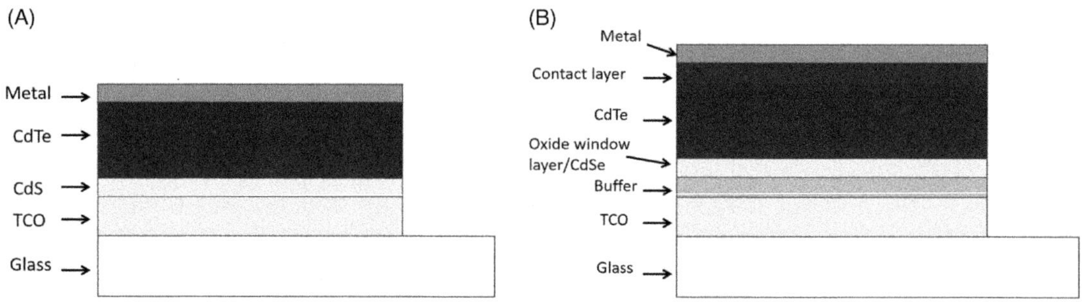

FIGURE 10.2 Typical CdTe solar cell structure showing. (A) The simplest implementation of device stack and (B) the current state of the art device architecture incorporating and oxide buffer/window layers, CdSe incorporation, and a back-contacting layer.

had reached 16.7% on the lab scale (Fig. 10.1) and despite industrial uptake remained there for over a decade, leading to a widespread belief that the technology had reached a plateau. This has because proven to be a false fear with efficiencies jumping past 22% in recent years with predictions that efficiencies in excess of 25% will be reached in the coming years.

The following sections discuss each of the key device layers shown in Fig. 10.2 in turn, detailing their development, deposition, and impact on cell performance.

10.2.1 Transparent Conductive Oxide (TCO)

For CdTe solar cells the choice of the TCO front-contact, requires careful consideration because, unlike many other technologies, it employs the superstrate structure. Therefore, as well as the usual TCO considerations of transparency and conductivity [5], as it is the first layer deposited it also has the requirement to be stable to all subsequent deposition processes. The processes that are to follow can therefore necessitate a particular choice of TCO. Two of the more common TCOs used for thin film PV are SnO_2:In_2O_3 (ITO) and ZnO:Al (AZO), both having low sheet resistance and high optical transparency. Both of these layers can be problematic for CdTe cell fabrication though. AZO has a tendency to break down at temperatures >400°C, losing conductivity, while ITO can similarly break down diffusing indium (Fig. 10.3), an *n*-type dopant, into the CdTe layer. For high deposition temperatures >500°C the standard choice is to use commercially available SnO_2:F (FTO), which has a sufficient figure of merit [5] and is highly stable. Alternative bespoke and high performance TCOs have been reported for CdTe, such as cadmium stannate Cd_2SnO_4 [6], but in general little of note is reported in this area and FTO has become the "go to" TCO.

10.2.2 The Window Layer

As the early development of the technology the standard heterojunction structure has been *n*-CdS/*p*-CdTe. A CdS "window" layer is widely used for a number of the thin film technologies (e.g., CIGS, CZTS) as it is nascently *n*-type, easy to deposit via a variety of routs and has a tolerably large 2.4 eV bandgap. CdS deposition for CdTe cells has been demonstrated through varied routes such as close space sublimation (CSS) [8] thermal evaporation, metal-organic chemical vapor deposition (MOCVD) [9], RF sputtering [10], or chemical batch deposition [11]. All of these routes have been demonstrated as capable of producing

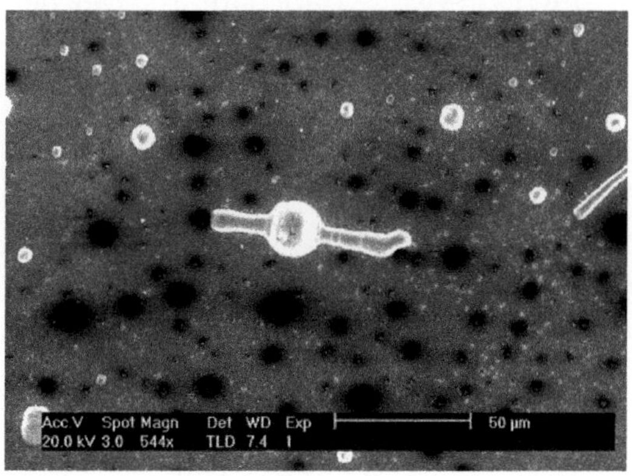

FIGURE 10.3 SEM image of a CdTe/ITO structure following CSS deposition and showing the breakdown of the ITO layer with indium "blobs" appearing on the sample surface [7].

device quality layers, but there are additional considerations that need to be factored in with the CdS layer. Due to the superstrate nature of CdTe solar cells, the CdTe layer is deposited on top of the CdS layer, often at higher temperatures than at which the CdS was deposited. This leads to recrystallization of the CdS [12] and intermixing of the CdS and CdTe layers to form $CdS_{1-x}Te_x$ and CdS_yTe_{1-y} phases [13]. The CdS layer must therefore be stable enough to this recrystallization and intermixing so as to not disintegrate, but overall this intermixing process is beneficial to cell performance. There is a high degree of lattice mismatch, and thus induced interfacial strain, between the CdS and CdTe [14], which is partly relieved by the intermixing. Although some optical losses occur due to the semimetallic quality of these intermixed phases [15], these are relatively minor compared to the overall improvement in junction quality. The degree of intermixing between the two junction layers is controlled by a number of factors such as the CdS and CdTe deposition methods and any postgrowth processing [13] (Fig. 10.4), meaning its optimization is nontrivial.

Although CdS has been demonstrated as a suitable choice for the "window" layer in CdTe devices it does have inherent limitations. Because of the one-sided nature of the junction (CdS is considerably more highly doped than CdTe) only carriers generated in the CdTe layer are effectively collected. This means that optical absorption in the CdS layer is essentially parasitic and results in performance loss for optical wavelengths of <520 nm [17]. The typical workaround to this problem has been to make the CdS as thin as possible [18], thereby minimizing any such losses. This only works to a point however. Reducing the CdS thickness too much, generally below around 100 nm, results in significant losses in fill factor and open circuit voltage. The reason behind this is a question of microstructure, when the CdS is too thin voids, or pinholes as they are commonly referred to, begin to form in the CdS layer [19] (Fig. 10.4). This can allow regions where the CdTe layer contacts directly to the underlying oxide layer, weakening the average quality of the junction, and bringing down the cell performance [20]. This may be overcome to an extent by the incorporation of a "buffer" or "highly resistive transparent" layer between the CdS and TCO

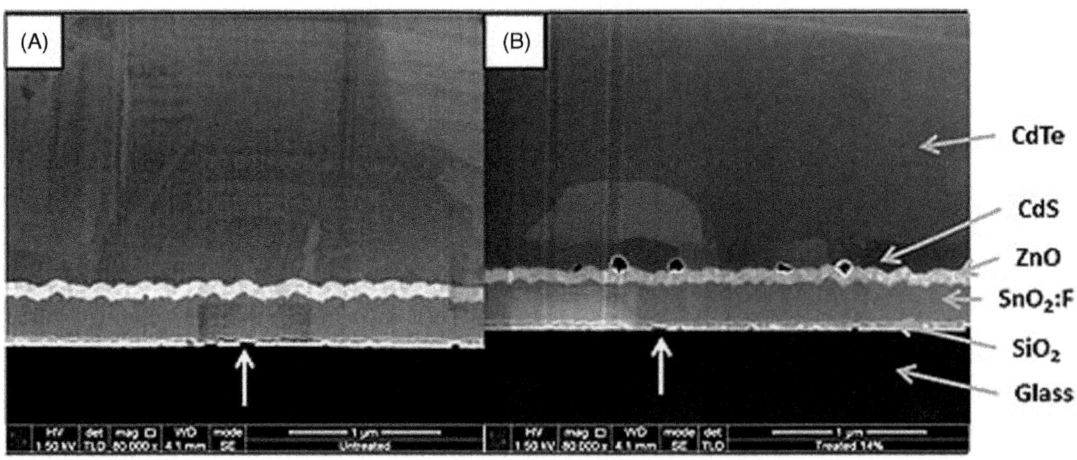

FIGURE 10.4 High-resolution secondary electron images of CdTe/CdS interface regions for. (A) As grown and (B) $CdCl_2$ treated devices [16].

layers [21] (Fig. 10.4). A number of different materials have been used such as a buffer, primarily ZnO [22], or SnO_2, but also a number of other such as $Zn_{1-x}Sn_xO$ [23]. These buffer layers have demonstrated a route to greater CdS thickness reduction without compromise of performance [21]. How the buffer layer achieves this has never been comprehensively established and it may in fact have more than one role. Clearly there is some involvement in minimizing the deleterious effects of CdS pinholes and associated localized reduction in diode quality, but it additionally appears it may be able to aid the band alignment between the TCO and CdS. An oxide/CdS bilayer has therefore become more standard as the window layer structure for CdTe cells than a CdS layer in isolation.

Despite the improvements offered by the introduction of a buffer layer, recent developments have shown CdTe solar cells evolve away from CdS layers all together, toward something with a higher bandgap, and thus greater degree of transparency. The first initial move in this direction was by the use of an oxidized CdS:O layer, which has a greatly increased bandgap compared to CdS owing to quantum confinement effects in the nanometer scale grain structure [24]. Despite the success of this method it never resulted in device improvement beyond 16.7%. The initial instinct to eliminate the CdS would be to simply replace it with a high bandgap oxide, with the likes of ZnO or SnO_2 seeming obvious choices. Unfortunately owing to the lattice mismatch problems such layers are inherently unsuited and in isolation produce devices with low performance. Recent work has identified $Mg_{1-x}Zn_xO$ (MZO) as being a suitable alternative [25] and the research field is quickly moving toward adopting this as a widespread alternative to CdS, coupled to the use of a CdSe interfacial layer to effect bandgap grading [26] while this may or may not establish itself as new standard window layer, as the technology continues to develop, we will see evolution away from the traditional CdS layer.

10.2.3 CdTe Absorber Layer

CdTe is a polycrystalline thin film which can be doped either *n* or *p* type. The deposited layers tend to be weakly *p*-type, with doping being increased via key postgrowth processing such as the chloride treatment (Section 10.2.4) or the copper back-contacting step (Section 10.2.5). The quality of the absorber layer can be highly dependent on the deposition technique used, with lower temperature deposition routes producing smaller grained and typically lower performing material (although recrystallization following chloride treatment will modify the grain structure significantly). Numerous deposition routes have been established, electrodeposition [27], sputtering [28], thermal evaporation [29], MOCVD [30], and CSS [31], all of which have been demonstrated of forming functional devices. Among these techniques it is CSS, which has been most readily adopted at the industrial level and indeed the majority of recent "champion" cells have used this technique. Irrespective of the deposition technique used however the CdTe layer itself is between 1 and 8 μm thick. Cells with layers thinner than 1 μm show reduced performance owing to incomplete optical absorption [32] or shunting due to pinholes [33], while layers thicker than this show performance loss owing to additional resistivity of the overly thick absorber. One would not expect to encounter losses at such a relatively low thickness such as 8 μm, when dealing with Si for example (which is typically >100 μm thick), but CdTe suffers from low carrier lifetimes of typically <10 ns [34]. Indeed even determining the lifetime of CdTe to be this large has proved problematic owing to particularly high surface recombination. Standard time resolved photoluminescence measurements of the CdTe free surface give values dominated by the surface recombination, leading to a large number of reports implying the carrier lifetime was <1 ns. This can be overcome to an extent by taking measurements through the glass [35] but for a truly accurate analysis of the carrier lifetime a two-photon technique is required [36], which gives a more accurate carrier lifetime value in the tens of nanosecond range [34]. The low carrier lifetimes are in part due to the overriding issue with CdTe thin films, the behavior of grain boundaries, which have been repeatedly linked to low performance via issues such as low carrier lifetime [35]. It is widely believed these interstices between the grains act as dominant recombination centers due to the presence of dangling bonds. One of the recent primary challenges for CdTe solar cells has been understand the role of the CdTe grain boundaries within functioning devices. The polycrystalline nature of the films and the misorientation between neighboring grains, has been aptly demonstrated by techniques such as electron back scattered diffraction [37], meaning one would indeed anticipate a high defect density at the grain boundaries and for them to act as preferential recombination centers. Although this thesis is supported by techniques such as cathodoluminescence, other techniques such as electron beam induced current (EBIC) have offered contradictory evidence. For more information on the role of grain boundaries in CdTe solar cells the reader is referred to the following review article on the subject [38]. Discussion of the CdTe layer in isolation is problematic as the layer itself is never used in isolation. All working devices require CdTe postgrowth treatments hence the discussion of the CdTe layer is, in essence, continued in Sections 10.2.4 and 10.2.5 Fig. 10.5.

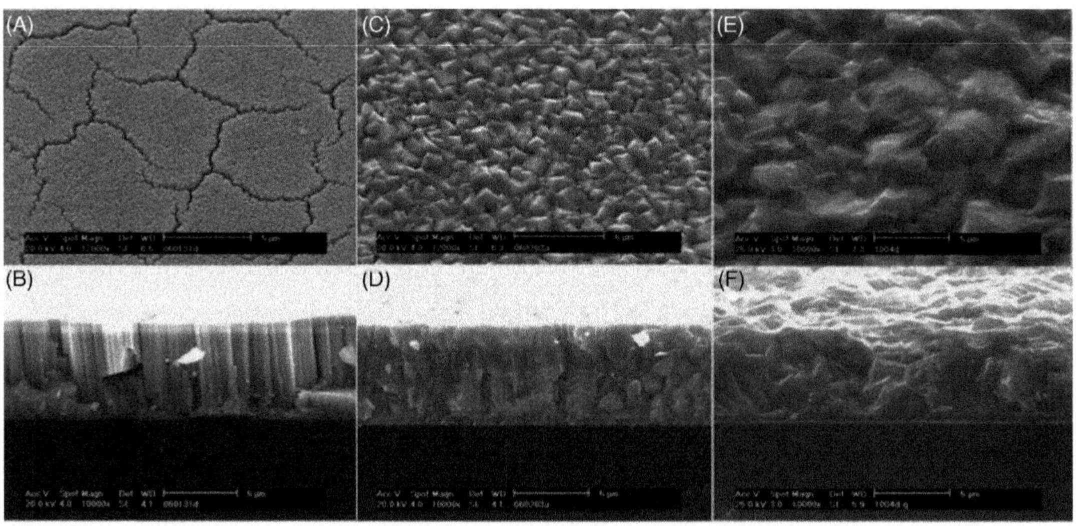

FIGURE 10.5 SEM micrographs (surfaces and cross-sections) from the three CdTe films with different CSS growth regimes. (A) +, (B) first regime (Tsub = 255°C), (C) +, (D) second regime (Tsub = 387°C), (E) +, (F) third regime (Tsub = 523°C) [39].

10.2.4 The Chloride Process

What is widely referred to as the chloride treatment step is typically essential to the production of functional CdTe solar cells. Indeed, barring single crystal cell exceptions [40] (which may to an extend be considered a separate technology entirely), no cell reported has exceeded 10% without some form of chloride treatment. It was first developed by Basol [41] using an electrochemical process, and the use of chlorine appears to have emerged from its prior use to photosensitize CdS films. Otherwise it is hard to understand the logical leap to applying Cl which, being a group VII element, one would initially assume would be an n-type dopant, to achieve p-type character while alternative treatments based of $MgCl_2$ [42] or CHF_2Cl [43] have been identified, cadmium chloride ($CdCl_2$) has long been established, and as such remains the research and industrial standard process. There a number of methodological variations in the manner of application but the principle remains the same; the free CdTe "back surface" is coated with a thin layer of $CdCl_2$, typically deposited via either thermal evaporation [44] or from a solution via spray or drop casting [45]. The stack structure is then annealed somewhere in the 380–450°C temperature range, usually in an air or oxygen containing ambient [46] (although some oxygen free processes have been reported as successful [30]). Following this annealing the cell is typically rinsed in water to remove any excess $CdCl_2$ remaining on the surface prior to whatever contacting procedure is being applied while the practical application of the $CdCl_2$ treatment was quickly established, the understanding of what the $CdCl_2$ treatment was actually doing to the device has taken longer to develop and has changed in recent times. This is primarily due to the multifaceted nature of its influences being hard to disentangle. On a structural level it has been widely demonstrated to mediate recrystallization in the CdTe and CdS layers [47], the level of recrystallization being partly dependent upon the starting grain structure of the films. For CdTe films deposited by low temperature methods, such as thermal evaporation or sputtering, which have a small as-deposited grain structure, $CdCl_2$ treatment induces near-complete recrystallization of the film to a significantly larger final grain structure [48] (Fig. 10.6). For higher temperature methods such as CSS, the as-deposited grain structure is large and thus more thermodynamically stable meaning recrystallization is only seen at the near CdS interface region where the grain structure is smaller and more defective [49]. Another standard result has been to observe an increase in carrier concentration following chloride treatment [42]. This has been widely attributed to the formation of the chlorine A-center V_{Cd}-Cl_i while the chloride treatment does undoubtedly have an effect on the doping level seen in measured devices, more recent work suggests its primary role may be to pacify grain boundaries. It has been demonstrated via high-resolution electron microscopy that the incorporated chlorine is predominately located at the grain boundaries, with their being little incorporated in the grain interiors [50]. This in turn has been shown to have a pronounced effect on the grain boundaries electrical behavior when analyzed by techniques such as EBIC [50]. Hence it may be considered that the process is in effect a passivation treatment rather than a doping step. There have also been suggestions that the chloride treatment may have inherent limits and that alternative processes need to be developed to overcome the current voltage limited performance of the technology (Section 10.3.2).

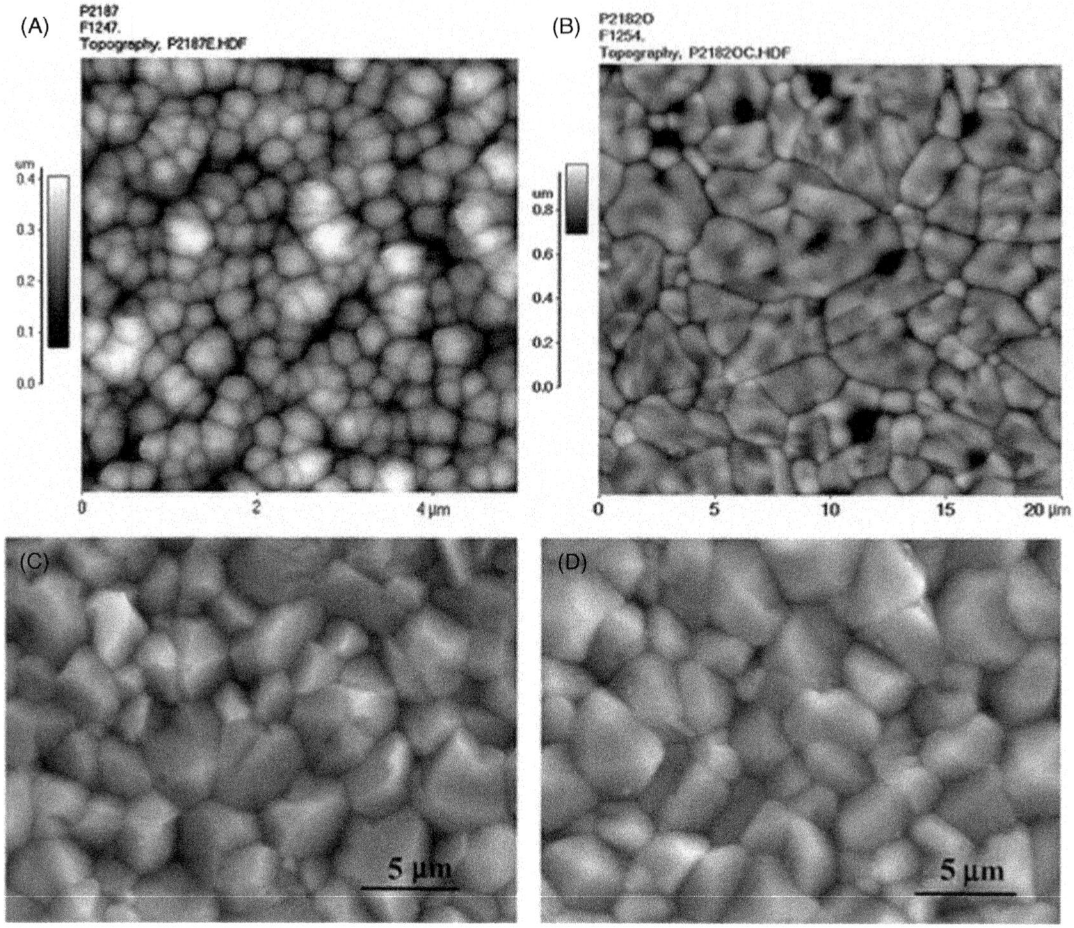

FIGURE 10.6 Structure of CdTe films before/after $CdCl_2$ treatment for different deposition temperatures. (A) Low temperature CdTe, as-deposited, (B) Low temperature CdTe, after $CdCl_2$ treatment at 440°C, (C) High temperature CdTe, as-deposited, (D) High temperature CdTe, after $CdCl_2$ treatment at 400°C [51].

10.2.5 Back-Contacting

One of the long-standing challenges of CdTe solar cell fabrication has been the back-contact. Owing to the high ionization potential of CdTe a metal with a work function of >5.9 eV is required to form an ohmic contact. Most metals do not have such a work function, and those that come close such as platinum and nickel are unsuitable, as they react with tellurium to form unwanted phases [52]. Other lower work function metals create a Schottky junction at the CdTe-metal interface resulting in the formation a rectifying contact. This generates a potential barrier which acts as a second diode opposing the main junction diode (Fig. 10.7A) and owing to the resultant barrier Φ_b (Fig. 10.7B) causes the phenomenon of "rollover" observed in first quadrant of a current voltage curve [53] (Fig.10.7C). A high forward bias lowers the main junction potential while simultaneously

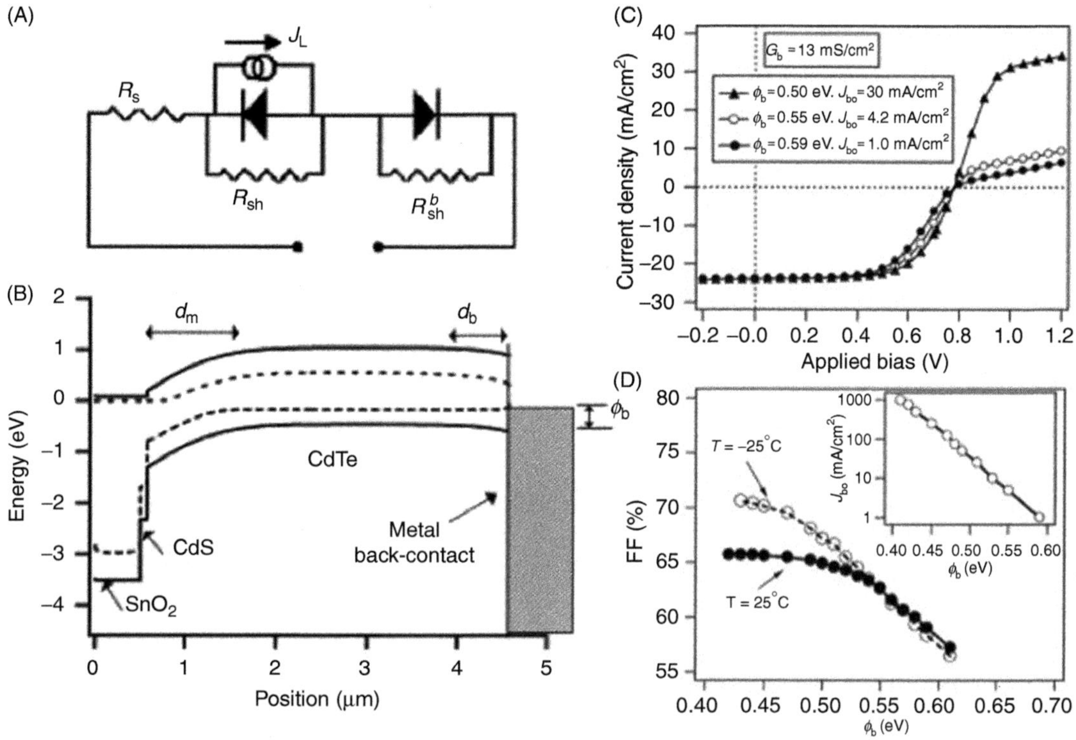

FIGURE 10.7 (A) Two-diode equivalent circuit model for a CdTe solar cell, (B) band diagram of two noninteracting diodes in the light at zero bias, (C) The effect of barrier height on the current–voltage characteristics (simulated), and (D) reduction of fill-factor with increase in barrier height [53].

lowering the back-contact barrier, the back-contact barrier dominates and current through the solar cell is limited, appearing as a flattening of the current voltage curve at high forward bias ($V > V_{OC}$). This can result in lowering fill factor values with increased barrier height (Fig. 10.7D).

Significant research has been undertaken to develop a way to form an ohmic contact between CdTe and a suitable metal. One such a way to overcome this challenge is to allow charge to tunnel through the Schottky barrier by heavily doping the CdTe layer at the back surface so that it is effectively "p^+", forming a pseudo-ohmic contact. This is not as simple as one might anticipate as achieving such high doping densities in CdTe is tricky because Fermi level pinning and self-compensation act against the introduction of acceptors in high concentrations. A fairly common technique is to modify the interface between CdTe and the metal by chemically etching the surface of CdTe to modify the surface. Nitric phosphoric acid [54] and bromine methanol solutions [55] are routinely used as etchants to preferentially remove Cd atoms at the back surface of CdTe before contacting, leaving behind a tellurium rich layer which has greater *p*-type character as a result of cadmium vacancies. However, excessive etching can cause pinholes in the CdTe film, which act as shunting pathways lowering performance. The introduction of copper at the back-contact as a *p*-type dopant has been most successful contacting technique and as a result the most widespread, reaching hole densities of 10^{15} cm^{-2}. However, due to the fast-diffusing nature of copper in CdTe this has also been widely demonstrated to lead to long-term stability issues. A heat treatment is often required for copper to form the desired Cu_2Te complexes but this can aid migration of copper atoms from the back-contact toward the main junction along grain boundaries, resulting in shunting pathways that reduce long-term performance. Upon reaching the main junction, copper can dope the CdS layer, increasing its resistivity, and is known to contribute defect states that aid recombination and lower performance [56]. This is not solely a problem at the point of contact fabrication as numerous studies have shown Cu-contacted devices to be less stable even under standard operating conditions. Additional layers placed between the CdTe and metal contact have also been used to reduce the contact resistance by choosing an appropriate material that forms an ohmic contact to each adjacent layer. ZnTe has demonstrated the effectiveness of this method [57] and is now used in commercial modules due to its similar lattice parameter and favorable band positions. When doped with copper it forms a pseudo-ohmic contact; however, doping densities must be kept low to avoid excessive copper migration but the overall stability of the contact is improved. Other interface layers that have been studied that could achieve higher doping densities while remaining stable include Sb_2Te_3 [58], FeS_2 [59,60], MoO_3 [61], NiP [62], and As_2Te_3 [63], which have shown an ability to form ohmic contacts to varying degrees. The current state of the art when it comes to contacting is now a copper doped layer of either ZnTe or Te coupled to an aluminum contact. This apparently gives sufficient ohmic contact coupled to stability suitable enough for module production where a 20+ year lifespan is essential.

10.2.6 General CdTe Solar Cell Production Notes

There are numerous subtleties to CdTe solar cell fabrication that, while useful to know, are never reported in journal articles as they are not particularly novel or exciting. They are however essential to producing working solar cells, this section is intended to give a brief selection of useful tips for fabrication intended to help CdTe solar cell novices.

- It is preferable to use a commercial TCO-coated glass, in particular FTO as this is stable to all but the most extreme of deposition processes. ITO may be applicable for some processes but for others it will break down.
- Deposit CdS at a thickness of >200 nm. This will be thick enough, irrespective of deposition technique to ensure good coverage. Some optical losses will result but the thickness can be reduced in subsequent depositions.
- The thickness of the CdTe required will depend on the deposition method used. For techniques such as sputtering, buried junctions (i.e., depletion region located away from the CdS interface) may occur for thicknesses much greater than1.5 μm so layers will need to be ~1 μm. For higher temperature deposition techniques due to the increased grain size thicker films may be required to ensure good coverage of the substrate and thus minimize shunting.

- For initial optimization of the chloride treatment, ensure an excess of the chloride (>100 nm) being used is present on the back surface. The treatment temperature will almost always fall in the 380–450°C range, and the time can simply be optimized by tracking cell performance as a function of treatment.
- Following chloride treatment wash the surface thoroughly in deionized water to remove any excess. Typically there will be oxide or chloride phases remaining at the surface [64], which this washing will not remove. Etching can aid removal of these phases with either nitric–phosphoric acid [65] or bromine methanol being standard.
- Numerous contact process of varying complexity can be employed and these will have a significant influence on the performance. However, a good basic contact is a few nanometer of Cu followed by a gold back-contact being applied. This will give consistently good results for most cells.
- To make a good front-contact to the solar cell the CdTe layer can be easily removed by manual scribing with a scalpel blade or similar. The CdS layer also needs to be removed, and this can be done by swabbing with HCl which will remove the CdS but not damage the TCO (unless it is ZnO based e.g., ZnO:Al).
- Contact size is an overlooked issue but must be considered carefully. Owing to the sheet resistance of TCO larger contacts will significantly reduce the FF of the measured cell. The temptation to use particularly small contacts should be avoided however as this can lead to significant error in estimation of the current density via even minor variations in the contact size. This size variation can occur readily if the back surface is conductive following doping. Defining the contact area by manual scribing or measuring using an optical aperture is advisable. A suitable contact size is considered to be in the 0.20–1.0 cm^2 range.
- Additional FF losses may also be seen if there is a large physical distance between the front and back-contacts, again due to the nonnegligible resistance of the TCO. This can be compensated for by the use of bus-bars across the cell which can be as simple as strips of conductive paint which will minimize parasitic shunting.

10.3 Looking Forward—Voltage, Doping, and Substrate Cells

Despite CdTe becoming established as an industrial technology, there remain a number of key areas still requiring development. The aim is not only to continue the upward trajectory of CdTe solar cells in an attempt to more closely match the 30% efficiency potential, but also to investigate alternative device structures such as the substrate cell, as this may have additional benefits for production. This section will discuss the development of the alternative substrate geometry solar cells and the current voltage limitation problems.

10.3.1 Substrate Cells

One of the more radical approaches to CdTe solar cell development has been the idea of inverting the cell structure to use the "substrate" geometry (Fig. 10.8). This mimics the

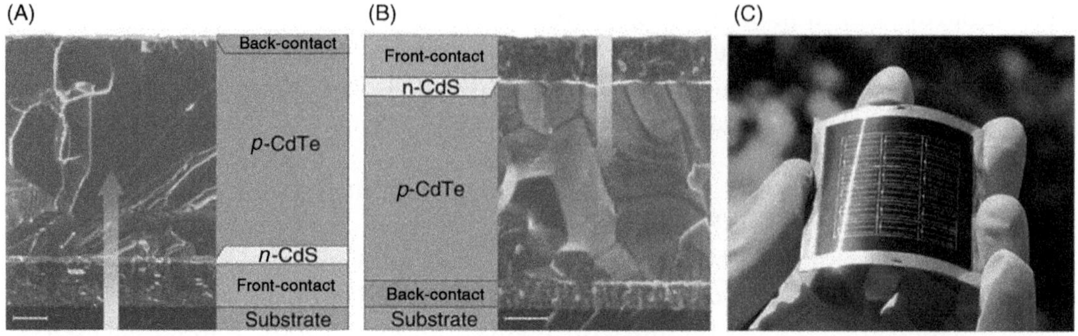

FIGURE 10.8 Scanning electron micrograph and schematic of the cross-section of a CdTe solar cell in the conventional superstrate configuration (A) and the substrate configuration (B) which allows the use of opaque substrates such as metal foils. In substrate configuration Mo/MoO_x and i-ZnO/ZnO:Al are used as electrical back and front contact, respectively. The scale bars correspond to 1 μm. The yellow arrows show the direction of illumination. (C) Photograph of a sample with several CdTe solar cells on flexible metal foil [67].

structure of other thin film technologies such as CIGS or CZTS, where the absorber layer is deposited prior to the *n*-type and TCO layers. An obvious question of course is, why bother? The use of glass as a basis for CdTe PV modules has several disadvantages such as it represents both 98% of the module weight and over a third of module cost. It also rather limits the potential for use in building integrated photovoltaics (BIPV) [66], something that thin film technologies should be highly suited for. CdTe in the substrate geometry, where CdTe is the first layer deposited, allows for nontransparent substrates to be used such as metal foils to be used. By using cheap metals such as stainless steel and aluminum it would allow roll-to-roll processing of flexible cells and has the potential to significantly reduce both module costs and weight, a significant advantage for BIPV [67]. From a scientific point of view the substrate configuration also allows for a finer control of junction formation, the chlorine activation, intermixing, and *p*-type doping of the CdTe [68] as the CdTe layer can be treated prior to addition of the partner layers. Despite these advantages though the performance of substrate CdTe devices is significantly lower than that which has been achieved for the superstrate geometry, with only a few reports of over 10% being achieved [69] and nothing in excess of 13.6%. The primary reason for the reduced performance is the increased difficulty of forming an efficient back-contact [70]. As well as the inherent work function issues discussed in Section 10.2.5, for substrate cells the contact must also be self-forming, similar to CIGS and CZTS [71], and stable to all subsequent deposition processes. The standard Cu process therefore becomes problematic due to Cu diffusion during device fabrication leading to degradation and compensation via excess Cu diffusion [72]. Molybdenum is typically the metal of choice for substrate cells due to its high purity and similar thermal expansion coefficient to CdTe [66] and is also a known quantity to an extent given the volume of work in the CIGS and CZTS fields. However, given the molybdenum work function of 4 eV a catastrophically large Schottky barrier is present which hinders performance [70]. As a result the majority of work that has been carried out on substrate CdTe PV thus far is focused on forming an ohmic contact via inclusion of carefully selected buffer layers. This has largely been using the approaches for superstrate cells with highly doped layers at the back surface such as Cu, Cu_xTe, and Sb_2Te_3 or layers that have a higher work function than CdTe to promote favorable band bending at the interface such as MoO_3 and WO_3 [73]. To date the most successful back-contact has utilized a combination of MoO_3 as the high work function buffer layer (6.5 eV) and Te as the back-contact buffer layer with 10% [70] being achieved for Cu free devices and over 13% with controlled Cu doping [67].

Despite the reduced performance there have been a few studies, which show that advantage of using the substrate configuration to analyze cell processing such as the role of Cu in degrading CdTe PV and the influence Cl processing the cells at different stages of cell fabrication. To date though the amount of research undertaken on substrate CdTe devices is vastly inferior to that of their superstrate counterpart. However, with recent developments in efficiencies, the potential for in depth analysis of the key step related to processing and the potential industrial advantages means that substrate CdTe may yet generate a substantial amount of research interest in future years.

10.3.2 Open Circuit Voltage Limitations

The impressive uplift in CdTe solar cell efficiency of recent years (Fig. 10.1) has been predicated on increased current, to the extent that the current champion cells now have short circuit current values close to the theoretical limit. This has been achieved by maximizing the optics of the cell; removing the CdS layer in favor of an oxide such as $Mg_{1-x}Zn_xO$ [25] to improve short wavelength collection, coupled to grading the CdTe bandgap via selenium incorporation to form lower bandgap $CdTe_{1-x}Se_x$ phases and improved long wavelength collection [26]. As the current is effectively maxed out any further improvement will therefore need to come via improvements in the open circuit voltage, something that has proved far more challenging to achieve. In the prior 20 years there has been less than 30 mV increase in the V_{OC} of champion cells. Improving beyond the current limit of 876 mV [3] (Fig. 10.9) to closer to the theoretical limit of $>$1100 mV seems a challenge as would require an increase in both minority carrier lifetime and doping density of the material [74], a considerable challenge. Recent work on single crystal CdTe absorber cells has demonstrated what may be possible though while the use of single crystals is obviously impractical for mass production owing to cost and deposition time considerations, they do provide tremendous insight into what may be possible. Work from National Renewable Energy Laboratory demonstrated that by using arsenic doped single crystal absorbers voltages in excess of 1 V were achievable [40]. These devices also removed the chloride and copper steps, which are considered standard practice. Similarly work which utilized single crystal absorbers but in a double heterojunction arrangement with cadmium magnesium telluride also yielded voltages in excess of 1 V [75]. The path toward similar voltage levels for polycrystalline equivalent cells would therefore appear to be centered around trying to replicate similar materials properties. This may mean abandoning the established chloride treatments process in search of alternatives which can yield higher doping, focusing

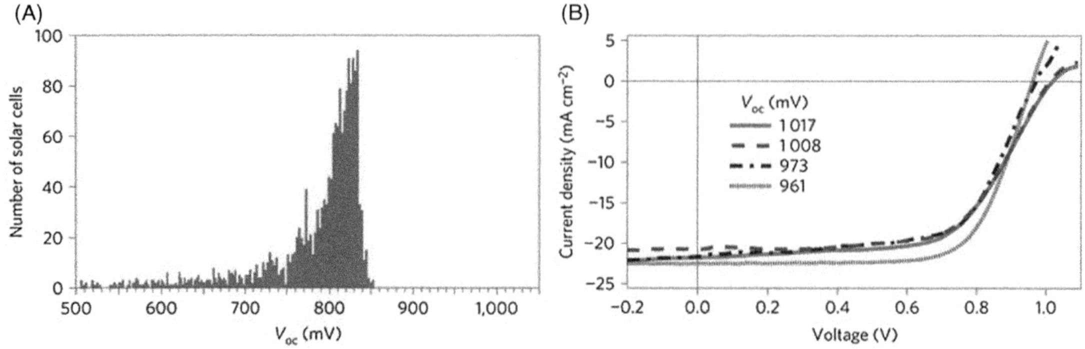

FIGURE 10.9 (A) Histogram of V_{OC} values for more than 2400 standard polycrystalline CdTe solar cells fabricated at NREL in the past decade. (B) Current density–voltage curves for As-doped single crystal CdTe solar cells showing higher open circuit voltage [40].

on the production of higher quality/purity polycrystalline CdTe layers or even adopting a different junction design to improve carrier lifetimes. It may ultimately prove though that polycrystalline material simply cannot match the quality of the single crystal equivalent, this has certainly proven true for silicon, but there are still a number of fundamental questions about both the material and device structure that need to be addressed. This is, at the time of writing, one of the areas the research community is particularly focused on.

10.4 Conclusion

The recent progress of CdTe solar cells should not be taken lightly. Having stagnated for over a decade recent breakthroughs leading to improved device performances have reinvigorated the research field leading to a raft of new cell designs and processes. The sudden emergence of alternative window layers such as MZO, use of absorber bandgap grading with Se, and breaking of 1 V with single crystal As-doping have all shown that this remains a technology in development, even at over 40 years old. Despite still being a work in progress though it has already firmly established itself as the module technology most directly competitive with silicon. There is also tremendous scope for further module cost reduction through either performance improvements, process refinement, or simple economies of scale. As a result CdTe seems to be a technology that will remain at the forefront of PV research and manufacture for the foreseeable future.

References

[1] Philipps, S. Warmuth, W. photovoltaics report—Fraunhofer ISE; 2016. Available from: https://www.ise.fraunhofer.de/en/publications/studies/photovoltaics-report.html.

[2] Madelung O: *Semiconductors: data handbook*, Berlin, 2004, Springer.

[3] Green MA, Emery K, Hishikawa Y, Warta W, Dunlop ED, Levi DH, et al: Solar cell efficiency tables (version 49), *Prog Photovolt Res Appl* 25:3–13, 2017.

[4] Bonnet D, Rabenhorst H. New results on the development of a thin film p-CdTe/n-CdS heterojunction solar cell. In: 9th IEEE photovoltaic specialists conference; 1972:129–133.

[5] Granqvist CG: Transparent conductors as solar energy materials: a panoramic review, *Sol Energy Mater Sol Cells* 91:1529–1598, 2007.

[6] Wu XZ: High-efficiency polycrystalline CdTe thin-film solar cells, *Sol Energy* 77:803–814, 2004.

[7] Major JD. CdTe solar cells: growth phenomena and device [PhD Thesis]. Durham University; 2008.

[8] Romeo N, Bosio A, Canevari V. The role of CdS preparation method in the performance of CdTe/CdS thin film solar cell. In: 3rd world conference on photovoltaic energy conversion; 2003:469–470.

[9] Zoppi G, Durose K, Irvine SJC, Barrioz V: Grain and crystal texture properties of absorber layers in MOCVD-grown CdTe/CdS solar cells, *Semicond Sci Technol* 21:763–770, 2006.

[10] Paudel NR, Compaan AD, Yan Y: Sputtered CdS/CdTe solar cells with MoO_{3-x}/Au back contacts, *Sol Energy Mater Sol Cells* 113:26–30, 2013.

[11] Lee J: Comparison of CdS films deposited by different techniques: effects on CdTe solar cell, *Appl Surf Sci* 252:1398–1403, 2005.

[12] McCandless BE, Moulton LV, Birkmire RW: Recrystallization and sulfur diffusion in CdCl2-treated CdTe/CdS thin films, *Prog Photovolt* 5:249–260, 1997.

[13] Taylor AA, Major JD, Kartopu G, Lamb D, Duenow J, Dhere RG, et al: A comparative study of microstructural stability and sulphur diffusion in CdS/CdTe photovoltaic devices, *Sol Energy Mater Sol Cells* 141:341–349, 2015.

[14] Castro-Rodrigueza R, Martel A, Mendez-Gamboa J, Pena JL: Nucleation and growth mechanism of CdTe cluster grown on CdS films, *J Cryst Growth* 306:249–253, 2007.

[15] Sites JR: Quantification of losses in thin-film polycrystalline solar cells, *Sol Energy Mater Sol Cells* 75:243–251, 2003.

[16] Major JD, Bowen L, Durose K: Focused ion beam and field emission gun-scanning electron microscopy for the investigation of voiding and interface phenomena in thin-film solar cells, *Prog Photovolt Res Appl* 20:892–898, 2012.

[17] Dhere R, Rose D, Albin D, Asher S, Al-Jassim M, Cheong H, et al. Influence of CdS/CdTe interface properties on the device properties. In: 26th photovoltaic specialists conference; 1997:435–438.

[18] Treharne RE, Seymour-Pierce A, Durose K, Hutchings K, Roncallo S, Lane D: Optical design and fabrication of fully sputtered CdTe/CdS solar cells, *J Phys Conf Ser* 286:12038, 2011.

[19] Archbold MD, Halliday DP, Durose K, Hase TPA, Boyle DS, Mazzamuto S, et al: Development of low temperature approaches to device quality CdS: a modified geometry for solution growth of thin films and their characterisation, *Thin Solid Films* 515:2954–2957, 2007.

[20] Koishiyev GT, Sites JR, Effect of weak diodes on the performance of CdTe thin-film modules. In: 34th IEEE photovoltaic specialists conference; 2009:79–82.

[21] Ferekides CS, Mamazza R, Balasubramanian U, Morel DL: Transparent conductors and buffer layers for CdTe solar cells, *Thin Solid Films* 480:224–229, 2005.

[22] Fedorenko YG, Major JD, Pressman A, Phillips LJ, Durose K: Modification of electron states in CdTe absorber due to a buffer layer in CdTe/CdS solar cells, *J Appl Phys* 118:165705, 2015.

[23] Major JD, Durose K. Co-sputtered $Zn_{1-x}Sn_xO$ buffer layers for CdTe thin film solar cells. In: 39th IEEE photovoltaic specialists conference; 2013:1146–1149.

[24] Wu X, Dhere RG, Yan Y, Romero MJ, Zhang Y, Zhou J, et al. High-efficiency polycrystalline CdTe thin-film solar cells with an oxygenated amorphous CdS (a-CdS:O) window layer. In: 29th IEEE photovoltaic specialists conference; 2002:531–534.

[25] Kephart JM, McCamy JW, Ma Z, Ganjoo A, Alamgir FM, Sampath WS: Band alignment of front contact layers for high-efficiency CdTe solar cells, *Sol Energy Mater Sol Cells* 157:266–275, 2016.

[26] Paudel NR, Yan Y: Enhancing the photo-currents of CdTe thin-film solar cells in both short and long wavelength regions, *Appl Phys Lett* 105:183510, 2014.

[27] Kim D, Pozder S, Zhu Y, Trefny JU. Polycrystalline thin film CdTe solar cells fabricated by electrodeposition, 1st world conference on photovoltaic energy conversion; 1994:334–337.

[28] Nowell MM, Scarpulla MA, Paudel NR, Wieland KA, Compaan AD, Liu X: Characterization of sputtered CdTe thin films with electron backscatter diffraction and correlation with device performance, *Microsc Microanal* 21:927–935, 2015.

[29] Kranz L, Perrenoud J, Pianezzi F, Gretener C, Rossbach P, Buecheler S, et al: Effect of sodium on recrystallization and photovoltaic properties of CdTe solar cells, *Sol Energy Mater Sol Cells* 105:213–219, 2012.

[30] Barrioz V, Irvine SJC, Jones EW, Rowlands RL, Lamb DA: In situ deposition of cadmium chloride films using MOCVD for CdTe solar cells, *Thin Solid Films* 515:5808–5813, 2007.

[31] Major JD, Proskuryakov YY, Durose K, Zoppi G, Forbes I: Control of grain size in sublimation-grown CdTe, and the improvement in performance of devices with systematically increased grain size, *Sol Energy Mater Sol Cells* 94:1107–1112, 2010.

[32] Paudel NR, Wieland KA, Compaan AD. Improvements in ultra-thin CdS/CdTe solar cells. In: 37th photovoltaic specialists conference; 2011:4–6.

[33] Major JD, Phillips LJ, Al Turkestani M, Bowen L, Whittles TJ, Dhanak VR, et al: P3HT as a pinhole blocking back contact for CdTe thin film solar cells, *Sol Energy Mater Sol Cells* 172:1–10, 2017.

[34] Kanevce A, Levi DH, Kuciauskas D: The role of drift, diffusion, and recombination in time-resolved photoluminescence of CdTe solar cells determined through numerical simulation, *Prog Photovolt* 22:1138–1146, 2014.

[35] Major JD, Al Turkestani M, Bowen L, Brossard M, Li C, Lagoudakis P, et al: In-depth analysis of chloride treatments for thin-film CdTe solar cells, *Nat Commun* 7:13231, 2016.

[36] Kanevce A, Kuciauskas D, Gessert TA, Levi DH, Albin DS. Impact of interface recombination on time resolved photoluminescence (TRPL) decays in CdTe solar cells (numerical simulation analysis). In: 38th IEEE photovoltaic specialists conference; 2012:848-853.

[37] Nowell MM, Scarpulla MA, Compaan AD, Liu X, Paudel NR, Kwon D, et al. Electron backscatter diffraction and photoluminescence of sputtered CdTe thin films. In: 37th IEEE photovoltaic specialists conference; 2011:001327–001332.

[38] Major JD: Grain boundaries in CdTe thin film solar cells: a review, *Semicond Sci Technol* 31:93001, 2016.

[39] Luschitz J, Lakus-Wollny K, Klein A, Jaegermann W: Growth regimes of CdTe deposited by close-spaced sublimation for application in thin film solar cells, *Thin Solid Films* 515:5814–5818, 2007.

[40] Burst JM, Duenow JN, Albin DS, Colegrove E, Reese MO, Aguiar JA, et al: CdTe solar cells with open-circuit voltage breaking the 1 V barrier, *Nat Energy* 1:16015, 2016.

[41] Basol BM, Ou SS, Stafsudd OM: Type conversion, contacts and surface effects in electroplated CdTe-films, *J Appl Phys* 58:3803–3813, 1985.

[42] Major JD, Treharne RE, Phillips LJ, Durose K: A low-cost non-toxic post-growth activation step for CdTe solar cells, *Nature* 511:334–337, 2014.

[43] Mazzamuto S, Vaillant L, Bosio A, Romeo N, Armani N, Salviati G: A study of the CdTe treatment with a Freon gas such as CHF_2Cl, *Thin Solid Films* 516:7079–7083, 2008.

[44] Major JD, Proskuryakov YY, Durose K: Impact of CdTe surface composition on doping and device performance in close Space sublimation deposited CdTe solar cells, *Prog Photovolt* 21:436–443, 2013.

[45] Potlog T, Ghimpu L, Antoniuc C: Comparative study of CdS/CdTe cells fabricated with and without evaporated Te-layer, *Thin Solid Films* 515:5824–5827, 2007.

[46] Regalado-Perez E, Reyes-Banda MG, Mathew X: Influence of oxygen concentration in the CdCl2 treatment process on the photovoltaic properties of CdTe/CdS solar cells, *Thin Solid Films* 582:134–138, 2015.

[47] Moutinho HR, Dhere RG, Al-Jassim MM, Mayo B, Levi DH, Kazmerski LL: Induced recrystallization of CdTe thin films deposited by close-spaced sublimation, *Ncpv Photovolt Prog Rev* 462:517–523, 1999.

[48] Moutinho HR, Dhere RG, Romero MJ, Jiang CS, To B, Al-Jassim MM. Recrystallization of PVD CdTe thin films induced by $CdCl_2$ treatment - a comparison between vapor and solution processes. In: 33rd IEEE photovoltaic specialists conference; 2008:1–5.

[49] Major JD, Bowen L, Durose K: Focused ion beam and field emission gun–scanning electron microscopy for the investigation of voiding and interface phenomena in thin-film solar cells, *Prog Photovolt Res Appl* 20:892–898, 2012.

[50] Li C, Wu Y, Poplawsky J, Pennycook TJ, Paudel N, Yin W, et al: Grain-boundary-enhanced carrier collection in CdTe solar cells, *Phys Rev Lett* 112:156103, 2014.

[51] Quadros J, Pinto AL, Moutinho HR, Dhere RG, Cruz LR: Microtexture of chloride treated CdTe thin films deposited by CSS technique, *J Mater Sci* 43:573–579, 2008.

[52] Demtsu SH, Albin DS, Pankow JW, Davies A: Stability study of CdS/CdTe solar cells made with Ag and Ni back-contacts, *Sol Energy Mater Sol Cells* 90:2934–2943, 2006.

[53] Demtsu SH, Sites JR: Effect of back-contact barrier on thin-film CdTe solar cells, *Thin Solid Films* 510:320–324, 2006.

[54] Yan Y, Jones KM, Wu X, Al-Jassim MM: Microstructure of CdTe thin films after mixed nitric and phosphoric acids etching and (HgTe, CuTe)-graphite pasting, *Thin Solid Films* 472:291–296, 2005.

[55] Moutinho HR, Dhere RG, Jiang CS, Al-Jassim MM, Kazmerski LL: Electrical properties of CdTe/CdS solar cells investigated with conductive atomic force microscopy, *Thin Solid Films* 514:150–155, 2006.

[56] Corwine C, Pudov AO, Gloeckler M, Demtsu SH, Sites JR: Copper inclusion and migration from the back contact in CdTe solar cells, *Sol Energy Mater Sol Cells* 82:481–489, 2004.

[57] Gessert TA, Sheldon P, Li X, Dunlavy D, Niles D, Sasala R, et al. Studies of ZnTe back contacts to CdS/CdTe solar cells. In: 26th IEEE photovoltaic specialists conference; 1997:419–422.

[58] Hu S, Zhu Z, Li W, Feng L, Wu L, Zhang J, et al: Band diagrams and performance of CdTe solar cells with a Sb_2Te_3 back contact buffer layer, *AIP Adv* 1:42152, 2011.

[59] Bhandari KP, Koirala P, Paudel NR, Khanal RR, Phillips AB, Yan Y, et al: Iron pyrite nanocrystal film serves as a copper-free back contact for polycrystalline CdTe thin film solar cells, *Sol Energy Mater Sol Cells* 140:108–114, 2015.

[60] Turner AK, Woodcock JM, Ozsan ME, Summers JG, Barker J, Binns S, et al: Stable, high efficiency thin film solar cells produced by electrodeposition of cadmium telluride, *Sol Energy Mater* 23:388–393, 1991.

[61] Yang R, Wang D, Jeng M, Ho K, Wang D: Stable CdTe thin film solar cells with a MoO_x back-contact buffer layer, *Prog Photovolt Res Appl* 24:59–65, 2016.

[62] Viswanathan V, Tetali B, Selvaraj P, Jagannathan S, Morel DL, Ferekides CS, et al. Ni_2P-a promising candidate for back contacts to CdS/CdTe solar cells. In: 28th IEEE photovoltaic specialists conference; 2000:587–590.

[63] Barrioz V, Proskuryakov YY, Jones EW, Major JD, Irvine SJC, Durose K, et al. Highly arsenic doped CdTe layers for the back contacts of CdTe solar cells. In: Materials research society symposia proceedings; 2007:367–372.

[64] Waters D, Niles DW, Gessert TA, Albin D, Rose D, Sheldon P. Surface analysis of CdTe after various precontact treatments. In: 2nd world conference on photovoltaic energy conversion; 1995:1031–1034.

[65] Major JD, Al Turkestani M, Durose K. Structural and electrical characterisation of $MgCl_2$-treated CdTe solar cells, 42nd IEEE photovoltaic specialists conference; 2015:1–5.

[66] Aliyu MM, Islam MA, Hamzah NR, Karim MR, Matin MA, Sopian K, et al: Recent developments of flexible CdTe solar cells on metallic substrates: Issues and prospects, *Int J Photoenergy*:6–9, 2012.

[67] Kranz L, Gretener C, Perrenoud J, Schmitt R, Pianezzi F, La Mattina F, et al: Doping of polycrystalline CdTe for high-efficiency solar cells on flexible metal foil, *Nat Commun* 4:2306, 2013.

[68] Kranz L, Gretener C, Perrenoud J, Jaeger D, Gerstl SS, Schmitt R, et al: Tailoring impurity distribution in polycrystalline cdte solar cells for enhanced minority carrier lifetime, *Adv Energy Mater* 4:1301400, 2014.

[69] Kranz L, Schmitt R, Gretener C, Perrenoud J, Pianezzi F, Uhl AR, et al. Progress toward 14 % efficient CdTe solar cells in substrate configuration. In: 39th IEEE photovoltaic specialists conference; 2013:1644–1648.

[70] Gretener C, Perrenoud J, Kranz L, Kneer L, Schmitt R, Buecheler S: CdTe/CdS thin film solar cells grown in substrate configuration, *Prog Photovolt Res Appl* 21:1590–1598, 2013.

[71] Lin YC, Hsieh YT, Lai CM, Hsu HR: Impact of Mo barrier layer on the formation of $MoSe_2$ in Cu(In,Ga) Se_2 solar cells, *J Alloys Compd* 661:168–175, 2016.

[72] Gretener C, Perrenoud J, Kranz L, Cheah E, Dietrich M, Buecheler S, et al: New perspective on the performance stability of CdTe solar cells degradation of CdTe solar cell performance, *Sol Energy Mater Sol Cells* 146:51–57, 2016.

[73] Gretener C, Perrenoud J, Kranz L, Baechler C, Yoon S, Romanyuk YE, et al: Development of MoO_x thin films as back contact buffer for CdTe solar cells in substrate configuration, *Thin Solid Films* 535:193–197, 2013.

[74] Hsiao KJ, Sites JR: Electron reflector to enhance photovoltaic efficiency: application to thin-film CdTe solar cells, *Prog Photovolt* 20:486–489, 2012.

[75] Zhao Y, Boccard M, Liu S, Becker J, Zhao XH, Campbell CM, et al: Monocrystalline CdTe solar cells with open-circuit voltage over 1 V and efficiency of 17%, *Nat Energy* 1:16067, 2016.

11

An Overview of Hybrid Organic–Inorganic Metal Halide Perovskite Solar Cells

Khagendra P. Bhandari, Randy J. Ellingson

CENTER FOR PHOTOVOLTAICS INNOVATION AND COMMERCIALIZATION, UNIVERSITY OF TOLEDO, TOLEDO, OH, UNITED STATES

Khagendra.bhandari@rockets.utoledo.edu

11.1 Introduction

Solar generated electricity is the world's fasted growing renewable energy with net solar generation increasing by an average of 8.3% per year [1]. The highest growth rates have been noted in China, followed by Japan, and USA in second and the third place, followed by three European counties such as United Kingdom (fourth), Germany (fifth), and France (sixth) [2]. Seven commercial technologies make up 98% of the current world market share with the thin film industry making up only about 13% of the total. However, the growth in thin film industry is faster than that in the crystalline silicon industry [3–5]. This 13% market share is divided into several established and emerging PV technologies including polycrystalline CdTe thin films, CIGS, amorphous silicon, organic, dye-sensitive, and quantum dots solar cells. To increase the market share, alternative technologies have to provide a desirable combination of high power conversion efficiency, low manufacturing costs, and excellent stability. Recently developed hybrid organic–inorganic metal halide perovskite, methylammonium halide perovskite $CH_3NH_3MX_3$ or $MAMX_3$(MA = CH_3NH_3, M = Pb or Sn, X = Cl, Br, and I) or simply perovskite, solar cells have a great potential to become one of the leading technologies in PV industry due to their high efficiency and low manufacturing costs. As a result of intensive research efforts across the world over the past 8 years, perovskite-based solar cell performances are now comparable to silicon-based solar cells, at least at the laboratory scale [6–8].

German mineralogist Gustav Rose discovered calcium titanate, also known as calcium titanium oxide ($CaTiO_3$), in 1839. As a mineral, it is called perovskite, named after Russian mineralogist, Lev A. Perovski (1792–856) [9]. In hybrid organic–inorganic metal halide perovskite ($CH_3NH_3MX_3$), $CH_3NH_3^+$ is an organic cation, M is a divalent metal cation (Pb^{2+} or Sn^{2+}), and X is a monovalent halide anion (Cl^{-1}, Br^{-1}, or I^{-1}). Because of their excellent optoelectronic properties and potential solution-processed synthesis [10–12], these materials have been studied with the aim of developing new materials for organic light

A Comprehensive Guide to Solar Energy Systems. http://dx.doi.org/10.1016/B978-0-12-811479-7.00011-7

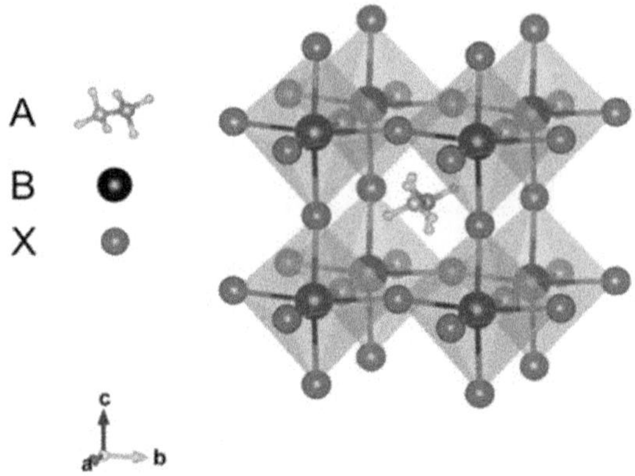

FIGURE 11.1 Crystal structure of hybrid organic-inorganic metal halide perovskites with the generic chemical formula *ABX*$_3$; Organic cations occupy position *A* [light gray/dark gray/gray *(green/blue/orange in the web version)*] and the metal cations and halides occupy the *B* (black) and X [*(purple in the web version)*] positions, respectively. *Reproduced with permission from Wang D, Wright M, Elumalai NK, Uddin A. Stability of perovskite solar cells. Solar Energy Mat Solar Cells 2016;147:255–75, copyright 2016, Solar Energy Materials and Solar Cells.*

emitting diodes [13] and field effect transistors [14]. Fig. 11.1 shows a schematic diagram of the generic ABX_3 perovskite crystal structure for a hybrid organic–inorganic metal halide perovskite [15]. The A position contains an organic cation ($CH_3NH_3^+$), B is a metal cation (Pb^+ or Sn^+), and X is a halide anion (Cl^- or Br^- or I^-).

The hybrid organic–inorganic lead halide perovskite compound was first used as visible-light sensitizers for photovoltaic cells in 2009 with the efficiency of 3.8% for X = Br and 3.1% for X = I, respectively at one sun illumination [16]. Perovskite was also used as a sensitizer in quantum dot-sensitized solar cells in 2011 with an efficiency of 6.5% [17]. These two performances indicated the potential of using perovskite for solar cells even though they were not very stable due to the presence of a liquid electrolyte. The first stable and solid-state perovskite was reported in 2012 with an efficiency of 9.7% [18] and in the same year another report on solid-state perovskite was published in science recording an efficiency of 10.9% [19]. Since then, perovskite-based solar cell performance has rapidly progressed with a best efficiency record of 22.1% in 2015 [6]. Progress in the perovskite solar cells has been remarkable within a short time period and is considered as the biggest scientific breakthrough in the PV industry [20–22].

Solution-based and vapor-based depositions are the two main deposition methods for the fabrication of high quality perovskite thin films. The solution-based deposition technique is cost effective and compatible with the fabrication process, which includes flexible substrates [23–25]. On the other hand, the vapor-based deposition technique is an industrial production technique with a potential for the commercialization of perovskite solar cells [26,27]. In either case, the deposition methods are relatively rapid and also consume very small amount of materials. These are some of the reasons why scientific communities are attracted to the perovskite solar cells industry.

The hybrid organic–inorganic metal halide perovskite-based materials exhibit several outstanding optical and electrical properties, which are ideal for photovoltaic applications. The absorption coefficient of perovskite ($\alpha > 10^5$ cm^{-1}) is higher than for existing PV materials such as CIGS, Si, GaAs, and CdTe with minimum Urbach energy and utilizes all the radiations higher than the band gap energy and provides higher short circuit current density (J_{SC}) even from ~300 nm thick films [28,29]. The Urbach energy of methylammonium tri-iodide is ~15 meV, very close to the Urbach energy of high quality GaAs [30]. The band gap of methylammonium lead tri-iodide is 1.5 eV and can be tuned with the addition of other halide ions ranging from 1.5 to 2.3 eV [31]. A very high open circuit voltage (V_{OC}) is achieved from perovskite solar cells relative to its band gap and also in comparison to other thin film solar cells [32,33]. Wolf et al. [30] found very sharp absorption edge with negligible deep defect states and minimum nonradiative recombination loss. The perovskite materials possess three important properties such as high electron and hole mobilities in the range of 10–60 $cm^2 V^{-1} s^{-1}$ [34–36]; long carrier lifetime as high as 1.07 μs [37]; and long diffusion lengths (>1 μm) [38]. Chen et al. [39] found still longer carrier lifetimes and diffusion lengths with the values of 30 μs and 23 μm for polycrystalline films and up to 3 ms and 650 μs for single crystals, respectively. All these factors are responsible for the high V_{OC} values exhibited by perovskite solar cells.

The photo-conversion efficiency achieved by perovskite solar cells is very high, but the stability of the solar cells in ambient environments is still a serious problem and improvements are relatively very slow. The stability is the main issue preventing perovskite solar cells from commercialization. Since the organic materials that make up perovskite solar cells are volatile, the organic species not only can easily escape from perovskite films and interact with moisture, but they are also not sustainable at even slightly raised temperatures. When completing current–voltage measurement, performance depends on which direction bias voltage is applied in solar cells. This does not happen in standard inorganic solar cells such as Si, CdTe, CIGS, and GaAs. Finally, there are environmental concerns about using hazardous materials such as lead in perovskite solar cells.

This chapter focuses on general information related to hybrid organic–inorganic metal halide perovskite thin film and solar cells. For a detailed analysis, the readers need to follow several other review papers or journal articles [15,40,41]. This chapter provides discussions on device fabrication, device architecture, progress made so far, and details about the stability of solar cells.

11.2 Thin Film Fabrication/Formation

The performance of perovskite solar cells or any other solar cells is mainly judged by the absorber layer film quality even though other layers are also equally important. There are several factors that determine the fabrication of high quality film and high performance solar cells such as controlled morphology, thickness uniformity, high surface coverage with no or minimum pinholes, material phase purity, and high crystallinity. For example, high crystallinity of the film determines the charge separation efficiency, charge transport, and the diffusion length of charge carriers [42,43]. To establish these criteria, engineering of material composition, film fabrication method, control crystallization, desired substrate selection, and solvent/additives selection are conducted with care. The nature of the substrate surface morphology determines the quality of the film formed on it. The perovskite film is of high quality when the substrates of interest are mesoporous, irrespective of TiO_2, Al_2O_3, or NiO, and it provides excellent device performance [44]. The crystal size of the mixed halide perovskite ($MAPbI_{3-x}Cl_x$) reduces to <100 nm in mesoporous Al_2O_3 compared to ~500 nm in planar system, resulting in an increase J_{SC} of the solar cells [45]. There are various deposition techniques for high quality perovskite thin films such as single-step solution deposition [19], two-step solution deposition [23], two-step vapor-assisted deposition [25], and thermal vapor deposition [26].

11.2.1 Single Step Deposition

In the one-step deposition process, both organic and inorganic compounds are codeposited either through solution or thermal evaporation processes as described elsewhere [46–48]. In the single-step solution process, a mixture of MX_2 (M = Pb or Sn and X = I, Br, and Cl) and methylammonium iodide (MAI) or formamidinium iodide (FAI) is dissolved in an organic solvent and the solution mixture is spun coated onto the substrate of interest. The film so obtained is annealed at 100–150°C to produce the final perovskite phase. In the deposition process used to fabricate high performance solar cells, the composition variation has been made from MAI poor to MAI rich in PbI_2 in the ratio of 1:2–3:1 [49,50]. When the precursor composition is changed, there needs to be a change in processing temperature and time to maintain desired crystal structure, phase purity, and morphology [43,51,52].

11.2.2 Two Step Sequential Deposition

In the two-step solution process, a MX_2 seed layer is first spun coated onto the substrate and the substrate is then dipped into MAI or FAI/isopropanol solution or another spin coating is made for MAI or FAI onto a MX_2 network to form a hybrid organic–inorganic metal halide perovskite [23,53]. The two-step process has been shown to be better as it provides a more uniform and controlled film and has been extensively used in solar cell fabrications [6,23,44]. The two-step process, in contrast, has some problems such as incomplete perovskite conversion and surface roughness in some cases but with the introduction of new techniques these problems have been reduced. One-step and two-step solution deposition processes are described in Fig. 11.2 [47].

11.2.3 Two Step Vapor Assisted Deposition

The two-step solution deposition method has been modified and described by Chen et al. [25], here the MAI is introduced into the MX_2 layer via a vapor deposition technique in order to control the morphology and grain size of the perovskite film in a better manner. This deposition method is time consuming as it takes a long time to fully convert to the perovskite and furthermore, the performance is not efficient [25].

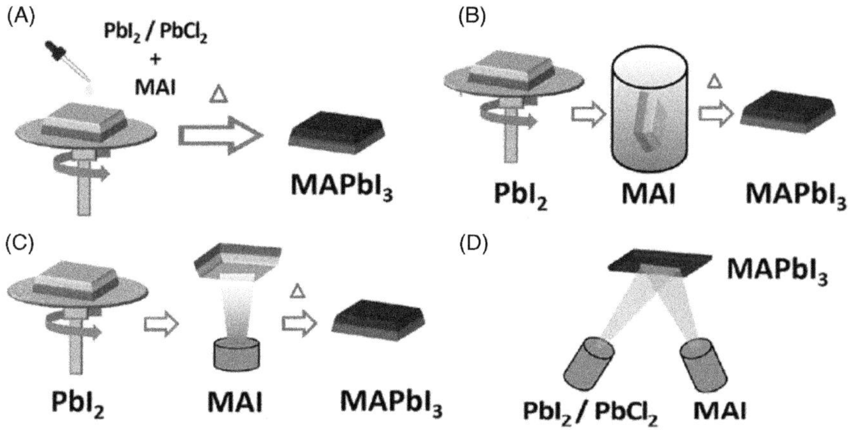

FIGURE 11.2 Deposition methods for perovskite thin films, including (A) single-step solution deposition, (B) two-step solution deposition, (C) two-step hybrid deposition, and (D) thermal vapor deposition. *Reproduced with the permission from Song Z, Watthage SC, Phillips AB, Heben MJ. Pathways toward high-performance perovskite solar cells: review of recent advances in organo-metal halide perovskites for photovoltaic applications. J Photon Energy 2016;6:022001, copyright 2016.*

11.2.4 Thermal Vapor Deposition

The thermal evaporation method utilizes a dual source for MX_2 and MAI or FAI with different heating elements to form perovskite films [26]. This method provides high-quality perovskite thin film with a uniform thickness and of pin-hole-free composition. The first planar heterojunction $MAPbI_{3-x}Cl_x$ solar cell was fabricated in 2013 using thermal vapor deposition technique with efficiency of $>15\%$ [26]. One of the disadvantages of thermal vapor deposition technique is the requirement of precise control of temperature during deposition as both precursor sources and the products have low thermal stability. The two-step vapor-assisted deposition and thermal vapor deposition methods are described in Fig. 11.2.

11.3 Perovskite Solar Cell Device Structure

The hybrid organic–inorganic metal halide perovskite thin film solar cell device structure is just a modified version of the dye-sensitized solar cell (DSSCs) device structure [16,17]. In the DSSC process, a porous TiO_2 film is deposited onto SnO_2:F which is then coated with dye molecules and the system is dipped into a liquid electrolyte kept in metal electrode such as platinum (Pt) [54]. For the metal electrode, a separate metal (Pt) plate is made with a thin layer of iodine electrolyte spread over a conducting sheet. The two metal plates are then joined and sealed together to prevent the electrolyte from leaking. In 2012, Kim et al. [18] fabricated the first solid-state perovskite solar cells by depositing $MAPbI_3$ onto sub-micron thick mesoporous TiO_2 film and completed the solar cells by depositing a hole-transport layer spiro-MeOTAD and back contact metal, Au. This work by Kim et al. [18] was followed by another similar solar cell fabrication technique by Lee et al. [19] with an efficiency of 9.7%. These two demonstrations not only drew the attention of scientific communities to perovskite solar cells but also achieved several milestones in device performance using mesoporous structures [23,44,50,53,55]. Mesoporous solar cells require a high annealing temperature to fabricate TiO_2 films that could be time and energy consuming and also costly. As discussed in the Introduction section, $MAPbX_3$ has a relatively long charge carrier diffusion length indicating that both electrons and holes can easily be transported to their electrodes without the need of mesoporous TiO_2 [42]. Therefore, instead of being limiting to mesoporous structures, researchers started working in planar device structures. In a planar structure, the perovskite absorber layer is deposited onto a compact electron transport layer (ETL) instead of depositing onto a mesoporous layer. By a dual source vapor deposition process, planar perovskite structures achieved an efficiency of 15.4% in 2013; the present day highest efficiency from this type of structure is close to 20% [26,56]. In both types of device structures, solar cells can be illuminated either through ETL layer or the hole transport layer (HTL). If the solar cell is illuminated through the ETL, the device structure is called an n-i-p structure whereas if the solar cell is illuminated through the HTL, the device structure is called a p-i-n structure.

11.3.1 Mesoporous Scaffold Structure

Mesoporous structures can be divided into two categories depending on which direction the light incidents on the solar cells. The first use of perovskite solar cell was based on mesoporous n-i-p structure and is still widely used to fabricate high performance solar cells. In the mesoporous n-i-p structure, a layer of about 50–70 nm thick compact ETL (TiO_2) is deposited on the fluorine-doped tin oxide (FTO: SnO_2:F) coated glass substrate. The ETL is covered by a 150–300 nm thick mesoporous metal oxide (TiO_2 or Al_2O_3) filled with perovskite. About 300 nm perovskite is continuously deposited before depositing a 150–200 nm HTL (spiro-MeOTAD) and back contact metal (Au) as shown in Fig. 11.3A. An inverted p-i-n mesoporous structure in perovskite solar cells can also

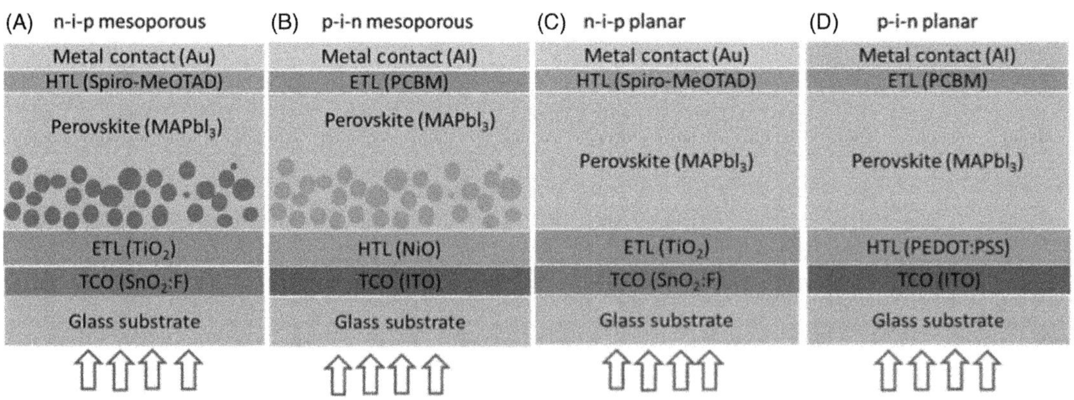

FIGURE 11.3 Schematic diagrams of perovskite solar cells in the (A) n-i-p mesoporous, (B) p-i-n mesoporous (C) n-i-p planar, and (D) p-i-n planar structures.

be achieved [57]. A typical device structure of mesoporous inverted p-i-n consists of the following arrangement, FTO/compact NiO_x/nanocrystal NiO/perovskite/PCBM/electrode (Fig. 11.3B).

In the mesoporous n-i-p structure, mesoporous and relative conductive TiO_2 or insulating Al_2O_3 scaffolds are used to facilitate electron transport between the perovskite absorber and the FTO electrode [19,58]. A complete pore-filling in the mesoporous structure is important in order to (1) prevent direct leakage of current between two contacts, (2) increase absorption of photons due to light scattering, and (3) enhance carrier collection [59]. For the support of light absorption with minimum shunting pathways, thin perovskite capping layer is always desired on top of the mesoporous structure. Also, if the mesoporous structure is thicker, the perovskite materials confined within the pores does not have enough space for the sufficient grain growth, thus ultimately reducing the device performance and lowering the V_{OC} and J_{SC} [60,61]. Therefore, the thickness of mesoporous TiO_2 not only determines the pore filling fraction and perovskite grains but also determines charge transport rate and collection efficiencies at the perovskite/TiO_2 interface. The development of the n-i-p mesoporous structure used to fabricate perovskite solar cells has been responsible for the increase in the solar cell efficiency from 3.8% to over 22%, which has taken place in just 8 years [6,16,44,50]. The device cross-section image, and current density–voltage (J–V) characteristics of a typical mesoporous structure is provided in Fig. 11.4 along with J–V parameters [62].

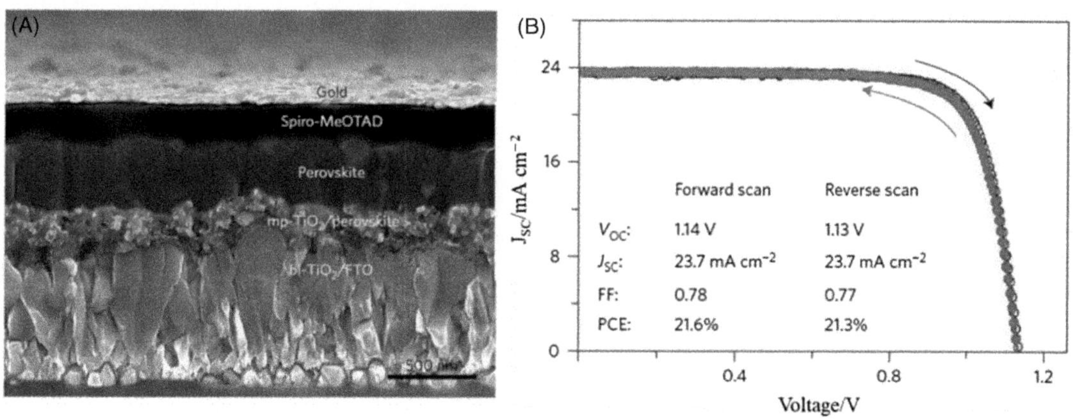

FIGURE 11.4 (A) A colored high-resolution cross-section Scanning electron microscopy (SEM) image of a complete solar cell fabricated by the polymer-templated nucleation and growth (PTNG) method, where the mesoporous TiO_2 layer (mp-TiO_2) is fully covered by a smooth and compact perovskite capping layer, (B) J–V curves of the champion solar cell prepared by PTNG method measured in both reverse and forward directions. *Adapted with permission from Bi D, Yi C, Luo J, Décoppet J-D, Zhang F, Zakeeruddin SM, Li X, Hagfeldt A, Grätzel M. Polymer-templated nucleation and crystal growth of perovskite films for solar cells with efficiency greater than 21%. Nat Energy 2016;1:16142, copyright 2016, Nature Publications.*

11.3.2 Planar Structure

In planar structure of perovskite solar cells, the mesoporous scaffold layer is removed and only the perovskite absorber layer is kept between the ETL and HTL. This planar structure can be considered as an evolution of the mesoporous structure. Its simplified device structure has attracted the interest of researchers working in the area of thin film PV cells. In the n-i-p planar structure, the solar cell is illuminated through ETL side whereas in the p-i-n planar structure, the solar cell is illuminated through HTL side as shown in Fig. 11.3C and D. The first successful demonstration of a planar structure only had an efficiency of 4% due to the inferior film quality and inadequate absorption of the perovskite film [63]. Today, the planar structure shows a similar efficiency performance to the mesoporous structure. The actual solar cell structure in the two cases, consists of glass/TCO/ETL/perovskite/HTL/metal and glass/TCO/HTL/perovskite/ETL/metal, respectively [26,43,47,63,64]. The efficiency of planar n-i-p solar cells has been increasing continuously with new developments and the present day highest efficiency is about 19% [50,65]. The J–V curves in perovskite solar cells do not coincide while scanning from forward to reverse and reverse to forward unlike in other regular solar cells such as CdTe, CGTS, and Si. This behavior is more severe in the n-i-p planar structure.

The inverted p-i-n planar structure resembles typical organic solar cells. The traditional organic transport layers such as [poly(3,4-ethylenedioxythiophene) polystyrene sulfonate] (PEDOT:PSS) and fullerence derivative [[6,6]-phenyl-C61-butyric acid methyl ester (PCBM)] are directly implemented as in the HTL and ETL layers of n-i-p perovskite solar cells. Similarly, instead of using FTO, tin-doped indium oxide (ITO) is preferred in the p-i-n structure as shown in Fig. 11.3. The efficiency of the p-i-n planar structure has been improving with the better selection of fullerence derivatives from the initial 3.9% efficiency to present day 18.9% [63,66–68]. The commonly used HTLs in the p-i-n structure are PEDOT: PSS, PTAA (poly-triarylamine), and NiO_x and ETLs are PCBM, $PC_{61}BM$, C_{60}, ZnO, and combinations of them [66,69–72]. The current–density voltage characteristics with J–V parameters and external quantum efficiency (EQE) data of perovskite solar cells in planar structure are shown in Fig. 11.5 [67].

11.4 Device Optimization

In the development of these solar cells, device optimization is an ongoing process and for efficient performance, each layer needs to be optimized. Since some of the layers in perovskite solar cells are already very well optimized, emphasis has been focused on the optimization of the three main layers such as the ETL, the perovskite absorber layer, and the HTL. This has achieved an efficiency of more than 22% over a short time period. This is a significant gain especially when compared to the progress made in other thin film technologies. As far as the optimization of the absorber layer is concerned, the focus has been on controlling the precursor solution, solution and film processing, perovskite composition, and the interface properties with the aim of obtaining smooth, pin-hole free perovskite films consisting of large grains with good crystallinity. The following optimization has been done to improve the absorber layer in perovskite solar cells.

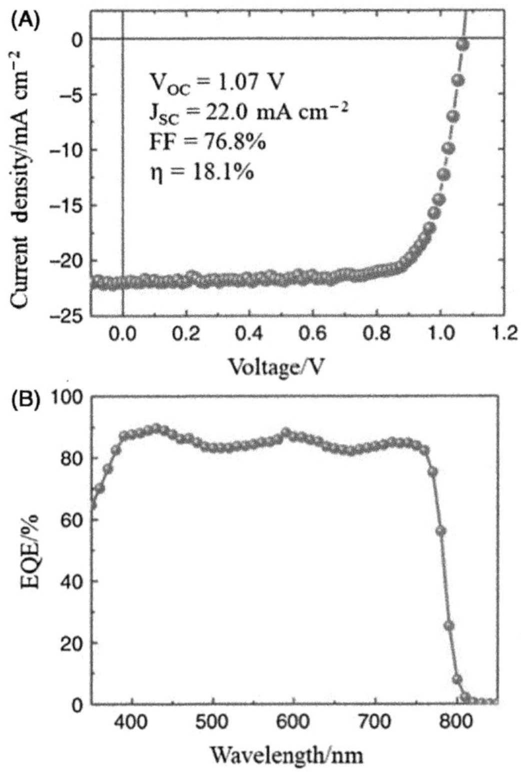

FIGURE 11.5 (A) The photocurrent of the $MAPbI_3$ solar cells in planar device structure under 1-sun illumination, (B) EQE of the solar cell with the active layer of $MAPbI_3$. *Adapted with the permission of Bi C, Wang Q, Shao Y, Yuan Y, Xiao Z, Huang J. Nonwetting surface-driven high-aspect-ratio crystalline grain growth for efficient hybrid perovskite solar cells. Nat Commun 2015;6, copyright 2015.*

11.4.1 Solvent to Film Optimization

The solvent engineering strategy of using mixed solvents and/or anti-solvents has become an effective but simple technique for realizing favorable perovskite film morphology in order to achieve high-performance perovskite solar cells [73,74]. In the perovskite film fabrication process, the precursor solution is prepared by dissolving in a mixture of dimethylsulfoxide (DMSO) and γ-butyrolactone (GBL) and the solution is spun coated by using one-step or two-step approach. DMSO is one of the commonly used solvent due to its strong coordination with MX_2 to form metal-halide complexes [44,75]. DMSO is not the only solvent that can produce high quality perovskite film as demonstrated by Wu et al. [76]. Instead of using DMSO/GBL combination, Kim et al. [77] used a combination of *N,N*-dimethylformamide (DMF) and GBL to prepare perovskite film with improved film quality. In the film fabrication process, application of antisolvent such as toluene is used to extract excess solvent such as DMSO and DMF for the rapid precipitation of the perovskite film before significant growth takes place [78]. The incorporation of toluene in film fabrication process produces small-grain and dense perovskite films.

Xiao et al. [79] demonstrated fast-deposition crystallization (FDC) procedure to yield highly uniform perovskite film using a spin coating of a DMF solution of $MAPbI_3$ followed immediately by the addition of chlorobenzene to induce crystallization. In this method, a rapid reduction of solubility of $MAPbI_3$ occurs with a fast nucleation and growth as a result of the addition of the second solution. The film fabricated by FDC produces a very large grain and full surface coverage in contrast to incomplete coverage obtained by conventional spin coating. A new concept of solvent–solvent extraction was proposed to fabricate high quality perovskite films at room temperature [80]. In this method, perovskite film is spun coated in a high boiling point solvent like *N*-methyl-2-pyrrolidone (NMP) and the wet film is immediately transferred into a low boiling point solvent like diethyl ether for the crystallization of smooth perovskite films. Other antisolvents used in fabricating high quality film are benzene and xylene [79]. Fig. 11.6 demonstrates morphological and structural characterization of $MAPbI_3$ films prepared by several deposition techniques.

FIGURE 11.6 Morphology control of hybrid organic–inorganic metal halide perovskite films: SEM images of $CH_3NH_3PbI_3$ obtained from (A) one-step spin coating, (B) two-steps spin coating, (C) one-step spin coating with toluene treatment, (D) thermal annealing of toluene treated film, (E) solvent–solvent extraction method, (F) vapor assisted deposition method, (G) vapor deposition method, (H) fast deposition crystallization method. *(D) Reprinted with permission from Lin Q, Armin A, Burn PL, Meredith P. Organohalide perovskites for solar energy conversion. Account Chem Res 2016:49:545–553, copyright (2016), American Chemical Society, (E) reprinted with permission from Zhou Y, Yang M, Wu W, Vasiliev AL, Zhu K, Padture NP. Room-temperature crystallization of hybrid-perovskite thin films via solvent–solvent extraction for high-performance solar cells. J Mater Chem A 2015;3:8178–84, copyright (2015) Royal Society of Chemistry, (F) Reprinted with permission of Chen Q, Zhou H, Hong Z, Luo S, Duan H-S, Wang H-H, Liu Y, Li G, Yang Y. Planar heterojunction perovskite solar cells via vapor-assisted solution process. J Am Chem Soc 2013; 136:622–5, copyright (2013) American Chemical Society, (G) reprinted with permission from Liu M, Johnston MB, Snaith HJ. Efficient planar heterojunction perovskite solar cells by vapour deposition. Nature 2013;501:395–8, copyright (2013) Nature Publishing Group, and (H) Reprinted with permission from Xiao M, Huang F, Huang W, Dkhissi Y, Zhu Y, Etheridge J, Gray-Weale A, Bach U, Cheng Y-B, Spiccia LA. Fast deposition-crystallization procedure for highly efficient lead iodide perovskite thin-film solar cells. Angew Chem Int Ed 2014;53:9898–903, copyright 2014, WILEY-VCH.*

The post-deposition annealing process serves to remove residual solvent from the solution-process, assist in perovskite formation from its precursors, and enhance crystallization and grain growth. A mixed halide perovskite like $MAPbI_{3-x}Cl_x$ has to be annealed longer, to accomplish full conversion, compared to the single halide system. These perovskite films are more stable at room temperature. The surface coverage of perovskite films decreases when the annealing takes a long time (e.g., more than 30 min at 110°C) at higher temperature due to the decomposition of the perovskite phase [51]. Annealing at higher temperatures (higher than 80°C) results in MAI loss and increases the relative PbI_2 content resulting in a decrease in device performance [43,82]. In general, annealing is done either in dry air or in an inert atmosphere like a nitrogen environment. Annealing of the perovskite films can also be conducted in pyridine or MAI vapor with the achievement of enhanced luminescence and carrier lifetime [83,84]. Similarly, Xiao et al. [85] conducted solvent annealing with DMF achieving improve crystallinity of perovskite films and enhancement of performance.

The annealing of the perovskite films discussed above is known as hotplate annealing under different atmospheres. Optical annealing approaches have also been reported for perovskite films using either a halogen lamp (near-infrared radiation) or a xenon lamp (photonic flash and intense pulse light) sources [86–88]. Optical annealing process is considered effective because heat light/radiation is absorbed by absorber layer as well as FTO layers. However, it was found that the power conversion efficiencies of perovskite solar cells are not as good as hot plate annealing processes. This might be because the temperature cannot be measured accurately in optical annealing method. Saliba et al. [89] introduced a concept of "flash" annealing, which was particularly effective for planar architecture cells, where the sample was annealed at high to a low temperature range at different time intervals.

11.4.2 Band Gap Optimization

It has been shown that chemical modification of the anions in organic–inorganic metal halide perovskite can tune the band gap over a wide range of the solar spectrum. For example, the band gap of the perovskite can be tuned to cover almost the entire visible range from 1.5 eV for $MAPbI_3$ to 2.3 eV for $MAPb(Br_xI_{1-x})_3$ when $x = 1$, with a significant modulation in photo-conversion efficiency (PCE) of perovskite solar cells [31,90]. The introduction of Br not only changes the band gap of the material but also makes the material more stable in humid condition. The band gap of perovskite can also be changed by the modification of the cation such as the replacement of methylammonium with a formamidinium ion (FA) to form formamidinium lead halide [$(HC(NH_2)_2PbI_3$] perovskite structure having a band gap of 1.48 eV [31,91,92]. The reduction in band gap in $FAPbI_3$ allows for absorption of photons over a wide range of the spectrum providing for a greater current collection. Instead of complete replacement, only a partial replacement of cation can also be done to change the band gap of the perovskite. Pellet et al. worked on an alloy of $MA_xFA_{1-x}PbI_3$ to extend the absorption to a longer spectral range and to enhance the thermal stability [93].

Similarly, Yang et al. fabricated high efficient solar cells using intramolecular exchange with the absorber: $(MAPbBr_3)_x(FAPb_3)_{1-x}$ [6].

Another way of changing the band gap of perovskite is by replacing toxic divalent metal (Pb) with a more environmentally friendly tin (Sn) in $MAMX_3$ [94,95]. Tin-based perovskites have promising photovoltaic properties like narrower optical band gap (~1.0 eV) and higher carrier mobility [96,97]. Germanium-based perovskite solar cell is equally promising. Mathews et al. [91] using a theoretical calculation showed that Ge can be an alternative replacement for Pb. Ternary bismuth halides perovskite material has also been reported as a replacement for toxic Pb-based perovskite but their results in PV applications are not yet available [98,99].

11.4.3 Electron and Hole Transporting Materials Optimization

Efficient charge extraction to the outer electrodes in perovskite solar cells is determined by the perovskite/ETL and perovskite/HTL interfaces. Proper ETL and HTL materials maintain a low surface and interface charge recombination which have high degree of charge selectivity. The most common metal oxide ETL essential for high efficient perovskite solar cells is mesoporous and planar TiO_2. The charge transfer rates from the perovskite absorber layer to TiO_2 ETL layer is very fast. At the same time, the electron recombination rates are also high in TiO_2 due to the low mobility and transport properties [100]. Zinc oxide, ZnO, nanorods and nanoparticles have been introduced as ETL layers in perovskite solar cells resulting in efficiencies as high as 11.1% and 15.9%, respectively [101,102]. Other metal oxides that have been utilized in mesoporous ETL layers in perovskite solar cells include Al_2O_3 [19,45], SiO_2 [103], ZrO_2 [104,105], and $SrTiO_3$ [106]. Zinc oxide has been used as a compact ETL layer in planar structure with an efficiency of 15.7% [24]. Baena et al. [107] fabricated high efficient planar perovskite solar cells of 18% efficiency using SnO_2 as ETL. Other ETL materials that have been included in a planar n-i-p structure format are CdS [108], CdSe [109], and TiO_2-graphene [110]. These are inorganic ETL materials for perovskite solar cells. ETL materials from organic solar cells are also widely used as ETLs in perovskite solar cells. Fullerence (C_{60}) and its derivatives such as [6,6]-phenyl-C61-butyric acid methyl ester ($PC_{61}BM$), indene-C60 bisadduct (ICBA), and $PC_{71}BM$ are ideal candidates as efficient electron extraction materials due to their low temperature fabrication, suitable energy level alignment and decent carrier mobility [63,111,112].

The first solid state perovskite solar cells were fabricated by the substitution of liquid electrolyte with 2,2′,7,7′-tetrakis(*N*,*N*-di-p-methoxyphenylamine)-9,9′spirobifluorene (spiro-OMeTAD) as the HTL, achieving an efficiency of 10% in the mesoporous structure [18,19]. Since then, this small molecular organic material has become the most commonly used HTL in high efficient perovskite solar cells. The preparation of the spirobifluorene core in spiro-OMeTAD molecule requires an extensive synthetic process and is responsible for an increase in the production costs. In the process of finding cost effective HTL materials for efficient carrier transport in perovskite solar cells, researchers have used the organic polymer, poly(triarylamine) (PTAA) [6], 3,4-ethylenedioxythiophene (EDOT) [113], and many other organic materials; the full list is provided elsewhere [114]. Even though these organic HTL materials have good transport properties, they unfortunately have poor stability. Inorganic HTL materials, on the other hand, are cost effective and demonstrate long-term stability although the device efficiency is lower than using organic HTLs. Examples of inorganic HTL materials are CuI [115], Cu:NiOx [116], NiO [117], CuSCN [118,119], and iron pyrite [120].

11.5 Stability Issues and Challenges of Perovskite Solar Cells

11.5.1 Stability Issues

The present research on perovskite solar cells mainly focused on material design, novel cell structures, and the underlying mechanisms. The issues of degradation of perovskite and the stability of the devices are huge challenges to the PV communities. It is very urgent to address these challenges to achieve good reproducibility and long lifetime for solar cells. Organo-metallic halide perovskite undergoes series of chemical reactions even under ambient atmospheric conditions and either decomposes into their components or the film can directly degrade into other chemicals. Niu et al. [121] identified four factors responsible for the degradation of perovskite films such as oxygen and moisture, UV light, solution processing, and temperatures.

Since the material is quite sensitive to oxygen and moisture, most of the fabrication processes are conducted in an inert atmosphere in a glove box. Significant degradation occurs in the solar cells during their testing under ambient conditions. The report by Seok et al. [90] indicates that degradation of perovskite film starts at a humidity of 55% and higher, displaying a color change from dark brown to yellow. This degradation prevents perovskite solar cells for outdoor applications. The degradation of perovskite film caused by oxygen and moisture is irreversible [121]. Niu et al. [121] found that absorption of $TiO_2/MAPbI_3$ film in the spectral range of 530–800 nm is greatly reduced after exposure to air with a humidity of 60% at 35°C for 18 h and the material's X-ray diffraction (XRD) peaks had completely disappeared. Leijtens et al. [61] demonstrated that the cause of degradation of perovskite solar cells is due to the degradation of TiO_2 in UV light. Thermal stability was tested for semi-finished perovskite solar cells by annealing them at 85°C for 24 h; they found only the PbI_2 remaining when analysed using XRD [122].

Even with these challenges, efficient and stable solar cells have been demonstrated in recent years with materials and interface engineering [8,105,123–125]. Kim et al. [18] reported stable performances of solar cells over 500 h for devices stored in air at room temperature and occasional exposed to air mass 1.5 global (AM1.5G) light illumination. The stability of perovskite solar cells under high humidity and temperature conditions has been improved by employing a moisture-resistant layer to prevent water entrance [104,126,127]. Leijtens et al. [128] demonstrated encapsulation techniques using glass sealing or laminate plastic films to improve device stability to over 125 days at 60°C under simulated sunlight. Stability of perovskite solar cells could also be improved by compositional engineering of the films [31,90,129].

11.5.2 J–V Hysteresis

Standard solar cells such as CdTe and Si, do not change the nature of the current density–voltage (J–V) curves while scanning from forward bias to reverse bias voltage and vice versa, but perovskite solar cells display anomalous J–V hysteresis phenomenon [130]. The variation in J–V characteristics depends on the direction and the rate of the voltage sweep. This lack of accuracy in J–V measurement poses a great concern in scientific community as it leads to inaccurate device efficiencies. The origin of hysteresis is still open for discussion although ferroelectricity [131,132], ion migration [133,134], and unbalanced charge collection rates [135] have been proposed as possible reasons. The application of mesoporous TiO_2 considerably weakens the hysteresis effects, while the equivalent planar devices struggle with severe hysteresis due to the induce dipole polarization in perovskite films as shown in Fig. 11.7 [44,130,136]. However, the degree of hysteresis depends on the thickness of the mesoporous TiO_2 layer [130]. The charge transfer between the perovskite and planar TiO_2 is not efficient in comparison with mesoporous TiO_2 layer. The hysteresis in

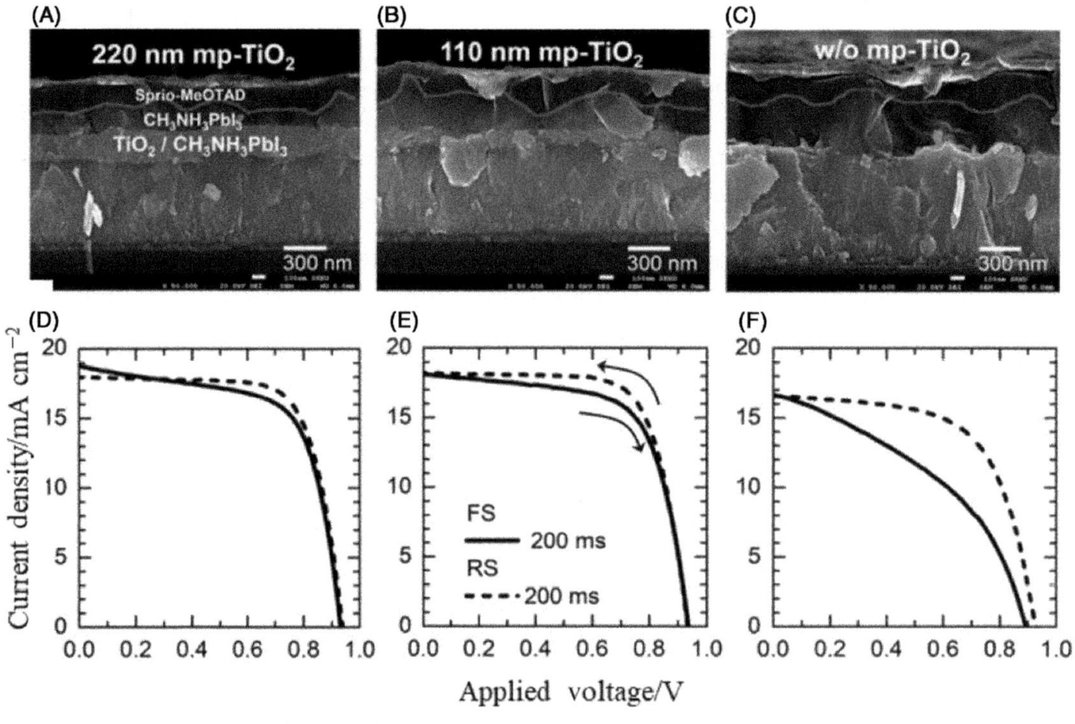

FIGURE 11.7 SEM Cross-sectional images of mesoporous structured solar cells with (A) 220 nm, (B) 110 nm thick mp-TiO_2, (C) planar structured solar cell without mp-TiO_2. J–V characteristics measured at forward bias scan (FS) and reverse bias scan (RS) for mesostructured devices, (D) 220 nm, (E) 110 nm thick mp-TiO_2, and (F) planar structured solar cell without mp-TiO_2. The solar cells were illuminated with AM1.5G illumination at a voltage setting time of 200 ms. *Adapted with permission from Kim H-S, Park N-G. Parameters affecting I–V hysteresis of $CH_3NH_3PbI_3$ perovskite solar cells: effects of perovskite crystal size and mesoporous TiO_2 layer. J Phys Chem Lett 2014;5:2927–9, copyright 2014, American Chemical Society.*

planar devices can be reduced by passivating the surface defects of the perovskite layer with treatments of pyridine or thiophene [111]. Surprisingly, hysteresis was not observed for inverted planar perovskite solar cells based on PEDOT:PSS/perovskite/$PC_{60}BM$ with high efficiencies [85,111,137]. However, substantial J–V hysteresis was observed when the temperature of hysteresis-free inverted planar solar cells was reduced to 175 K [138] or when the devices were older than a week [133]. From the discussion, we conclude that hysteresis in perovskite solar cells does not depend on the type of device architecture, but it can be minimized by improving the stability of the perovskite and the engineering of the interfaces. The device cross-sectional imaging and corresponding J–V characteristics of perovskite solar cells showing hysteresis behavior are presented in Fig. 11.7 [136].

11.6 Summary

The film formation process of perovskite plays a critical role in determining device performances. Perovskite thin films can be prepared by various techniques; the main techniques include one-step solution, two-step sequential deposition processes, vapor deposition, and vapor assisted solution processing. The perovskite film morphology is crucially important for device performance. It can be optimized by various approaches such as the use of additives, thermal annealing, solvent annealing, atmospheric control, and solvent engineering. During film fabrication, control over crystallization, and growth of perovskites is required to achieve uniform film with full surface coverage, large crystal size, and even good stability. Improved processing of film fabrication with proper morphological control has already created great success in perovskite PV, and it is believed that this will lead to a significant breakthrough in developing useful and successful perovskite solar cells in the future.

High performance solar cells are based on mesoporous oxide scaffolds because of their good film quality. However, the high temperature annealing processes required in mesoporous film formation increases processing complexity and cost. The fabrication of perovskite solar cells using mesoporous structures might mitigate against the compatibility of implementing a high performance flexible product with the integration of tandem cells into commercialized existing technologies. However, the planar device configuration approach is considered an appropriate technological path for the fabrication of high performance flexible solar cells. It is believed that numerous research efforts within the perovskite PV community will, in the near future, solve the prevailing stability issues resulting in the practical commercialization of perovskite solar cells.

References

[1] International Energy Outlook 2016. US Energy Information Administration, 2016.

[2] Schmela M: *Global market outlook for solar power/2016-2020*, Brussels, Belgium, 2016, Solar power Europe.

[3] Bhandari KP, Collier JM, Ellingson RJ, Apul DS: Energy payback time (EPBT) and energy return on energy invested (EROI) of solar photovoltaic systems: a systematic review and meta-analysis, *Renew Sust Energy Rev* 47:133–141, 2015.

[4] Song Z, McElvany CL, Phillips AB, Celik I, Krantz PW, Watthage SC, et al: A technoeconomic analysis of perovskite solar module manufacturing with low-cost materials and techniques, *Energy Environ Sci* 10(6), 2017.

[5] Wallace SK, Mitzi DB, Walsh A: The steady rise of kesterite solar cells, *ACS Energy Lett* 2:776–779, 2017.

[6] Yang WS, Noh JH, Jeon NJ, Kim YC, Ryu S, Seo J, et al: High-performance photovoltaic perovskite layers fabricated through intramolecular exchange, *Science* 348:1234–1237, 2015.

[7] Chiang C-H, Nazeeruddin MK, Grätzel M, Wu C-G: The synergistic effect of H_2O and DMF towards stable and 20% efficiency inverted perovskite solar cells, *Energy Environ Sci* 10:808–817, 2017.

[8] Tan H, Jain A, Voznyy O, Lan X, García de Arquer FP, Fan JZ, et al: Efficient and stable solution-processed planar perovskite solar cells via contact passivation, *Science* 355:722–726, 2017.

[9] Schmidt LJ. Tracking down the truth of perovski. In: 38th Rochester mineralogical symposium program notes; 2011, p. 31–2.

[10] Mitzi DB: Synthesis, crystal structure, and optical and thermal properties of $(C_4H_9NH_3)_2M_{I4}$ (M = Ge, Sn, Pb), *Chem Mater* 8:791–800, 1996.

[11] Mitzi DB, Prikas M, Chondroudis K: Thin film deposition of organic–inorganic hybrid materials using a single source thermal ablation technique, *Chem Mater* 11:542–544, 1999.

[12] Mitzi DB, Dimitrakopoulos CD, Kosbar LL: Structurally tailored organic–inorganic perovskites: optical properties and solution-processed channel materials for thin-film transistors, *Chem Mater* 13:3728–3740, 2001.

[13] Chondroudis K, Mitzi DB: Electroluminescence from an organic–inorganic perovskite incorporating a quaterthiophene dye within lead halide perovskite layers, *Chem Mater* 11:3028–3030, 1999.

[14] Mitzi DB, Chondroudis K, Kagan CR: Organic-inorganic electronics, *IBM J Res Dev* 45:29–45, 2001.

[15] Wang D, Wright M, Elumalai NK, Uddin A: Stability of perovskite solar cells, *Sol Energy Mater Sol Cells* 147:255–275, 2016.

[16] Kojima A, Teshima K, Shirai Y, Miyasaka T: Organometal halide perovskites as visible-light sensitizers for photovoltaic cells, *J Am Chem Soc* 131:6050–6051, 2009.

[17] Im J-H, Lee C-R, Lee J-W, Park S-W, Park N-G: 6.5% efficient perovskite quantum-dot-sensitized solar cell, *Nanoscale* 3:4088–4093, 2011.

[18] Kim H-S, Lee C-R, Im J-H, Lee K-B, Moehl T, Marchioro A, et al: Lead iodide perovskite sensitized all-solid-state submicron thin film mesoscopic solar cell with efficiency exceeding 9%, *Sci Rep* 2:591, 2012.

[19] Lee MM, Teuscher J, Miyasaka T, Murakami TN, Snaith HJ: Efficient hybrid solar cells based on meso-superstructured organometal halide perovskites, *Science* 338:643–647, 2012.

[20] Newcomer Juices Up the Race to Harness Sunlight. Science 2013;342:1438–9.

[21] Shin SS, Yeom EJ, Yang WS, Hur S, Kim MG, Im J, et al: Colloidally prepared La-doped $BaSnO_3$ electrodes for efficient, photostable perovskite solar cells, *Science* 356:167–171, 2017.

[22] Ummadisingu A, Steier L, Seo J-Y, Matsui T, Abate A, Tress W, et al: The effect of illumination on the formation of metal halide perovskite films, *Nature* 545(7653), 2017, Advance online publication.

[23] Burschka J, Pellet N, Moon S-J, Humphry-Baker R, Gao P, Nazeeruddin MK, et al: Sequential deposition as a route to high-performance perovskite-sensitized solar cells, *Nature* 499:316–319, 2013.

[24] Liu D, Kelly TL: Perovskite solar cells with a planar heterojunction structure prepared using room-temperature solution processing techniques, *Nat Photonics* 8:133–138, 2014.

[25] Chen Q, Zhou H, Hong Z, Luo S, Duan H-S, Wang H-H, et al: Planar heterojunction perovskite solar cells via vapor-assisted solution process, *J Am Chem Soc* 136:622–625, 2013.

[26] Liu M, Johnston MB, Snaith HJ: Efficient planar heterojunction perovskite solar cells by vapour deposition, *Nature* 501:395–398, 2013.

[27] Chen C-W, Kang H-W, Hsiao S-Y, Yang P-F, Chiang K-M, Lin H-W: Efficient and uniform planar-type perovskite solar cells by simple sequential vacuum deposition, *Adv Mater* 26:6647–6652, 2014.

[28] Yin W-J, Shi T, Yan Y: Unique properties of halide perovskites as possible origins of the superior solar cell performance, *Adv Mater* 26:4653–4658, 2014.

[29] Solomon M, Johnson A: New research in solar cells: urbach tails and open circuit voltage, *Elem Boston Coll Undergrad Res J*:11, 2015.

[30] De Wolf S, Holovsky J, Moon S-J, Löper P, Niesen B, Ledinsky M, et al: Organometallic halide perovskites: sharp optical absorption edge and its relation to photovoltaic performance, *J Phys Chem Lett* 5:1035–1039, 2014.

[31] Eperon GE, Stranks SD, Menelaou C, Johnston MB, Herz LM, Snaith HJ: Formamidinium lead trihalide: a broadly tunable perovskite for efficient planar heterojunction solar cells, *Energy Environ Sci* 7:982–988, 2014.

[32] Li Y, Ding B, Chu Q-Q, Yang G-J, Wang M, Li C-X, et al: Ultra-high open-circuit voltage of perovskite solar cells induced by nucleation thermodynamics on rough substrates, *Sci Rep* 7:46141, 2017.

[33] Tress W, Marinova N, Inganäs O, Nazeeruddin MK, Zakeeruddin SM, Graetzel M: Predicting the open-circuit voltage of $CH_3NH_3PbI_3$ perovskite solar cells using electroluminescence and photovoltaic quantum efficiency spectra: the role of radiative and non-radiative recombination, *Adv Energy Mater* 5, 2015, 1400812.

[34] Wehrenfennig C, Eperon GE, Johnston MB, Snaith HJ, Herz LM: High charge carrier mobilities and lifetimes in organolead trihalide perovskites, *Adv Mater* 26:1584–1589, 2014.

[35] Leijtens T, Stranks SD, Eperon GE, Lindblad R, Johansson EMJ, McPherson IJ, et al: Electronic properties of meso-superstructured and planar organometal halide perovskite films: charge trapping, photodoping, and carrier mobility, *ACS Nano* 8:7147–7155, 2014.

[36] Shi D, Adinolfi V, Comin R, Yuan M, Alarousu E, Buin A, et al: Low trap-state density and long carrier diffusion in organolead trihalide perovskite single crystals, *Science* 347:519–522, 2015.

[37] Kiermasch D, Rieder P, Tvingstedt K, Baumann A, Dyakonov V: Improved charge carrier lifetime in planar perovskite solar cells by bromine doping, *Sci Rep* 6:39333, 2016.

[38] Stranks SD, Eperon GE, Grancini G, Menelaou C, Alcocer MJ, Leijtens T, et al: Electron-hole diffusion lengths exceeding 1 micrometer in an organometal trihalide perovskite absorber, *Science* 342:341–344, 2013.

[39] Chen Y, Yi H, Wu X, Haroldson R, Gartstein Y, Rodionov Y, et al: Extended carrier lifetimes and diffusion in hybrid perovskites revealed by Hall effect and photoconductivity measurements, *Nat Commun*:7, 2016.

[40] Salim T, Sun S, Abe Y, Krishna A, Grimsdale AC, Lam YM: Perovskite-based solar cells: impact of morphology and device architecture on device performance, *J Mater Chem A* 3:8943–8969, 2015.

[41] Shi Z, Guo J, Chen Y, Li Q, Pan Y, Zhang H, et al: Lead-free organic–inorganic hybrid perovskites for photovoltaic applications: recent advances and perspectives, *Adv Mater* 29, 2017, 1605005.

[42] Xing G, Mathews N, Sun S, Lim SS, Lam YM, Grätzel M, et al: Long-range balanced electron-and hole-transport lengths in organic-inorganic CH3NH3PbI3, *Science* 342:344–347, 2013.

[43] Eperon GE, Burlakov VM, Docampo P, Goriely A, Snaith HJ: Morphological control for high performance, solution processed planar heterojunction perovskite solar cells, *Adv Funct Mater* 24:151–157, 2014.

[44] Jeon NJ, Noh JH, Kim YC, Yang WS, Ryu S, Seok SI: Solvent engineering for high-performance inorganic–organic hybrid perovskite solar cells, *Nat Mater* 13:897–903, 2014.

[45] Ball JM, Lee MM, Hey A, Snaith HJ: Low-temperature processed meso-superstructured to thin-film perovskite solar cells, *Energy Environ Sci* 6:1739–1743, 2013.

[46] You J, Hong Z, Yang YM, Chen Q, Cai M, Song T-B, et al: Low-temperature solution-processed perovskite solar cells with high efficiency and flexibility, *ACS Nano* 8(2):1739–1780, 2014.

[47] Song Z, Watthage SC, Phillips AB, Heben MJ: Pathways toward high-performance perovskite solar cells: review of recent advances in organo-metal halide perovskites for photovoltaic applications, *J Photon Energy* 6:022001–122001, 2016.

[48] Yu Y, Wang C, Grice CR, Shrestha N, Chen J, Zhao D, et al: Improving the Performance of formamidinium and cesium lead triiodide perovskite solar cells using lead thiocyanate additives, *ChemSusChem* 9:3288–3297, 2016.

[49] Roldan-Carmona C, Gratia P, Zimmermann I, Grancini G, Gao P, Graetzel M, et al: High efficiency methylammonium lead triiodide perovskite solar cells: the relevance of non-stoichiometric precursors, *Energy Environ Sci* 8:3550–3556, 2015.

[50] Zhou H, Chen Q, Li G, Luo S, Song T-b, Duan H-S, et al: Interface engineering of highly efficient perovskite solar cells, *Science* 345:542–546, 2014.

[51] Song Z, Watthage SC, Phillips AB, Tompkins BL, Ellingson RJ, Heben MJ: Impact of processing temperature and composition on the formation of methylammonium lead iodide perovskites, *Chem Mater* 27:4612–4619, 2015.

[52] Wang Q, Shao Y, Dong Q, Xiao Z, Yuan Y, Huang J: Large fill-factor bilayer iodine perovskite solar cells fabricated by a low-temperature solution-process, *Energy Environ Sci* 7:2359–2365, 2014.

[53] Im J-H, Jang I-H, Pellet N, Grätzel M, Park N-G: Growth of $CH_3NH_3PbI_3$ cuboids with controlled size for high-efficiency perovskite solar cells, *Nat Nanotechnol* 9:927–932, 2014.

[54] O'regan B, Grfitzeli M: A low-cost, high-efficiency solar cell based on dye-sensitized, *Nature* 353:737–740, 1991.

[55] You J, Yang Y, Hong Z, Song T-B, Meng L, Liu Y, et al: Moisture assisted perovskite film growth for high performance solar cells, *Appl Phys Lett* 105:183902, 2014.

[56] Bai Y, Dong Q, Shao Y, Deng Y, Wang Q, Shen L, et al: Enhancing stability and efficiency of perovskite solar cells with crosslinkable silane-functionalized and doped fullerene, *Nat Commun* 7:12806, 2016.

[57] Wang K-C, Jeng J-Y, Shen P-S, Chang Y-C, Diau EW-G, Tsai C-H, et al: p-Type mesoscopic nickel oxide/organometallic perovskite heterojunction solar cells, *Sci Rep* 4:4756, 2014.

[58] Grätzel C, Zakeeruddin SM: Recent trends in mesoscopic solar cells based on molecular and nanopigment light harvesters, *Mater Today* 16:11–18, 2013.

[59] Leijtens T, Lauber B, Eperon GE, Stranks SD, Snaith HJ: The importance of perovskite pore filling in organometal mixed halide sensitized TiO_2-based solar cells, *J Phys Chem Lett* 5:1096–1102, 2014.

[60] Choi JJ, Yang X, Norman ZM, Billinge SJ, Owen JS: Structure of methylammonium lead iodide within mesoporous titanium dioxide: active material in high-performance perovskite solar cells, *Nano Lett* 14:127–133, 2013.

[61] Leijtens T, Eperon GE, Pathak S, Abate A, Lee MM, Snaith HJ: Overcoming ultraviolet light instability of sensitized TiO_2 with meso-superstructured organometal tri-halide perovskite solar cells, *Nat Commun*:4, 2013.

[62] Bi D, Yi C, Luo J, Décoppet J-D, Zhang F, Zakeeruddin SM, et al: Polymer-templated nucleation and crystal growth of perovskite films for solar cells with efficiency greater than 21%, *Nat Energy* 1:16142, 2016.

[63] Jeng JY, Chiang YF, Lee MH, Peng SR, Guo TF, Chen P, et al: $CH_3NH_3PbI_3$ perovskite/fullerene planar-heterojunction hybrid solar cells, *Adv Mater* 25:3727–3732, 2013.

[64] Jeng JY, Chen KC, Chiang TY, Lin PY, Tsai TD, Chang YC, et al: Nickel oxide electrode interlayer in $CH_3NH_3PbI_3$ perovskite/PCBM planar-heterojunction hybrid solar cells, *Adv Mater* 26:4107–4113, 2014.

[65] Li H, Shi W, Huang W, Yao E-P, Han J, Chen Z, et al: Carbon quantum dots/TiOx electron transport layer boosts efficiency of planar heterojunction perovskite solar cells to 19%, *Nano Lett* 17:2328–2335, 2017.

[66] Heo JH, Han HJ, Kim D, Ahn TK, Im SH: Hysteresis-less inverted $CH_3NH_3PbI_3$ planar perovskite hybrid solar cells with 18.1% power conversion efficiency, *Energy Environ Sci* 8:1602–1608, 2015.

[67] Bi C, Wang Q, Shao Y, Yuan Y, Xiao Z, Huang J: Non-wetting surface-driven high-aspect-ratio crystalline grain growth for efficient hybrid perovskite solar cells, *Nat Commun*:6, 2015.

[68] Dong Q, Yuan Y, Shao Y, Fang Y, Wang Q, Huang J: Abnormal crystal growth in $CH_3NH_3PbI_{3-x}Cl_x$ using a multi-cycle solution coating process, *Energy Environ Sci* 8:2464–2470, 2015.

[69] Nie W, Tsai H, Asadpour R, Blancon J-C, Neukirch AJ, Gupta G, et al: High-efficiency solution-processed perovskite solar cells with millimeter-scale grains, *Science* 347:522–525, 2015.

[70] You J, Meng L, Song T-B, Guo T-F, Yang YM, Chang W-H, et al: Improved air stability of perovskite solar cells via solution-processed metal oxide transport layers, *Nat Nanotechnol* 11:75–81, 2016.

[71] Liang PW, Chueh CC, Williams ST, Jen AKY: Roles of fullerene- based interlayers in enhancing the performance of organometal perovskite thin- film solar cells, *Adv Energy Mater*:5, 2015.

[72] Park JH, Seo J, Park S, Shin SS, Kim YC, Jeon NJ, et al: Efficient $CH_3NH_3PbI_3$ perovskite solar cells employing nanostructured p- type NiO electrode formed by a pulsed laser deposition, *Adv Mater* 27:4013–4019, 2015.

[73] Rong Y, Tang Z, Zhao Y, Zhong X, Venkatesan S, Graham H, et al: Solvent engineering towards controlled grain growth in perovskite planar heterojunction solar cells, *Nanoscale* 7:10595–10599, 2015.

[74] Chen H, Wei Z, He H, Zheng X, Wong KS, Yang S: Solvent engineering boosts the efficiency of paintable carbon-based perovskite solar cells to beyond 14%, *Adv Energy Mater* 6, 2016, 1502087.

[75] Li W, Fan J, Li J, Mai Y, Wang L: Controllable grain morphology of perovskite absorber film by molecular self-assembly toward efficient solar cell exceeding 17%, *J Am Chem Soc* 137:10399–10405, 2015.

[76] Wu Y, Islam A, Yang X, Qin C, Liu J, Zhang K, et al: Retarding the crystallization of PbI_2 for highly reproducible planar-structured perovskite solar cells via sequential deposition, *Energy Environ Sci* 7:2934–2938, 2014.

[77] Kim H-B, Choi H, Jeong J, Kim S, Walker B, Song S, et al: Mixed solvents for the optimization of morphology in solution-processed, inverted-type perovskite/fullerene hybrid solar cells, *Nanoscale* 6:6679–6683, 2014.

[78] Sakai N, Pathak S, Chen H-W, Haghighirad AA, Stranks SD, Miyasaka T, et al: The mechanism of toluene-assisted crystallization of organic–inorganic perovskites for highly efficient solar cells, *J Mater Chem A* 4:4464–4471, 2016.

[79] Xiao M, Huang F, Huang W, Dkhissi Y, Zhu Y, Etheridge J, et al: A fast deposition-crystallization procedure for highly efficient lead iodide perovskite thin-film solar cells, *Angew Chem Int Ed* 53:9898–9903, 2014.

[80] Zhou Y, Yang M, Wu W, Vasiliev AL, Zhu K, Padture NP: Room-temperature crystallization of hybrid-perovskite thin films via solvent–solvent extraction for high-performance solar cells, *J Mater Chem A* 3:8178–8184, 2015.

[81] Lin Q, Armin A, Burn PL, Meredith P: Organohalide perovskites for solar energy conversion, *Acc Chem Res* 49:545–553, 2016.

[82] Dualeh A, Tétreault N, Moehl T, Gao P, Nazeeruddin MK, Grätzel M: Effect of annealing temperature on film morphology of organic–inorganic hybrid pervoskite solid- state solar cells, *Adv Funct Mater* 24:3250–3258, 2014.

[83] de Quilettes DW, Vorpahl SM, Stranks SD, Nagaoka H, Eperon GE, Ziffer ME, et al: Impact of microstructure on local carrier lifetime in perovskite solar cells, *Science* 348:683–686, 2015.

[84] Tosun BS, Hillhouse HW: Enhanced carrier lifetimes of pure iodide hybrid perovskite via vapor-equilibrated re-growth (VERG), *J Phys Chem Lett* 6:2503–2508, 2015.

[85] Xiao Z, Dong Q, Bi C, Shao Y, Yuan Y, Huang J: Solvent annealing of perovskite- induced crystal growth for photovoltaic- device efficiency enhancement, *Adv Mater* 26:6503–6509, 2014.

[86] Troughton J, Charbonneau C, Carnie MJ, Davies ML, Worsley DA, Watson TM: Rapid processing of perovskite solar cells in under 2.5 seconds, *J Mater Chem A* 3:9123–9127, 2015.

[87] Troughton J, Carnie MJ, Davies ML, Charbonneau C, Jewell EH, Worsley DA, et al: Photonic flash-annealing of lead halide perovskite solar cells in 1 ms, *J Mater Chem A* 4:3471–3476, 2016.

[88] Lavery BW, Kumari S, Konermann H, Draper GL, Spurgeon J, Druffel T: Intense pulsed light sintering of $CH_3NH_3PbI_3$ solar cells, *ACS Appl Mater Interfaces* 8:8419–8426, 2016.

[89] Saliba M, Tan KW, Sai H, Moore DT, Scott T, Zhang W, et al: Influence of thermal processing protocol upon the crystallization and photovoltaic performance of organic–inorganic lead trihalide perovskites, *J Phys Chem C*, 2014.

[90] Noh JH, Im SH, Heo JH, Mandal TN, Seok SI: Chemical management for colorful, efficient, and stable inorganic–organic hybrid nanostructured solar cells, *Nano Lett* 13:1764–1769, 2013.

[91] Koh TM, Fu K, Fang Y, Chen S, Sum TC, Mathews N, et al: Formamidinium-containing metal-halide: an alternative material for near-IR absorption perovskite solar cells, *J Phys Chem C* 118:16458–16462, 2014.

[92] Stoumpos CC, Malliakas CD, Kanatzidis MG: Semiconducting tin and lead iodide perovskites with organic cations: phase transitions, high mobilities, and near-infrared photoluminescent properties, *Inorg Chem* 52:9019–9038, 2013.

[93] Pellet N, Gao P, Gregori G, Yang T-Y, Nazeeruddin MK, Maier J, et al: Mixed-organic-cation perovskite photovoltaics for enhanced solar-light harvesting, *Angew Chem Int Ed* 53:3151–3157, 2014.

[94] Hao F, Stoumpos CC, Cao DH, Chang RPH, Kanatzidis MG: Lead-free solid-state organic-inorganic halide perovskite solar cells, *Nat Photonics* 8:489–494, 2014.

[95] Noel NK, Stranks SD, Abate A, Wehrenfennig C, Guarnera S, Haghighirad A-A, et al: Lead-free organic-inorganic tin halide perovskites for photovoltaic applications, *Energy Environ Sci* 7:3061–3068, 2014.

[96] Hao F, Stoumpos CC, Guo P, Zhou N, Marks TJ, Chang RPH, et al: Solvent-mediated crystallization of $CH_3NH_3SnI_3$ films for heterojunction depleted perovskite solar cells, *J Am Chem Soc* 137:11445–11452, 2015.

[97] Umari P, Mosconi E, De Angelis F: Relativistic GW calculations on $CH_3NH_3PbI_3$ and $CH_3NH_3SnI_3$ perovskites for solar cell applications, *Sci Rep*:4, 2014.

[98] Mel'nikova SV, Zaitsev AI: Ferroelectric phase transition in Cs_3Bi_{2I9}, *Phys Solid State* 39:1652–1654, 1997.

[99] Ivanov YN, Sukhovskii AA, Lisin VV, Aleksandrova IP: Phase transitions of Cs_3Sb_{2I9}, Cs_3Bi_{2I9}, and Cs-$_3Bi_2Br_9$ crystals, *Inorg Mater* 37:623–627, 2001.

[100] Gubbala S, Chakrapani V, Kumar V, Sunkara MK: Band- edge engineered hybrid structures for dye-sensitized solar cells based on SnO_2 nanowires, *Adv Funct Mater* 18:2411–2418, 2008.

[101] Son D-Y, Im J-H, Kim H-S, Park N-G: 11% efficient perovskite solar cell based on ZnO nanorods: an effective charge collection system, *J Phys Chem C* 118:16567–16573, 2014.

[102] Zhang R, Fei C, Li B, Fu H, Tian J, Cao G: Continuous size tuning of monodispersed ZnO nanoparticles and its size effect on the performance of perovskite solar cells, *ACS Appl Mater Interfaces* 9:9785–9794, 2017.

[103] Yun J, Ryu J, Lee J, Yu H, Jang J: SiO_2/TiO_2 based hollow nanostructures as scaffold layers and Al-doping in the electron transfer layer for efficient perovskite solar cells, *J Mater Chem A* 4:1306–1311, 2016.

[104] Mei A, Li X, Liu L, Ku Z, Liu T, Rong Y, et al: A hole-conductor–free, fully printable mesoscopic perovskite solar cell with high stability, *Science* 345:295–298, 2014.

[105] Hashmi SG, Martineau D, Li X, Ozkan M, Tiihonen A, Dar MI, et al: Air processed inkjet infiltrated carbon based printed perovskite solar cells with high stability and reproducibility, *Adv Mater Technol* 2:1600183, 2017.

[106] Bera A, Wu K, Sheikh A, Alarousu E, Mohammed OF, Wu T: Perovskite oxide $SrTiO_3$ as an efficient electron transporter for hybrid perovskite solar cells, *J Phys Chem C* 118:28494–28501, 2014.

[107] Baena JPC, Steier L, Tress W, Saliba M, Neutzner S, Matsui T, et al: Highly efficient planar perovskite solar cells through band alignment engineering, *Energy Environ Sci* 8:2928–2934, 2015.

[108] Liu J, Gao C, Luo L, Ye Q, He X, Ouyang L, et al: Low-temperature, solution processed metal sulfide as an electron transport layer for efficient planar perovskite solar cells, *J Mater Chem A* 3:11750–11755, 2015.

[109] Wang L, Fu W, Gu Z, Fan C, Yang X, Li H, et al: Low temperature solution processed planar heterojunction perovskite solar cells with a CdSe nanocrystal as an electron transport/extraction layer, *J Mater Chem C* 2:9087–9090, 2014.

[110] Wang JT-W, Ball JM, Barea EM, Abate A, Alexander-Webber JA, Huang J, et al: Low-temperature processed electron collection layers of graphene/TiO_2 nanocomposites in thin film perovskite solar cells, *Nano Lett* 14:724–730, 2013.

[111] Chiang C-H, Tseng Z-L, Wu C-G: Planar heterojunction perovskite/$PC_{71}BM$ solar cells with enhanced open-circuit voltage via a (2/1)-step spin-coating process, *J Mater Chem A* 2:15897–15903, 2014.

[112] Docampo P, Ball JM, Darwich M, Eperon GE, Snaith HJ: Efficient organometal trihalide perovskite planar-heterojunction solar cells on flexible polymer substrates, *Nat Commun*:4, 2013.

[113] Li H, Fu K, Hagfeldt A, Grätzel M, Mhaisalkar SG, Grimsdale AC: A simple 3,4-ethylenedioxythiophene based hole-transporting material for perovskite solar cells, *Angew Chem Int Ed* 53:4085–4088, 2014.

[114] Yu Z, Sun L: Recent progress on hole- transporting materials for emerging organometal halide perovskite solar cells, *Adv Energy Mater*:5, 2015.

[115] Christians JA, Fung RC, Kamat PV: An inorganic hole conductor for organo-lead halide perovskite solar cells. Improved hole conductivity with copper iodide, *J Am Chem Soc* 136:758–764, 2013.

[116] Jung JW, Chueh CC, Jen AKY: A low-temperature, solution-processable, Cu-doped nickel oxide hole-transporting layer via the combustion method for high-performance thin-film perovskite solar cells, *Adv Mater* 27:7874–7880, 2015.

[117] Kwon U, Kim B-G, Nguyen DC, Park J-H, Ha NY, Kim S-J, et al: Solution-Processible crystalline NiO nanoparticles for high-performance planar perovskite photovoltaic cells, *Sci Rep*:6, 2016.

[118] Qin P, Tanaka S, Ito S, Tetreault N, Manabe K, Nishino H, et al: Inorganic hole conductor-based lead halide perovskite solar cells with 12.4% conversion efficiency, *Nat Commun*:5, 2014.

[119] Ye S, Sun W, Li Y, Yan W, Peng H, Bian Z, et al: CuSCN-based inverted planar perovskite solar cell with an average PCE of 15.6%, *Nano Lett* 15:3723–3728, 2015.

[120] Huckaba AJ, Sanghyun P, Grancini G, Bastola E, Taek CK, Younghui L, et al: Exceedingly cheap perovskite solar cells using iron pyrite hole transport materials, *ChemistrySelect* 1:5316–5319, 2016.

[121] Niu G, Li W, Meng F, Wang L, Dong H, Qiu Y: Study on the stability of $CH_3NH_3PbI_3$ films and the effect of post-modification by aluminum oxide in all-solid-state hybrid solar cells, *J Mater Chem A* 2:705-710, 2014.

[122] Conings B, Drijkoningen J, Gauquelin N, Babayigit A, D'Haen J, D'Olieslaeger L, et al: Intrinsic thermal instability of methylammonium lead trihalide perovskite, *Adv Energy Mater*:5, 2015.

[123] Chiang CH, Nazeeruddin MK, Gratzel M, Wu C-G: The synergistic effect of H_2O and DMF towards stable and 20% efficiency inverted perovskite solar cells, *Energy Environ Sci* 10:808–817, 2017.

[124] Hashmi SG, Tiihonen A, Martineau D, Ozkan M, Vivo P, Kaunisto K, et al: Long term stability of air processed inkjet infiltrated carbon-based printed perovskite solar cells under intense ultra-violet light soaking, *J Mater Chem A* 5:4797–4802, 2017.

[125] Aitola K, Domanski K, Correa-Baena J-P, Sveinbjörnsson K, Saliba M, Abate A, et al: High temperature-stable perovskite solar cell based on low-cost carbon nanotube hole contact, *Adv Mater* 29, 2017, 1606398.

[126] Habisreutinger SN, Leijtens T, Eperon GE, Stranks SD, Nicholas RJ, Snaith HJ: Carbon nanotube/polymer composites as a highly stable hole collection layer in perovskite solar cells, *Nano Lett* 14:5561–5568, 2014.

[127] Wang F, Shimazaki A, Yang F, Kanahashi K, Matsuki K, Miyauchi Y, et al: Highly efficient and stable perovskite solar cells by interfacial engineering using solution-processed polymer layer, *J Phys Chem C*, 2017.

[128] Leijtens T, Eperon GE, Noel NK, Habisreutinger SN, Petrozza A, Snaith HJ: Stability of metal halide perovskite solar cells, *Adv Energy Mater* 5:1500963, 2015.

[129] Jeon NJ, Noh JH, Yang WS, Kim YC, Ryu S, Seo J, et al: Compositional engineering of perovskite materials for high-performance solar cells, *Nature* 517:476–480, 2015.

[130] Snaith HJ, Abate A, Ball JM, Eperon GE, Leijtens T, Noel NK, et al: Anomalous hysteresis in perovskite solar cells, *J Phys Chem Lett* 5:1511–1515, 2014.

[131] Wei J, Zhao Y, Li H, Li G, Pan J, Xu D, et al: Hysteresis analysis based on the ferroelectric effect in hybrid perovskite solar cells, *J Phys Chem Lett* 5:3937–3945, 2014.

[132] Frost JM, Butler KT, Walsh A: Molecular ferroelectric contributions to anomalous hysteresis in hybrid perovskite solar cells, *APL Mater* 2:081506, 2014.

[133] Tress W, Marinova N, Moehl T, Zakeeruddin S, Nazeeruddin MK, Grätzel M: Understanding the rate-dependent J–V hysteresis, slow time component, and aging in $CH_3NH_3PbI_3$ perovskite solar cells: the role of a compensated electric field, *Energy Environ Sci* 8:995–1004, 2015.

[134] Eames C, Frost JM, Barnes PR, O'regan BC, Walsh A, Islam MS: Ionic transport in hybrid lead iodide perovskite solar cells, *Nat Commun*:6, 2015.

[135] Bergmann VW, Weber SA, Ramos FJ, Nazeeruddin MK, Grätzel M, Li D, et al: Real-space observation of unbalanced charge distribution inside a perovskite-sensitized solar cell, *Nat Commun*:5, 2014.

[136] Kim H-S, Park N-G: Parameters affecting I–V hysteresis of $CH_3NH_3PbI_3$ perovskite solar cells: effects of perovskite crystal size and mesoporous TiO_2 layer, *J Phys Chem Lett* 5:2927–2934, 2014.

[137] Xiao Z, Bi C, Shao Y, Dong Q, Wang Q, Yuan Y, et al: Efficient, high yield perovskite photovoltaic devices grown by interdiffusion of solution-processed precursor stacking layers, *Energy Environ Sci* 7:2619–2623, 2014.

[138] Bryant D, Wheeler S, O'Regan BC, Watson T, Barnes PR, Worsley D, et al: Observable hysteresis at low temperature in "hysteresis free" organic–inorganic lead halide perovskite solar cells, *J Phys Chem Lett* 6:3190–3194, 2015.

12

Organic Photovoltaics

Ross A. Hatton
WARWICK UNIVERSITY, COVENTRY, UNITED KINGDOM
Ross.Hatton@warwick.ac.uk

12.1 Introduction

As part of a global effort to curb CO_2 emissions the past decade has seen a dramatic acceleration in the deployment of crystalline silicon PVs, spurred by unprecedented reductions in manufacturing costs and state subsidies [1,2]. In many parts of the world the cost of electricity produced by large-scale silicon PV installations is now competitive with that produced using conventional fossil fuels, which is transforming the global energy generating sector [1,2]. However, for expansion of the range of PV applications for integration into buildings, transportation, and consumer electronics there is a need for PV technologies that are compatible with light weight and flexible substrates and whose color can be engineered to match the intended application, the latter of which is particularly important for consumer acceptance.

In conventional inorganic semiconductors such as silicon, the atoms are held together by strong covalent bonds and the thickness of semiconductor needed for PV applications is of the order of 100 μm [3,4]. For crystalline and multi-crystalline silicon PVs, which are the dominant PV technologies of today [1], the optimal semiconductor thickness is ~150 μm, which is comparable to the thickness of a piece of A4 paper. For silicon this large thickness is needed due to its relatively weak absorption of near-infra red light, which makes up a large proportion of the useful solar spectrum for PV applications [4]. Consequently, PVs based on conventional semiconductors are inherently brittle and must be supported on rigid flat plate substrates (e.g., glass), which renders them heavy and unsuitable for use in many important emerging application areas. For example, the electrification of transport systems across the world is progressing rapidly [5], and the roofs of cars and lorries are ideal platforms for PV modules provided the energy required to transport the extra weight is small compared to the electrical energy generated. For automotive applications the PV module must also be very low profile and conformal to the contours of the vehicle, so as not to increase fuel consumption due to increased air flow drag, but also for consumer acceptance. These requirements make silicon PVs wholly unsuitable for this application space. Similarly for portable consumer electronics, for which there is a strong case for integration of PVs as a source of auxiliary power to reduce the time between battery recharges and to enable the relentless demand for increased functionality and computing power.

A Comprehensive Guide to Solar Energy Systems. http://dx.doi.org/10.1016/B978-0-12-811479-7.00012-9

Again, for consumer acceptance the PV module must not add significantly to the weight of the device or detract from the ergonomics or its aesthetics. A further critically important constraint for both of the aforementioned application areas is the need for the PV module to be comprised of nontoxic materials, as consumers and regulatory bodies demand that catastrophic failure of the device does not present a significant risk to the user or natural environment.

The dominant commercial PV technologies of today are fabricated using batch-to-batch processes, which are slow compared to continuous production processes such as roll-to-roll printing. Conventional PVs are also fabricated with at least one high temperature-processing step, which limits the choice of supporting substrate and is energy intensive. The latter is important because it is a key determinant of the energy payback time (EPBT), which is the time taken for the PV module to produce the energy that goes into its production, and therefore its CO_2 mitigation potential [6]. For example, silicon production requires a processing temperature of $\sim$1500°C and so this energy intensive processing step, combined with the relatively thick layer of semiconductor used and slow batch-to-batch fabrication process, means that its EPBT is typically 1.5–2.7 years [7,8].

In contrast, PVs based on organic semiconductors (organic photovoltaics—OPVs), use a semiconductor thickness of less than 1 μm, which is less than 1% of the thickness used in conventional inorganic PVs. Additionally, OPVs are processed at temperatures below 150°C, making them compatible with low cost flexible plastic substrates, such as poly(ethylene terephthalate) [9], and continuous roll-to-roll processing by printing from solution [10] or vacuum evaporation [11,12] (Fig. 12.1). Consequently, OPV modules can be extremely light weight at <0.5 kg m^{-2} [13,14] and with a thickness of <1 mm [13,14] making them very well matched to applications in transportation and portable consumer electronics. While the fabrication cost is likely to be lowest for wholly printed OPVs (since all processes are performed at ambient pressure) both printing and vacuum evaporation

FIGURE 12.1 Vacuum processed small molecule OPVs produced by Heliatek GmbH (photographer: Tim Deussen) (left); solution processed polymer/small molecule OPVs produced by InfinityPV ApS (right: upper and lower).

are low temperature, rapid fabrication methods that are scalable to large area, and so offer a large cost advantage over the high temperature batch-to-batch processes used to fabricate convention PV [15].

An important consequence of the low processing temperature for OPVs is that the energy required to fabricate an OPV module is expected to be ~5% of that needed to produce a multicrystalline silicon PV module. Additionally, due to the very low thickness of semiconductor needed, less than 1 g of organic semiconductor is sufficient to produce 1 m^2 of OPV cells. As a result, OPVs have an intrinsically low carbon footprint both in terms of the manufacturing process and the materials used, and can return the energy used in their fabrication within a few months of installation [15]. For example, in 2013 an EPBT of ~180 days was reported for a large-scale installation of solution-processed OPVs with a power conversion efficiency of only 1.6%–1.8% when operated in a southern European setting [16], and so it is realistic to expect that an EPBT of <100 days is reachable with today's high performance OPVs. An EPBT of <100 days would be the lowest of any renewable source of electricity, being comparable to that of land-based wind turbines [8,16]. Remarkably, a fundamental analysis suggests that an EPBT of the order of 1 day could ultimately be achieved for solution-processed OPVs, through innovations in materials and manufacture [17].

Outside of the physical sciences the term *organic* is very often associated with living things, although in the context of OPVs this prefix refers to the use of carbon-based semiconducting molecules as the light-harvesting element, in place of conventional inorganic semiconductors. Interestingly however, two of the most important molecules for life; heme and chlorophyll are both based on the porphyrin macrocycle, which is a type of organic semiconductor. Indeed, porphyrin derivatives were used in early stages of the development of OPVs owing to their strong absorption in the visible spectrum [18]. Importantly, like heme and chlorophyll, the organic semiconductors used in high performance OPVs contain no toxic heavy metals, such as lead or cadmium, or rare earth metals such as ruthenium, which are essential elements of other types of emerging PVs [19,20]. For practical applications the possibility of catastrophic failure of the materials separating the active electronic materials from the environment must be considered. The use of toxic elements such as lead and cadmium, which are used in lead halide perovskite [19] and quantum dot PVs [20] respectively, presents a significant barrier to market entry for these PV technologies in many parts of the world, not least because contamination of the natural environment by toxic metals is well recognized as a major global problem [21]. Indeed, in Europe and the USA, regulations on lead already restrict the shipping of lead perovskite PVs in these areas. In contrast OPVs are almost certainly the most environmentally sustainable class of emerging PV technology in terms of the toxicity and sustainability of the materials used in their production. At the end of life, OPV modules can also be safely incinerated and the metals used in the electrodes recovered, which is made easy by simple deinstallation to form a compact roll [16]. Since the thickness of the organic semiconductor layer used in an OPV is less than 1 μm, the total CO_2 produced by incineration per unit area of PV module is also very small [8,16].

It is anticipated that OPVs will make their largest contribution as a source of renewable electricity in the context of building integrated PV (e.g., in architectural glass, windows, and roof tiles) where their light weight and the possibility of engineering the color gives OPVs a major advantage over other classes of PV (Fig. 12.2). Early stage large-scale deployment will most likely be in the form of semitransparent shaded glass, which uses the energy absorbed providing shade to generate electricity. Advantageously, in this context the PV is incorporated in the inert gas filled gap between the glass planes and so the module is well insulated from the environment. OPVs are particularly well matched to this large area application because, in contrast most other types of PV, the efficiency with which light is converted to electricity improves as the temperature increases up to an optimal operating temperature of 40°C [22], which is easily achieved in the cavity between the planes of double glazing, even in northern climates. In addition, the organic semiconductors used in OPVs can be optimized to harvest photons outside of the visible spectrum, enabling the realization of highly transparent OPVs. To this authors' knowledge, Heliatek GmbH are currently leading in this application space, offering OPVs achieving 6% efficiency with up to 50% transparency that are also suitable for integration into vehicles.

Another early stage application for OPVs is in consumer electronics (Fig. 12.2—top left) and micro-generators for indoor wireless smart devices in the home and in a retail environment to keep track of stock. Unlike most other types of PV the power conversion efficiency of OPVs improves at light levels below 1 sun intensity [22,23] making them ideal for indoor applications where the light intensity is typically less than 5% of 1 sun intensity. Additionally, the light harvesting organic semiconductor in OPVs can be engineered to make optimal use of the light emitted by incandescent, fluorescent, and LED lights used in commercial and residential environments.

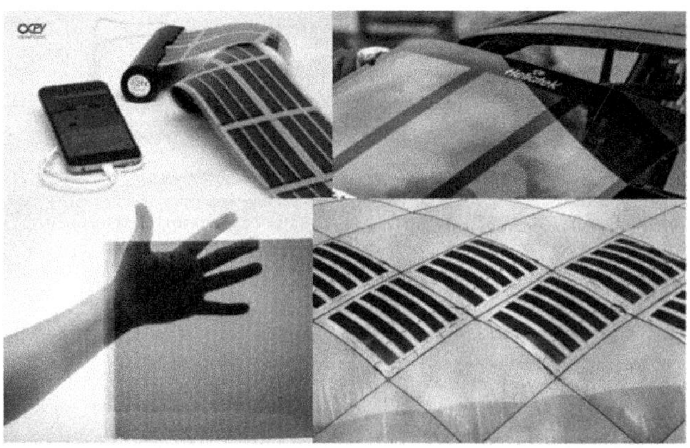

FIGURE 12.2 Commercially available OPVs: InfinityPV HeLi-on compact solar charger (top left); Heliatek's Heliafilm incorporated into a car roof (in collaboration with Webasto) (top right and bottom) and on top of an inflatable building. Bottom left photograph credited to André Wirsig.

12.2 Operating Principles

The basic physical processes that give rise to the photovoltaic effect in an OPV [24–26] are different to those which underpin the operation of conventional inorganic PVs due to a fundamental difference between the structure of crystalline inorganic semiconductors and molecular semiconductors: In the former, such as silicon, all atoms form strong covalent bonds with their neighbors, and are part of a continuous periodic lattice that extends in three dimensions [3]. Strong coupling of a very large number of equivalent atomic orbitals on adjacent atoms gives rise to the formation of bands of closely spaced electron states which, in a semiconductor at ambient temperature, are almost completely filled or completely empty corresponding to the valence band and conduction band respectively. The high density of closely spaced electron states in a band allows charge carriers to move easily through the crystal lattice under the influence of an applied electric field. Charge carriers can be vacancies in the valence band (i.e., holes) or electrons in the conduction band. While it is often useful to consider charge carriers in semiconductors as particles, in crystalline semiconductors they can also be described as waves, which extend over several crystal lattice sites. The delocalized nature of charge carriers in crystalline inorganic semiconductors helps to ensure that the coulombic force of attraction between electrons and holes is very small. Additionally, the small background population of free electrons and holes that always exists in semiconductors at ambient temperature, also helps to screen the interaction between charge carriers by polarization of the crystal lattice in the space between them, a property that is embodied in the relative permittivity, or dielectric constant (ε_r), of the material. Due to the combined effect of the delocalization of charge carriers and the large dielectric constant, the binding energy between a photo-generated electron and hole in a crystalline inorganic semiconductor is typically much smaller than the thermal energy of the charge carriers at room temperature: ~25 meV. Consequently, absorption of a photon of light with sufficient energy to excite an electron from the valence band to the conduction band gives rise directly to the production of a free electron in the conduction band and free hole in the valance band. Since these charges are not coulombically bound to one another they can easily be extracted to the external circuit using a moderate electric field, such as that which exists across the interface between *n* and *p* doped semiconductors. For this reason, conventional PVs are based on a *p-n* junction [3], in which the same semiconductor doped *p* and *n*-type forms the junction across which photo-generated charge carriers are separated. Across a *p-n* junction there is no discontinuity (step) in the energy of the conduction and valence band edges across the interface between *p* and *n*-type regions.

In contrast, photon absorption in organic semiconductors does not result in the direct formation of a free electron and hole, because the coulombic energy of attraction between them is large: 0.2–1 eV [24–26]. This large binding energy is a consequence of the electronic structure of organic semiconductors in the solid state, which is dominated by that of the constituent molecules. Coupling between the highest occupied molecular orbital (HOMO) and lowest unoccupied molecular orbital (LUMO) of adjacent molecules is weak because the molecules are bound together only by weak van der Waals-type interactions.

For this reason, organic semiconductors are often described as van der Waals solids, and the energy levels that transport charge are highly localized on individual molecules. Consequently, the properties of organic semiconductors are typically understood in terms of an array of weakly interacting potential wells, as depicted in Fig. 12.3. Since the integrity of the molecule is preserved in organic semiconductors, a molecular description of the orbitals involved in charge transport is most widely used, in which the HOMO is equivalent to the valence band and the LUMO to the conduction band in a crystalline inorganic semiconductor. In some circumstances, such as at low temperature, a band model can be used to explain the behavior of some highly ordered organic semiconductors (e.g., pentacene), although even at low temperature the bandwidth is very narrow (<0.1 eV) due to the very weak interaction between adjacent molecules.

Upon photon absorption, the photo-excited electron is localized in the LUMO of the same molecule as the hole, which for small molecule organic semiconductors corresponds to a separation of the order of <1 nm. While semiconducting polymers can have dimensions hundreds of times greater than this, the electronic and optical properties are dominated by that of much smaller segments of the polymer chain known as the *conjugation length*. This is because, in real polymer films, twists and kinks in the polymer chain disrupt the strong coupling of the *p*-atomic orbitals that gives rise to the HOMO and LUMO [18]. As a result, a photo-excited electron in the LUMO and the hole formed in the HOMO are spatially confined to the same small segment of the polymer chain. This very close proximity of oppositely charged charge carriers, combined with the relatively low dielectric constant of typical of organic semiconductors, results in a strong coulombic interaction between the photo-generated electron and hole. Consequently light absorption in organic semiconductors does not result in direct formation of free charge carriers, but of tightly bound electron-hole pairs, which can be described in terms of a charge neutral quazi-particle called a Frenkel exciton (Fig. 12.4 process 1) [24,26]. Since excitons have no net charge

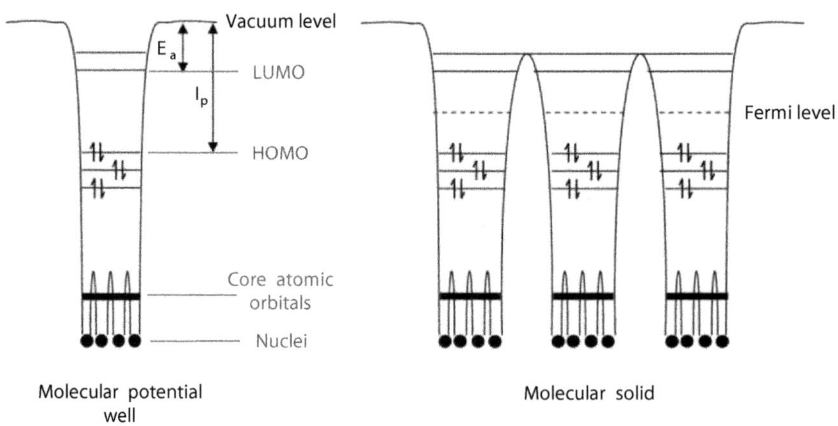

FIGURE 12.3 Schematic diagram illustrating the potential well for an isolated molecule (left) and an aggregation of molecules (right) held together by van der Waals interactions (i.e., a molecular solid). The electron affinity (E_a) and ionization potential (I_p) are also shown.

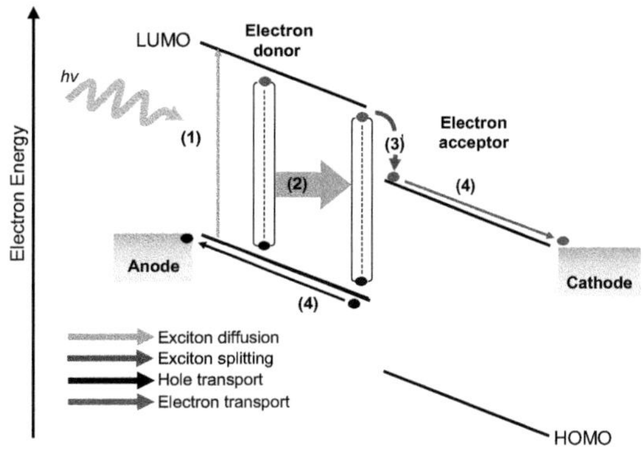

FIGURE 12.4 Simplified energy level diagram depicting a donor-acceptor heterojunction and the four key processes for photocurrent generation: (1) photon absorption to form an exciton; (2) exciton diffusion to the organic heterojunction; (3) exciton splitting; and (4) charge carrier extraction.

they are not influenced by the built-in electric field that exists across the semiconductor layers in an OPV as a result of the difference in work function between the two electrodes. Excitons do however diffuse randomly, visiting hundreds of individual molecules before relaxation to the ground state (Fig. 12.4 process 2) with the emission of light, or nonradiatively by dissipating their energy as heat (i.e. quanta of thermal energy called phonons [3]). In an OPV excitons are split at the junction between two dissimilar molecules having offset HOMO and LUMO orbitals as illustrated in Fig. 12.4 (process 3). The heterojunction provides the thermodynamic driver for the spontaneous splitting of the exciton, as there is both a favorable enthalpy change and entropy change when a single exciton is dissociated to form two charge carriers.

It is important that the heterojunction in an OPV is carefully engineered to ensure that the potential energy step is just enough to dissociate the exciton but no more, since the maximum potential difference across an OPV is determined by the difference in energy between a hole in the HOMO of the molecule that has been oxidized (i.e., the electron donor) and the electron in the LUMO of the molecule that has been reduced (i.e., the electron acceptor). In practice, the minimum energy offset required at the heterojunction is 0.2–0.3 eV. Once the Frenkel exciton has been split the electron and hole in adjacent molecules are still coulombically bound to one another, although much less strongly than when on the same molecule, so a moderate electric field in conjunction with the chemical potential gradient that results from the concentration gradient of charge carriers at the heterojunction, is sufficient to ensure efficient extraction of the charge carriers to the external circuit. Indeed, the efficiency with which photons can be converted to electrons in the external circuit in an OPV can approach 100% provided the excitons are formed within less than one exciton diffusion length of the heterojunction in both the donor and acceptor phases [27,28]. The donor-acceptor heterojunction in an OPV is equivalent to the *p-n* junction in a conventional inorganic PV, in that it is the part of the device that generates free charge carriers (Fig. 12.5).

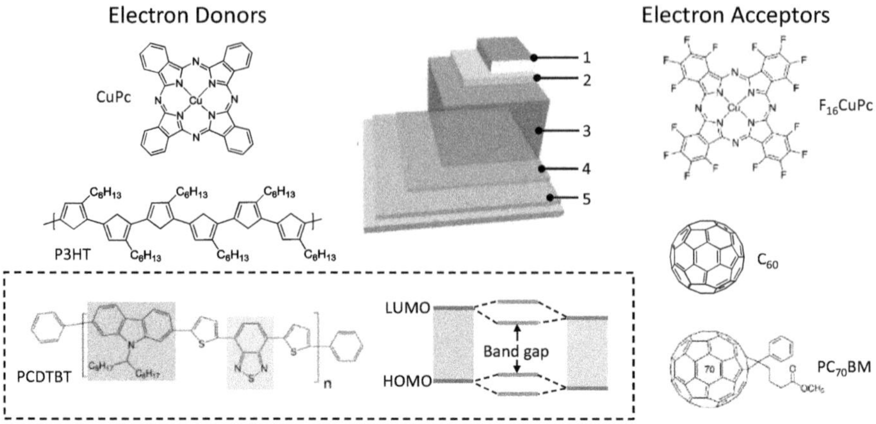

FIGURE 12.5 Key components of a conventional architecture OPV device: (1) opaque metal electrode; (2) electron extraction layer; (3) donor-acceptor heterojunction; (4) hole extraction layer; (5) conducting oxide coated glass transparent electrode. Also shown is an illustrative selection of electron donor and electron acceptor type small molecule and polymeric organic semiconductors.

Unfortunately the exciton diffusion length in most organic semiconductors is of the order of 5–40 nm and so the donor and acceptor phases must be structured on this scale for all of the absorbed light to be converted to electricity [24–26].

The efficiency of most types of PV device decreases strongly with increasing temperature in the range 20–60°C, which coincides with the important temperature range for practical application, as even in northern climates the standard operating temperature of a PV module is well above 20°C and most typically in the range 30–60°C. For example, the efficiency of crystalline silicon and copper indium gallium (di)selenide (CIGS) PV is ~15% lower at 60°C [29]. Consequently, the standard test condition for PVs of 25°C favors the dominant PV technologies of today, because these types of PVs exhibit a negative temperature coefficient. OPVs are a rare example of a class of PVs that exhibit a positive temperature coefficient, and have an optimal efficiency at 40–50°C, which is well-matched to the normal operating temperature of a PV device in the field. The increase in efficiency with temperature results from an increase in the amount of light harvested, due to spectral broadening of the absorption spectrum [22], and reduced device series resistance due to improved charge carrier mobility. Although the relative importance of these two mechanisms is the subject of debate [22], it is intuitive that charge transport in molecular semiconductors is thermally activated process because charge carriers must hop from one potential well to the next (Fig. 12.3) [18].

12.3 Device Structure

Organic semiconductors can absorb light very strongly, although most typically only over a narrow range of wavelengths, which renders them brightly colored. It is for this reason that many small molecule organic semiconductors used in OPVs are closely related to dyes used as paint pigments [18]. In the context of an OPV device it is important that the donor and acceptor materials have complimentary absorption spectra, to ensure good coverage of the

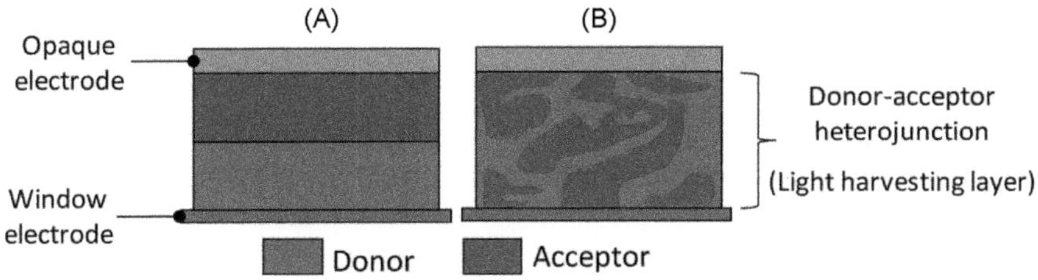

FIGURE 12.6 Schematic illustration of a bilayer (A) and bulk heterojunction (B) OPV device architecture. The thickness of the organic heterojunction in a bilayer device architecture is constrained to <80 nm by the *exciton diffusion bottle neck*. The thickness of the bulk-heterojunction (BHJ) can exceed 300 nm.

solar spectrum. However, in a simple bilayer device architecture (Fig. 12.6) the thickness of the donor and acceptor materials on either side of the heterojunction is constrained to much less than needed to absorb all of the photons at any given wavelength, because the exciton diffusion length in most organic semiconductors is limited to less than 40 nm [25,26]. Excitons formed at distances greater than the exciton diffusion length from the heterojunction relax to the ground state before arriving at the heterojunction and so do not contribute to photocurrent generation. This constraint is known as the *exciton diffusion bottleneck* [25,26] and is most effectively circumvented using the bulk-heterojunction (BHJ) film morphology, in which the donor and acceptor phases are codeposited and spontaneously phase separate into a complex interpenetrating network of donor and acceptor phases as schematically illustrated in Fig. 12.6B. By carefully engineering the molecular structure and film deposition conditions it is possible to ensure that the dimensions of both phases are comparable to twice the exciton diffusion length, and so all excitons formed throughout the thickness of the films are within one exciton diffusion length of a heterojunction. Despite the complexity of this layer, it works remarkably well given that free charge carriers must move along narrow winding tracks of each phase to the electrodes without recombining. Remarkably, it has now been shown for a number of different BHJ materials that almost 100% of absorbed photons can be converted to excitons in the external circuit; see for example [27] and [28], and it is possible to increase the BHJ thicknesses to greater than 300 nm for some material systems without adversely affecting device series resistance. What is critically important when scaling to film thickness beyond ~100 nm is that the electron and hole mobility in each phase is closely matched to avoid the accumulation of one charge carrier type, which results in increased electron-hole recombination [30,31].

12.4 Challenges and Opportunities for Improved Performance

Over the past decade the certified power conversion efficiency of laboratory scale OPVs has increased rapidly from ~3% to over 13% (Fig. 12.7), to a level comparable to that of amorphous silicon PV. This impressive progress has, to a very significant extent, been driven by commercial organizations that recognize the potential of OPVs to meet the needs of the application areas outlined earlier in this chapter. These organizations include

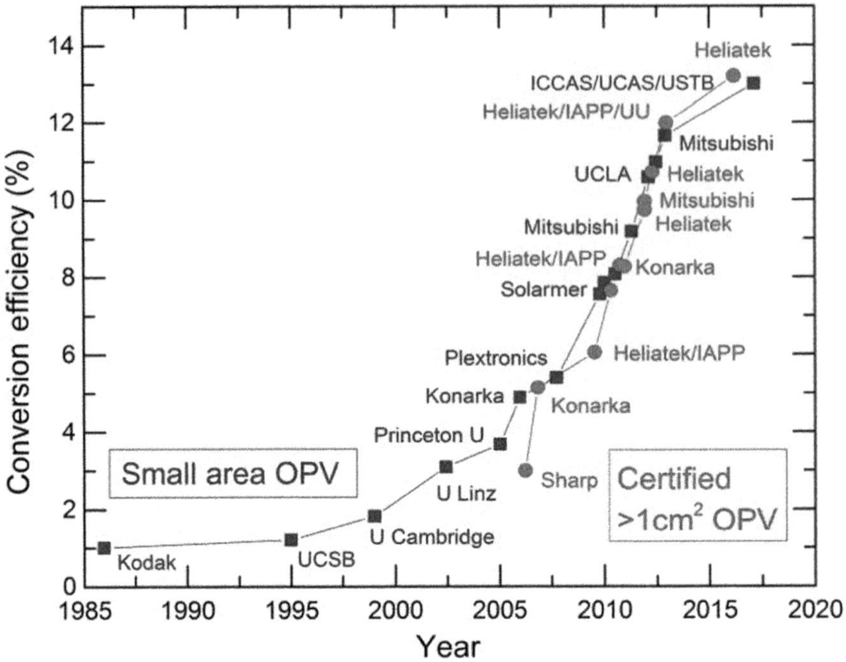

FIGURE 12.7 Graph of OPV power conversion efficiency under 1 sun simulated solar illumination. Figure adapted from www.orgworld.de record chart. Figure credited to Professor Karl Leo.

InfinityPV ApS [32], who to the author's knowledge are the only company currently selling OPVs directly to the public (Fig. 12.2, top left), although other companies are selling business-to-business, including Heliatek GmbH [33] and Belectric OPV GmbH (becoming OPVIUS GmbH) [34]. Other companies seeking to commercialize OPVs include Eight19, Next Energy Technology, Mitsubishi Chemical Corporation, Toshiba, although this not an exhaustive list.

To date, OPV power conversion efficiency above 11% has only been achieved for OPVs with multijunction device architecture, in which two or three individual heterojunctions are fabricated directly on top of one another, and electrically connected in series by internal electrodes [35] (Fig. 12.8). The advantage of this approach is that it allows each heterojunction to be optimized to harvest a particular part of the incident solar spectrum. Consequently, optimized multijunction OPVs perform 20%–30% [35,36] better than single junction devices, which justifies the added complexity in device architecture. At the time of writing, the certified record power conversion efficiency for an OPV device was held by Heliatek GmbH [33], for a triple heterojunction small molecule OPV that harvested light of wavelength 450 nm to 950 nm. Notably however, the efficency of solution processed double junction OPVs is very close behind at 12.8%–13.0% [37,38].

An interesting feature of multijunction OPVs is that the upper limit for the open-circuit voltage (V_{oc}) is the sum of the individual junctions that make up the stack, although this is only achieved by closely matching the hole and electron currents produced in adjacent

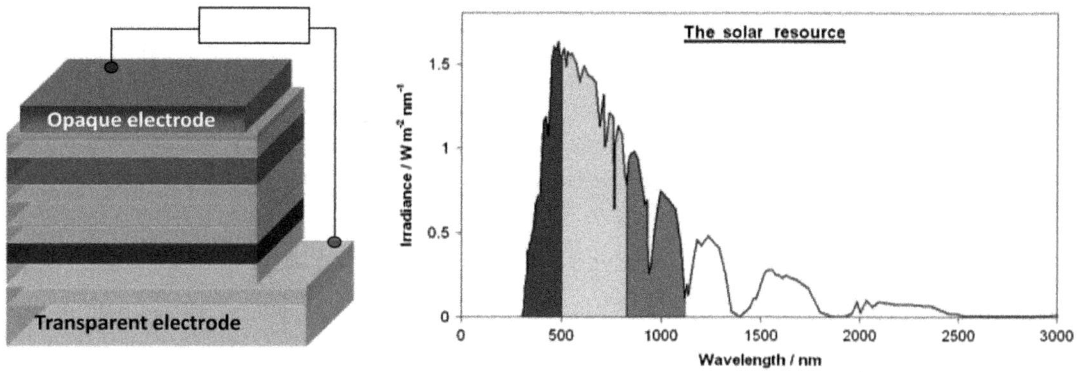

FIGURE 12.8 Schematic diagram of the structure of a three heterojunction OPV, with junctions optimized to harvest the blue, green, and red parts of the solar spectrum (left). The solar spectrum at ground level highlighting the approximate range of wavelengths harvested by each organic heterojunction in a triple junction OPV (right).

devices [35,39]. The potential to achieve much greater output voltages than possible for a single junction OPV is well illustrated by the work of Sullivan et al. [39] who demonstrated that a V_{oc} exceeding 5 V could be achieved for a device comprising five individual heterojunctions. In that work the motivation was to achieve an output voltage sufficient to recharge a battery for consumer electronics application, which requires a threshold potential for recharging of 4.0–4.2 V. The advantage of that seemingly unconventional approach, over simply connecting five individual OPV cells in series to achieve the same voltage, is that a PV module comprising high voltage multijunction OPVs connected in parallel can continue to charge the battery even if in partial shade (Fig. 12.9), which is particularly useful for portable consumer electronic aplications.

Now that the power conversion efficiency of laboratory scale OPVs is very close to threshold for large-scale applications, the research effort in industry and academia is increasingly focusing on addressing challenges around scale-up of materials and processes to enable low cost production of modules, as well as developing approaches to improving module stability toward the fluctuations in light intensity and temperature experienced in real world applications. The following sections highlight a number of areas from which significant advances in OPV performance are likely to stem in the coming years. The power conversion of a PV device is proportional to the product of the open-circuit voltage (V_{oc}), the fill factor (FF), and the short circuit-current density (J_{sc}) [3], and so in the following part of this chapter each point is discussed in terms of the effect on one or more of these device parameters.

12.4.1 Increasing Power Conversion Efficiency

For single junction OPVs empirical considerations suggest that a power conversion efficiency of 12% is within reach [36,40], and for multijunction device architectures 15% is technically feasible using refinements of existing device structures and materials [35,41]. A laboratory scale device efficiency of 15% is an important benchmark because it is

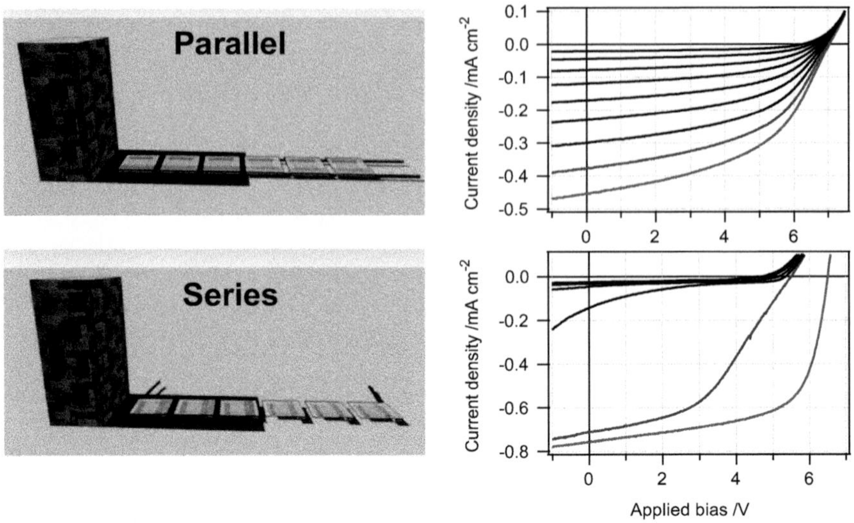

FIGURE 12.9 Diagrams (left) and current-voltage characteristics (right) showing how a module of multijunction OPV cells reported by Sullivan et al. [39] maintains a high operating voltage even in partial shade. Diagram (left) credited to Dr. Paul Sullivan.

reasonable to expect that this would translate to a 10%–12% module efficiency, which would be sufficient for a number of the aforementioned application areas. Importantly, fundamental descriptions indicate that the upper efficiency limit for a single junction OPV may be as high as 20%–24% [36,42] and so multijunction OPVs may yet achieve a module efficiency above 15%, although this will require a much better understanding of the ideal heterojunction morphology as well as the development of materials with improved optical and electrical properties. Structural and chemical impurities in organic semiconductors limit the efficiency of today's high performance OPVs by serving as electron-hole recombination centers and impeding the transport of free charges to the electrodes, which constrains the thickness of the organic semiconductor that can be used. The density of such impurity states in organic semiconductors is typically many orders of magnitude higher than the impurity level in PV grade inorganic semiconductors, and so it is likely that considerable improvements in the OPV performance can be achieved simply by reducing the density of these structural and chemical impurities.

For commercial scale OPVs based on vacuum processable organic semiconductors, simple small molecules are favored because they are available in high volume, have no insulating parts, and form highly crystalline phases. There are also many examples of small molecule semiconductors that strongly absorb near infra-red light, in addition to those that absorb strongly in the ultraviolet and visible parts of the solar spectrum. Examples of small molecule organic semiconductors that have proved particularly useful for OPVs are metal phthalocyanines (particularly zinc and copper phthalocyanine) and the cage fullerenes (particularly C_{60} and C_{70}), some examples of which are given in Fig. 12.5. The well-defined size, rigidity, and high symmetry of small molecule organic semiconductors favors crystallization in the condensed phase. A high degree of crystallinity in both donor and accept phases is beneficial for two reasons: it helps to maximize the free charge carrier mobility (velocity per unit field), which enables efficient charge carrier extraction to the external circuit. Electrical losses due to recombination are more prevalent in disordered organic semiconductor films because free charge carriers are delayed as they move toward the electrodes, which increases the likelihood of electron-hole recombination, thus eroding the J_{sc} and device *FF*. A high degree of crystallinity in both donor and acceptor phases also helps to minimize losses in V_{oc}, because it reduces the energy needed to split the exciton. One widely used means of improving the degree of phase separation and crystallinity in both donor and accept phases in a bulk heterojunction, is to anneal the film at low temperature (<150°C) either during film deposition (in the case of vacuum deposited OPV) or post-film deposition (for solution processed OPV), which also helps to ensure that the film morphology has achieved equilibrium.

What is particularly special about organic semiconductors (both small molecule and polymer) as a class of semiconducting materials for electronic devices, is their amenability to facile engineering of the HOMO and LUMO energies to match the needs of the application. This can be achieved by chemical modification of the conjugated core, or by attaching electronegative or electropositive groups at the periphery of the conjugated core, which modify the HOMO and LUMO energies via the inductive effect [18,43,44]. Like conventional semiconductors, organic semiconductors can be doped using *n* or *p* type dopants to increase their conductivity. In the context of an OPV device, doping is however only useful for minimizing the barrier to charge carrier extraction by the electrodes, since dopants in the photoactive region operate as electron-hole recombination centers, degrading the J_{sc}. However, organic semiconductors are sometimes described as having either *p* or *n* type character even when undoped because they conduct one charge carrier type more efficiency than the other. An intrinsic preference for the conduction of electrons (holes) reflects a combination of the accessibility of the LUMO (HOMO) for injection or extraction of charge carriers to the external circuit, and the high electron (hole) mobility. Consequently, the character of an organic semiconductor can be switched from *n* to *p*-type without altering the optical properties by tuning the energy of the orbitals responsible for charge transport to make them more or less accessible for the injection/extraction of charge carriers by the electrode. To illustrate the extent to which the properties of organic semiconductors can be engineered by simple chemical modification it is useful to consider the case of copper phthalocyanine, which was first recognized for its semiconducting properties by Eley and Vartanyan (independently) in 1948 [45] and for decades was used as the electron donor in small molecule OPVs [26]. By replacing the hydrogen atoms at the periphery of the macrocycle with electronegative atoms such as fluorine, the electron affinity and ionization potential can be increased by approximately the same amount of ~1 eV, making the LUMO more accessible for electron injection/extraction without significantly changing the optical properties. Using this approach Yang et al. [46] demonstrated that fluorinated copper phthalocyanine can serve an electron acceptor in OPVs.

It is possible to change the HOMO-LUMO gap in organic semiconductors by altering the extent of conjugation. In the case of copper phthalocyanine the HOMO-LUMO gap is most easily reduced by extending the degree of conjugation to make the corresponding napthalocyanine [43]. Similar strategy can be applied to conjugated polymers, although for polymeric semiconductors the HOMO-LUMO gap is most often reduced by incorporating into the polymer backbone alternating electron-rich (i.e., donor type) and electron-poor (i.e., acceptor type) units [47]. For example, the small HOMO-LUMO gap polymer PCDTBT (Fig. 12.5) has carbazole electron rich units [*light gray* (*red* in the web version)] and benzothiadiazole electron deficient [*dark gray* (*blue* in the web version)] units bridged by a conjugated thiophene unit. The reduction in HOMO-LUMO gap results from mixing of the orbitals of these electron rich and electron poor units as illustrated in Fig. 12.5. Using this approach polymeric semiconductors have been developed for OPVs that are capable of absorbing light with wavelengths beyond 900 nm, which has been a key driver of recent improvements in OPV power conversion efficiency [48].

To increase the efficiency of OPVs to 15% and beyond, there is a need for the development of narrow band gap organic semiconductors offering higher electron and hole mobilities, when part of a bulk heterojunction blend, than is currently achieved ($<10^{-2}$ cm^2 V^{-1}s^{-1}). A consequence of the slow transport of photo-generated charge carriers out of the device is the accumulation of charge in the photoactive material which results in recombination losses and reduced J_{sc}. Given that the charge carrier mobility in single phase organic semiconductors can exceed 10 cm^2 V^{-1}s^{-1} [18] there is considerable scope for improvement in this area, even when the complex interpenetrating morphology of the bulk heterojunction is taken into account. Notably, due to the relatively low charge carrier mobility in most organic semiconductors used in OPVs (several orders of magnitude below that of conventional inorganic semiconductors) it has, until recently, been a widely held view that the thickness of a bulk-heterojunction could not exceed 100–200 nm without incurring excessive electrical losses [49,50]. However, it is now recognized that much thicker photoactive layers can be used, even with charge carrier mobilities at their current levels, provided the electron mobility in the electron acceptor and hole mobility in the electron donor phases are closely matched [49,50]. A significant mismatch in the electron and hole mobility results in an accumulation of one charge carrier type in the device that reduces the built-in electric field needed to facilitate extraction of photo-generated charge carriers [49–51]. Using thicker photoactive layers simultaneously increases the proportion of light harvested and improves the manufacturability, as processing films with a thickness of 100 nm over large area is challenging, not least because it is very difficult to achieve a high degree of uniformity of film thickness, which limits the range of high throughput deposition techniques suitable for manufacturing [50].

One of the most effective ways to improve the efficiency of OPVs is to increase the J_{sc} by making better use of the solar spectrum in the wavelength range 700–1000 nm (i.e., near infra-red light). While C_{60} and its derivatives are widely used as the electron acceptor in OPVs, they only strongly absorb short wavelength light ($\lambda < 500$ nm), and so there has been growing interest in the development of nonfullerene electron acceptors that strongly absorb visible and near-IR light [52,53]. An alternative approach to increasing the efficiency of light harvesting is to use two different donors with complimentary absorption spectra (or two acceptors with complimentary absorption spectra) in the same bulk heterojunction [51,54–56]. Despite the additional complexity in the morphology of the bulk heterojunction layer, this approach has proved to be remarkably effective.

Even for today's best performing OPVs, it is recognized that there is considerable scope for reducing losses in V_{oc}, not only by engineering the energy offset at the organic heterojunction to the absolute minimum required for Frenkel exciton dissociation [44], but also by improving the degree of crystallinity in both donor and acceptor phases to minimize the energy cost in dissociating the charge transfer exciton [42]. A different approach to minimizing the energy cost of splitting the exciton and recombination losses is to increase the dielectric constant (ε_r) of both the donor and acceptor materials, which reduces the coulombic energy of attraction between the electron and hole and thus the energy needed to separate them. Computational analysis predicts that a power conversion efficiency of ~20% is within reach if the ε_r can be increased from current values of 2–4 up to ~7.5 [42,57]. Notably, increasing ε_r to ~9 would reduce the exciton binding energy to an insignificant level such that a donor/acceptor heterojunction would no longer be required [42]. One way to increase the dielectric constant, well matched to solution processable organic semiconductors, is to modify the solubilizing side chains so that they include polar bonds or dipolar terminating moieties that are free to rotate in response to an electric field [57,58].

Beyond engineering the photoactive layer in OPVs, improvements in device efficiency will also result from advances in antireflective coatings and high-performance substrates and electrodes with refractive indices matched to that of the organic semiconductors. Another rapidly advancing area is in the development of universal strategies for maximizing light absorption in OPVs by light trapping using nanophotonic or plasmonic electrode structures [59]. One approach that has received relatively little attention to date, but is actually very well matched to OPV device architectures, is to trap the incident light as plasmonic excitations (i.e., collective oscillations of the conduction band electrons) at the surface of a nanostructured metal electrode [60,61] (Fig. 12.10). Similar to metal nanoparticles, nano-holes in a metal film can have a very large absorption cross-section, which enables strong coupling with incident light, while having the advantage of being confined to the plane of the electrode and so not undermining the electrical integrity of the diode. Due to the very large absorption coefficient of organic semiconductors used to harvest light in OPVs, light trapped as plasmonic excitations at the surface of one (or both) electrodes in an OPV can be used to directly excite electrons from the HOMO to the LUMO in an adjacent organic semiconductor, before the plasmonic excitations dissipate their energy as heat in the metal. While research in this area is at an early stage, this approach is very attractive because it offers a universally applicable means of increasing J_{sc} without increasing the complexity of device fabrication. For commercial viability, the process used to nanostructure the electrode must be low cost and easily scalable, ruling out the use of conventional lithographic methods for the formation of a high density of nanoapertures.

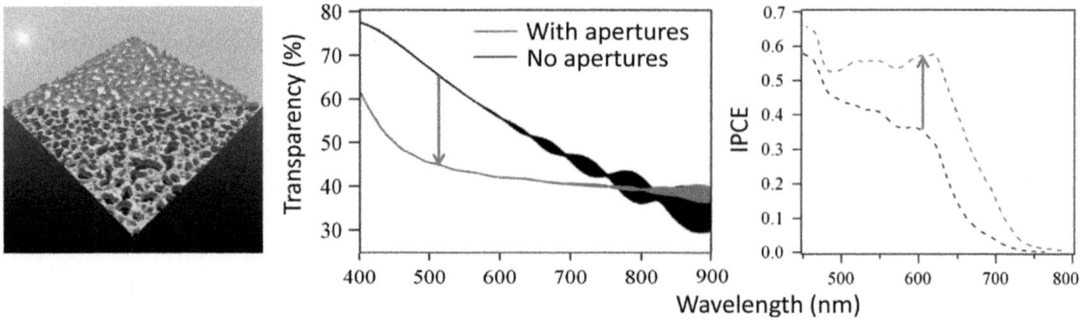

FIGURE 12.10 An ultra-thin noble metal film with a dense array of suboptical wavelength apertures formed by thermal annealing and imaged using an atomic force microscope (left); transparency of an 11 nm thick silver film electrode supported on plastic with and without a dense array of suboptical wavelength apertures (middle); incident-photon-to-charge-efficiency (IPCE) spectra for an OPV device using a silver film electrode with and without apertures (right). These data show how the incorporation of suboptical wavelength apertures into a silver window electrode used in an OPV, device reduce the amount of light entering the device by direct transmission but increase photocurrent generation due to light trapping as plasmonic excitations at the surface of the electrode. The data shown is adapted from reference [60].

One approach has been to use a random array of suboptical wavelength apertures formed by rapid thermal annealing of an optically thin metal film, which offers the advantage of a broad band optical response, scalability, and simplicity [60,61].

12.4.2 Improving Long-Term Stability

The stability of OPVs is already suitable for many consumer electronics applications and for indoor devices with lifetimes of several years (e.g., smart electronic tags). However, for building integration and automotive applications OPVs must have long-term stability toward: (1) ultra-violet light—which is a significant proportion of the solar spectrum at ground level; (2) temperatures up to 60°C—which can be achieved even in relatively temperate climates; and (3) oxygen and moisture ingress into the device—which is inevitable given sufficient time. For the latter, the encapsulation materials are critically important and so a great deal of research effort has focused on the development of flexible materials with excellent barrier properties toward water and oxygen, based on widely available raw materials that can also be processed at low cost [62]. One innovative approach has been the development of ultra-thin glass, which can be bent through a radius of curvature small enough to be compatible with roll-to-roll printing [63]. If this type of glass can be produced at low enough cost it will be very attractive as a substrate for OPVs, as glass offers outstanding water and oxygen barrier properties even when very thin.

The stability of OPV devices depends not only on the intrinsic chemical stability of the organic semiconductors used, but also on their morphological stability and the stability of the interfaces they make with the electrode materials, as well as the stability of the electrode materials themselves. One way to extend device lifetime is to integrate into the electroactive part of the device a material whose function does not change significantly when oxidized, such that it serves as a sink for water and oxygen [64]. An important advance in OPV research came with the discovery that thin films of the partially reduced transition metal oxides WO_{3-x}, MoO_{3-x}, or V_2O_{5-x}, deposited either by vacuum evaporation or solution processing, can be used as high performance hole-extracting materials at the interface between the light harvesting organic semiconductors and the hole-extracting electrode [65,66]. What is special about these materials is that they are intrinsically *n*-type due to oxygen array vacancies, with a large work function and wide bandgap. Consequently, these oxides are transparent and can be used to ensure optimized alignment of the electrode Fermi level with the HOMO of the donor phase in a bulk heterojunction. The high work function is important for OPV stability, because it enables the use of organic semiconductors with a high ionization potential (i.e., a deep lying HOMO), which offer improved stability toward oxidation in air.

Accelerated stability testing of Heliatek's [33] double junction cells (efficiency ~7.7%) using conventional foil encapsulates have shown less than 10% degradation after 3000 h in damp heat (85°C/85% relative humidity) under 1 sun continuous illumination. This remarkable stability, particularly given the complexity of the device, bodes very well for the prospects of achieving OPV module lifetimes suitable for building integration applications in the near future.

12.4.3 Minimizing the Cost of Materials and Device Fabrication

To realize the full cost advantage and potential environmental benefits of OPVs over other types of emerging PV technologies, there are several areas in which developments are needed. It is essential that the organic semiconductors used in an OPV module can be synthesized using inexpensive, easily accessible starting materials in a small number of synthetic steps, as the production cost (and embodied energy) scales rapidly with number of synthesis steps. For example, P3HT (Fig. 12.5), which was the work horse of OPV research for many years, can be produced in as few as three synthetic steps using low cost precursor materials [67]. Given that less than 1 g of organic semiconductor is needed to fabricate 1 m^2 of OPV devices, the volume of materials required for large-scale production of OPVs is relatively small, which gives scope for flexibility in the synthetic methods used for scale-up [68].

An important aspect that has received surprisingly little attention to date, is the need to move away from using chlorinated solvents for deposition of solution-processed organic semiconductors [69,70]. Chlorinated solvents are toxic and unsustainable, and so the precautionary and containment measures needed to enable their use for large-scale roll-to-roll production will inevitably add to the production cost of solution-processed OPVs while also undermining the green credentials of this type of OPV. Notably, this is not a problem that needs to be addressed for small molecule OPVs fabricated by vacuum processing.

A key difference between the design of silicon PVs and OPVs, is that the latter are typically fabricated on a glass (or plastic) substrate coated with a wide bandgap conducting oxide which serves as the transparent electrode. Conversely, in a silicon PV the light enters the device through the top-surface, which is printed with a grid of closely spaced silver wires. The silver grid lines are opaque, so light enters the device through the gaps in the metal grid. The conducting oxide electrodes used in OPVs comprise a 100–200 nm thick layer of a heavily doped oxide such as F:SnO_2 or Sn:In_2O_3 (ITO) supported on glass, and offer a sheet resistance ~10 Ω per square combined with an average transparency of ~80% over the wavelength range 400–900 nm. (Note: The electrical resistance of a square sample of thin film conductor is independent of substrate area because the resistance is proportional to the length and inversely proportional to the width, so the resistance of thin film conductors is most often given *per unit square* without defining the area). Conducting oxide electrodes are inherently brittle and are produced using energy intensive batch-to-batch manufacturing processes, which renders them poorly matched to OPVs. Consequently it is widely recognized that a low-cost alternative is needed that is compatible with flexible substrates [9,71,72]. A number of solutions to this complex problem have been proposed, including electrodes based on conducting polymers, carbon nanotubes, graphene, and metal nanowires [71]. However, all have significant disadvantages either in terms of environmental concerns relating to the toxicity of the materials used (as for nano-Ag and Cd), the complexity of the fabrication process, unacceptably high surface roughness or impractically high sheet resistance, and so no clear leader has emerged. For large area (practical) applications the electrode sheet resistance cannot realistically exceed 15 Ω per square without incurring excessive resistive losses, and so it is a false economy to increase the electrode transparency at the expense of sheet resistance beyond 15 Ω per square [73]. A reduction in the sheet resistance of the transparent electrode to ~1 Ω per square would open the door to new electrode geometries beyond the narrow-strip designs currently used, which would reduce module manufacturing complexity and costs [73]. To date random arrays of Ag wires with a diameter of the order of 20–100 nm and printed silver grid electrodes offer the best performance in terms of sheet resistance and transparency [71]. However, for application in OPVs these electrodes must be used in conjunction with a heavily doped conducting polymer or doped small molecule layer to span the gaps between wires/grid lines and smooth the surface, which inevitably complicates the fabrication process. Further serious drawbacks are instability resulting from localized heating at the contacts between silver nanowires [74], combined with the susceptibility of nano-silver toward oxidation particularly by airborne sulfur containing compounds.

In recent years there has been growing interest in using unpatterned silver films with a thickness of 6–12 nm fabricated by vacuum evaporation as the transparent electrode in OPVs [9,64,71,72,76,77]. Large area roll-to-roll deposition of metals by vacuum evaporation is an established industrial process for the manufacture of low cost food packaging and insulation foils, and offers excellent control over metal thickness and uniformity: For example, the 20 nm thick aluminum film used for crisp packaging is deposited at $\sim 10^{-2}$ Pa at a rate of 400–800 m min^{-1}. In the most rigorous economic model for the large-scale manufacture of OPVs published to date, vacuum deposition of the metal electrode is shown to represent a tiny percentage of the total cost of a solution processed OPV [75]. This method of electrode production is therefore well matched to OPVs, and could be fully integrated into the production of vacuum-processed OPVs. For unpatterned metal film window electrodes, silver is the metal of choice because it has the lowest extinction coefficient and highest electrical conductivity among metals. While silver is relatively expensive metal (comparable to that of indium used in ITO glass) it has been argued that it could easily be recovered and recycled by incinerating the OPV device at the end of its useful life [16]. To maximize the far-field transparency, silver electrodes are invariably sandwiched between two oxide layers—a concept borrowed from the low emissivity glass industry. This triple-layer architecture has been shown to perform as well as ITO glass in OPVs [9,71].

In recent years there has been growing interest in using copper as the base metal for a window electrode in OPVs, because this metal has comparable electrical conductivity to Ag but is ~1% of the cost, and is already widely used in the microelectronics industry because it offers high resistance to electro-migration. In general copper has received relatively little attention as an electrode material for PV applications due to its higher susceptibility to oxidation than silver. However, it has recently been shown that ultra-thin copper films can be made much more stable toward oxidation in air either by partial oxidation during deposition [64] or by capping with a sub-1 nm film of aluminum [64,77,78] without deteriorating its optical or electrical properties (Fig. 12.11), and so copper may yet emerge as the window electrode material of choice for the large-scale production of OPVs and other classes of emerging thin film PV.

As explained earlier in this chapter, the widespread utilization of OPVs in transportation and for building integrated PV will depend on whether suitable flexible barrier materials are forthcoming at low enough cost. High performance water and oxygen barrier layers invariably comprise multiple laminated layers of materials, and so achieving the necessary degree of flexibility for roll-to-roll processing is nontrivial. It is likely that the cost of the encapsulation materials for OPV will exceed that of the OPV module itself, and so advances in this area are needed.

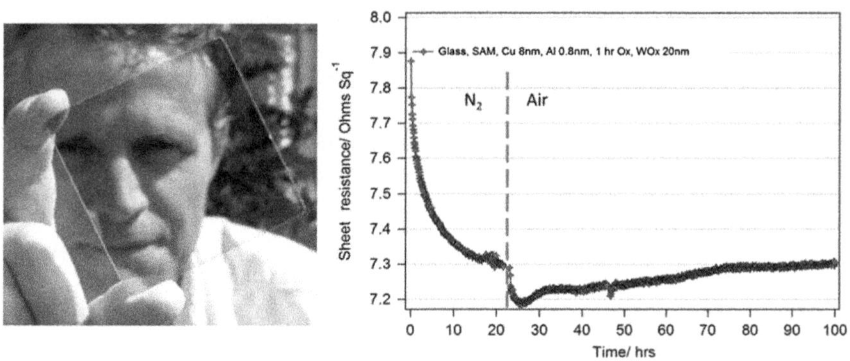

FIGURE 12.11 Photograph of a robust 8 nm (~60 atom) thick copper film supported on chemically modified glass (left); evolution of sheet resistance of an 8 nm thick Cu electrode capped with an 0.8 nm aluminium passivation layer and 20 nm thick tungsten oxide hole transport layer. The electrode was bought into ambient air after 23 h testing in a nitrogen atmosphere (right). Photograph credited to Dr. Ross A. Hatton. Right hand figure credited to Dr. Oliver S. Hutter [78].

12.5 Conclusion

OPVs are extremely light weight, low profile, and offer tuneable color, and so are very well matched to a range of important application areas for which other PV technologies are much less well-suited, including architectural glass, automotive applications, and in consumer electronics. Given the rapid pace of advancement in OPVs, combined with the commercial success of large area displays based on organic light emitting diodes—which demonstrates that scale-up of organic optoelectronic devices for commercial applications is possible—it is now highly likely that OPVs will become a significant player in the PV sector in the coming decade. Furthermore, if low cost encapsulation materials are forthcoming, and roll-to-roll fabrication of high efficiency solution processed OPVs can be scaled-up, OPVs will offer a large cost advantage over other types of PV technology combined with the lowest energy payback time. The low energy outlay, combined with the absence of heavy metals or rare earth elements will also almost certainly make OPVs the most sustainable PV technology.

References

[1] Rekinger M, Thies F, Masson G, Orland S. Global Market Outlook For Solar Power/2015-2019. European Photovoltaic Industry Association Report.

[2] Sussams L, Leaton J. Expect the Unexpected: The Disruptive Power of Low-carbon Technology. Carbon Tracker and the Grantham Institute at Imperial College London Report. February 2017.

[3] Hook JR, Hall HE: *Solid state physics*, 2nd ed., John Wiley & Sons, 1991.

[4] Fonash SJ. Solar Cell Device Physics (Second Edition). Elsevier.

[5] Astarloa B, Kaakeh A, Lombardi M, Scalise J, et al. Editor David Sims. The future of electricity: new technologies transforming the grid edge: World Economic Forum in collaboration with Bain & Company. March 2017.

[6] Nelson J, Gambhir A, Ekins-Daukes N. Grantham Institute for Climate Change: Solar power for CO_2 mitigation 2014;11:1–15.

[7] Burger B, Kiefer K, Kost C, Nold S, Philipps S, Preu R, Rentsch J, Schlegl T, Stryi-Hipp G, Willeke G, Wirth H, Brucker I, Häberle A, Warmuth W. Photovoltaics Report Fraunhofer Institute for Solar Energy Systems, ISE with support of PSE AG. Freiburg, July 12, 2017.

[8] Lizin S, Van Passel S, De Schepper E, Maes W, Lutsen L, Manca J, Vanderzande D: Life cycle analyses of organic photovoltaics: a review, *Energy Environ Sci* 6:3136–3149, 2013.

[9] Jungheum Y: Ultrathin metal films for transparent electrodes of flexible optoelectronic devices, *Adv Mater* 18:1606641, 2017.

[10] Søndergaard R, Hösel M, Angmo D, Larsen-Olsen TT, Krebs FC: Roll-to-roll fabrication of polymer solar cells, *Mater Today* 15:36–49, 2012.

[11] Taylor DM: Vacuum-thermal-evaporation: the route for roll-to-roll production of large-area organic electronic circuits, *Semicond Sci Technol* 30:054002, 2015.

[12] Uhrich CL, Falkenberg C, Rabe J, Heimke B, Klemet M, Wilde C, Wichtendahl R, Pfeiffer M, Organic solar cells: from lab to roll-to-roll production. Proc SPIE. 9184, Organic Photovoltaics XV, 918415. 2014). doi: 10.1117/12.2061897.

[13] Eritt M, Challenges of vacuum roll-to-roll processing of organic solar cells and encapsulation processes AIMCAL Conference. Dresden May 30–June 2, 2016. Heliatek GmbH.

[14] Technical specifications of polymer solar module produced by InfinityPV. Available from: www.infinitypv.com/images/infinityPV_OPV_organic_solar_cells.pdf.

[15] Griffith MJ, Cooling NA, Vaughan B, Elkington DC, Hart AS, Lyons AG, Quereshi S, Belcher WJ, Dastoor PC: Combining printing, coating, and vacuum deposition on the roll-to-roll scale: a hybrid organic photovoltaics fabrication, *IEEE J Sel Top Quantum Electron* 22:4100614, 2015.

[16] Krebs FC, Espinosa N, Hösel M, Søndergaard RR, Jørgensen M: 25th anniversary article: rise to power—OPV-based solar parks, *Adv Mater* 26:29–39, 2014.

[17] Espinosa N, Hösel M, Angmo D, Krebs FC: Solar cells with one-day energy payback for the factories of the future, *Energy Environ Sci* 5:5117–5132, 2012.

[18] Sun, SS, Dalton, LR, editors. Introduction to organic electronic and optoelectronic materials and devices. CRC Press, page 144.

[19] Babayigit A, Ethirajan A, Muller M, Conings B: Toxicity of organometal halide perovskite solar cells, *Nat Mater* 15:247–251, 2016.

[20] engül H, Theis TL: An environmental impact assessment of quantum dot photovoltaics (QDPV) from raw material acquisition through use, *J Clean Prod* 19:21–31, 2011.

[21] Su C, Jiang LQ, Zhang WJ: A review on heavy metal contamination in the soil worldwide: situation, impact and remediation techniques, *Environ Skep Crit* 3:24–38, 2014.

[22] Burlingame Q, Zanotti G, Ciammaruchi L, Katz EA, Forrest SR: Outdoor operation of small molecule organic photovoltaics, *Org Electron* 41:274–279, 2017.

[23] Hadipour A, de Boer B, Blom PWM: Organic tandem and multi-junction solar cells, *Adv Funct Mater* 18:169–181, 2008.

[24] Gregg BA: Excitonic solar cells brian, *Phys Chem B* 107:4688–4698, 2003.

[25] Kippelen B, Brédas J-L: Organic photovoltaics, *Energy Environ Sci* 2:251–261, 2009.

[26] Forrest SR: The limits to organic photovoltaic cell efficiency, *MRS Bull* 30:28–32, 2005.

[27] Kyaw AKK, Wang DH, Gupta V, Leong WL, Ke L, Bazan GC, Heeger AJ: Intensity dependence of current–voltage characteristics and recombination in high-efficiency solution-processed small-molecule solar cells, *ACS Nano* 7:4569–4577, 2013.

[28] Park SH, Roy A, Beaupré S, Cho S, Coates N, Moon JS, Moses D, Leclerc M, Lee K, Heeger AJ: *Nat Photon* 3:297–302, 2009.

[29] www.easac.eu/fileadmin/docs/Low_Carbon/KVA_workshop/Renewables/2013_09_Easac_Stockholm_Leo.pdf.

[30] Albrecht S, et al: Quantifying charge extraction in organic solar cells: the case of fluorinated PCPDTBT, *J Phys Chem Lett* 5:1131–1138, 2014.

[31] Kirchartz T, Agostinelli T, Campoy-Quiles M, Gong W, Nelson J: Understanding the thickness-dependent performance of organic bulk heterojunction solar cells: the influence of mobility, lifetime, and space charge, *J Phys Chem Lett* 3:3470–3475, 2012.

[32] www.infinitypv.com.

[33] www.heliatek.com.

[34] www.opvius.com (formally Belectric OPV).

[35] Ameri T, Lia N, Brabec CJ: Highly efficient organic tandem solar cells: a follow up review, *Energy Environ Sci* 6:2390–2413, 2013.

[36] Janssen RAJ, Nelson J: Factors limiting device efficiency in organic photovoltaics, *Adv Mater* 25:1847–1858, 2013.

[37] Qin Y, Chen Y, Cui Y, Zhang S, Yao H, Huang J, Li W, Zheng Z, Hou J: Achieving 12.8% efficiency by simultaneously improving open-circuit voltage and short-circuit current density in tandem organic solar cells, *Adv Mater* 29:1606340, 2017.

[38] Cui Y, Yao HF, Gao BW, Qin YP, Zhang SQ, Yang B, He C, Xu BW, Hou JH: Fine-tuned photoactive and interconnection layers for achieving over 13% efficiency in a fullerene-free tandem organic solar cell, *J Am Chem Soc* 139:7302–7309, 2017.

[39] Sullivan P, Schumann S, Da Campo R, Howells T, Duraud A, Shipman M, Hatton RA, Jones TS: Ultra-high voltage multijunction organic solar cells for low-power electronic applications, *Adv Energy Mater* 3:239–244, 2013.

[40] Scharber MC: On the efficiency limit of conjugated polymer: fullerene-based bulk heterojunction solar cells, *Adv Mater* 28:1994–2001, 2016.

[41] Dennler G, Scharber MC, Ameri T, Denk P, Forberich K, Waldauf C, Brabec CJ: Design rules for donors in bulk-heterojunction tandem solar cells-towards 15 % energy-conversion efficiency, *Adv Mater* 20:579–583, 2008.

[42] Koster LJA, Shaheen SE, Hummelen JC: Pathways to a new efficiency regime for organic solar cells, *Adv Energy Mater* 2:1246–1253, 2012.

[43] Leznoff, CC, Lever ABP, editors. Phthalocyanines: properties and applications. VCH publishers Inc.

[44] Li S, Ye L, Zhao W, Zhang S, Mukherjee S, Ade H, Hou J: Energy-level modulation of small-molecule electron acceptors to achieve over 12% efficiency in polymer solar cells, *Adv Mater* 28:9423–9429, 2016.

[45] Eley DD. Phthalocyanines as semiconductors. Nature 1948;162:819 and Vartanyan AT. J Phys Chem U.S.S.R. 1948;22:769.

[46] Yang JL, Schumann S, Hatton RA, Jones TS: Copper hexadecafluorophthalocyanine (F16CuPc) as an electron accepting material in bilayer small molecule organic photovoltaic cells, *Org Electron* 11:1399–1402, 2010.

[47] Mazzio KA, Luscombe CK: The future of organic photovoltaics, *Chem Soc Rev* 44:78–90, 2015.

[48] Li W, Furlan A, Hendriks KH, Wienk MM, Janssen RAJ. Efficient tandem and triple-junction polymer solar cells. J Am Chem Soc 2013;135:5529–5532.

[49] Armin A, Yazmaciyan A, Hambsch M, Li J, Burn PL, Meredith P: Electro-optics of conventional and inverted thick junction organic solar cells, *ACS Photonics* 2:1745–1754, 2015.

[50] Armin A, Hambsch M, Wolfer P, Jin H, Li J, Shi Z, Burn PL, Meredith P: Efficient, large area, and thick junction polymer solar cells with balanced mobilities and low defect densities, *Adv Energy Mater* 5:1401221, 2015.

[51] Gasparini N, Lucera L, Salvador M, Prosa M, Spyropoulos GD, Kubis P, Egelhaaf H-J, Brabec CJ, Ameri T: High-performance ternary organic solar cells with thick active layer exceeding 11% efficiency, *Energy Environ Sci* 10:885–892, 2017.

[52] Sullivan P, Duraud A, Hancox I, Beaumont N, Mirri G, Tucker JHR, Hatton RA, Shipman M, Jones TS: Halogenated boron subphthalocyanines as light harvesting electron acceptors in organic photovoltaics, *Adv Energy Mater* 1:352–355, 2011.

[53] Yao H, Chen Y, Qin Y, Yu R, Cui Y, Yang B, Li S, Zhang K, Hou J: Design and synthesis of a low bandgap small molecule acceptor for efficient polymer solar cells, *Adv Mater* 28:8283–8287, 2016.

[54] Chen C-C, Chang WH, Yoshimura K, Ohya K, You J, Gao J, Hong Z, Yang Y: An efficient triple-junction polymer solar cell having a power conversion efficiency exceeding 11%, *Adv Mater* 26:5670–5677, 2014.

[55] Nam M, Cha M, Lee HH, Hur K, Lee K-T, Yoo J, Han IK, Kwon SJ, Ko D-H: Long-term efficient organic photovoltaics based on quaternary bulk heterojunctions, *Nat Commun* 8:14068, 2017.

[56] An Q, Zhang F, Zhang J, Tang W, Deng Z, Hu B: Versatile ternary organic solar cells: a critical review, *Energy Environ Sci* 9:281–322, 2016.

[57] de Gier HD, Jahani F, Broer R, Hummelen JC, Havenith RWA: Promising strategy to improve charge separation in organic photovoltaics: installing permanent dipoles in PCBM analogues, *J Phys Chem A* 120:4664–4671, 2016.

[58] Cho N, Schlenker CW, Knesting KM, Koelsch P, Yip H-L, Ginger DS, Jen AKY: High-dielectric constant side-chain polymers show reduced non-geminate recombination in heterojunction solar cells, *Adv Energy Mater* 4:1301857, 2014.

[59] Chueh CC, Crump M, Jen AKY: Optical enhancement via electrode designs for high-performance polymer solar cells, *Adv Funct Mater* 26:321–340, 2016.

[60] Stec HM, Hutter OS, Hatton RA: Plasmon-active nano-aperture window electrodes for organic photovoltaics, *Adv Energy Mater* 3:193–199, 2013.

[61] Pereira HJ, Hutter OS, Dabera GDMR, Rochford LA, Hatton RA: Copper light-catching electrodes for organic photovoltaics, *Sus Energy Fuels* 1:859–865, 2017.

[62] Giannouli M, Drakonakis VM, Savva A, Eleftheriou P, Florides G, Choulis SA: Methods for improving the lifetime performance of organic photovoltaics with low-costing encapsulation, *Chem Phys Chem* 16:1134–1154, 2015.

[63] www.youtube.com/watch?v=uXQEpVRtGtw ultra-thin glass – coring willow glass.

[64] Hutter OS, Stec HM, Hatton RA: An indium-free low work function window electrode for organic photovoltaics which improves with in situ oxidation, *Adv Mater* 25:284–288, 2013.

[65] Xie F, Choy WCH, Wang C, Li X, Zhang S, Hou J: Low-temperature solution-processed hydrogen molybdenum and vanadium bronzes for an efficient hole-transport layer in organic electronics, *Adv Mater* 25:2051–2055, 2013.

[66] Yin Z, Wei J, Zheng Q: Interfacial materials for organic solar cells: recent advances and perspectives, *Adv Sci* 3:1500362, 2016.

[67] Osaka I, McCullough RD: Advances in molecular design and synthesis of regioregular polythiophenes, *Acc Chem Res* 41:1202–1214, 2008.

[68] Bannock JH, Krishnadasan SH, Nightingale AM, Yau CP, Khaw K, Burkitt D, Halls JJM, Heeney M, de Mello JC: Continuous synthesis of device-grade semiconducting polymers in droplet-based microreactors, *Adv Func Mater* 23:2123–2129, 2012.

[69] Zhao J, Li Y, Yang G, Jiang K, Lin H, Ade H, Ma W, Yan H: Efficient organic solar cells processed from hydrocarbon solvents, *Nat Energy* 1:15027, 2016.

[70] McDowell C, Bazan GC: Organic solar cells processed from green solvents, *Curr Opin Green Sustainable Chem* 5:49–54, 2017.

[71] Cao W, Li J, Chen H, Xue J: Transparent electrodes for organic optoelectronic devices: a review, *J Photon Energy* 4:040990, 2014.

[72] Jeong S, Jungn S, Kang H, Lee D, Choi S-B, Kim S, Park B, Yu K, Lee J, Lee K: Role of polymeric metal nucleation inducers in fabricating large-area, flexible, and transparent electrodes for printable electronics, *Adv Funct Mater* 27:1–8, 2017.

[73] Jin H, Pivrikas A, Lee KH, Aljada M, Hambsch M, Burn PL, Meredith P: Factors influencing the efficiency of current collection in large area, monolithic organic solar cells, *Adv Energy Mater* 2:1338, 2012.

[74] Lagrange M, Sannicolo T, Muñoz-Rojas D, Lohan BG, Khan A, Anikin M, Jiménez C, Bruckert F, Bréchet Y, Bellet D: Understanding the mechanisms leading to failure in metallic nanowire-based transparent heaters, and solution for stability enhancement, *Nanotech* 28:055709, 2017.

[75] Mulligan CJ, Wilson M, Bryant G, Vaughan B, Zhou X, Belcher J, Dastoor PC: A projection of commercial-scale organic photovoltaic module costs, *Sol Energ Mat Sol C* 120:9–17, 2014.

[76] Zhao G, Wang W, Bae T-S, Lee S-G, Mun CW, Lee S, Yu H, Lee GH, Song M, Yun J: Stable ultrathin partially oxidized copper film electrode for highly efficient flexible solar cells, *Nat Commun* 6:8830, 2015.

[77] Hutter OS, Hatton RA: A hybrid copper:tungsten suboxide window electrode for organic photovoltaics, *Adv Mater* 27:326–331, 2015.

[78] Hutter OS. Nanostructured copper electrodes for organic photovoltaics [PhD Thesis]. University of Warwick; 2015.

13

Upconversion and Downconversion Processes for Photovoltaics

Aruna Ivaturi*, Hari Upadhyaya**

*UNIVERSITY OF STRATHCLYDE, GLASGOW, UNITED KINGDOM; **WOLFSON CENTRE FOR MATERIALS PROCESSING, INSTITUTE OF MATERIALS AND MANUFACTURING, DEPARTMENT OF MECHANICAL, AEROSPACE AND CIVIL ENGINEERING, BRUNEL UNIVERSITY, LONDON, UXBRIDGE, UNITED KINGDOM

hari.upadhyaya@brunel.ac.uk, aruna.ivaturi@strath.ac.uk

13.1 Introduction

In all photovoltaic (PV) devices, apart from carrier recombination and parasitic resistance related losses, there are primarily two main loss mechanisms arising due to the absorption threshold of the absorber material [1,2]. All the incident photons with energy less than this threshold are not absorbed and hence do not significantly contribute to the generation of electron-hole pairs. These losses are called the *sub-bandgap or transmission losses*. For example, in the case of crystalline silicon solar cells, about 20% of the sun's energy (AM1.5 solar spectrum) is lost owing to these losses (see Fig. 13.1) [3]. On the other hand, all the incident photons with energy greater than the absorption threshold give rise to *lattice thermalization losses* because of the excess energy that is transformed into heat. This loss mechanism accounts for approximately 35% of the sun's energy for a crystalline silicon device [3]. Besides these two primary losses related to the intrinsic properties of the absorber material, there are other losses more related to the electronic properties of solar cells: (1) contact voltage losses, (2) recombination losses due to poor interface or material quality, (3) junction losses, and (4) reflection losses from interfaces. All these fundamental losses directly lead to an efficiency limit of ~30% for single-junction PV devices under nonconcentrated AM1.5 illumination—this is the so-called *Shockley–Queisser theoretical efficiency limit* (*S-Q limit*) [4].

One generic approach to address the fundamental losses arising from the mismatch between the incident photon energy and the absorber bandgap (or the absorption threshold) is via manipulating the sunlight prior to conversion also termed as *photon conversion*. The *sub-bandgap or transmission losses* can be addressed via a process called *upconversion* (UC) whereas the *lattice thermalization losses* can be addressed via *downconversion* (DC) [1].

This chapter gives a brief overview of the upconversion and downconversion concepts, materials, and integrated PV devices reported in the literature for performance enhancement.

A Comprehensive Guide to Solar Energy Systems. http://dx.doi.org/10.1016/B978-0-12-811479-7.00013-0

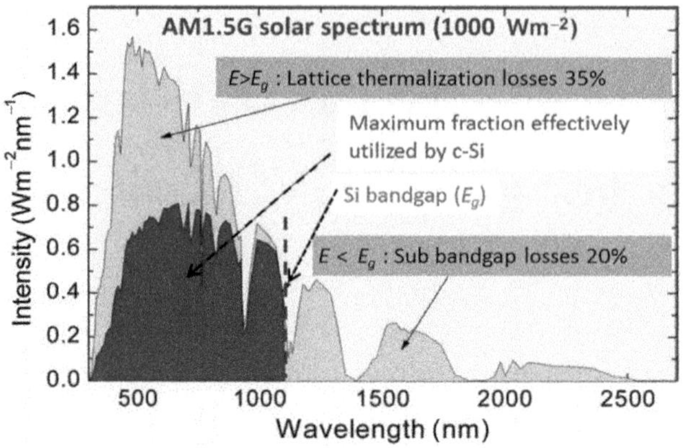

FIGURE 13.1 Spectral distribution of the AM1.5G solar irradiance (shaded in gray (green in the web version)) with the maximum fraction effectively utilized by c-Si solar cell (shaded in dark gray (red in the web version)). *Dotted line* gives the bandgap E_g of c-Si. All photons with $E < E_g$ give rise to sub-bandgap or transmission losses and with $E > E_g$ give rise to thermalization losses.

13.2 Upconversion

Upconversion refers to an anti-Stokes type nonlinear optical emission process in which one higher-energy photon is emitted for every two or more absorbed lower-energy photons (see Fig. 13.2) [5].

Upconversion of low-energy photons from a noncoherent radiation source like the sun is the most frequently a multistep process—the ground state absorption of low-energy photons populates the metastable energy level. Then energy transfer between the two excited ions or molecules occurs followed by excitation of higher-energy level. This is followed by a nonradiative relaxation and then spontaneous emission of the desired higher-energy photon due to radiative transition from the higher-excited energy level to the ground state [6]. Since the first experimental demonstration in 1966 [7], this effect has received renewed interest due to its ever-expanding application base, for example in, lasing [8], laser cooling [9], temperature sensing [10], biomedical imaging and therapy [11,12], 3D displays [13] and for broadening the spectral response of PV devices [6]. In 2002, Trupke and Green

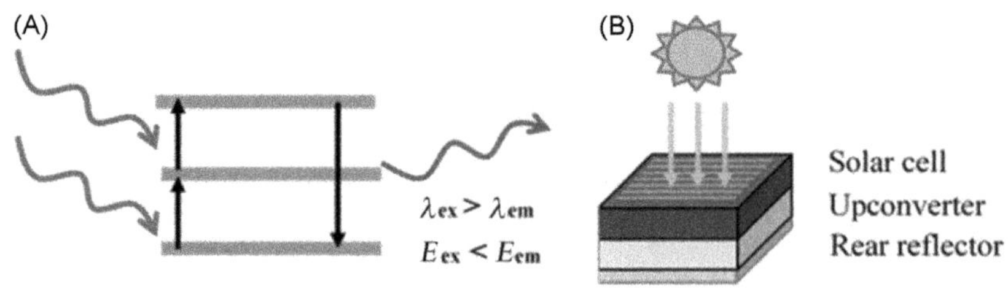

FIGURE 13.2 (A) Schematic showing upconversion process. (B) Frequently used configuration of integrating upconverter layer at the rear of the solar cell for addressing transmission losses. The rear reflector aids in harvesting the upconverted photons.

reported the first theoretical study of upconversion in the context of photovoltaics and showed that a system consisting conventional single-junction bifacial solar cells with a single bandgap of 2 eV and an ideal upconverter at its rear can achieve a PCE of 47.6% under nonconcentrated sunlight and 63.2% for concentrated sunlight [14]. Later in 2012 Johnson and Conibeer reported that the theoretical limit for efficiency of a Si solar cell with bandgap of 1.2 eV and an ideal upconverted illuminated by nonconcentrated light was 40%, as compared to efficiency of 33.25% for the solar cell alone [15]. In contrast to the theoretical predictions, several reported experimental studies have shown, however, that the expected efficiency increase in real devices are rather low, though significant enhancements are possible with materials with high upconversion quantum yields [6]. The reasons for such discrepancy and the ways to address these are discussed later in the chapter.

13.2.1 Upconversion Materials

Both Lanthanide (Ln)-based upconverters as well as organic upconverters have been explored in the literature to enhance the NIR response of PV devices. Lanthanides belong to the group of rare-earth (RE) elements, along with yttrium and scandium. The lanthanides include all the elements from lanthanum (with the atomic number 57 and an orbital configuration $5s^25p^64f^05d^16s^2$) to lutetium (with the atomic number 71 and an orbital configuration $5s^25p^64f^{14}5d^16s^2$). Upconversion is observed in materials based on the trivalent lanthanide ions (Ln^{3+}) (where 4f shell is partially filled and all $5d^1$ and $6s^2$ electrons are removed, except in lanthanum with an empty 4f shell and lutetium with completely filled 4f shell). Ln^{3+} ions exhibit unique optical properties due to many possible radiative transitions between the energy levels of the partially filled 4f shell, which are shielded by the completely filled outer lying 5s and 5p shells. As a result, the optical transitions in the 4f shell are only marginally affected by the surroundings and appear at nearly the same energies for different host materials [16]. Although the energetic position of an energy level is mainly undisturbed by the surroundings, the precise nature of the energy levels (especially the width and strength of the different transitions) is determined by the host material. The crystal field of the host material at the position of the Ln^{3+} results in a splitting of the energy levels into crystal field components (the so-called Stark levels) leading to an effectively broader absorption spectrum. A broader absorption spectrum is very much desired for applications in photovoltaics, as a larger fraction of the solar spectrum can be used for upconversion and, consequently, be utilized by the solar cell. Also, the combination of the doping level of the Ln^{3+} in the host and the host crystal structure determines the average distance d between the Ln^{3+}, which in turn has a strong impact on the upconversion performance. The host material of the Ln^{3+} also determines the likelihood of nonradiative losses such as multiphonon relaxation (MPR). The transition probability for MPR depends on the phonon energies of the host material and the energy gap between the considered energy levels [17]. A large variety of different host materials with low phonon energies to suppress nonradiative losses have been proposed to be suitable host materials for Ln^{3+} [5,16,18–25]. There are many reviews available that provide good overviews on available rare earth based upconverter materials [5,18,25–28]. To date, high internal upconversion quantum yield values at low irradiance suitable for photovoltaics have been reported only for monocrystalline and multicrystalline materials [6]. Upconversion nanomaterials, unfortunately are typically a factor of 10^3–10^5 less efficient than their bulk counter parts [29], but have the advantage and possibility of being combined with concepts such as photonics and plasmonics for enhancing their upconversion performance [30–35].

Organic upconverters utilized for addressing the sub-bandgap losses are based on triplet–triplet annihilation (TTA). Here, a sensitizer species absorbs light and crosses to a triplet state, the energy of which is rapidly and efficiently transferred to a second species, the emitter. The emitter triplet states are roughly half the energy of its first excited singlet state, so when two such triplets, $|T1\rangle|T1\rangle$, combine to create a supramolecular singlet, internal conversion to the $|S1\rangle|S0\rangle$ state occurs to yield subsequent upconverted fluorescence [36]. Though a quadratic process at low efficiency, once a certain triplet concentration is reached, the triplet decay is dominated by bimolecular reactions and the response on illumination density changes to linear [37–40]. In comparison to lanthanide-based upconverting materials, a higher absorptance can be achieved for thinner layers in organic upconverters, as the absorption transition is a singlet–singlet transition and not a partially forbidden transition like within the lanthanide's 4f shell. However, photostability and chemical lifetime are major challenges for organic upconverter materials in the context of their application in photovoltaics. Several reviews on TTA-based upconversion have been published, which discuss the available materials and underlying processes in detail [28,41,42].

The theoretical quantum efficiency of upconversion process is limited to 50%. The upconversion layer is usually placed at the rear of a PV device to capture the transmitted photons and thus it is possible to independently optimize the layer for enhanced device performance. This requires firstly the PV device to be bifacial such that it efficiently transmits the low-energy photons and secondly it efficiently utilizes the upconverted photons emitted by the upconverter. For efficient utilization of external radiation that impinges on the rear of the PV device, a rear reflector is placed behind the upconverter unit to enhance the performance of both the solar cell and upconverter.

13.2.2 PV Devices With Upconverters

In 1983, Saxena was the first to suggest the application of upconverters (terbium-doped lanthanum fluoride and thulium-doped calcium tungstate materials) for PV devices. However, the actual measurements with the solar cells were not reported [43]. The following subsections give a brief overview of the reported results on the upconversion-aided enhancement of IR response of PV devices. In the GaAs, c-Si as well as a-Si solar cells, upconverter is attached to the rear (in most cases) or front (proof-of-concept) of the solar cells. Whereas in dye-sensitized solar cells (DSSCs), organic solar cells, as well as perovskite solar cells (PSCs), additional concepts for upconverters internally integrated into the device have also been explored.

13.2.2.1 GaAs Solar Cells

In 1995, the first experimental demonstration of a proof-of-concept of upconversion enhanced response for any PV system was reported by Gibart et al. Using a 100 μm-thick Er^{3+}- and Yb^{3+}-doped vitroceramic at the rear of ultrathin GaAs solar cells, an external quantum efficiency due to upconversion (EQE_{UC}) of 2.5% under monochromatic laser excitation of 891 nm and ~25 W cm^{-2} irradiance was reported (corresponding to a normalized value of 10^{-3} $cm^2 W^{-1}$) [43]. Since then considerable progress has been made in the field of application of upconversion to PV devices.

13.2.2.2 Crystalline Silicon Solar Cells

The most frequently investigated materials for addressing the sub-bandgap losses of crystalline Si solar cells are the Ln^{3+}-doped hexagonal sodium rare-earth tetrafluoride (β-$NaREF_4$) [44–47]. In addition to these, other alternatives such as Ln^{3+} doped rare-earth oxide (RE_2O_3), oxysulfide (RE_2O_2S) [47–49], glasses, and glass ceramics have also been explored and reported with the corresponding enhancement in solar cell response [50,51].

In 2003 Shalav et al. reported for the first time the application of microcrystalline β-$NaYF_4$:20%Er^{3+} upconverter to a bifacial crystalline silicon solar cell [52]. In 2005 the same group used the upconverter mixed in an acrylic adhesive medium [53] and reported EQE_{UC} of the combined silicon solar cell upconverter system of 2.5 % under monochromatic excitation of 1523 nm at an irradiance ~0.2 W cm^{-2} (normalized EQE_{UC} value < 0.125 $cm^2 W^{-1}$). By improving the device and changing the matrix to white oil with rubberizer matrix, the same group reported in 2007 an increased EQE_{UC} of 3.4% at 1523 nm, however at increased irradiance of 2.4 W cm^{-2}, resulting in a lower-normalized EQE_{UC} of 0.014 cm^2 W^{-1} [54]. In 2010 Fischer et al. reported an EQE_{UC} of 0.34% at a much lower irradiance of 1.09 W cm^{-2} at 1523 nm for β-$NaYF_4$:20%Er^{3+} upconverter powder filled into a powder cell that was attached to the rear of a silicon solar cell using an index-matching liquid. This corresponds to a normalized EQE_{UC} of 0.03 cm^2 W^{-1} [55]. Using an optimized bifacial silicon solar cell [56] on the other hand, an EQE_{UC} of 1.69% was reported (normalized EQE_{UC} of 0.15 cm^2 W^{-1}) and a further increase in EQE_{UC} with increase in the Er^{3+} doping concentration to 25% as well as increasing the concentration of phosphor was reported in 2014 [57–59]. For example, using 84.9 w/w% of β-$NaYF_4$:20%Er^{3+} embedded in perfluorocylobutane (PFCB), an EQE_{UC} as high as 5.72% was achieved under 1523 nm excitation with an irradiance of 0.45 W cm^{-2}, corresponding to a normalized EQE_{UC} of 0.126 cm^2 W^{-1} [58]. In 2013, Martin-Rodriguez et al. explored microcrystalline Gd_2O_2S:10%Er^{3+} as a potential alternative to β-$NaYF_4$:20%Er^{3+} [48]. Under monochromatic excitation at 1511 nm with an irradiance of 0.50 W cm^{-2}, an EQE_{UC} of 4.09% was achieved for the microcrystalline material filled in a powder cell and around 7.5% for the material embedded in PFCB with a concentration of 84.9 w/w% [58]. In 2013 another potential alternative the monocrystalline BaY_2F_8:30%Er^{3+} with even higher EQE_{UC} was reported by Boccolini et al. [16]. As proof-of-concept the upconverter was mounted in front of a conventional silicon solar cell, an EQE_{UC} of 6.5% at a high irradiance of 8.5 W cm^{-2} under illumination at 1557 nm was reported [16]. Later in 2015 by using optimized bifacial silicon solar cells and using the monocrystalline BaY_2F_8:30%Er^{3+}, a record value of EQE_{UC} of 8.0% was reported at a higher irradiance of 0.45 W cm^{-2} by Fischer et al. This value corresponds to normalized EQE_{UC} value of 0.177 cm^2 W^{-1} [60]. Another group of upconversion materials that have been applied on top of the crystalline silicon solar cells are fluorozirconate glass with 9.1% Er^{3+} doping and fluoroindate glass with 2.25% Er^{3+} and 0.1% Yb^{3+} co-doping [61,62]. In 2009 Henke et al. using fluorozirconate glass with 9.1% Er^{3+} doping, reported an EQE_{UC} of 1.6% achieved at 1540 nm, under a very high irradiance, ~107 W cm^{-2}, resulting in rather low normalized value around 10^{-9} cm^2 W^{-1} [61].

Later in 2013 Hernandez-Rodriguez et al. reported an EQE_{UC} of 0.4% (1480 nm excitation, 67 W cm^{-2} irradiance) corresponding to a normalized EQE_{UC} of 6.0×10^{-5} cm^2 W^{-1} for fluoroindate glass with 2.25% Er^{3+} and 0.1% Yb^{3+} co-doping [62]. Pellé et al. on the other hand in 2011 reported EQE_{UC} of 1.4% and 2.4% respectively for ZBLAN:4.7%Er^{3+} fluoride glass and disordered CaF_2–YF_3 crystal, under 1540 nm excitation with 141 W cm^{-2} irradiance for upconverter material attached at the rear of a bifacial silicon solar cell [63].

It is important to note that though for application in photovoltaics broadband illumination is more relevant, most of the early investigations as discussed above were carried out under monochromatic excitation. In 2011, Goldschmidt et al. first reported experiments applying broadband illumination to upconverter solar cell devices. Using 98 w/w% of β-$NaYF_4$:20%Er^{3+} in zapon varnish, they showed an EQE_{UC} of 0.81% averaged over the broadband excitation region from 1460 nm to 1600 nm with a photon flux equivalent to 458 suns (in the considered spectral region) [64]. For the assessment of materials under broadband illumination, a more meaningful figure-of-merit is considered to be the additional short-circuit current density $\Delta j_{SC,UC}$ that can be generated due to upconversion of sub-band-gap photons under illumination with the full solar spectrum onto the upconverting solar cell device than the averaged EQE_{UC} [63]. In 2014 Fischer et al. applied 75.7 w/w% of β-$NaYF_4$:25%Er^{3+} embedded in the polymer PFCB with to an optimized bifacial silicon solar cell and reported a $\Delta j_{SC,UC}$ of 3.4 mA cm^{-2} and 13.3 mA cm^{-2} for the equivalent solar concentration levels of 78 suns and 207 suns. By increasing the concentration of β-$NaYF_4$:25%Er^{3+} embedded in PFCB to 84.9 w/w%, the additional current due to upconversion $\Delta j_{SC,UC}$ was reported to be increased to 8.1 mA cm^{-2} at 70 suns concentration [58]. On the other hand, for 84.9 w/w% of Gd_2O_2S:10%Er^{3+} embedded in PFCB a $\Delta j_{SC,UC}$ of 5.9 mA cm^{-2} was reported under 67 suns concentration, revealing the superior performance of the β-$NaYF_4$:25%Er^{3+} under broadband illumination compared to the Gd_2O_2S:10%Er^3 [47]. In 2015 Fischer et al. also reported record enhancements in using monocrystalline BaY_2F_8:30%Er^{3+} to optimize bifacial silicon solar cells, under an equivalent solar concentration of 94 suns, an additional short-circuit current density due to upconversion of 17 mA cm^{-2} was achieved. This corresponds to a record relative enhancement of 0.55% (40-fold increase compared to first values published in 2011) in the additional short-circuit current density generated due to upconversion [64].

13.2.2.3 Amorphous Silicon Solar Cells

Due to the higher-absorption threshold (around 730 nm), the potential gain due to upconversion for hydrogenated amorphous silicon (a-Si:H) solar cells is even higher than for crystalline silicon solar cells (absorption threshold around 1100 nm). Both lanthanum-based upconverters and organic upconverters have been applied to a-Si:H solar cells.

In 2009 de Wild et al. first reported the application of β-$NaYF_4$:18%Yb^{3+},2%Er^{3+} embedded in PMMA between a ZnO:Al 0.5% rear contact and a white paint rear reflector to a-Si:H solar cells [65]. An EQE_{UC} of approximately 0.02% under laser illumination at 980 nm with 3 W cm^{-2} irradiance was reported (resulting in a normalized EQE_{UC} of 0.7×10^{-4} $cm^2 W^{-1}$) [66]. In 2013, using a similar configuration with Gd_2O_2S:10% Yb^{3+},5%Er^{3+} upconverter, de Wild et al. reported an EQE_{UC} of 0.06% under laser illumination at 981 nm with 0.2 W cm^{-2} irradiance (a higher normalized EQE_{UC} of 0.003 $cm^2 W^{-1}$) and a $\Delta j_{SC,UC}$ of 0.1 mA cm^{-2} under broad band illumination of 20 suns concentration [67]. In 2012, Chen et al. reported the application of β-$NaYF_4$:25%Yb^{3+},1%Ho^{3+} microprisms to both a standard p–i–n a-Si and an a-Si/a-Si tandem solar cell and observed that the relative enhancement of the current is roughly the same for both the solar cells [68]. For the tandem solar cell, they reported current enhancements equivalent to EQE_{UC} of up to 0.007% under laser illumination at 980 nm with 0.5 W cm^{-2} irradiance (i.e., normalized EQE_{UC} of 1.4×10^{-4} $cm^2 W^{-1}$). By replacing Ho^{3+} by Er^{3+}, an improved EQE_{UC} of 0.015% under similar laser excitation was reported (i.e., normalized EQE_{UC} of 2.88×10^{-4} $cm^2 W^{-1}$) [69].

In 2012, Cheng et al. first reported application of the organic TTA upconverting system of PQ4PDNA/rubrene [nitroaminopalladiumtetrakis porphyrin (PQ4PdNA) as sensitizer and rubrene as the emitter] to an a-Si:H silicon solar cell. The cuvette with PQ4PDNA/rubrene dissolved in toluene was optically coupled to the rear of the solar cells with an index-matching liquid. A $\Delta j_{SC,UC}$ of 0.30 mA cm^{-2} under irradiation with an equivalent concentration of 48 suns was reported [70]. Shulze et al. increased the absorption and the outcoupling of upconverted photons by adding 100 μm-diameter Ag-coated glass spheres to the cuvettes with PQ4PDNA/rubrene in toluene, resulting in $\Delta j_{SC,UC}$ of 0.275 mA cm^{-2} under irradiation with an equivalent concentration of 19 suns [71]. The current was not measured directly, but calculated based on EQE_{UC} measurements covering the relevant spectral range. Later in 2014, Schulze and Schmidt reported an $\Delta j_{SC,UC}$ of 4.7×10^{-3} mA cm^{-2} under a very low irradiation of 1.4 suns, using the upconverter system of PQ4PDNA and rubrene/9,10-bis-phenylethynylanthracene in cuvettes additionally filled with Ag-coated glass spheres [28].

13.2.2.4 Dye-Sensitized Solar Cells

The sub-bandgap losses in DSSCs strongly depend on the absorption threshold of the dye species used. A relatively large number of studies have been published in literature reporting the use of Ln^{3+}-based upconverter materials to enhance the device performance of DSSCs. However, many of these results must be considered with care as the enhancements cannot be clearly attributed to upconversion. In 2017, Liu et al. developed a composite photoanode of Nb_2O_5 coated TiO_2 NWAs/UC-EY-TiO_2 NPs on flexible Ti mesh for fully flexible DSSC and studied the effects Er^{3+}-Yb^{3+} doping concentration on the upconversion luminescence intensity and corresponding photovoltaic performance. When the Er^{3+} and Yb^{3+} doping amounts were controlled at 2.0 Er^{3+}-1.0 Yb^{3+}, the PCE of the flexible DSSCs reached about 7.29%, which was enhanced about 52% for the DSSCs of undoped TiO_2 composites. The introduction of Nb_2O_5 coating further enhanced the PCE from 7.29 to 8.10% [72]. In another study, Han et al. synthesized upconversion Er, Yb-CeO_2 hollow spheres using carbonaceous spheres as removable template via hydrothermal method for improving the efficiency of dye-sensitized solar cells (DSSCs). PCE was enhanced by 27% due to the upconversion effect, scattering effect, and higher dye loading capacities of the modified devices with Er,Yb-CeO_2 [73]. In 2016 Meng et al. mixed plasmon-enhanced upconversion composite Y_2O_3:Er/Au@TiO_2 having three-dimensional star-like morphology with submicron TiO_2 (200 nm) nanoparticles to form a multifunctional scattering layer in TiO_2-based DSSCs. The system showed extended light-absorbing range to near-infrared region and improved light-scattering ability, leading to an improved photovoltaic performance of DSSCs. With the optimum mixing ratio of 23%, a PCE of 8.62% was attained as compared to 6.77% without the upconversion composite which corresponds to an improvement of 27.6% [74]. In another study, Luoshan et al. synthesized and incorporated novel multi-shell-coated and Au-nanoparticle decorated hexagonal submicron β-$NaYF_4$:Yb^{3+}, Er^{3+}@SiO_2@Au/@TiO_2 prisms into the photoanodes of DSSCs. The optimal PCE obtained in a typical DSSC containing the modified photoelectrode was 7.79%, corresponding to 28.1% higher than that of the DSSC with a pure TiO_2 photoanode [75]. Later, Yu et al. designed up-conversion YbF_3-Ho/TiO_2 nanoheterostructures and explored them as photoelectrode material to yield DSSCs with enhanced performance. When integrating the UC/TiO_2 nanoheterostructures into DSSCs, an overall PCE of 8.0% was achieved—this corresponded to 23% enhancement in the overall conversion efficiency [76]. Liao et al. on the other hand synthesized upconversion/semiconductor submicron hollow spheres composed of inner $Na_xGdF_yO_z$:Yb/Er shell and outer TiO_2 shell by exploiting colloidal carbon spheres as the scaffold. The hollow spheres were then incorporated into the TiO_2 nanoparticle film photoanode to yield DSSCs with improved performance. Compared to the PCE of 6.81% for the pristine device, the DSSC assembled with the introduction of 8 wt% $Na_xGdF_yO_z$:Yb/Er@TiO_2 hollow spheres in the photoanode exhibited an optimal PCE of 7.58%, corresponding to 11.31% performance enhancement [77]. Bai et al. utilized 1D upconversion CeO_2:Er, Yb nanofibers synthesized by electrospinning method as scattering layer on top of TiO_2 photoanode in DSSC and observed an enhancement of 14% in the PCE due to the upconversion and scattering effects [78]. In 2014 Yuan et al. developed and explored hexagonal core–shell-structured β-$NaYbF_4$:Er^{3+}(2%)/$NaYF_4$:Nd^{3+}(30%), as an NIR energy relay material for DSSCs. These UCNPs were shown to harvest light energy in multiple NIR regions, and subsequently convert the absorbed energy into visible light where the DSSCs strongly absorb. The NIR-insensitive DSSCs show compelling photocurrent enhancement through binary upconversion under NIR light illumination either at 785 or 980 nm, substantiating efficient energy relay by these UCNPs. The overall conversion efficiency of the DSSCs was improved from 4.95% to 5.24% with the introduction of UCNPs under simulated AM 1.5 solar irradiation [79]. In another study, Zhou et al. synthesized core–shell structured β-$NaYF_4$:Er^{3+}, Yb^{3+}@SiO_2 upconversion nanoparticles with similar diameters to TiO_2 nanoparticles and mixed them in TiO_2 photoelectrode, which eliminated the size-dependent light scattering effect in the light harvesting process, leading to an enhancement of 6% in the PCE (from 5.96% to 6.34%). In 2011, Xie et al. doped TiO_2 nanoparticles with Er^{3+} and Yb^{3+} ions, which were then directly incorporated for the electrodes of the DSSC. The Er^{3+}/Yb^{3+} acted not only as upconverter, but also as dopant, which improved the electrical properties. Furthermore, the addition of upconverter also modified the scattering properties, as discussed in the later works on upconverters integrated into the TiO_2 photoanode [80] or the rear reflector [81]. In 2012 Yuan et al. intermixed colloidal nanocrystals of β-$NaYF_4$:20%Yb^{3+},2%Er^{3+} with the Z907 dye for DSSC. Under monochromic excitation with 980 nm and 8 W cm^{-2}, an EQE_{UC} of 0.011% was observed, which is equivalent to a normalized EQE_{UC} of 0.132×10^{-4} cm^2 W^{-1}. Both fluorescence resonance energy transfer and luminescence-mediated energy transfer were discussed as potential routes for the migration of the energy from the upconverter to the dye [82]. In another report, in 2011 Liu et al. attached $Y_3Al_5O_{12}$ transparent ceramic co-doped with 3.0% Yb^{3+} and 0.5% Er^{3+} to the rear of a DSSC, which showed a $\Delta j_{SC,UC}$ of 0.2 mA cm^{-2} under 980 nm laser excitation of 500 mW power [83]. In 2013 Miao et al. applied microcrystalline Yb_2O_3 and β-$NaYF_4$:15%Yb^{3+},3%Er^{3+} on top of DSSC and observed under 980 nm laser illumination of ~474 W cm^{-2}, EQE_{UC} of only 0.029% for the device containing the former and 0.015% for the device containing latter—resulting in normalized EQE_{UC} values of 6.1×10^{-7} cm^2 W^{-1} and 3.3×10^{-7} cm^2 W^{-1}, respectively [84]. In 2013 Nattestad et al. first report the application of organic upconverter in a DSSC. A degassed solution of the PQ4PDNA/rubrene upconverter system was placed in a cavity integrated between the DSSC and rear reflector. For an illumination equivalent to 3 suns, $\Delta j_{SC,UC}$ of 2.25×10^{-3} mA cm^{-2} was observed [85]. The same group in 2014 observed an improvement by using an upconverting system PQ4PDNA with the hybrid emitter rubrene/9,10-bis-phenylethynylanthracene. An additional short-circuit current density of 4.05×10^{-4} mA cm^{-2} under an extremely low irradiation equivalent to 0.3 suns was observed [28].

13.2.2.5 Organic Solar Cells

In organic solar cells, sub-bandgap losses amount to more than 70% of the incident photons and more than 50% of the incident power [6]. In 2011 Wang et al. used layer of commercial upconversion phosphor based on a yttrium fluoride host doped with ytterbium and erbium to P3HT:PCBM solar cell. An increase in photocurrent density of ~0.0135 mA cm^{-2} due to upconversion was reported under 975 nm laser diode illumination (25 mW cm^{-2}) [86]. Later in 2012 the same group integrated upconversion MoO_3:Yb^{3+}/Er^{3+} layer as a buffer layer between the active layer and the electrode into a P3HT:PCBM-based organic solar cell and reported an increase in photocurrent density to upconversion under 975 nm laser diode illumination under different excitation intensity [87]. In 2012 Schulze et al. applied PQ4PDNA/rubrene-based upconverter materials (as described for the a-Si:H solar cells, in Section 13.2.2.3) to two organic solar cells with PCDTBT:$PC_{71}BM$ and P3HT:ICBA respectively as active materials and with 3.8% and 3.1% efficiencies under AM1.5G irradiation. An additional current due to upconversion of 0.048 mA cm^{-2} was achieved at an equivalent concentration of 17.3 suns for the solar cell made of P3HT:ICBA. In another report a layer of β-$NaYF_4$:Yb^{3+},Er^{3+} nanoparticles deposited on top of a glass/ITO/PEDOT:PSS/P3HT:PCBM/Ca/Al solar cell structure under irradiation with a 980 nm laser with 4.9 W cm^{-2} was reported to achieve an EQE_{UC} of 0.004% (a normalized value of 8.9×10^{-6} cm^2 W^{-1}) [88]. In 2017 Lin et al. reported a solid-state organic interband solar cell that shows enhanced photocurrent derived from TTA that converts sub-bandgap light into charge carriers. Femtosecond resolution transient absorption spectroscopy and delayed fluorescence spectroscopy were used to provide evidence for the triplet sensitization and upconversion mechanisms, while external quantum efficiency measurements in the presence of a broadband background light were used to demonstrate that sub-bandgap performance enhancements achievable in the device [89].

13.2.2.6 Perovskite Solar Cells

PSCs (e.g. based on $CH_3NH_3PbI_3$) are usually unable to utilize light beyond the visible region limited by their intrinsic bandgap, which accounts for 58% of the total solar energy. Consequently, upconversion has been recently explored as a promising way to harvest of this region of the solar spectrum and transform it into available visible light to enhance the IR response of PSCs. In 2016, Chen et al. integrated $LiYF_4$:Yb^{3+},Er^{3+} single crystal in front of PSC and demonstrated efficiency enhancement [90]. In another study Roh et al. introduced hydrothermally grown $NaYF_4$:Yb^{3+},Er^{3+} nanoprisms as upconverting centers to the TiO_2 mesoporous layer in PSCs and showed an increase in PCE due to upconversion [91]. Later in 2017, Wang et al. demonstrated performance enhancement by using hydrothermally grown 3% Er^{3+} and 6% Yb^{3+} co-doped TiO_2 nanorod arrays as electron transfer material in PSCs as compared to those based on undoped TiO_2 [92]. However, in all these studies, the device characteristics under laser irradiation or under higher solar concentrations (to clearly attribute the contributions as only due to upconverison) were not reported. In another study in 2016, He et al. incorporated $NaYF_4$:Yb^{3+}, Er^{3+} upconversion nanoparticles as the mesoporous electrode for $CH_3NH_3PbI_3$ based PSCs. The incorporation of $NaYF_4$:Yb^{3+}, Er^{3+} nanoparticles as the mesoporous electrode led to a short-circuit current density of 0.74 mA cm^{-2} upon excitation with 980 nm laser with 28 W cm^{-2} [93]. More recently in 2017, Zhou et al. demonstrated that semiconductor plasmon-sensitized nanocomposites ($mCu_{2-x}S@SiO_2@Er_2O_3$), which could efficiently convert broadband infrared light (800–1600 nm) to visible benefiting from the Localized Surface Plasmon Resonance (LSPR) of $Cu_{2-x}S$ when mixed with TiO_2 paste and applied as electron extraction layer in a PSCs led to an expanded response in the range of 800–1000 nm. The short-circuit current density of the device was observed to increase with the power density of the 980 nm excitation laser with >0.85 mA cm^{-2} obtained for 45 W cm^{-2} [94].

13.2.3 Approaches to Increase Upconversion Performance Enhancement

In contrast to the promising potential of upconversion and the steady progress that has been made in terms of application to PV over the last few years, the overall efficiency enhancement achieved upon application of upconversion are still low. This necessitates the important means to be explored to increase upconverting solar cell device performance. There are two approaches that could potentially help in addressing this issue: (1) enhancing the upconverter properties and (2) by modifying the upconverter environment.

13.2.3.1 Material Optimization

13.2.3.1.1 LN^{3+}-BASED UPCONVERTERS

For enhancing the upconverter properties especially in case of lanthanum-based upconverters, the choice and concentration of dopant material and the properties of the host are important and need to be optimized for highest device performance. In the context of photovoltaics, the additional desirable properties for the host material and upconverter are chemical and optical stability, strong crystal fields at the position of the upconverting material for high radiative transition probabilities and broad absorption, high transparency outside the active region and low scattering to avoid parasitic optical losses and finally, a high refractive index to reduce optical losses.

13.2.3.1.2 ORGANIC UPCONVERTERS

In addition to synthesis of even more efficient TTA upconverter molecules, optimizing the concentration of these materials in their surroundings is an important factor to be considered. Increasing the concentration of active species has been mentioned as the major strategy to increase the UCQY of TTA-based upconverter materials—especially increasing the steady-state concentration of emitter molecules in their excited state [28]. As simply increasing the concentration of emitter molecules in the solution is limited by the solution limit, increasing the concentration of sensitizer molecules as well as the lifetime of the excited triplet state of the emitter is considered a promising approach [6].

13.2.3.2 Material Environment

13.2.3.2.1 PLASMONICS AND PHOTONICS

Dielectric and metal photonic nanostructures are powerful tools to enhance upconversion performance, because they can enhance both the absorption and emission processes. Photonic nanostructures can increase the local irradiance on the upconverter, thus increasing absorption and, due to the nonlinearity of upconversion, also the UCQY. The photonic nanostructures on the other hand change the local photonic density of states (LDOS), which can be used to amplify desired radiative transitions and suppress others [31–34].

It is important to note that both the plasmonic and the dielectric photonic concepts to enhance upconversion performance are not compatible with mono- and microcrystalline upconverter materials. Instead, they require the use of either implanted lanthanide ions or the use of nanocrystalline or organic TTA-based upconverter materials.

13.2.3.2.2 SPECTRAL CONCENTRATION

The spectral range, in which upconverter efficiently converts incident photons to higher energies, is quite narrow (typically <100 nm). The limited spectral range can be extended by either co-doping or in combination with a sensitizer [95], or by using a second luminescent material that absorbs in a wide spectral range and emits in the absorption range of the upconverter. The latter process called the *spectral concentration* has the advantage of extending spectral range and enhancing the photon flux in the absorption range of the upconverter, which then leads to a higher UCQY [96,97].

13.3 Downconversion

Downconversion (also known as "quantum cutting") is the opposite process to upconversion where one high-energy photon is "cut" into two lower-energy photons (see Fig. 13.3). This process can reduce thermalization losses of hot charge carriers after the absorption of a high-energy photon. If both lower-energy photons can be absorbed by the solar cell, current doubling is achieved for the region of the solar spectrum that consists of photons with energies exceeding 2 E_g [98,99].

In 1957 Dexter first proposed the idea to obtain quantum yields >100% by creating multiple photons through "cutting" a single photon into two lower-energy photons [100]. The mechanism he proposed involved the simultaneous energy transfer from a donor to two acceptors, each accepting half the energy of the excited donor. In 1974 the first experimental evidence for quantum yields >100% was reported for $YF_3{:}Pr^{3+}$. The mechanism was not the one proposed by Dexter, but involved two sequential emission steps from the high-energy $^1S_0$ level of Pr^{3+} ($^1S_0 \rightarrow {}^1I_6$ followed by relaxation to the $^3P_0$ level and emission of a second visible photon from $^3P_0$) [101,102]. Later, quantum cutting via two sequential energy transfer steps in the Gd^{3+}–Eu^{3+} couple was discovered and, based on the analogy with the two-step energy transfer process leading to upconversion, it was called "downconversion" [103]. The potential of downconversion for increasing the efficiency of solar cells was

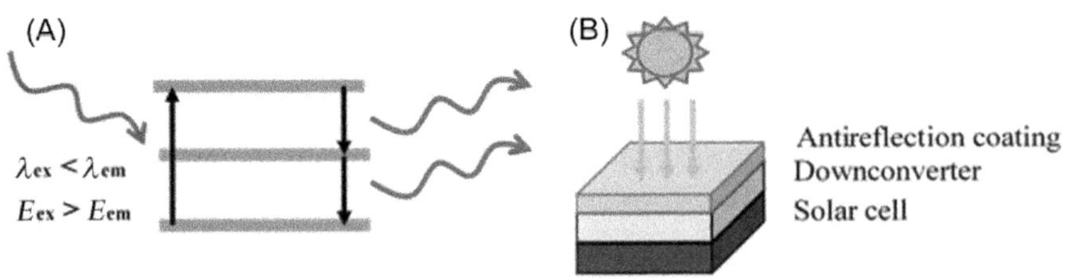

FIGURE 13.3 (A) Schematic showing downconversion process (B) Frequently used configuration of integrating downconverter layer at the rear of the solar cell for addressing transmission losses. The antireflection coating aids in harvesting the upconverted photons.

realized soon afterwards [98]. The first experimental demonstration of downconversion for solar cells involved the Tb^{3+}–Yb^{3+} couple where quantum cutting was achieved through cooperative energy transfer from Tb^{3+} to two Yb^{3+}-ions, via the same mechanism that was suggested by Dexter [104].

13.3.1 Downconversion Materials

Most of the early work on downconversion for solar cell applications is dominated by lanthanide ions due to their useful optical properties. Recently, nanomaterials have also been explored as potential downconverters. To maximize benefits obtained through downconversion, the host material and lanthanide ions must be carefully chosen accordingly with the type and design of the solar cell. Host materials must exhibit high transmittance, photostability, excitation energy, absorption strength, chemical and thermal stability, and low scattering [105–107]. The formation of defects and traps within the host material results in energy being absorbed inside the host instead of being transferred to the activator ion [105,106]. Therefore a highly crystalline host is recommended. Main criteria to choose the appropriate lanthanide ion are high emission lifetime and good chemical and electrical stability [106]. A downconverter is usually placed on the top of a solar cell and one of the drawbacks of this is that the layer often is highly reflective which causes radiative losses. This can be avoided by an antireflective coating designed to reflect the downconverted emission back into the solar cell [105]. Of the various lanthanide ions that have been explored as downconverters for solar cells, a good experimental demonstration involved a system of a Tb^{3+}-Yb^{3+} couple in $(Y,Yb)PO_4$:Tb^{3+} [108]. Downconversion in Tb^{3+}-Yb^{3+} is also reported in $GdAl_3(BO_3)_4$ [109], $GdBO_3$ [110], Y_2O_3 [111], CaF_2 nanocrystals [112,113], and lanthanum borogermanate glass [114]. Good quantum yields desirables for solar cell applications have also been obtained with Gd^{3+} [115] and Eu^{3+} [116]. Other common downconversion lanthanides materials useful for solar cell integration are Pr^{3+}-Yb^{3+} co-doped KY_3F crystals [117], Nd:SrTiO thin films [118] and $LiGdF_4$:Eu^{3+} and $LiGdF_4$: Er^{3+}, Tb^{3+} [119], among others [105,106]. Nanophosphors such as an $SmPO_4$ NPs-doped TiO_2 [120], Eu^{2+}-doped barium silicate (Ba_2SiO_4:Eu^{2+}) [121] have also been explored. Recently, graphene quantum dots (QDs) were explored as an option for downconversion in commercial solar cells [122]. Hybrid nanostructures, for example, colloidal ZnS nanoparticles/Si-nanotips showed promising potential for efficient solar spectrum utilization in crystalline silicon solar cells [123]. Recently new kind of QDs based on inorganic perovskites have also been explored [124].

13.3.2 PV Devices With Downconverters

Since the pioneering work reported by Trupke and Richards in 2002, different materials such as oxides doped with lanthanide ions [125,126], silicon nanoparticles [127–129], and quantum dots [130–133], among others have been explored as downconversion layers for solar cells.

13.3.2.1 Silicon and GaAs-Based Solar Cells

In 2012 Chin-Lung et al. reported >2% increase in the conversion efficiency of a screen-printed monocrystalline silicon solar cells by using a layer of Eu^{3+}-doped $Y(OH)_3$ nanotubes [134]. In 2014 Chowdhury et al. reported an increase in J_{sc} from 2.33% to 5.16% due to the presence of Si NPs deposited onto a aSi:H (n+)/a-Si:H (i)/c-Si (p) HIT solar cell layer stack [129]. In another study, Gardelis and Nassiopoulou reported an increase of up to 37.5% in conversion efficiency of a Si-based solar cell using $CuInS_2$/ZnS core-shell quantum dots on the active area [130]. They attributed this improvement to the combined effect of downconversion and the antireflecting property of the dots. In 2015 Merigeon et al. successfully used Pr^{3+}–Yb^{3+} co-doped with ZBLA (family of glasses with a composition of ZrF_4-BaF_2-LaF_3-AlF_3-NaF) as downconverters on top of a commercial silicon solar cell [135]. In 2016 Meng-Lin Tsai et al. reported an increase in the efficiency of an n-type Si heterojunction solar cell using graphene quantum dots (GQDs) as downconverters [122]. In 2014 Chien-Chung et al. utilized a dual-layer of CdS QDs on a hybrid GaAs solar cell and reported >2% increase in the overall power conversion efficiency [136]. In general, QDs have been used as downconverters in GaAs-based solar cells. Some examples are layers of crystalline silicon nanopillar arrays and QDs of CdS and ZnS nanoparticles/Si nanotips [123] and QDs of CdS and CdSe/ZnS core-shell in colloidal solutions deposited on the top layer of a GaAs cell [137].

13.3.2.2 Dye-Sensitized Solar Cells

In DSSCs the downconverter can be integrated either with the photoelectrode and the dye inside the cell or an additional layer containing the downconverter can be placed in the front of the photoanode. One advantage of using an external layer is that, on the one hand, it allows using energy in the UV range owing to its conversion to visible light and on the other hand to improve stability on DSSCs because it prevents degradation of the dye or electrolyte caused by UV energy.

Recently lanthanide ions as Eu^{3+}, Sm^{3+} [103], $LiGdF_4$:Eu [138], $Ca_3La_{3(1-x)}Eu_{3x}(BO_3)_5$ [139], among others on TiO_2 films have been studied as photoanodes. The DSSCs were reported to have increased efficiencies in all cases. A successful example is a solar cell with a photoanode of TiO_2 doped with $Y_3Al_5O_{12}$:Ce proposed by Guang Zhu et al. with an efficiency of 7.91% which is higher than an efficiency of 6.97% of a solar cell without the downconversion layer [140]. Other typical oxides utilized in DSSCs such as SiO_2, ZnO, $SrTiO_3$, and Nb_2O_5 doped with phosphors for a downconversion approach were explored lately [141]. In 2014 Lim et al. [142] studied the possibility of using downconversion and upconversion materials (double composite layer) on TiO_2. The downconversion material was $Y_3Al_5O_{12}:Ce^{3+}$ while the upconversion material was $Gd_2O_3:Er^{3+}/Yb^{3+}$. These DSSCs were reported to have about 21% higher PCE than the solar cell without the inclusion of the double composite layer. Recently, Chander et al. demonstrated that by using a nanophosphor $YVO_4:Eu^{3+}$ on a TiO_2 photoanode, it is possible to improve the absorption and outdoor stability of a DSSCs [143].

13.3.2.3 Organic Solar Cells

As other solar cells, OPV cells also present the possibility to increase their efficiencies and UV stability through downconversion. In 2014 Li et al. utilized $SmPO_4$ nanophosphor as downconverter in a hybrid solar cell of TiO_2/P3HT bulk heterojunction and reported 3% increase in PCE [144]. In 2015 Liu et al., utilized $NaYF_4$:Yb^{3+}/Er^{3+} nanophosphors into the TiO_2 (cathode) and reported an increase in the short circuit current in a PCDTBT:PC_{71}BM heterojunction solar [145]. In another study, Na et al. reported a bifunctional layer of Eu^{3+}-doped ZnO for downconversion purposes in a P3HT:PC_{61}BM bulk heterojunction solar cell. The bilayer also served as an electron transporting layer. An increase in the short circuit current density by about 2% was reported [146]. In a similar scheme, Svrcek et al used silicon nanocrystals (Si-nc) with PTB7:PC_{71}BM bulk heterojunction solar cells and reported up to 24% photocurrent enhancement under concentrated sunlight. In this hybrid solar cell, the Si-nc was incorporated into a PEDOT:PSS thin film [147].

13.3.2.4 Perovskite Solar Cells

Downconversion has also been explored to improve both the UV photon harvesting and UV stability of PSCs [148–151]. In 2016 Hou et al. reported 29% enhancement in PCE by incorporating $ZnGa_2O_4$:Eu^{3+} into PSCs [148]. In 2017 Jin et al. introduced fluorescent carbon dots, which could effectively convert ultraviolet to blue light in the mesoporous TiO_2 (m-TiO_2) layer of the traditional PSCs and reported around 12% increase in the PCE. The devices were reported to maintain nearly 70% of the initial efficiency after 12 h of full sunlight illumination, while the bare devices maintain only 20% of the initial efficiency [149]. In another study, Jian et al. used a transparent luminescent downconverting layer of an Eu-complex (Eu–4,7-diphenyl-1,10-phenanthroline) in PSCs and reported an enhancement of 11.8% in short-circuit current density (J_{sc}) and 15.3% in PCE [150]. Similarly, Chen et al. reported enhanced photovoltaic performance (35% increase in PCE) and UV-stability of PSCs by applying downconversion CeO_2:Eu^{3+} nanophosphor—TiO_2 composite electrodes [151].

13.4 Conclusions

This chapter provided a comprehensive overview on photon conversion/spectral conversion as a potential approach to address the transmission and thermalization losses in PV devices ranging from conventional crystalline Si and amorphous Si solar cells to the emerging technologies including organic solar cells, dye-sensitized solar cells, and perovskite solar cells. Where as in the former the spectral conversion layers are integration on the front and/or rear of the PV device, the emerging technologies give the extra possibility of integration within the device. To exploit the full potential of photon conversion/spectral conversion concentrated research efforts toward optimizing the materials as well as devices is important.

References

[1] Shalav A. [Ph.D. Thesis]. University of New South Wales; 2006.

[2] Bagnall DM, Boreland M: *Energy Policy* 36:4390, 2008.

[3] Goldschmidt JC, Fischer S, Löper P, Krämer KW, Biner D, Hermle M, et al. In: Proc. 25th Eur. Photovoltaic Solar Energy, Conf. and Exhibition (25th EU PVSEC). Valencia, Spain; 2010. p. 229. doi:10.4229/25thEUPVSEC2010-1CO.6.5.

[4] Shockley W, Queisser HJ: *J Appl Phys* 32:510, 1961.

[5] Auzel F: *Chem Rev* 104:139, 2004.

[6] Goldschmidt JC, Fischer S: *Adv Optical Mater* 510:3, 2015.

[7] Auzel F. C R Acad Sci (Paris) 1966; 262(1016); 1966 263(819); Ovsyankin V, Feofilov PP. JETP Lett 1966;3(317); 1966;3(322).

[8] Mohageg M, Matsko AB, Maleki L: *Opt Express* 20:16704, 2012.

[9] Seletskiy DV, Melgaard SD, Bigotta S, Lieto AD, Tonelli M, Sheik-Bahae M. Nat Photon 2010;4:161 Quintanilla M, Cantelar E, Cusso F, Villegas M, Caballero AC. Appl Phys Express 2011;4:022601; Dong B, Cao B, He Y, Liu Z, Li Z, Feng Z. Adv Mater 2012;24:1987.

[10] Li Z, Zhang Y, Jiang S: *Adv Mater* 20:4765, 2008.

[11] Idris NM, Gnanasammandhan MK, Zhang J, Ho PC, Mahendran R, Zhang Y: *Nat Med* 18:1580, 2012.

[12] Wang F, Han Y, Lim CS, Lu Y, Wang J, Xu J, et al: *Nature* 463:1061, 2010.

[13] Trupke T, Green MA, Würfel P: *J Appl Phys* 92:4117, 2002.

[14] Johnson CM, Conibeer GJ: *J Appl Phys* 112:103108, 2012.

[15] Boccolini A, Faoro R, Favilla E, Veronesi S, Tonelli M: *J Appl Phys* 114:064904, 2013.

[16] (a)Weber MJ: *Phys Rev* 157:262, 1967. (b)Weber MJ: *Phys Rev* 171:283, 1968. (c)Riseberg LA, Moos HW: *Phys Rev* 174, 1968.

[17] Strümpel C, McCann M, Beaucarne G, Arkhipov V, Slaoui A, Švrc ek V, et al: *Solar Energy Mater Solar Cells* 91:238, 2007.

[18] Pollnau M, Gamelin DR, Lüthi SR, Güdel HU, Hehlen MP: *Phys Rev B* 61:3337, 2000.

[19] Suyver JF, Grimm J, van Veen MK, Biner D, Krämer KW, Güdel HU: *J Lumin* 117:1, 2006.

[20] Auzel F, Pecile D: *J Lumin* 11:321, 1976.

[21] Yocom P, Wittke J, Ladany I: *Metallurg Trans* 2:763, 1971.

[22] Johnson L, Guggenheim H, Rich T, Ostermayer F: *J Appl Phys* 43:1125, 1972.

[23] (a)Kano T, Yamamoto H, Otomo Y: *J Electrochem Soc* 119:1561, 1972. (b)Menyuk N, Dwight K, Pierce JW: *Appl Phys Lett* 21:159, 1972. (c)Sommerdijk JL: *J Lumin* 6:61, 1973.

[24] (a)Auzel FE: *Proc IEEE* 61:758, 1973. (b)Auzel F: *J Lumin* 31–32:759, 1984. (c)Suyver JF, Aebischer A, Garcia-Revilla S, Gerner P, Güdel HU: *Phys Rev B* 71:125123, 2005. (d)Heer S, Kömpe K, Güdel H-U, Haase M: *Adv Mater* 16:2102, 2004. (e)Krämer KW, Güdel HU: *Phys Rev B* 56:13830, 1997.

[25] van der Ende BM, Aarts L, Meijerink A: *Phys Chem Chem Phys* 11:11081, 2009.

[26] de Wild J, Meijerink A, Rath JK, van Sark WGJHM, Schropp REI: *Energy Environ Sci* 4:4835, 2011.

[27] Liu L, Huang D, Draper SM, Yi X, Wu W, Zhao J: *Dalton Trans* 42:10694, 2013.

[28] Schulze TF, Schmidt TW: *Energy Environ Sci* 8:103, 2014.

[29] Haase M, Schäfer H: *Angew Chem Int Ed* 50:5808, 2011.

[30] Schietinger S, Aichele T, Wang H-Q, Nann T, Benson O: *Nano Lett* 10:134, 2010.

[31] (a) Goldschmidt JC, Fischer S, Steinkemper H, Hallermann F, von Plessen G, Krämer KW, et al: *IEEE J Photovoltaics* 2:134, 2012. (b) Garcia-Adeva AJ: *J Lumin* 128:697, 2008.

[32] (a)Fischer S, Hallermann F, Eichelkraut T, von Plessen G, Krämer KW, Biner D, et al: *Opt Express* 21:10606, 2013. (b)Fischer S, Hallermann F, Eichelkraut T, von Plessen G, Krämer KW, Biner D, et al: *Opt Express* 20:271, 2012.

[33] Herter B, Wolf S, Fischer S, Gutmann J, Blasi B, Goldschmidt JC: *Opt Express* 21:A883, 2013.

[34] Verhagen E, Kuipers L, Polman A: *Opt Express* 17:14586, 2009.

[35] Wu DM, García-Etxarri A, Salleo A, Dionne JA: *J Phys Chem Lett* 5:4020, 2014.

[36] Gordon J, Ballato J, Smith DW Jr, Jin J: *J Opt Soc Am B* 22:1654, 2005.

[37] Smith DW Jr, Chen S, Kumar SM, Ballato J, Topping C, Shah HV, et al: *Adv Mater* 14:1585, 2002.

[38] Kraemer KW, Biner D, Frei G, G€udel HU, Hehlen MP, Luethi SR: *Chem Mater* 16:1244, 2004.

[39] Ehrmann PR, Campbell JH. J Am Ceram Soc 2002;85:1061; Feng X, Tanabe S, Hanada T. J Appl Phys 2001;89:3560; Xun ZY, Chen XX, Fen C, Hui LJ, Bo YG. Opt Electron Lett 2012;8:0273.

[40] Dexter DL, Schulman JH: *J Chem Phys* 22:1063, 1954.

[41] Simon YC, Weder C: *J Mater Chem* 22:20817, 2012.

[42] (a) Singh-Rachford TN, Castellano FN: *Coordin Chem Rev* 254:2560, 2010. (b) Zhao J, Ji S, Guo H: *RSC Adv* 1:937, 2011.

[43] Saxena VN: *Indian J Pure Appl Phys* 21:306, 1983.

[44] Boyer J-C, van Veggel FCJM: *Nanoscale* 2:1417, 2010.

[45] Ivaturi A, MacDougall SKW, Martin-Rodriguez R, Quintanilla M, Marques-Hueso J, Kramer KW, et al: *J Appl Phys* 114:013505, 2013.

[46] MacDougall SKW, Ivaturi A, Marques-Hueso J, Krämer KW, Richards BS: *Sol Energy Mater Sol C* 128:18, 2014.

[47] Fischer S, Martín-Rodríguez R, Fröhlich B, Krämer KW, Meijerink A, Goldschmidt JC: *J Lumin* 153:281, 2014.

[48] Martín-Rodríguez R, Fischer S, Ivaturi A, Froehlich B, Krämer KW, Goldschmidt JC, et al: *Chem Mater* 25:1912, 2013.

[49] Pokhrel M, Kumar G, Sardar D: *J Mater Chem A* 1:11595, 2013.

[50] Yeh DC, Sibley WA, Schneider I, Afzal RS, Aggarwal I: *J Appl Phys* 69:1648, 1991.

[51] Ivanova S, Pellé F: *J Optical Soc Am B* 26:1930, 2009.

[52] Shalav A, Richards BS, Trupke T, Corkish RP, Krämer KW, Güdel HU, et al. In: Presented at Proc. 3rd world conf. photovoltaic energy conversion, Osaka, Japan, May 11–18; 2003.

[53] Shalav A, Richards BS, Trupke T, Krämer KW, Güdel HU: *Appl Phys Lett* 86:13505, 2005.

[54] Richards BS, Shalav A: *IEEE Trans Electron Devices* 54:2679, 2007.

[55] Fischer S, Goldschmidt JC, Löper P, Bauer GH, Brüggemann R, Krämer K, et al: *J Appl Phys* 108:044912, 2010.

[56] Rüdiger M, Fischer S, Frank J, Ivaturi A, Richards BS, Krämer KW, et al: *Sol Energy Mater Sol C* 128:57, 2014.

[57] Fischer S, Ivaturi A, Frohlich B, Rudiger M, Richter A, Kramer KW, et al: *IEEE J Photovoltaics* 4:183, 2014.

[58] Fischer S. [Dissertation]. Albert-Ludwigs Universität Freiburg; 2014. ISBN:978–3–8396–0785–5.

[59] Arnaoutakis GE, Marques-Hueso J, Ivaturi A, Kraemer KW, Fischer S, Goldschmidt JC, et al: *Opt Express* 22:A452, 2014.

[60] Fischer S, Favilla E, Tonelli M, Goldschmidt JC: *Sol Energy Mater Sol C* 136:127, 2015.

[61] Henke B, Ahrens B, Johnson JA, Miclea PT, Schweizer S. In: Presented at SPIE – 7411, Nanoscale photonic and cell technologies for Photovoltaics II, San Diego, California, USA; 2009.

[62] Hernandez-Rodriguez MA, Imanieh MH, Martin LL, Martin IR: *Sol Energy Mater Sol C* 116:171, 2013.

[63] Pellé F, Ivanova S, Guillemoles J-F: *EPJ Photovoltaics*, 2011doi: 10.1051/epjpv/2011002.

[64] Goldschmidt JC, Fischer S, Löper P, Krämer KW, Biner D, Hermle M, et al: *Sol Energy Mater Sol C* 95:1960, 2011.

[65] (a) de Wild J, Meijerink A, Rath JK, van Sark WGJHM, Schropp REI. In: Presented at Proc. 24th European photovoltaic solar energy conference, Hamburg, Germany, September 21–25; 2009. (b) de Wild J, Meijerink A, Rath JK, van Sark WGJHM, Schropp REI. Sol Energy Mater Sol C 2010;94:1919.

[66] de Wild J, Rath JK, Meijerink A, van Sark WGJHM, Schropp REI: *Sol Energy Mater Sol C* 94:2395, 2010.

[67] de Wild J, Duindam TF, Rath JK, Meijerink A, van Sark WGJHM, Schropp REI: *IEEE J Photovoltaics* 3:17, 2013.

[68] Chen Y, Hao X, Zhou J, Jiao Y, He W, Wang H, et al: *Mater Lett* 83:49, 2012.

[69] Chen Y, Zhou J, Jiao Y, He W, Wang H, Hao X, et al: *J Lumin* 134:504, 2013.

[70] Cheng YY, Fueckel B, MacQueen RW, Khoury T, Clady RGCR, Schulze TF, et al: *Energy Environ Sci* 5:6953, 2012.

[71] Schulze TF, Czolk J, Cheng Y-Y, Fueckel B, MacQueen RW, Khoury T, et al: *J Phys Chem C* 116:22794, 2012.

[72] Liu W, Zhang H, Wang H-G, Zhang M, Guo M: *Appl Surf Sci* 422:304–315, 2017.

[73] Han G, Wang M, Li D, Bai J, Diao G: *Sol Energy Mater Sol Cells* 160:54–59, 2017.

[74] Meng F, Luo Yi, Zhou Y, Zhang J, Zheng Y, Cao G, et al: *J Power Sources* 316:207–214, 2016.

[75] Luoshan M, Bai L, Bu C, Liu X, Zhu Y, Guo K, et al: *J Power Sources* 307:468–473, 2016.

[76] Yu J, Yang Y, Fan R, Wang P, Dong Y: *Nanoscale* 8:4173–4180, 2016.

[77] Liao W, Zheng D, Tian J, Lin Z: *J Mater Chem A* 3:23360–23367, 2015.

[78] Bai J, Zhao R, Han G, Li Z, Diao G: *RSC Adv* 5:43328–43333, 2015.

[79] Yuan C, Chen G, Li L, Damasco JA, Ning Z, Xing H, et al: *ACS Appl Mater Interfaces* 6:18018–18025, 2014.

[80] (a) Jun Z, Haiou S, Wei G, Shunhao W, Chuntao Z, Fang X, et al: *J Power Sources* 226:47, 2013. (b) Yu J, Yang Y, Fan R, Liu D, Wei L, Chen S, et al: *Inorg Chem* 53:8045, 2014. (c) Zhou Z, Wang J, Nan F, Bu C, Yu Z, Liu W, et al: *Nanoscale* 6:2052, 2014.

[81] Ramasamy P, Kim J: *Chem Commun* 50:879, 2014.

[82] Yuan C, Chen G, Prasad PN, Ohulchanskyy TY, Ning Z, Tian H, et al: *J Mater Chem* 22:16709, 2012.

[83] Liu M, Lu Y, Xie ZB, Chow GM: *Sol. Energy Mater Sol C* 95:800, 2011.

[84] Miao C, Liu T, Zhu Y, Dai Q, Xu W, Xu L, et al: *Opt Lett* 38:3340, 2013.

[85] Nattestad A, Cheng YY, MacQueen RW, Schulze TF, Thompson FW, Mozer AJ, et al: *J Phys Chem Lett* 4:2073, 2013.

[86] Wang H-Q, Batentschuk M, Osvet A, Pinna L, Brabec CJ: *Adv Mater* 23:2675–2680, 2011.

[87] Wang H-Q, Stubhan T, Osvet A, Litzov I, Brabec CJ: *Sol Energy Mater Sol Cells* 105:196–201, 2012.

[88] Schulze TF, Czolk J, Cheng Y-Y, Fückel B, MacQueen RW, Khoury T, et al: *J Phys Chem C* 116:22794–22801, 2012.

[89] Lin Y-HL, Koch M, Brigeman AN, Freeman DME, Zhao L, Bronstein H, et al: *Energy Environ Sci* 10:1465–1475, 2017.

[90] Chen X, Xu W, Song H, Chen C, Xia H, Zhu Y, et al: *ACS Appl Mater Interfaces* 8:9071–9079, 2016.

[91] Roh J, Yu H, Jang J: *ACS Appl Mater Interfaces* 8:19847–19852, 2016.

[92] Wang X, Zhang Z, Qin J, Shi W, Liu Y, Gao H, et al: *Electrochim Acta* 245:839–845, 2017.

[93] He M, Pang X, Liu X, Jiang B, He Y, Snaith H, et al: *Angew Chem Int Ed* 55:4280–4284, 2016.

[94] Zhou D, Liu D, Jin J, Chen X, Xu W, Yin Z, et al: *J Mater Chem A* 5:16559–16567, 2017.

[95] (a) Zhang J, Shade CM, Chengelis DA, Petoud S: *J Am Chem Soc* 129:14834, 2007. (b) Zou W, Visser C, Maduro JA, Pshenichnikov MS, Hummelen JC: *Nat Photon* 6:560, 2012.

[96] Strümpel C, McCann M, del Canizo C, Tobías I, Fath P. In: Presented at the Proc. 20th European photovoltaic solar energy conference, Barcelona, Spain, June 6–10; 2005.

[97] Goldschmidt JC, Löper P, Fischer S, Janz S, Peters M, Glunz SW, et al. In: Presented at the Proc. Optoelectronic and microelectronic Materials and Devices (COMMAD), Sydney, Australia; 2008.

[98] Trupke T, Green M, Würfel P: *J Appl Phys* 92:1668, 2002.

[99] Richards B: *Sol Energy Mater Solar Cell* 90:1189, 2006.

[100] Dexter D: *Phys. Rev.* 108:630, 1957.

[101] Sommerdijk J, Bril A, de Jager A: *J Lumin* 8:341, 1974.

[102] Piper W, DeLuca J, Ham F: *J Lumin* 8:344, 1974.

[103] Vergeer P, Vlugt T, Kox M, den Hertog M, van der Eerden J, Meijerink A: *Phys Rev B* 71:014119, 2005.

[104] Wegh R, Donker H, Oskam K, Meijerink A: *Science* 283:663, 1999.

[105] Bünzli J-CG, Chauvin A-S: Chapter 261 – Lanthanides in solar energy conversion. In Jean-Claude GB, Vitalij KP, editors: *Handbook on the physics and chemistry of rare earths*, Elsevier, 2014, pp 169–281.

[106] Zhang QY, Huang XY: *Prog Mater Sci* 55:353, 2010.

[107] Huang X, Han S, Huang W, Liu X: *Chem Soc Rev* 42:173–201, 2013.

[108] van der Ende BM, Aarts L, Meijerink A: *Phys Chem Chem Phys* 11:11081–11095, 2009.

[109] Zhang QY, Yang GF, Jiang ZH: *Appl Phys Lett* 91:051903, 2007.

[110] Li Z, Zeng Y, Qian H, Long R, Xiong Y: *CrystEngComm* 14:3959, 2012.

[111] Jun-Lin Y, Xiao-Yan Z, Jing-Tai Z, Zhi-Jun Z, Hao-Hong C, Xin-Xin Y: *J Phys D Appl Phys* 41:105406, 2008.

[112] Ye S, Zhu B, Chen J, Luo J, Qiu JR: *Appl Phys Lett* 92:141112, 2008.

[113] van Pieterson L, Reid MF, Meijerink A: *Phys Rev Lett* 88:067405, 2003.

[114] Liu X, Ye S, Qiao Y, Dong G, Zhu B, Chen D, et al: *Appl Phys B* 96:51, 2009.

[115] Wegh RT, Donker H, Oskam KD, Meijerink A: *Science* 283:663, 1999.

[116] Yen-Chi C, Woan-Yu H, Teng-Ming C: *J Rare Earths* 29:907, 2011.

[117] Xu Y, Zhang X, Dai S, Fan B, Ma H, Adam J-l, et al: *J Phys Chem C* 15:13056, 2011.

[118] Fix T, Rinnert H, Blamire MG, Slaoui A, MacManus-Driscoll JL: *Sol Energy Mater Sol Cells* 102:71, 2012.

[119] Wegh RT, Donker H, van Loef EVD, Oskam KD, Meijerink A: *J Lumin* 87–89:1017, 2000.

[120] Li QH, Yuan YB, Wei TH, Li Y, Chen ZH, Jin X, et al: *Sol Energy Mater Sol Cells* 130:426, 2014.

[121] Chen JY, Huang CK, Hung WB, Sun KW, Chen TM: *SolEnergy Mater Sol Cells* 120A:168, 2014.

[122] Tsai M-L, Tu W-C, Tang L, Wei T-C, Wei W-R, Lau SP, et al: *Nano Lett* 16:309, 2016.

[123] Huang C-Y, Wang D-Y, Wang C-H, Chen Y-T, Wang Y-T, Jiang Y-T, et al: *ACS Nano* 4:5849, 2010.

[124] Wang Y, Li X, Song J, Xiao L, Zeng H, Sun H: *Adv Mater* 27:7101, 2015.

[125] Tai Y, Zheng G, Wang H, Bai J: *J Solid State Chem* 226:250, 2015.

[126] Gao G, Wondraczek L: *J Mater Chem C* 1:1952, 2013.

[127] Stupca M, Alsalhi M, Al Saud T, Almuhanna A, Nayfeh MH: *Appl Phys Lett* 91:063107, 2007.

[128] Carrillo-López J, Luna-López JA, Vivaldo-De la Cruz I, Aceves-Mijares M: *Sol Energy Mater Sol Cells* 100:39, 2012.

[129] Chowdhury FI, Alnuaimi A, Islam K, Nayfeh A. In: Proceedings of the IEEE 40th Photovoltaic Specialist Conference (PVSC); 2014, p. 2209–13.

[130] Gardelis S, Nassiopoulou AG: *Appl Phys Lett* 104:183902, 2014.

[131] Ahn Y, Kim J, Shin S, Ganorkar S, Kim Y-H, Kim Y-T, et al: *J Sol Energy Eng* 137:021011, 2014.

[132] Chen HC, Lin CC, Han HV, Chen KJ, Tsai YL, Chang YA, et al: *Sol Energy Mater Sol Cells* 104:92, 2012.

[133] Pi X, Li Q, Li D, Yang D: *Sol Energy Mater Sol Cells* 95:2941, 2011.

[134] Cheng CL, Yang JY: *IEEE Electron Device Lett* 33:697, 2012.

[135] Merigeon J, Maalej O, Boulard B, Stanculescu A, Leontie L, Mardare D, et al: *Opt Mater* 48:243, 2015.

[136] Chung C-C, Tran BT, Han H-V, Ho Y-T, Yu H-W, Lin K-L, et al: *Electron Mater Lett* 10:457, 2014.

[137] Han H-V, Lin C-C, Tsai Y-L, Chen H-C, Chen K-J, Yeh Y-L, et al: *Sci Rep* 4:5734, 2014.

[138] Kim HJ, Song JS, Kim SS: *JKorean Phys Soc* 45:609, 2004.

[139] Dai WB, Lei YF, Li P, Xu LF: *J Mater Chem A* 3:4875, 2015.

[140] Zhu G, Wang X, Li H, Pan L, Sun H, Liu X, et al: *Chem Commun* 48:958, 2012.

[141] Yao N, Huang J, Fu K, Deng X, Ding M, Xu X: Rare earth ion doped phosphors for dye-sensitized solar cells applications, *RSC Adv* 6:17546–17559, 2016.

[142] Lim MJ, Ko YN, Chan Kang Y, Jung KY: *RSC Adv* 4:10039, 2014.

[143] Chander N, Khan AF, Komarala VK: *RSC Adv* 5:66057, 2015.

[144] Li Q, Yuan Y, Wei T, Li Y, Chen Z, Jin X, et al: *Sol Energy Mater Sol Cells* 130:426, 2014.

[145] Liu C, Zhang X, Li J, He Y, Li Z, Li H, et al: *Synth Met* 204:65, 2015.

[146] Na W, Qun L, Xvsheng Q, Chang-Qi M: *Mater Res Express* 2:125901, 2015.

[147] Svrcek V, Yamanari T, Mariotti D, Mitra S, Velusamy T, Matsubara K: *Nanoscale* 7:11566, 2015.

[148] Hou X, Xuan T, Sun H, Chen X, Li H, Pan L: *Sol Energy Mater Sol Cells* 149:121, 2016.

[149] Jin J, Chen C, Li H, Cheng Y, Xu Lin, Dong B, et al: *ACS Appl Mater Interfaces* 9:14518, 2017.

[150] Jiang L, Chen W, Zheng J, Zhu L, Mo Li, Li Z, et al: *ACS Appl Mater Interfaces* 9:26958, 2017.

[151] Chen W, Luo Q, Zhang C, Shi J, Deng X, Yue L, et al: *J Mater Sci Mater Electron* 28:11346, 2017.

Further Reading

[152] Goldschmidt JC: *Novel solar cell concepts*, München, 2010, Verlag Dr. Hut.

14

Advanced Building Integrated Photovoltaic/Thermal Technologies

Fangliang Chen, Frank Pao, Huiming Yin
COLUMBIA UNIVERSITY, NEW YORK, NY, UNITED STATES
hy2251@columbia.edu

14.1 Introduction

Integrated technologies for harvesting solar energy in the building sector, such as building-integrated photovoltaic (BIPV) systems [1–3], building-integrated solar thermal systems [4–6], or building-integrated photovoltaic/thermal (BIPVT) systems [7–9], have evolved as viable technologies toward the nearly zero energy building scenario. Those integrated systems replace parts of the conventional building materials and the components in the climate envelope of buildings, such as facades and roofs, and simultaneously serve as both a building envelope material and power generator [10–12]. Compared with most conventional nonintegrated systems, in addition to the power supply function, the integrated system offers several advantages: (1) there is no need for the allocation of land or facilitation of the PV system, (2) it does not require additional assembly components such as brackets and rails, and (3) it thus achieves significant savings in terms of the total building material costs and associated labor fees [13, 14].

Today, most photovoltaic (PV) modules in production are based on crystalline silicon wafer technologies. The electricity conversion efficiency of silicon solar modules available for commercial application is about 12%–20% [15]. However, the majority of the incoming solar energy is either reflected or absorbed as heat [16]. Consequently, the working temperature of the solar cells increases considerably after prolonged operations. Solar panel temperature is one of the important factors that affects electricity conversion efficiency, yet most solar cells show a heat-related performance loss of about 0.4%–0.5% $(°C)^{-1}$ [17]. Without a cooling system, in-service surface temperatures can be up to 40–50°C above ambient temperature, resulting in 16%–25% reductions in electricity generation or malfunction beyond the operational temperature range [18]. The rise in PV temperature not only reduces electricity generation, but also shortens the life-span of the module itself. The BIPVT system appears as an exciting new technology as it merges PV and thermal systems, simultaneously harvesting both electrical and the thermal energy [19]. The most common BIPVT systems are realized through a heat transfer fluid in an open-loop (usually air) [20–22] or closed-loop (usually liquid) configuration [23–25], which are shown in Fig. 14.1A and B, respectively.

A Comprehensive Guide to Solar Energy Systems. http://dx.doi.org/10.1016/B978-0-12-811479-7.00014-2

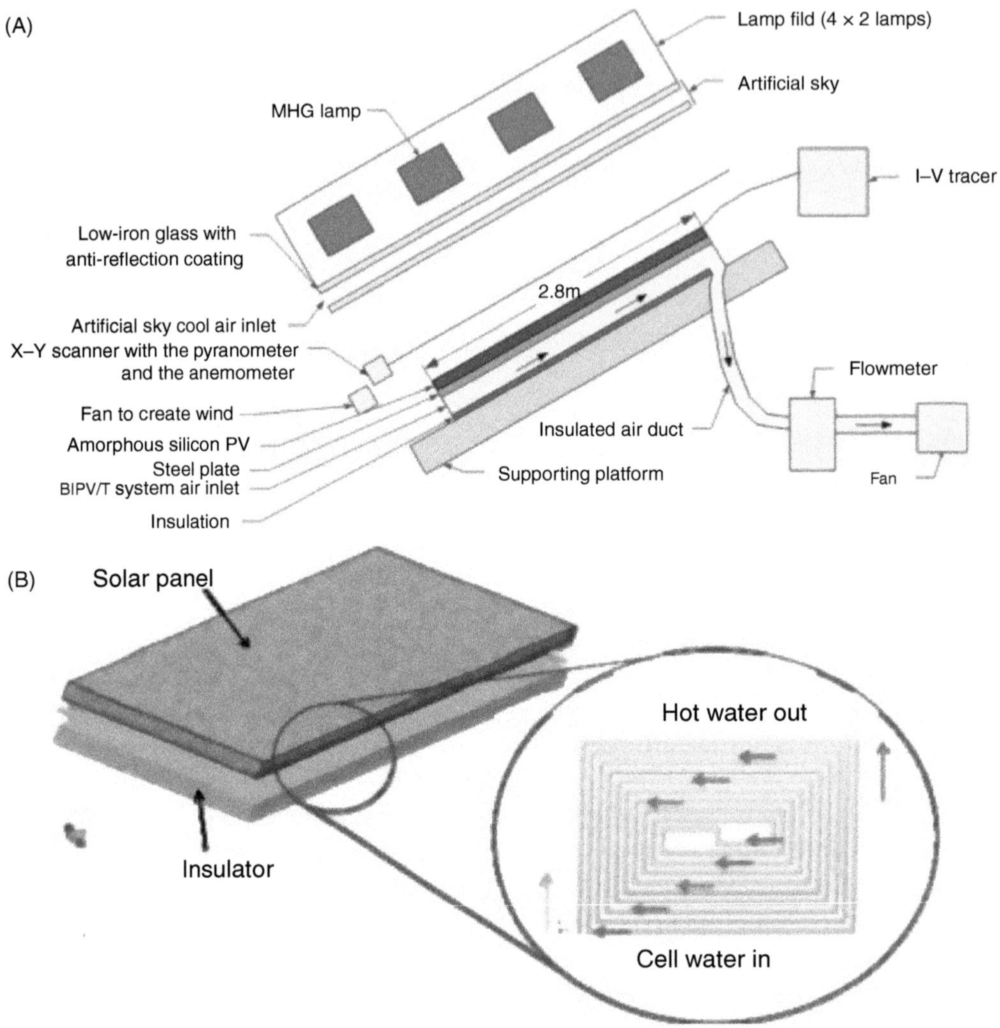

FIGURE 14.1 Schematic illustration of the BIPV/T system cooled by (A) open-loop [21], and (B) closed-loop [26].

Generally, the closed-loop configuration with liquid is more efficient than the open-loop with air as the heat transfer media due to the high thermophysical properties of liquid compared to air [27]. Most closed-loop BIPVT systems employ water tubes for cooling and thermal energy collection. Normally, the water tubes are embedded in insulation materials and covered by absorber materials in contact with the PV elements above [26]. However, such designs commonly exhibit poor heat conduction due to the small contact area between the absorber and water pipes, and thus not able to efficiently cool the PV cells to increase the PV efficiency. In this chapter, we present the state-of-the-art BIPVT system, and introduce a commercial building integrated thermal electric roofing system (BITERS) by Atlantis Energy Systems, Inc. [28] and a novel BIPVT solar roof panel developed by Columbia University [29–31]. Comparisons among typical PVTs or BIPVTs with water as the heat transfer fluid found in the literature demonstrate that significant energy conversion efficiency improvement can be achieved by the presented BIPVT roof panel to harvest solar energy in the form of PV electricity as well as heat energy through the collection of warm water. Overall, the performance indicates a very promising prospective of the new BIPVT multifunctional roofing panel.

14.2 Building Integrated Thermal Electric Roofing System

Advanced solar BIPV glass and solar PV SunSlates and TallSlates Roof Tiles are shown in Fig. 14.2. The SunSlates and TallSlates are both BIPV solar roofing products. The SunSlates system is one of the most advanced BIPV Class A roofing systems in the market and it offers a roof warranty and has a UL Class A fire certification. The TallSlates is a solar module that generates electricity and act as a roof slate as well, which is ideal for straight slope roof applications. The gutters on each side of the module act as drain channels as well as support for the mounting system.

It was reported from field experience that the SunSlate installation could be time consuming and prone to mistakes. The new hybrid product, Building integrated thermal electric roofing (BITER) system, and modular system could be hoisted preassembled onto an existing roof or a new construction roof with excellent workability for installation.

The BITER is a solar hybrid system that can harvest solar energy in the form of both thermal energy and electricity. The thermal component is located directly beneath the SunSlates or the TallSlates and as a result the solar modules convert the solar energy into DC power at a relatively high efficiency. All internal systems can be preinstalled in the home prior to the units' arrival, resulting in a dramatic labor and also insurance saving for the roofing contractor. The units are then lowered into place by crane onto preset cleats

FIGURE 14.2 Solar roof system made by AES. (A) SunSlates, and (B) TallSlates.

attached to the roof and securely locked down. They are then plumbed to the in-house connecting lines. Once installed a roof can either be finished with slates for the surrounding edges or left as is with an asphalt shingle roof.

The energy harvesting mechanism of the BITER is illustrated in Fig. 14.3, the liquid going through the thermal system is being pumped and circulated throughout the sloping roof. The thermal energy is being captured and transferred to the heat exchanger that can provide domestic hot water. Meanwhile, DC electricity is being generated by the SunSlates and feed-in to the power grid through the inverter or the batteries if the customer prefers a standalone system.

For a SunSlates system, once all strings are checked and if they reach the acceptable power, they will be connected to the inverter. Then the PV system would be turned on and it would start supplying AC power to the grid. In the case of a standalone system, the strings could be connected to the battery via a charge controller and start storing electricity that could be used to power the DC appliances and lighting applications.

One of the pilot projects of the BITERS is shown in Fig. 14.4. The BITERS thermal system uses a pump to circulate the thermal liquid throughout the system. As the intensity of the solar power increases so does the throughput of the liquid in the thermal tubing, resulting in the generation of more thermal energy. In addition, while heat is being extracted through the thermal tubing from the SunSlates. The SunSlates are being cooled thus preventing the loss of efficiency in converting sunlight to DC and ultimately AC electricity. Furthermore, as much thermal energy had been taken out from the attic, the house would be cooler which results in reduction on the air conditioning energy consumption. In case

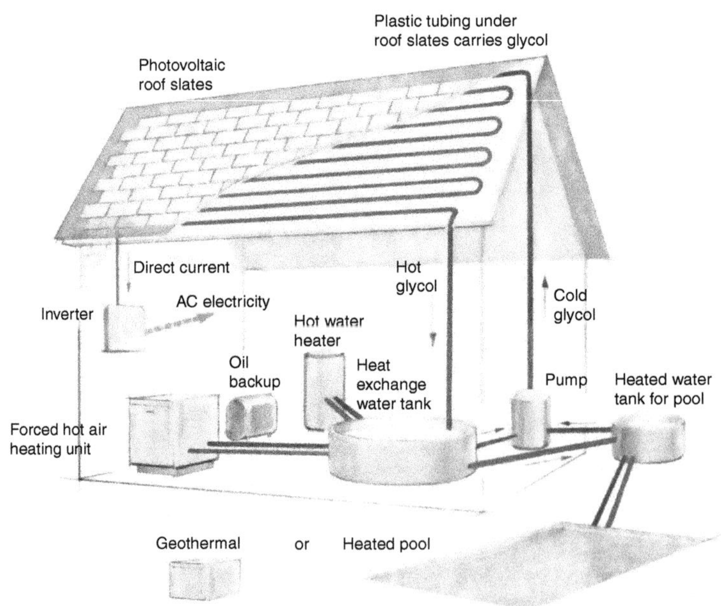

FIGURE 14.3 Operation of the BITER system in a building.

FIGURE 14.4 A pilot project of the BITER system.

there is an excess of thermal energy available on standard domestic hot water application, the heat could be stored for later use, or it could be used for a hot tub or for heating a pool or for geothermal use, which would be subject to locations.

Because of its high performance coupled with aesthetic design of the BITER and quality of all the BIPV installations, AES received the Frost & Sullivan "2009 World Building-Integrated Photovoltaic Niche Player of the Year Award" [32]. The award was in recognition of AES's successful initiatives in entering a niche market with innovative and user friendly solutions.

14.3 BIPVT Solar Roof

Recent studies [30, 31] have shown that, by integrating functionally graded materials (FGMs) as a cooling substrate, a novel BIPVT roofing panel is able to efficiently harvest solar energy. The purpose of the proposed FGM layer is to create a lightweight layer of solar roofing panel with a varying thermal conductivity in the thickness direction. The FGM layer involves a gradation of material phases from metal dominated layer to polymer materials. Water tubes are embedded in the top part of the FGM layer, where the high aluminum concentration creates high thermal conductivity so that heat flow from surface can be immediately transferred to water tubes in all directions, yet be insulated by the lower part composed of a polymer layer. The proposed PVT panel can be integrated into the building skin with water circulation, flow control, and heat storage and utilization systems. Solar energy is collected by BIPVT panels in the form of electricity and heat. Electricity can be used locally or transmitted to the grid, while the heat can be stored or directly used indoors for floor heating or clean drying, or used externally to supply heat to swimming pools.

14.3.1 Design and Manufacture of the Novel FGM Panel

The details of design and fabrication procedure of the FGM panel can be found in Ref. [30]. The FGM panel with embedded water tubes was fabricated by a combined vibration and sedimentation approach. The purpose of the designed double serpentine shape of the water tubes as shown in Fig. 14.5A is to remove the heat within the panel more uniformly and thus keeping the temperature of the panel surface much more uniform. When the FGM panel was cured and taken out from the vacuum oven, it was seamlessly glued to a structural substrate, such as fire-retardant plywood, or lightweight concrete through a moisture resistant epoxy adhesive as shown in Fig. 14.5B. The integrated FGM panel and substrate were then sprayed by a thin-layer of moisture resistant resin, which reacts chemically with the natural moisture in the wood, creating a highly moisture resistant substance similar to polyurethane. Due to the gradual change of the proportion of materials, only a small amount of aluminum powder is needed. This reduces the cost of the panel, as the aluminum material is relatively expensive. The heat absorbing layer and insulation layer in traditional PVTs can be replaced by one FGM layer that integrates the high thermal conductivity in the top part and low thermal conductivity in the lower part. As the volume fraction of aluminum particles continuously varies in the thickness direction, the thermomechanical property distribution changes smoothly, and avoids the thermal stress concentration across layers and increases the structural integrity and durability of the panels.

14.3.2 Assembling of the BIPVT

After polishing the metal rich surface of the FGM, 16 monocrystalline silicon PV cells (with a rated efficiency of 18.8% and rated power of 2.88 W) with dimensions of 127 × 127 mm were mounted onto the FGM layer by a thermal conductive adhesive. To generate a

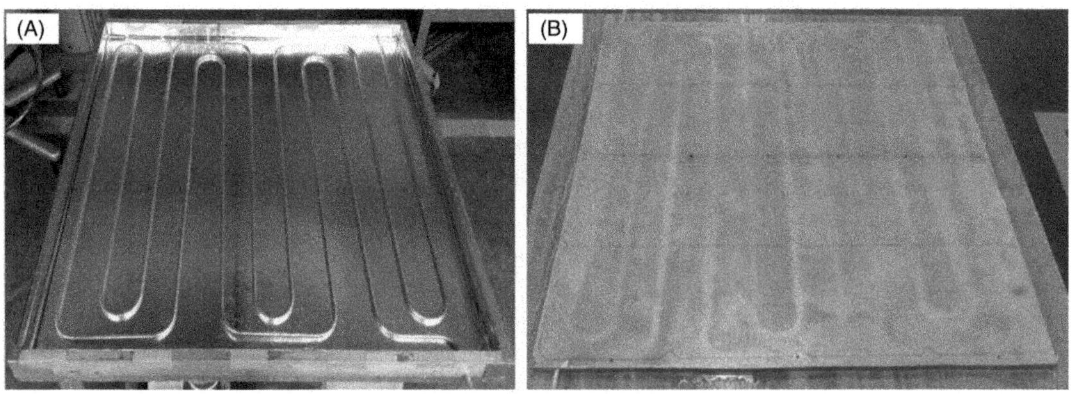

FIGURE 14.5 Embedment of water tube and integration of substrate. (A) Double serpentine shape of water tubes in the casting mold, and (B) assembling of FGM panel with structural substrate.

uniform bond surface between the FGM and the PV cells and avoid any residual adhesive overflow to the cell surfaces, 16 adhesive regions with dimensions of 101.6 × 101.6 mm were prepared by peeling off the preattached tapes from the uniform one that was applied on the FGM panel (Fig. 14.6A). The 16 PV cells were gently attached onto the adhesive and a slight uniform pressure was applied on each PV cell. When the adhesive was cured after 24 h, the applied pressure was removed and the mounted PV module was complete as shown in Fig. 14.6B.

For field applications, a transparent protective waterproofing layer could be further mounted on the BIPVT panel to protect the power generating elements and underlying building materials from external environmental distress such as moisture migration, surface wear and impact from dust, wind, storm, and so on. A schematic illustration of the developed BIPVT panel is shown in Fig. 14.6C, where the FGM layer gradually transits material phases from a well-conductive side (aluminum dominated) attached to a PV solar cell, to a highly insulated side (polymer materials) bonded to a structural substrate. The water tubes were embedded in the top part of the FGM layer, where the high aluminum concentration creates a high thermal conductivity so that heat can be immediately transferred to the water tubes from all directions, yet they ought to be insulated by the bottom part of FGM layer and the thermal insulation plywood. The substrate provides support for mechanical loading and functions as thermal insulation for the building envelope. The multilayered solar panel was designed in such a way that

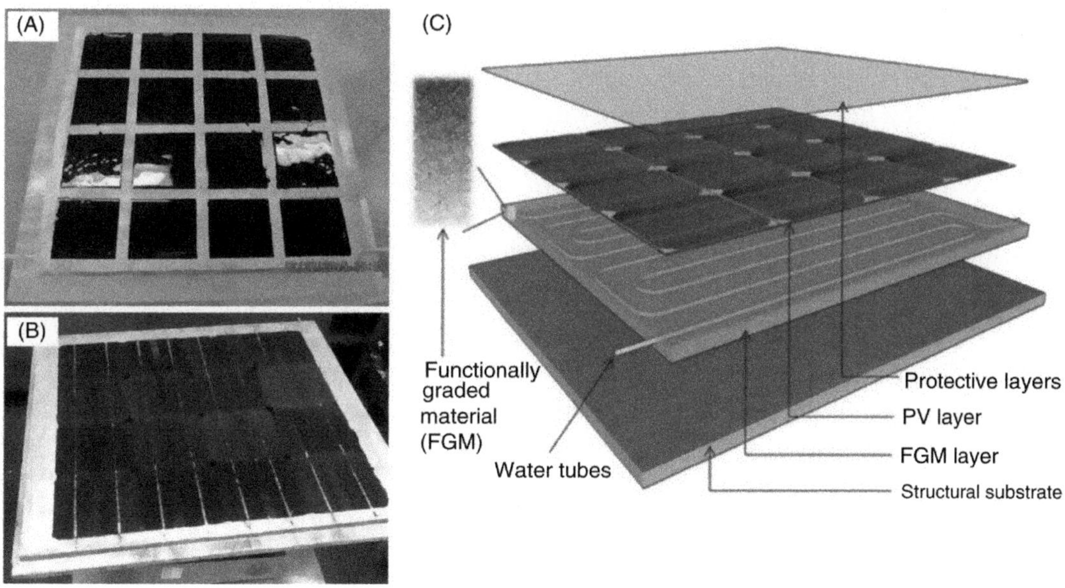

FIGURE 14.6 A novel building-integrated photovoltaic-thermal (BIPVT) roofing panel. (A) Layering of conductive adhesive, (B) integrated solar panel, and (C) the BIPVT layers for assembly.

layers with potentially shorter life expectancies can be easily replaced or removed from the design based on the sustainability measurement criteria, which can be applicable to other sustainable building materials and systems for building construction.

14.3.3 Integration of a Multifunctional Roofing System

A multifunctional roofing system can be easily assembled by using the proposed BIPVT panel, which is sized to be easily integrated with the industry standard structural framing spacing of 304.8 mm on center. Custom panels can also be easily developed for longer spans. The length of the panel can be either 1219 or 2438 mm, based on the weight and thickness requirements (1219 mm is illustrated for demonstration). The prototype design references the International Building Code (IBC 2015) [33] for the roof assembly requirements for weather protection, such as flashings to prevent moisture penetrations at all roof interruptions and terminations. The proposed panelized system will employ a field-applied joint system while gasketed design alternatives can also be developed. Both electrical and fluid connections will be made between the individual panels.

The manufacture of this panel will not be more complex than the existing combination of a traditional roof and PV panels. The proposed BIPVT panel can be integrated into a building's skin with relevant system components such as water circulation, flow control, and heat storage. The innovations of the proposed BIPVT technology are summarized as follows:

- Due to the ability to control temperatures through the water flow, the PV modules can work at lower temperatures in the summer resulting in a higher PV efficiency.
- The collected hot water can be directly utilized by water heating systems for indoor or for external use.
- Due to the temperature control on the roof, better thermal comfort in the building can be achieved in the summer time, a cooler room temperature can be obtained, and the energy demand for cooling can be reduced.
- In winter, a warm water flow can be used to remove frost or ice on the roof, clean solar panels, and thus restore and enhance solar energy utilization.

Traditional solar panels are rebuilt on the rooftop or attached to structural elements of the building skin, which is less than optimal. Firstly, the power generating elements of such panels are typically adhered to a structural substrate, supported by a structural framework that penetrates the building's water proofing system. Secondly, traditional solar panels cannot shield the building skin from wind loads and conventional configurations necessitate redundancies as the panel substrate and frame must be designed to resist the same wind and snow loads as the building envelope. However, the present BIPVT is not a simple superposition of individual structural component and PV module, but provides a viable solution to significantly increase overall energy utilization efficiency while alleviating the disadvantages of a traditional approach.

14.4 Modeling Procedures and Performance Evaluation of the Multifunctional BIPVT Panel

14.4.1 Laboratory Testing Setup

The performance of the prototype BIPVT panel was investigated at Columbia university's Carleton Laboratory via a solar simulator. The solar simulator was established in a custome-built room with a temperature-controlled exhaust system, which cools the room by forced-air ventilation as shown in Fig. 14.7A. A special metal-halide MHG lamp was assembled in the solar simulator, and provided an irradiance of up to 4 kW m^{-2}. The solar simulator was able to: (1) offer high irradiance efficiency and superior spatial uniformity on the test area, (2) produce a dense multiline spectrum of the rare earths that is comparable to a continuous spectrum, and (3) provide a spectral distribution very close to natural sunlight. During the test, the BIPVT panel was located under the solar lamp with a fixed distance of 1422.4 mm from the top surface of the panel to the bottom edge of the lamp. The irradiance intensity of the metal lamp was controlled by adjusting the power output from the simulator controller (Fig. 14.7B), which ranges from 620 to 1250 W m^{-2} at

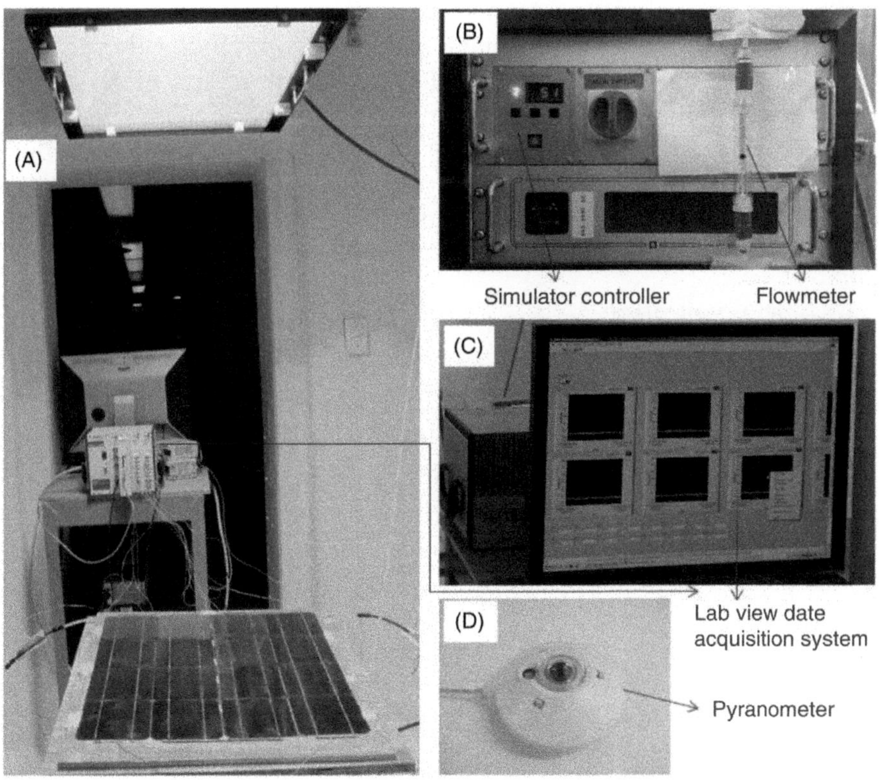

FIGURE 14.7 Performance evaluation of the BIPVT Panel. (A) Test setup, (B) simulator controller and flowmeter, (C) data acquisition system, and (D) Pyranometer.

the selected height. A water source was connected to the inlet of the panel to cool down the surface temperature of the FGM panel as well as the solar cell. A laboratory flowmeter (Fig. 14.7B) with a flow range of 2–250 mL min^{-1} and accuracy of 2% was used to control the flowrate inside the water tube.

Fourteen thermocouples were installed to measure the temperature distribution throughout the whole panel surface. Among those 14 thermocouples, 12 (six located at the two edges of the panel and other six were distributed at the center line on the solar panel) were attached on the FGM panel; while two were attached on the inlet (No. 1) and outlet (No. 15) aluminum water tubes. The detailed distributions of the thermocouples were shown in Fig. 14.8. The temperature readings from the thermocouples were acquired by the data acquisition system with a sampling frequency of 2 Hz.

14.4.2 Estimation of Heat Collection

The test consists of two stages. In the first stage, the BIPVT panel was exposed to solar radiation without water flowing inside until a stable state was reached; while in the second stage, a stable water flow was introduced and controlled by the flowmeter until another stable state was reached. The thermal distributions of the FGM panel under different irradiance levels (620, 800, and 1000 W m^{-2}) at different water flows (30, 60, 90, and 120 mL min^{-1} for 620 W m^{-2}; and 30, 60, 90, and 150 mL min^{-1} for 800 and 1000 W m^{-2}) were monitored. It was found that the temperature evolution trends for different flow rates under different irradiance levels are quite similar. For this reason, only two typical ones corresponding to an irradiance level of 620 W m^{-2} are shown in Fig. 14.9. T_1 and T_{15} are the measured temperatures of the water in the inlet and outlet tubes, respectively, which are static in the first stage. It shows that the temperature of the BIPVT panel gradually increases in the first stage as continuous

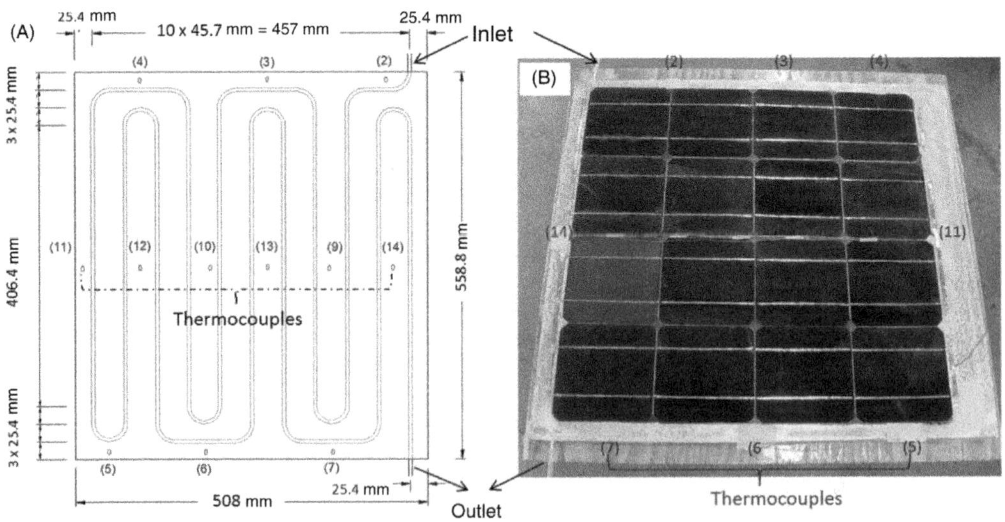

FIGURE 14.8 Thermal couple distributions. (A) Schematic illustration, and (B) real BIPVT panel.

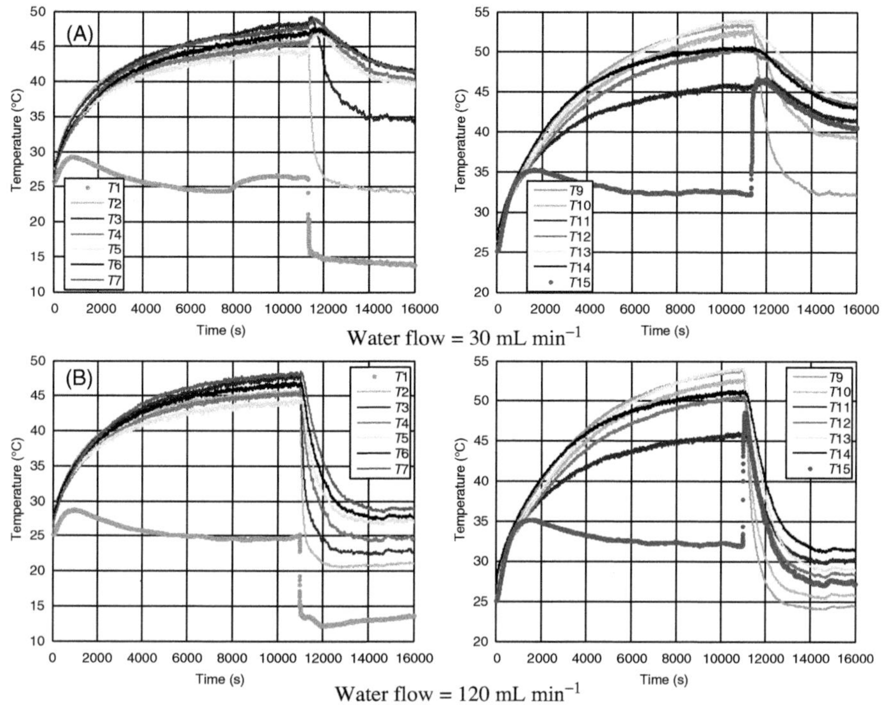

FIGURE 14.9 Thermal distribution of the BIPVT panel at an irradiance level = 620 W m^{-2}.

irradiance energy is absorbed and accumulates in the panel. Under the selected irradiance level (620 W m^{-2}), it took about 2.5 h for the BIPVT panel to reach a relatively stable peak temperature. As shown in Fig. 14.9, the temperature of the BIPVT panel during this stage increases from room temperature (~25°C) to a peak value (~51.5°C). It was also observed that the peak temperatures at the two edges are lower than those at the central line of the panel. As the water circuit was closed in this stage, the temperature of both the inlet and outlet water initially increased and then reached to a stable state in a much shorter period of time compared with those on the surface of the BIPVT panel. When the temperature of the panel reaches a stable peak value, the water valve was turned on and a stable flow of water was introduced and controlled by the flowmeter, which initiates the second stage.

In the second stage, the recorded surface temperatures of the BIPVT panel dropped rapidly due to the water flowing through the tubes embedded in the FGM panel, until another stable state was reached. Once the water flow was introduced, the temperature on the inlet water tube (T_1) suddenly droped as cool water flowed into the circuit; while in contract, the temperature on the outlet water tube (T_{15}) suddenly increased as heated water flows out from the outlet water tube. Thereafter, the outlet water temperature gradually decreased as the water continued flowing in the circuit until it reached the second stable state. Fig. 14.9 shows that the lower the water flow rate, the longer is the time required to cool down the BIPVT panel and the higher the temperature of the second stable state.

The thermal distributions of the BIPVT panel at other irradiance levels (800 and 1000 W m^{-2}) with different water flow rates are similar with those shown in Fig. 14.9. Table 14.1 summarizes the temperature increase of the outlet water compared with the inlet water (ΔT_{water}, which is the temperature difference between the outlet water (T_{15}) and that of the inlet water (T_1) in the second stable stage and indicates the harvested heat energy collected by the warm water) and the temperature decrease of the BIPVT surface (ΔT_{panel}, which is the average temperature difference of all thermocouples attached on the BIPVT panel surface between the first and second stages and reflects the enhanced electricity generation efficiency of the solar cell). Table 14.1 illustrates the increase in temperature of the collected water from the BIPVT panel can be as high as 37.5°C at an irradiance of 1000 W m^{-2} with a relatively low flow rate of 30 mL min^{-1}. As shown in the table, the decrease in surface temperature of the BIPVT can be as high as 32°C when the flow rate is increased to 150 mL min^{-1}. It is expected that, as if the water flow rate was further increased, the BIPVT panel surface temperature would decrease further.

To quantify the thermal performance, the collected heat gain by the present BIPVT panel and its corresponding thermal conversion efficiency are evaluated. According to the definition of instantaneous thermal efficiency [34], the useful collected heat Q_{water} gain by the BIPVT panel at different water flowing rates can be calculated by Eq. (14.1):

$$Q_{water} = \dot{m}_{water} \times C_{water} \times \Delta T_{water} \tag{14.1}$$

where $\dot{m}_{water}$ is the mass flow rate of the water flowing in the water pipe, and $C_{water} = 4.19$ kJ kg^{-1} K^{-1} is the specific heat capacity of water.

The thermal efficiency of the BIPVT panel $\eta_{thermal}$ is a ratio of the collected thermal energy to the irradiance energy absorbed by the panel, which can be expressed as [21, 29]:

Table 14.1 Overall Thermal Performance of the BIPVT

Solar Radiation (W m^{-2})	Water Flow Rate (mL min^{-1})	ΔT_{water} (°C)	ΔT_{panel} (°C)	Q_{water} (W)	$\eta_{thermal}$ (%)
620	30	26.5	−12.7	55.52	31.54
	60	22.94	−18.6	93.86	53.33
	90	17.6	−21.8	110.62	62.85
	120	14.1	−23.9	119.83	67.14
800	30	32.2	−15.7	67.46	29.71
	60	25.6	−20.8	107.26	47.23
	90	20.3	−24.2	127.59	56.18
	150	13.6	−28.1	142.46	62.73
1000	30	37.5	−14.3	78.56	27.68
	60	29.5	−22.9	123.61	43.54
	90	23.7	−27.8	148.95	52.47
	150	16.1	−32	168.65	59.41

$$\eta_{\text{thermal}} = \frac{Q_{\text{water}}}{E_{\text{IN}}} = \frac{Q_{\text{water}}}{I_R \cdot A} \tag{14.2}$$

where $E_{\text{IN}} = I_R \cdot A$ is the absorbed irradiance by the BIPVT panel, which is the product of the irradiance intensity (I_R) and the total area of the BIPVT panel and the frame. $A = 0.28\ \text{m}^2$ (508×558.8 mm) for the prototype BIPVT panel in this study. It should be mentioned that the thermal efficiency defined in Eq. (14.2) does not take into account the energy required to pump the water into the panel; this required energy is also not accounted for in the definition of the electric efficiency.

14.4.3 Estimation of Electricity Generation

The generated voltages of the BIPVT panel under different solar irradiances with different introduced water flow were collected by the LabVIEW data acquisition system. The electrical power produced by the cell can be easily determined along the I–V sweep by the equation $P = IV$. The corresponding I–V and power–voltage (P–V) curves for the BIPVT panel at an irradiance of 620 W m^{-2} with different water flow rates are shown in Fig. 14.10. At the I_{SC} and V_{OC} points, the output powers are zero, while the maximum power can be easily identified at the peak point from the P–V curve. At the maximum power point the voltage and current are denoted as V_{MP} and I_{MP} respectively, which is illustrated in Fig. 14.10A.

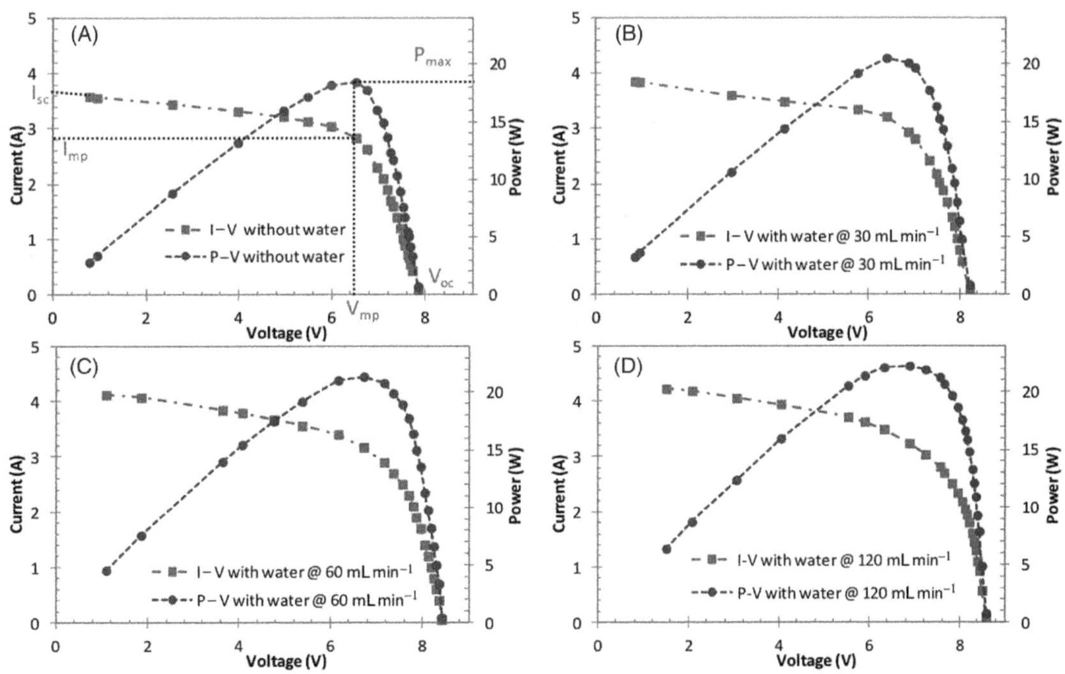

FIGURE 14.10 I–V curve and output power of the BIPVT with different flowing water under irradiance of 620 W m^{-2}. (A) Without water, (B) water flowing rate = 30 mL min^{-1}, (C) water flowing rate = 60 mL min^{-1}, and (D) water flowing rate = 120 mL min^{-1}.

From the I–V and P–V curves, some basic parameters such as fill factor (FF) and electricity conversion efficiency (η_{pv}) can be determined for the BIPVT under different irradiances with different water flowrates by following some fundamental studies in the literature [35]. The FF is essentially an index of the BIPVT panel quality. It is calculated in Eq. (14.3) by comparing the maximum output electricity (E_{MAX}) to the theoretically calculated vale (E_T) that would be generated at both the open circuit voltage and short circuit current together. A larger FF is desirable, which corresponds to an I–V sweep that is more square-like.

$$\mathrm{FF} = \frac{F_{\mathrm{MAX}}}{E_T} = \frac{I_{\mathrm{MPP}} \cdot V_{\mathrm{MPP}}}{I_{\mathrm{SC}} \cdot V_{\mathrm{OC}}} \tag{14.3}$$

The electricity conversion efficiency (η_{pv}) is defined as the ratio of the output electricity E_{pv} with respect to the absorbed solar irradiance E_{IN} by the BIPVT panel, which is represented in Eq. (14.4). E_{pv} can be taken as E_{MAX} because the solar cell can be operated up to its maximum energy output to get the maximum efficiency. In determining the efficiency, the dimensions of the BIPVT frames were also considered in the calculations, as they are necessary components that take up area though they are there for structural purposes.

$$\eta_{\mathrm{pv}} = \frac{E_{\mathrm{pv}}}{E_{\mathrm{IN}}} = \frac{E_{\mathrm{MAX}}}{I_R A} \tag{14.4}$$

The determined parameters of the BIPVT panel under different irradiances with different water flow rates are summarized in Table 14.2. Under a solar irradiance of 620 W m^{-2}, compared with the case in which no water flows, the output electricity energy increases from 18.45 to 22.65 W and the electric efficiency raises from 10.5% to 12.9% after a water flow of 120 mL min^{-1} was introduced. Both E_{pv} and η_{pv} increased as the solar irradiance and water flow rates increased. With a water flow rate of 150 mL min^{-1}, E_{pv} reaches 32.96 and 44.91 W and η_{pv} reaches 14.51% and 15.82% under solar irradiances of 800 and 1000 W m^{-2} respectively. The PV efficiency is enhanced by 21.1% and 24.0% for those two solar irradiances respectively, comparing with the case in which no water flows. Based on the thermal and electric performances provided in Tables 14.1 and 14.2, the total energy efficiencies $E_T = \eta_{thermal} + \eta_{pv}$ of the BIPVT panel under different solar irradiances and water flowing rates can be estimated. The maximum η_T evaluated in this study under the solar irradiance of 620 W m^{-2} with water flowing at a rate of 120 mL min^{-1} is approximately 79.8%. With a solar irradiance of 800 W m^{-2} and a water flowrate of 150 mL min^{-1} the total efficiency drops to 77.3%, and with a solar irradiance of 1000 W m^{-2} and a water flowrate of 150 mL min^{-1}, and the value is approximately 75.2%.

14.4.4 Overall Efficiency and Comparisons With Other Relevant PVT Collectors

For customers who want to utilize a larger amount of hot water through the BIPVT, they may need to decrease the water flow rate and thus sacrifice the electricity gain, and vice

Table 14.2 Electric Performance of the BIPVT

Solar Irradiance (W m^{-2})	Water Flow Rate (mL min^{-1})	I_{SC}/A (ADC)	V_{OC}/V (VDC)	I_{MP}/A (ADC)	V_{MP}/V (VDC)	E_{pv}/W	FF (%)	η_{pv} (%)
620	0	3.69	7.86	2.83	6.52	18.45	63.62	10.48
	30	3.92	8.23	3.21	6.39	20.51	63.58	11.65
	60	4.17	8.41	3.17	6.73	21.32	60.83	12.11
	90	4.2	8.54	3.19	6.88	21.96	61.19	12.48
	120	4.24	8.6	3.29	6.89	22.65	62.11	12.87
800	0	4.69	9.28	3.64	7.47	27.21	62.51	11.98
	30	5.09	9.47	3.98	7.62	30.35	62.96	13.36
	60	5.27	9.56	4.09	7.67	31.37	62.27	13.81
	90	5.31	9.57	4.13	7.72	31.88	62.74	14.04
	150	5.39	9.71	4.22	7.81	32.96	62.97	14.51
1000	0	5.68	10.02	4.52	8.01	36.22	63.61	12.76
	30	6.17	10.31	4.91	8.24	40.46	63.60	14.25
	60	6.26	10.42	5.09	8.32	42.35	64.92	14.92
	90	6.38	10.57	5.18	8.43	43.67	64.75	15.38
	150	6.62	10.83	5.24	8.57	44.91	62.64	15.82

versa. In this sense, a desired water flow rate can be adjusted based on a customer's need in terms of the harvested heat energy or the enhanced electricity generation efficiency. Fig. 14.11 provides a general comparison of the relative increase in the outlet water temperature compared to the inlet water temperature, thermal efficiency, and electric efficiency provided by the present BIPVT panel at different water flow rates under different irradiance intensities.

The comparison of the electric and thermal efficiencies of the present BIPVT roofing panel with other typical PVTs or BIPVTs with water as the heat transfer fluid is presented in Table 14.3. Note that none of the studies in Table 14.3 took into account the energy needed to operating pumps when calculating the efficiency levels. Overall, the literature contains many more performance evaluations of various PVT systems than that of BIPVT systems. Thus, only some of them with similar features to the presented BIPVT panel were collected in this comparison. Among the PVTs compared in the table, the PVT water collector proposed by Fudholi et al. [36], which consists of a combined PV module and a spiral flow absorber, exhibited the highest performance in the current literature. It was reported that this absorber produced a PVT efficiency of 68.4%, a PV efficiency of 13.8%, and a thermal efficiency of 54.6% at a solar radiation level of 800 W m^{-2} and mass flow rate of 0.041 kg s^{-1}.

One of the BIPVTs highlighted in the table was designed by Chow et al. [37] and involved a centralized PV and hot water collector wall system. They too conducted experimental studies under different operating modes at different seasons. Their test results showed that the thermal efficiency and the corresponding electricity conversion efficiency found by statistical analysis were 38.9% and 8.56%, respectively. Corbin and Zhai [24] developed a prototype BIPVT roof collector and conducted an experimentally

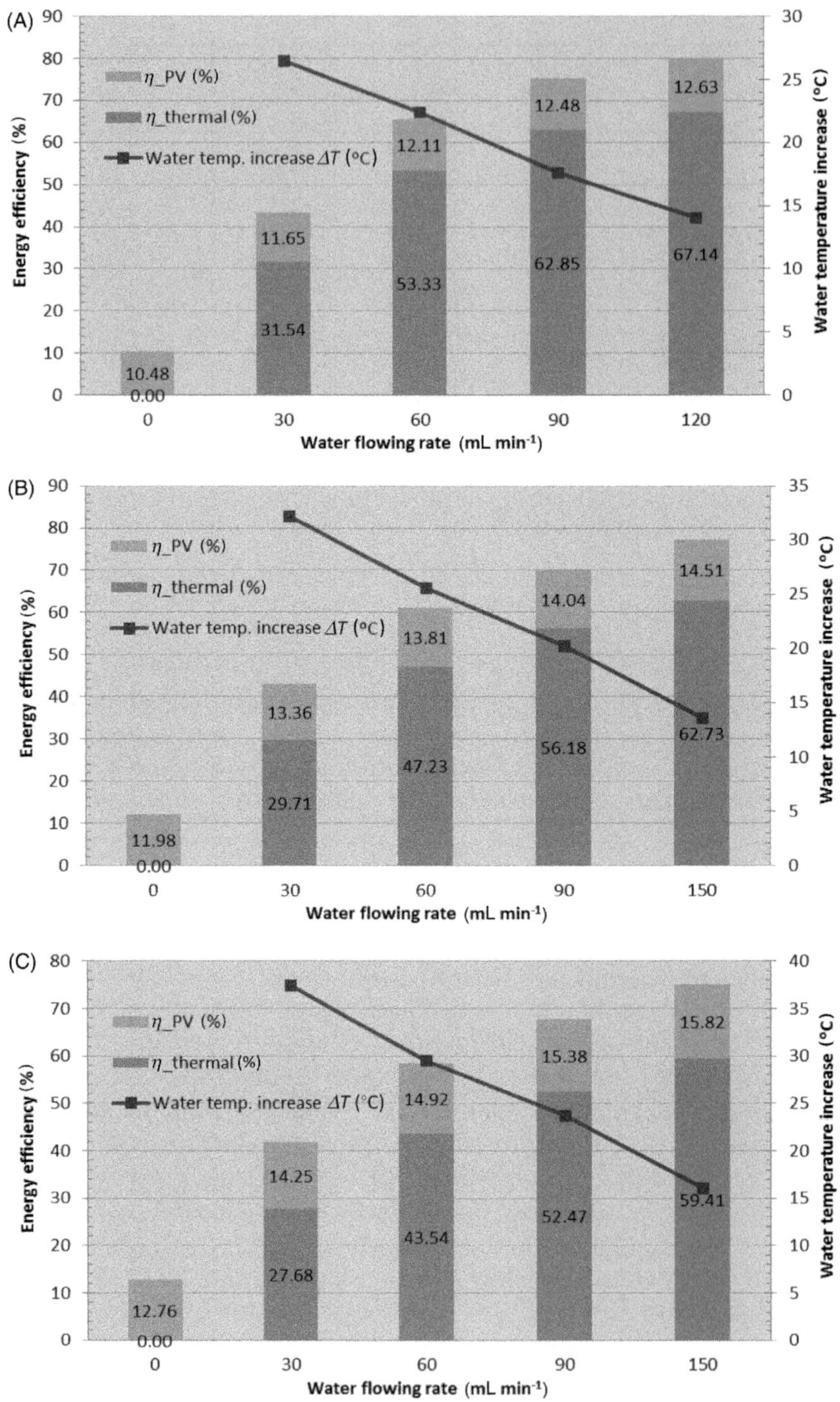

FIGURE 14.11 Increased water temperature, thermal and electricity conversion efficiencies of the BIPVT at different solar irradiance. (A) 620 W m^{-2}, (B) 800 W m^{-2}, and (C) 1000 W m^{-2}.

Table 14.3 Comparison of Panel Efficiencies Among Different Studies

Study	BIPVT (Yes/No)	Radiation: Daily (MJ m^{-2})	Radiation: Transient (W m^{-2})	η_{PV} (%)	$\eta_{thermal}$ (%)	η_{PVT} (%)
Huang et al. [39]	No	6.6 ~ 15.9	–	~9	12.2 ~ 44.5	21.2 ~ 53.5
He et al. [16]	No	10.5 ~ 19.8	–	3.9 ~ 5.4	28.8 ~ 52.0	32.7 ~ 57.4
Chow et al. [40]	No	12.8 ~ 22.2	–	9.9 ~ 10.2	45–48	54.9 ~ 58.2
Ji et al. [41]	No	9.3 ~ 21.2	–	9.4 ~ 10.3	34.7 ~ 56.9	40.6 ~ 63.1
Fudholi et al. [37]	No	–	800	13.8	54.6	68.4
Chow et al. [37]	Yes	8.3 ~ 12.8	–	8.56	38.9	47.46
Corbin and Zhai [24]	Yes	–	1000[a]	15.9	19	34.9
Anderson et al. [23]	Yes	Daily average	–	12.5–14	40–60[b]	–
Kim et al. [38]	Yes	Daily average	–	~17	~30	~47
Ibrahim et al. [34]	Yes	–	690	10.4–11.3	45–51	55–62
Present BIPVT	Yes	–	620	12.6	67.1	79.7
Present BIPVT	Yes	–	800	14.5	62.7	77.2
Present BIPVT	Yes	–	1000	15.8	59.4	75.2

BIPVT, Building integrated photovoltaic/thermal; *CFD*, computational fluid dynamics.
[a]Experimentally validated CFD simulation.
[b]A packing factor of 40% was applied to the collector.

validated computational fluid dynamics simulation. Their reported results showed that their BIPVT roof collector was able to provide thermal and combined (thermal plus electrical) efficiencies of 19% and 34.9%, respectively. Anderson et al. [23] designed a prototype BIPVT water collector that was integrated into the standing seam or toughed sheet roof. In calculating the thermal efficiency, a packing factor of 40% was applied, while the electrical efficiency was calculated by an indirect method which included an electrical efficiency loss factor of 0.5% (°C)$^{-1}$. Their results showed that the thermal and electric efficiencies of the BIPVT roof collector were within the ranges of 12.5% ~ 14% and 40% ~ 60%, respectively. Kim et al. [38] developed a water-type unglazed BIPVT collector. Based on expressions similar to Eqs. (14.2) and (14.4) in this study, the thermal and electrical efficiencies of their BIPVT under a clear day from 9:00 am to 3:00 pm were calculated, respectively. It was reported that the average thermal and electrical efficiencies were 30% and 17%, respectively.

Based on the spiral flow absorber designed in reference [36], Ibrahim et al. [34] developed a BIPVT roof system with the spiral flow copper absorber attached to the bottom of PV modules on the roof. The collector was a flat plate with single glazing sheet. The thermal performance of the PVT unit was evaluated by the Hottel–Whillier equations while the electrical efficiency was calculated by the indirect method with an electrical efficiency loss factor of 0.45% (°C)$^{-1}$. From the energy analysis for a mass flow rate of 0.027 kg s^{-1} and an average solar radiation of 690 W m^{-2}, it was found that the average electric and thermal efficiencies of the BIPVT roof were 10.8% and 48%, respectively.

From the comparisons shown in Table 14.3, the BIPVT panel presented in this chapter, is able to harvest solar irradiance more efficiently in form of electricity and heat than most PVT or BIPVT technologies currently available in the literature. Because of the ability to control temperatures through the water flow, the PV modules can work at lower temperatures in the summer time, leading to a higher efficiency for PV utilization. Furthermore, due to the temperature control on the roof, better thermal comfort in the building can be achieved and, therefore, the energy demand for cooling can be reduced in the summer time. Moreover, the warm water flow can be applied to remove frost or ice on the roof in wintertime, thus further enhancing solar energy utilization. In addition, as shown in Tables 14.1 and 14.2, the temperature increase of the collected water, the useful collected heat, and the increased PV efficiency vary from the water flowing rate and the irradiance levels. This provides customers with much flexibility to adjust the water flowing rate to meet their specific requirements.

14.5 Summary and Conclusions

In this chapter state-of-the-art BIPVT systems are discussed; namely a commercial BITERS system developed by Atlantis Energy Systems, Inc., and a novel BIPVT solar roof panel developed by Columbia University. As an essential component of the multifunctional building envelope, a novel FGM panel was successfully integrated into the BIPVT panel. The FGM layer gradually transits material phases from well-conductive side (aluminum dominated) to another highly insulated side (HDPE dominated). Due to the high thermal conductivity of the upper part of the FGM, the heat in the PV cells can be immediately transferred into the FGM and then captured by the cool water flowing in all directions through the embedded tubes. Thus the elevated operation temperature in the PV cells can be easily cooled down and as a consequence the PV efficiency can be considerably enhanced.

The developed BIPVT panel can be integrated into a building skin with relevant system components such as water circulation, flow control, and heat storage. It enables heat harvesting while improving the efficiency of the PV modules by controlling the temperature. The electricity can be directly transmitted to the grid, while the heat can be stored or directly used. This system can be used for the generation of electricity and heat for both residential and commercial buildings, which in turn reduces the heating and cooling costs. The performances of the prototype BIPVT panel in terms of its thermal efficiency and PV efficiency have also been evaluated in the laboratory. Test results showed that

- The increased temperature of the collected water by the BIPVT can be as high as 37.5°C at an irradiance of 1000 W m^{-2} with a relatively low flow rate at 30 mL min^{-1}, and that the decreased surface temperature of the BIPVT can be as high as 32°C when the flow rate is increased to 150 mL min^{-1}. It is expected that, as the water flow rate increases, the BIPVT surface temperature will further decrease.

- At a water flow rate of 150 mL min^{-1}, the output electricity energy E_{pv} reaches 32.96 and 44.91 W and the PV efficiency η_{pv} reaches 14.51% and 15.82% under solar irradiance of 800 and 1000 W m^{-2}, respectively. The PV efficiency is also enhanced by 21.1% and 24.0% for those two solar irradiances respectively, compared with the case when no water flow is introduced.
- At an introduced water flow rate of 150 mL min^{-1}, the total energy efficiency ($\eta_T = \eta_{thermal} + \eta_{pv}$) of the presented BIPVT roofing panel has been evaluated to be79.8%, 77.3%, and 75.2% under solar irradiances of 620, 800, and 1000 W m^{-2} respectively.

Overall, the test results demonstrate that significant energy conversion efficiency improvement can be achieved for both electricity generation and heat collection by the developed BIPVT roofing panel. From the comparisons of the present BIPVT with other relevant technologies, it was found that the developed BIPVT panel is able to harvest solar irradiance more efficiently in form of electricity and heat than most PVT or BIPVT technologies currently described in the literature. Due to the temperature control on the roof, better thermal comfort in the building can be achieved and, therefore, the energy demand for cooling can be reduced in the summer time. Moreover, the warm water flow can be applied to remove frost or ice on the roof in wintertime, thus further restore and enhance solar energy utilization. In addition, this BIPVT provides customers with great flexibilities to adjust the water flowing rate to meet each home's specific requirements.

Acknowledgment

This work is sponsored by the National Science Foundation CMMI 0954717, whose support is gratefully acknowledged. The authors appreciate the contributions from Mr. Yuda Sun, Mr. Xin He, Mr. Bin Li, Mr. Joao Silva, and Mr. Rodolfo Kusmaonthe on laboratory testing and the ongoing support from the Henry Mitchell Weitzner Research Fund.

References

[1] Zogou O, Stapountzis H: Energy analysis of an improved concept of integrated PV panels in an office building in central Greece, *Appl Energy* 88(3):853–866, 2011.

[2] Jelle BP, Breivik C: State-of-the-art building integrated photovoltaics, *Energy Proc* 20:68–77, 2012.

[3] Chae YT, Kim J, Park H, Shin B: Building energy performance evaluation of building integrated photovoltaic (BIPV) window with semi-transparent solar cells, *Appl Energy* 129:217–227, 2014.

[4] Archibald J. Building integrated solar thermal roofing systems-history, current status and future promise. In: Proceedings of the solar conference. American Solar Energy Society; American Institute Of Architects; 1999. p. 95–100.

[5] Lamnatou C, Mondol JD, Chemisana D, Maurer C: Modelling and simulation of building-integrated solar thermal systems: behaviour of the coupled building/system configuration, *Renew Sustain Energy Rev* 48:178–191, 2015.

[6] Buker MS, Riffat SB: Building integrated solar thermal collectors—a review, *Renew Sustain Energy Rev* 51:327–346, 2015.

[7] Pantic S, Candanedo L, Athienitis AK: Modeling of energy performance of a house with three configurations of building-integrated photovoltaic/thermal systems, *Energy Build* 42(10):1779–1789, 2010.

[8] Athienitis AK, Bambara J, O'Neill B, Faille J: A prototype photovoltaic/thermal system integrated with transpired collector, *Sol Energy* 85(1):139–153, 2011.

[9] Vats K, Tiwari GN: Energy and exergy analysis of a building integrated semitransparent photovoltaic thermal (BISPVT) system, *Appl Energy* 96:409–416, 2012.

[10] Gueymard CA, duPont WC: Spectral effects on the transmittance, solar heat gain, and performance rating of glazing systems, *Sol Energy* 83(6):940–953, 2009.

[11] Parida B, Iniyan S, Goic R: A review of solar photovoltaic technologies, *Renew Sustain Energy Rev* 15(3):1625–1636, 2011.

[12] Sun L, Lu L, Yang H: Optimum design of shading-type building-integrated photovoltaic claddings with different surface azimuth angles, *Appl Energy* 90(1):233–240, 2012.

[13] Chow TT: A review on photovoltaic/thermal hybrid solar technology, *Appl Energy* 87(2):365–379, 2010.

[14] Jelle BP: Building integrated photovoltaics: a concise description of the current state of the art and possible research pathways, *Energies* 9(1):21, 2015.

[15] Muller J, Hinken D, Blankemeyer S, Kohlenberg H, Sonntag U, Bothe K, Dullweber T, Kontges M, Brendel R: Resistive power loss analysis of PV modules made from halved 15.6 #x2009; #x00D7; #x2009;15.6 cm^2 silicon PERC solar cells with efficiencies up to 20.0 #x0025;, *IEEE J Photovolt* 5(1):189–194, 2015.

[16] He W, Chow T-T, Ji J, Lu J, Pei G, Chan L: Hybrid photovoltaic and thermal solar-collector designed for natural circulation of water, *Appl Energy* 83(3):199–210, 2006.

[17] Solar panel temperature—facts and tips. Available from: http://www.solar-facts-and-advice.com/solar-panel-temperature.html.

[18] Skoplaki E, Palyvos JA: On the temperature dependence of photovoltaic module electrical performance: a review of efficiency/power correlations, *Sol Energy* 83(5):614–624, 2009.

[19] Agrawal B, Tiwari GN: *Building integrated photovoltaic thermal systems: for sustainable developments*, London, 2011, Royal Society of Chemistry, Burlington House.

[20] Nagano K, Mochida T, Shimakura K, Murashita K, Takeda S: Development of thermal-photovoltaic hybrid exterior wallboards incorporating PV cells in and their winter performances, *Sol Energy Mater Sol Cells* 77(3):265–282, 2003.

[21] Yang T, Athienitis AK: A study of design options for a building integrated photovoltaic/thermal (BIPV/T) system with glazed air collector and multiple inlets, *Sol Energy* 104:82–92, 2014.

[22] Yang T, Athienitis AK: Experimental investigation of a two-inlet air-based building integrated photovoltaic/thermal (BIPV/T) system, *Appl Energy* 159:70–79, 2015.

[23] Anderson TN, Duke M, Morrison GL, Carson JK: Performance of a building integrated photovoltaic/thermal (BIPVT) solar collector, *Sol Energy* 83(4):445–455, 2009.

[24] Corbin CD, Zhai ZJ: Experimental and numerical investigation on thermal and electrical performance of a building integrated photovoltaic–thermal collector system, *Energy Build* 42(1):76–82, 2010.

[25] Ghani F, Duke M, Carson JK: Effect of flow distribution on the photovoltaic performance of a building integrated photovoltaic/thermal (BIPV/T) collector, *Sol Energy* 86(5):1518–1530, 2012.

[26] Ibrahim A, Othman MY, Ruslan MH, Mat S, Sopian K: Recent advances in flat plate photovoltaic/thermal (PV/T) solar collectors, *Renew Sustain Energy Rev* 15(1):352–365, 2011.

[27] Kumar A, Baredar P, Qureshi U: Historical and recent development of photovoltaic thermal (PVT) technologies, *Renew Sustain Energy Rev* 42:1428–1436, 2015.

[28] Pao F, Gottlieb R. Building integrated thermal electric hybrid roofing system. US Pattern number: 8215070; 2012.

[29] Yin HM, Yang DJ, Kelly G, Garant J: Design and performance of a novel building integrated PV/thermal system for energy efficiency of buildings, *Sol Energy* 87:184–195, 2013.

[30] Chen FL, He X, Yin HM: Manufacture and multi-physical characterization of aluminum/high-density polyethylene functionally graded materials for green energy building envelope applications, *Energy Build* 116:307–317, 2016.

[31] Chen F, Yin H: Fabrication and laboratory-based performance testing of a building-integrated photovoltaic-thermal roofing panel, *Appl Energy* 177:271–284, 2016.

[32] Frost & Sullivan. 2009. World building-integrated photovoltaic niche player of the year Award. Available from: http://www.frost.com/prod/servlet/press-release.pag?docid=191394073.

[33] Council, I.C. 2015. International Building Code, ICC; 2014.

[34] Ibrahim A, Fudholi A, Sopian K, Othman MY, Ruslan MH: Efficiencies and improvement potential of building integrated photovoltaic thermal (BIPVT) system, *Energy Convers Manag* 77:527–534, 2014.

[35] Chander S, Purohit A, Sharma A, Arvind Nehra SP, Dhaka MS: A study on photovoltaic parameters of mono-crystalline silicon solar cell with cell temperature, *Energy Rep* 1:104–109, 2015.

[36] Fudholi A, Sopian K, Yazdi MH, Ruslan MH, Ibrahim A, Kazem HA: Performance analysis of photovoltaic thermal (PVT) water collectors, *Energy Convers Manag* 78:641–651, 2014.

[37] Chow TT, He W, Ji J: An experimental study of façade-integrated photovoltaic/water-heating system, *Appl Therm Eng* 27(1):37–45, 2007.

[38] Kim J-H, Park S-H, Kang J-G, Kim J-T: Experimental performance of heating system with building-integrated PVT (BIPVT) collector, *Energy Proc* 4:1374–1384, 2014.

[39] Huang BJ, Lin TH, Hung WC, Sun FS: Performance evaluation of solar photovoltaic/thermal systems, *Sol Energy* 70(5):443–448, 2001.

[40] Chow TT, Ji J, He W: Photovoltaic-thermal collector system for domestic application, *J Sol Energy Eng* 129(2):205–209, 2006.

[41] Ji J, Lu J-P, Chow T-T, He W, Pei G: A sensitivity study of a hybrid photovoltaic/thermal water-heating system with natural circulation, *Appl Energy* 84(2):222–237, 2007.

15

Integration of PV Generated Electricity into National Grids

Graham Stein*, Trevor M. Letcher**,†

*NATIONAL GRID, WARWICK, UNITED KINGDOM; **UNIVERSITY OF KWAZULU-NATAL, DURBAN, SOUTH AFRICA; †LAUREL HOUSE, FOSSEWAY, STRATTON ON THE FOSSE, UNITED KINGDOM

trevor@letcher.eclipse.co.uk, graham.stein@nationalgrid.co.uk

15.1 Introduction: Rapid Growth of the Solar PV Industry

Solar cells have been used to produce electricity since the 1950s, but it is only recently, in the 2000s, that electricity-supplying companies have used solar energy as a source for producing grid-electricity. In this short time the photovoltaic industry has been a success story both from the decarbonisation of the electricity generating industry point of view and from the industry growth itself.

PV technology advances and deployment over past 20 years have been amazing as discussed in Chapter 1 and Chapter 28. The *raison d'etre* is of course the looming specter of global warming and climate change [1]. Growth in the United Kingdom was relatively slow as can be expected for a country not imbued with an overabundance of sunshine, after all the United Kingdom lies north of 52° N. In spite of that the United Kingdom now has the fourth largest solar PV capacity in Europe with 11% of the Europe's total (Chapter 28).

This chapter refers to the integration of solar PV into national power supply networks with particular relevance to the United Kingdom. Much of the information has been obtained from the National Grid Future Energy Scenarios [2], the National Grid Summer Outlook report [3], the Sheffield PV Monitoring Project [4], The National Grid control room twitter feed [5], and the National Grid System Operability Framework document [6]. Many of the statements in this chapter are taken directly from these documents.

The issues facing operators in integrating wind energy into grid networks have a close affinity to the problems facing PV. A detailed report into wind energy integration by Jurgen Weiss and T Bruce Tsuchida can be found in a related book [7].

A Comprehensive Guide to Solar Energy Systems. http://dx.doi.org/10.1016/B978-0-12-811479-7.00015-4

15.2 Why We Need to Integrate Solar Power into National Grids

Over the past century the world has become reliant on fossil fuel for most of its electricity. This has led to a build-up of CO_2 atmospheric pollution that has seen the levels of CO_2 rise from 250 parts per million (250 molecules per 10^6 molecules) in the 1800s to 413 ppm as measured at Mauna Loa Observatory on April 26, 2017 [8]. Fossil fuel burning power-stations are responsible for at least 25% of this CO_2 pollution. This increase in CO_2 is largely responsible for our present climate change and global warming [1].

Although the amount of energy reaching the surface of the Earth is enormous it is, at the same time, relatively dilute in that even in the hottest tropical areas the energy from the sun is of the order of 1000 W m^{-2}. Photovoltaic technology is one way of exploiting the sun's energy to produce electricity and of reducing our reliance on fossil fuel.

The development of photovoltaic cells over the past 30 years has led to a dramatic reduction in its costs and today electricity produced from PV panels is possibly cheaper per watt than fossil fuel [9]. This reduction in costs, combined with the introduction of specific policies supporting renewable generation development, has led to a huge growth in solar PV capacity in the United Kingdom which is set to continue. In the United Kingdom the solar PV capacity has increased from 9.3 GW in February 2016 to 11.7 GW in February 2017. It is expected that the growth will continue at a rate of 150 MW per month over the next 12 months, rising to 13.5 GW in February 2018 [10].

The growth in both the number of solar PV installations and their combined capacity is unprecedented in UK energy supply. Solar PV is now highly significant in the energy mix and its characteristics influence the way energy supply markets behave and how the networks used to deliver it perform.

The first feature most people would think of when asked about how much electricity a solar panel will generate is its dependency on the sun. Solar PV output varies as the sun comes and goes. Ideally, electricity companies need a steady production of electricity, which should exactly match their clients' demand. Most traditional power plants are designed to provide a steady base-load of energy and are equipped to ramp up or down to meet consumers' needs as they vary steadily throughout the day. Yet solar PV output varies. Does this really matter though?

The physical balancing of the available energy from nuclear, gas-fired and coal-powered electricity generation, and including solar PV (and also wind) with the demand is a knife-edge process. To start, let's look at how electricity demand varies in the United Kingdom and focus on summer months. This is taken from reference [6].

The hourly variation over a day is shown in Fig. 15.1 (taken from Fig. 2.3 of page 21 of reference [6]). The peak electricity demand in summer ranges from 39 GW in early April to 33 GW in mid-May. This is reasonably constant until mid-August when the demand rises again.

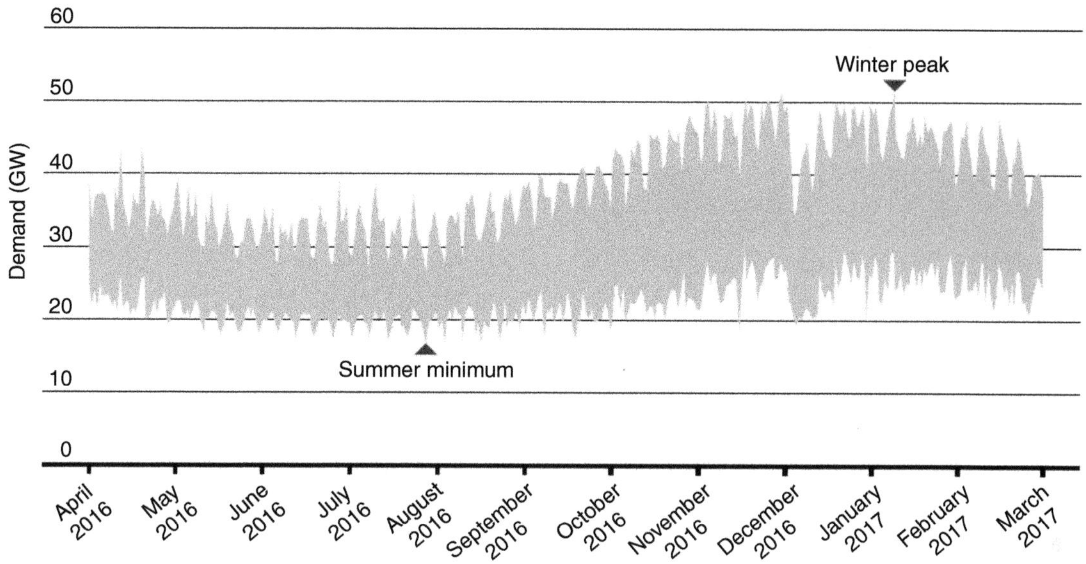

FIGURE 15.1 The hourly energy demand variation over a day for a year [6]. *Used with permission from the National Grid.*

15.3 How Solar PV Fits in

A key issue for the grid operator is the balancing of the various generating energy systems to meet the varying demand illustrated earlier. Year round modeling provides a good insight to the system and forms the basis for the balancing process.

Fig. 15.2 (taken from Fig. 2.8 page 26 of reference [6]) illustrates how generation is dispatched to meet a transmission demand on a particular day.

Looking from the bottom up, nuclear energy generation is assumed to run all day as a base-load. In this example the interconnector flow from Europe is next in line. The contribution from large-scale renewable energy (solar and wind) are next to be added before the various conventional generating energies are stacked on top—hydro, biomass, gas, and coal. If the demand has not been satisfied then storage units are brought in and if there is still a shortfall, then other balancing resources are used to make up the remaining deficit.

In the figures shown, distributed generation is placed over the transmission demand line, showing what the total demand for electricity will be. It is shown this way because distributed generation is not dispatched and doesn't play the same role in the market as the conventional transmission connected generation and is hence not "visible" in the same way. But it is illustrated to make clear its effect on the transmission demand and subsequently the generation required to balance the system. Fig. 15.3 (taken from Fig. 2.9 page 27 of reference [6]) shows what might happen in 2025.

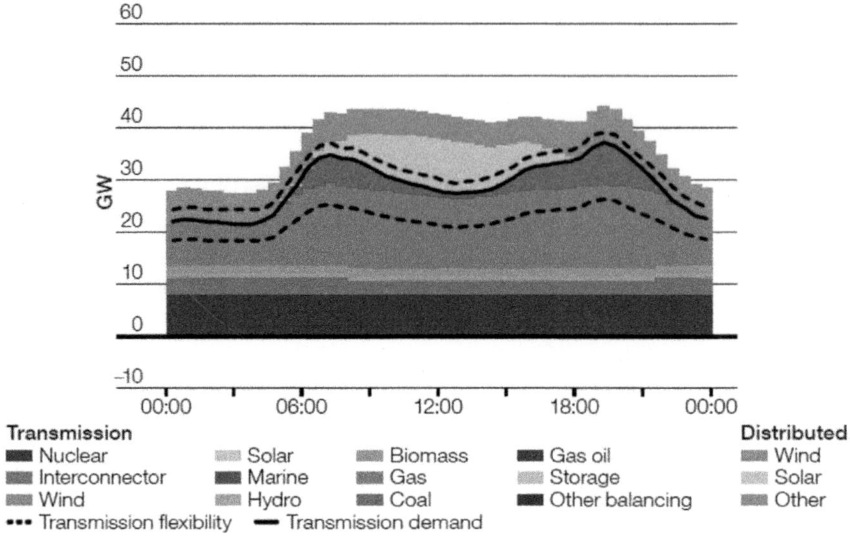

FIGURE 15.2 Energy generation on a particular day [6]. *Used with permission from the National Grid.*

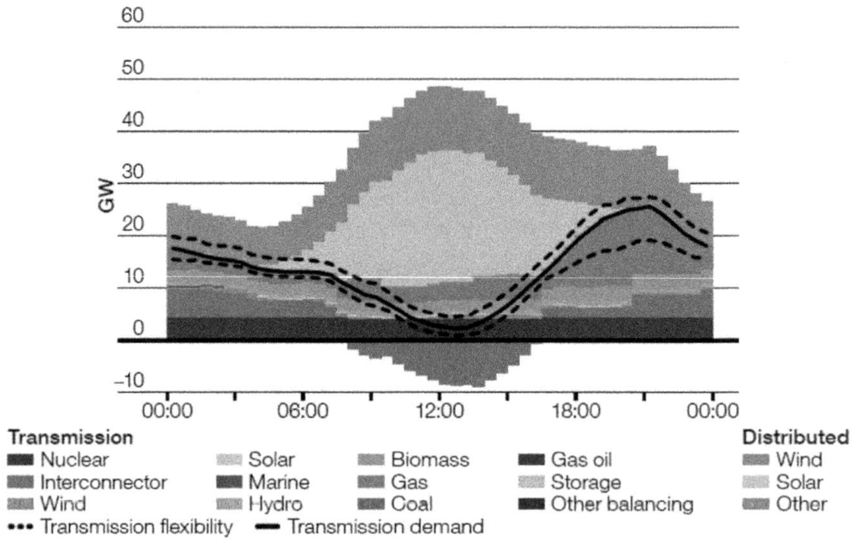

FIGURE 15.3 A prediction of energy generation on a day in 2025 [6]. *Used with permission from the National Grid.*

15.4 Is the Duck Relevant to Solar PV in United Kingdom?

One question raised by the growth of solar power is illustrated by the 'Duck Curve' described in Chapter 4 (Fig. 4.24). The figure highlights the experience of the California Independent System Operator showing the change in demand for grid electricity over the six years from 2011 [11a,11b].

The curve is the amount of energy that the traditional power plants have to supply on a typical day. The drop in the middle of the day is due to the amount of energy supplied by utility and roof top PV power supplies. As a result of the increased PV power from roof top systems and from solar farms from 2011 to 2016, this drop becomes more pronounced and the steepness of the 'neck' becomes greater.

This curve could well represent related systems of combined base-load power and solar PV power, anywhere in the world and you can see a similar effect in Figs. 15.2 and 15.3 earlier. During the day, when solar plants are most productive, demand for electricity from traditional power plants plummets. In the evening, the need for traditional power increases due to the combined effect of solar output decreasing and energy demand, increasing. This results in a steep rise (duck's neck) in demand, which must be met by the other sources power supply.

These other sources must be available at the right time and able to ramp up at the necessary rate. These could be the traditional power plants we're used to, but it might be prohibitively expensive to keep these open just to meet short steep rises in demand, in which case, we would see flexibility provided in other ways.

This "flexibility" could be consumers switching off their own demand in response to high prices or direct payments. Fast acting alternative sources of energy like batteries could perform the same function. Where the need for flexibility is great enough and transparent to those who might provide it, it's likely that we'll see investment in new facilities and the energy industry will develop new ways of buying and selling this capability.

So, how do we know that grid operators can make use of this flexibility and keep the lights on the way we're used to? Operators have a good track record of this, and perhaps the best example to refer to here is the solar eclipse in North America on Monday, August 21, 2017. The Californian system operator successfully negotiated a 5 GW drop off in solar output, and its rapid return at an average rate of 0.15 GW per minute. The National Grid has looked at how to cope in Great Britain when the same thing happens in 2026. Their report is entitled "A preliminary study of how a solar eclipse due in 2026 will affect GB when there is up to 26 GW of solar PV".

15.5 Effect of Growth in Small Distributed Installations

Another issue for the grid-electricity generating industry is the wide distribution of solar PV panels. PV works very well in small, distributed installations. Because they are small and numerous, there is very limited justification to add remote metering and control facilities. Roof top units are usually in the kilowatt range and the utility scale units (solar farms) can be anything from 10 MW to 200 MW and in the United Kingdom they are usually less than 20 MW.

Even for larger installations, the owner/operator may not want to control their output as it makes sense to sell all of the available output. After all, it is free when the sun shines.

The resulting shortcoming is the lack of information (called visibility here) of small-scale generators given to the grid operator who is responsible for ensuring that the right energy balance is achieved second by second. At present visibility is defined in terms of regulatory thresholds for small generators above which visibility is mandated. The values are 50 MW in England and Wales, 30 MW in Southern Scotland, and 10 MW in the North of Scotland.

For a view into the future, see Figs. 15.4 and 15.5 (Figs. 5.3 and 5.4 of reference [6]).

The results show the distribution of generation output from visible and invisible sources for 2016/17 and the prediction for 2025/27 in the "Consumer Power" and "Slow

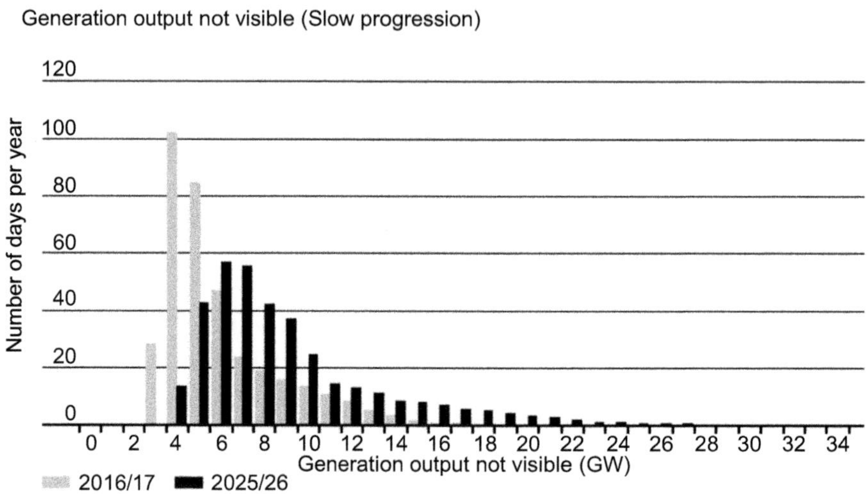

FIGURE 15.4 Generation output which is not visible; based on a slow progression scenario [6]. *Used with permission from the National Grid.*

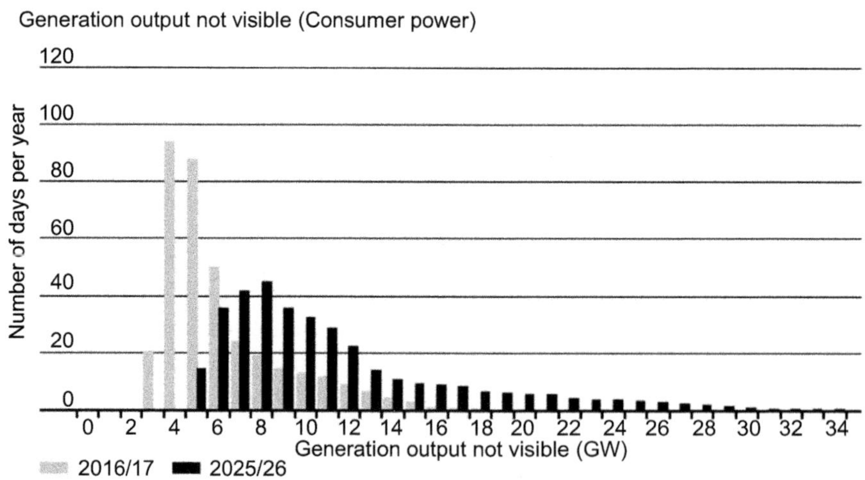

FIGURE 15.5 Generation output which is not visible; based on a consumer power scenario [6]. *Used with permission from the National Grid.*

Progression" scenarios (these are two of four scenarios used by National Grid to look at future gas and electricity demand and supply—in consumer power an empowered consumer makes choices and drives activity whilst in slow progression change is slow because of a lack of direction and ability to make investment choices). The results highlight the increased influence of distributed resources, which are not visible over time. In 2016/17 the highest total output from generators not visible is 17 GW. By 2025/26 it is predicted that this will rise to 34 GW in consumer power and 27 GW in slow progression.

Not only does this generated electricity need to go somewhere (someone needs to want to make use of it at the time it's produced), but it also has to be balanced with the output of other sources as described earlier. This means, we could either see a desire to limit the output of solar PV directly when it is particularly sunny and windy, or a growth in the flexibility which balances its output. We could find that much of the output is exported to the European continent. But it could also be used to charge batteries or other forms of energy storage. Of course, solar PV operators may start to limit their own output if they find that the price they receive is not satisfactory.

However, it is possible to overcome this shortcoming we have highlighted. After all, you know that solar PV output will be high if the sun is shining, and won't if it's not.

To improve the way PV generation output was forecast in the United Kingdom, a pilot project was started with Sheffield Solar [4]. The data has helped grid operators to better manage flows on the electricity grid network. To obtain the maximum solar PV output, it is necessary to estimate the availability of the solar resource. In the United Kingdom this is being done by the Met office at a number of sample sites, the radiance observations are being linked to solar power generation, and forecasting models are being developed.

This exercise becomes more challenging if you don't know the level of capacity which is behind the connection. Some designers size the panels and the converter rating differently, with the panels being oversized. Fig. 15.6 (see Fig. 5.2 in [6]) demonstrates the behavior of two solar installations with the same export capacity but with different generating capacities. Storage would add further complexities to the use of solar PV power as illustrated in Fig. 15.6 (see Fig. 5.2 of reference [6]). Equally, the additional storage could provide the flexibility required to compensate for some of the other issues touched on in this chapter (see reference [12]).

15.6 'Nonsynchronous' Inverter Type Generators Supporting the Network

Solar PV generators are linked to the grid by inverters which convert DC electrical power from panels into AC power suitable for injecting into the grid. Properly configured, a grid tie inverter (GTI) enables a home owner to use a power generation system such as solar system without extensive rewiring and without needing batteries to fill in peaks and troughs in output. If the power being produced is insufficient, the deficit will be sourced from the electricity grid.

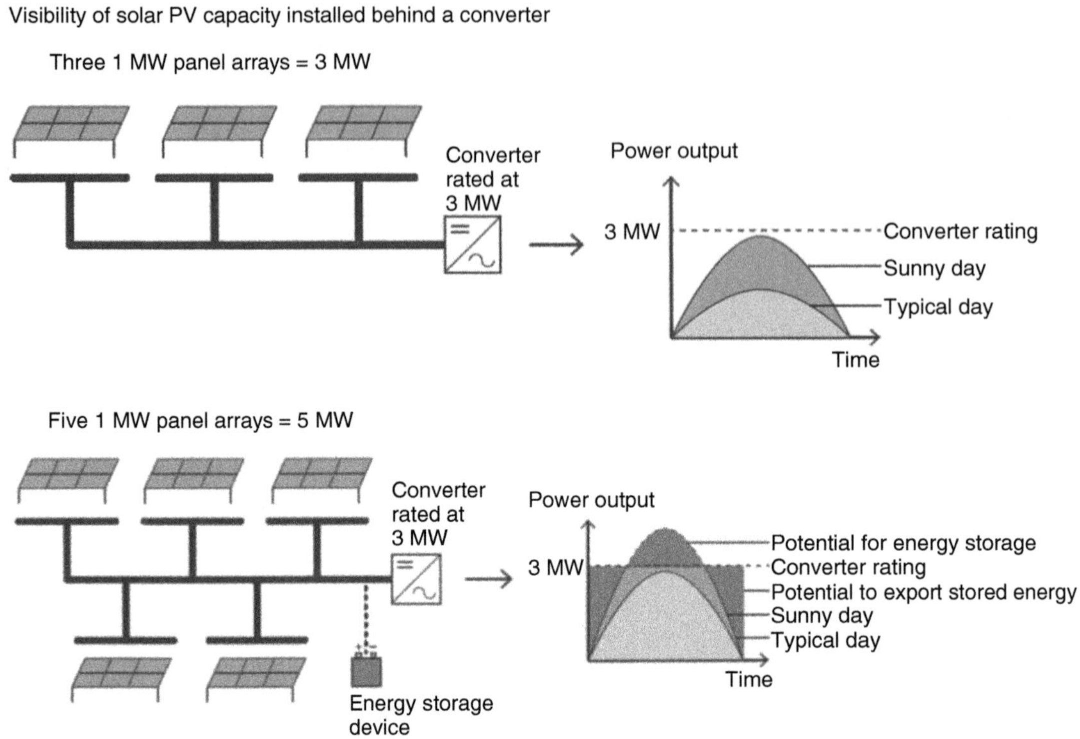

FIGURE 15.6 The behavior of two solar installations with the same export capacity but with different generating capacities [6]. *Used with permission from the National Grid.*

The GTI must match the phase of the grid voltage and maintain the output voltage slightly higher than the grid voltage at any instant. It does this by measuring the current AC grid waveform and manipulating its power electronic components to create an output with a voltage to match and therefore correspond with the grid. A high-quality modern GTI has a fixed unity power factor, which means its output voltage and current are perfectly lined up, and its phase angle is within 1 degree of the AC power grid.

The electricity grid voltage will fluctuate slightly as it is reconfigured as necessary in normal operation. Larger disturbances occur when electrical equipment failures happen and are dealt with. During a voltage disturbance the inverter needs to decide whether to keep generating an output matched to the network or to shut down and has to make this decision quickly based on the voltage it measures. This behavior needs to be carefully specified and can be tricky for manufacturers to design in to their inverters. It differs from a conventional "synchronous" generator, which has a inherent tendency to keep generating and supporting the network because it is electromagnetically coupled to the network.

The inverters used to link solar PV generators to the grid have not needed to provide support to the network thus far, as loss of their output hasn't been significant to the network's recovery from a disturbance. The growth in solar PV capacity means that this is being reconsidered. Changes are underway to implement European Network rules (see reference [13]), which could place requirements to withstand network disturbances on generators as small as 1 MW due to the increasing proportion of electricity being supplied by small scale generators.

Inverters can also be configured to operate with a phase angle, which is intentionally different to the local voltage it observes and is either fixed or varies in response to a need. This has the effect of raising or lowering local grid voltage. This feature is uncommon at the moment but is increasingly being asked for where lots of solar PV are connected and local voltages have the potential to rise to a level that network equipment is not designed to withstand: in other words, the "problem" created by the solar PV is being solved by the solar PV itself.

It is possible to use solar PV installations to support the networks beyond their local connection. Distributed energy resources such as solar PV could help stabilize the transmission network through a coordinated voltage control approach. This has been proposed by the National Grid and the electricity distribution company for London and the South East of England, UK Power Networks, through the Power Potential project [14]. The project proposes to investigate the implementation of a novel voltage control arrangement. The benefit here is that it is potentially cheaper and more competitive than conventional network reinforcement. For further details see references [6] and [14].

15.7 Converter Technology

The current generation of converter technology typically locks on to existing voltage reference signals using phase lock loops (PLL). This works fine while there is a strong system to connect to, acting as a reference. The idea is explored conceptually in Fig. 15.7 (see Fig. 4.21 reference [6]) using the analogy of a driver going through a tunnel. Further information on phase-locked loop technology can be found in references: [15–19].

As the electricity system becomes increasingly converter dominated, the system strength decreases and the reference is increasingly more susceptible to disturbances. It is possible to bolster system strength using either conventional "synchronous" generators or equipment known as synchronous condensers. These come at a cost, in terms of upfront investment, operating costs, and if fuel is burnt, carbon emissions. Consequently there is a growing hope and conviction among many engineers that the future is inverters, which behave much more like conventional generators. Virtual synchronous machine (VSM) technology is one way to approach this and the possibility has been discussed in various papers and industrial working groups. Other options using more sophisticated monitoring and control techniques which are capable at looking at and reacting to what is going on across the network are also being considered.

More information on VSM technology can be found in reference [20].

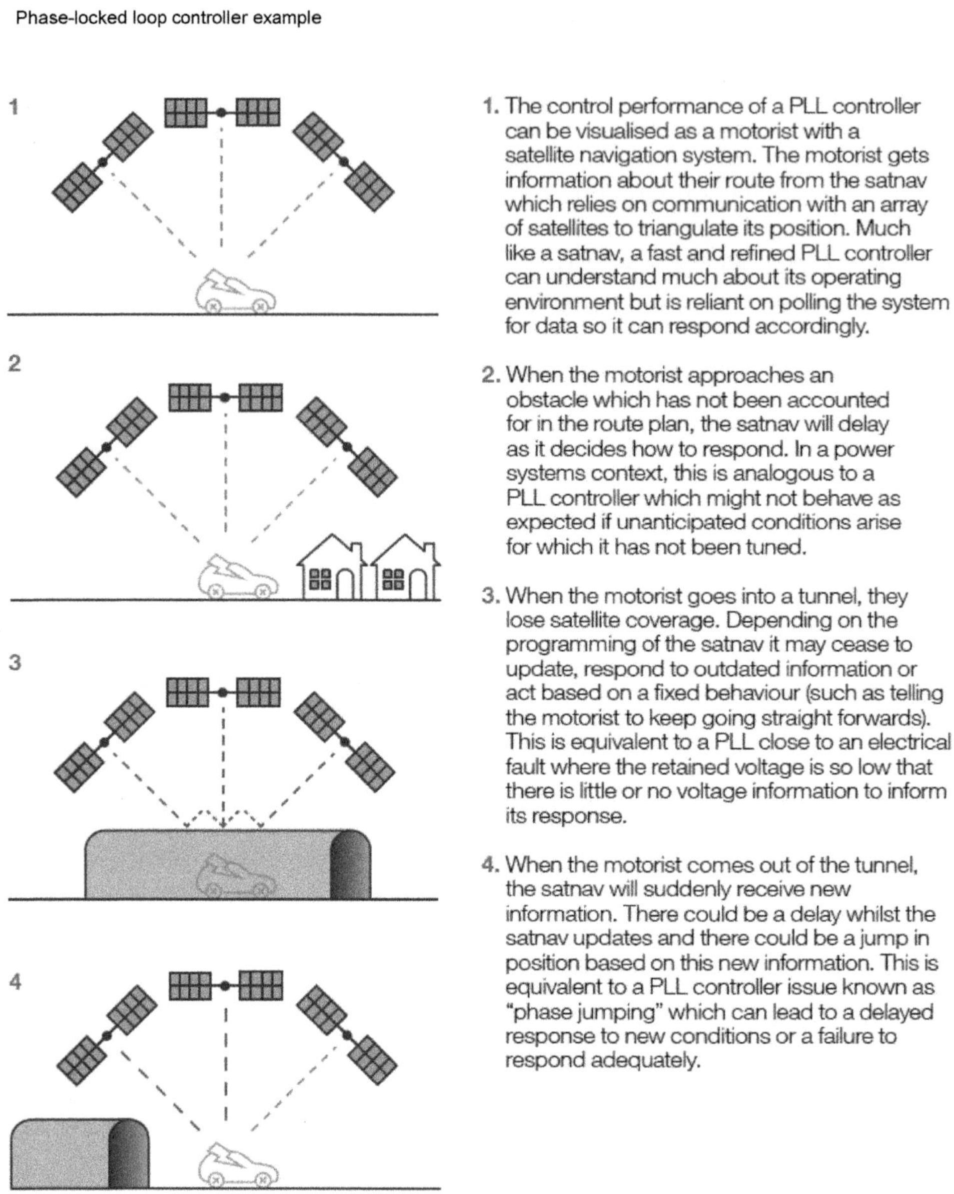

FIGURE 15.7 Phase-locked loop technology; analogous to a driver in a tunnel. *Used with permission from the National Grid.*

15.8 Conclusions

This chapter highlights a number of issues that grid operators are grappling with because of the rapid growth in PV. It doesn't cover the challenge of accommodating this new equipment in the local distribution network, which is a further topic in itself. However, all of the issues highlighted have a number of technical and administrative solutions which, depending on your point of view, would seem to be small in scale compared to the policy objectives they allow them to be achieved. Some of the potential solutions open up a whole new world of possibilities: domestic solar PVs combined with batteries and smart meters for example. Perhaps this will be the start of new era where energy consumers understand and decide exactly where their energy comes from and solar PV will be the technology that made the difference.

References

[1] Climate change: observed impacts on planet earth, 2nd ed., Editor Letcher TM, Elsevier, 2016.

[2] Available from: http://fes.nationalgrid.com/fes-document/ or http://www2.nationalgrid.com/uk/industry-information/future-of-energy/future-energy-scenarios/.

[3] Available from: http://www2.nationalgrid.com/UK/Industry-information/Future-of-Energy/FES/summer-outlook/.

[4] Available from: https://www.solar.sheffield.ac.uk/pvlive/.

[5] Available from: https://twitter.com/ngcontrolroom?lang=en.

[6] Future Energy Scenarios, National Grid; 2007. Available from: http://www2.nationalgrid.com/UK/Industry-information/Future-of-Energy/System-Operability-Framework/.

[7] Wind energy engineering: a handbook for onshore and offshore wind turbines, chapter 20, Editor Letcher TM, Elsevier, 2017.

[8] Available from: https://www.co2.earth/daily-co2.

[9] Available from: http://www.independent.co.uk/environment/solar-and-wind-power-cheaper-than-fossil-fuels-for-the-first-time-a7509251.htmlenergyinnovation.org/2015/02/07/levelized-cost-of-energy/.

[10] Available from: https://www.pv-magazine.com/2017/04/06/uk-150-mw-of-new-solar-pv-each-month-for-next-12-months-says-national-grid/.

[11a] Available from: https://www.greentechmedia.com/articles/read/the-california-duck-curve-is-real-and-bigger-than-expected.

[11b] Available from: http://www2.nationalgrid.com/WorkArea/DownloadAsset.aspx?id=8589941312.

[12] Storing energy: with special reference to renewable energy sources, Editor Letcher TM, Elsevier, 2016.

[13] Available from: http://www.entsoe.eu/major-projects/network-code-development/requirements for generators/.

[14] Available from: http://nationalgridconnecting.com/power-potential-ready-make-name/.

[15] Eren S, Bakhshai A, Jain P, Control of grid connected voltage source inverter with LCL filter, Institute of Electrical and Electronic Engineers DOI: 10.1109/APEC. 2012.6166021.

[16] Eren S, Karimi-Ghartemani M, Bakhshai A, Enhanced frequency-adaptive phase-locked loop for distributed power generation system applications, Queen's University Department of Electrical and Computer Engineering Kingston, Ontario, Canada. Available from: http://citeseerx.ist.psu.edu/viewdoc/download?doi=10.1.1.499.6870&rep=rep1&type=pdf.

[17] Limongi LR, Bojoi R, Pica C, Profumo F, Tenconi A, Politecnico di Torino, Department of Electrical Engineering, Torino – Italy. Analysis and comparison of phase locked loop techniques for grid utility applications, Institute of Electrical and Electronic Engineers. DOI: 10.1109/PCCON. 2007.373038.

[18] Xiao-Qiang GUO, Wei-Yang WU, He-Rong GU, Yanshan University. China. Phase locked loop and synchronization methods for gridinterfaced converters: a review. Przeglad Elektrotechniczny (review) ISSN 0033-2097,R.87 NR4/2011.

[19] Luna A, Citro C, Gavriluta C, Hermoso J, Candela I, Rodriguez P, Department of Electrical Engineering, SEER, Technical University of Catalonia, Spain. Advanced PLL structures for grid synchronization in distributed generation, Renewable Energies and Power Quality Journal 2012; 1: 1747–56.

[20] D'Arco S, Suul JA, Fosso OB: A virtual synchronous machine implementation for distributed control of power converters in SmartGrids, *Electric Power Syst Res* 122:180–197, 2015.

16

Small-Scale PV Systems Used in Domestic Applications

Nesimi Ertugrul
UNIVERSITY OF ADELAIDE, ADELAIDE, AUSTRALIA
nesimi.ertugrul@adelaide.edu.au

16.1 Introduction

Solar energy is greatly utilized by PV technology, and is simply limited by PV cell efficiency, availability of suitable area for installation, and associated power electronics converter technology. Commercial PV cells (Fig. 16.1A) commonly used in small domestic and industrial applications have a typical working photo-electrical efficiency of 10%–20% and have an open-circuit voltage of 600 mV, which is usually not a suitable level for practical applications [1]. To produce higher voltages and more power (between 10 and 300 W), PV cells are interconnected (about 40) to form larger units called panels or modules (Fig. 16.1B). Multiple modules are also installed on roofs of a building or at the ground level in a rack to form a PV array (Fig. 16.1C), which offer much higher voltage and power levels.

In addition to multiple PV cells, a typical PV module (or panel) consists of a number of auxiliary components which bring the module to usable form a transparent top surface (e.g., glass); an encapsulant (to hold together the top surface and rear surface of the cells); a rear layer (for sealing); and a metal frame around the outer edge for additional rigidity and ease of assembly.

Note that although Fig. 16.1 illustrates a type of PV technology, the structure of end products may vary to suit specific technology as well as target installations. For example, architecturally and structurally integrated PV arrays are emerging in the form of roof tiles and window glasses and with increased level of integrated power electronic circuits at cell and string levels. To enable higher packing density on roof tops, shade-tolerant design using static reflectors in tandem structures have also been implemented which can offer increased power output.

Although PV technologies can come in different forms, to utilize PV energy effectively, its DC output should be converted to match load characteristics of applications. A PV module/array is rarely connected to an electrical load directly (Fig. 16.2A) unless its voltage near the maximum power point (MPP) is higher than the operating voltage of the load. As classified in Fig. 16.2, complexity and topology of intermediate circuits (converters, filters

A Comprehensive Guide to Solar Energy Systems. http://dx.doi.org/10.1016/B978-0-12-811479-7.00016-6

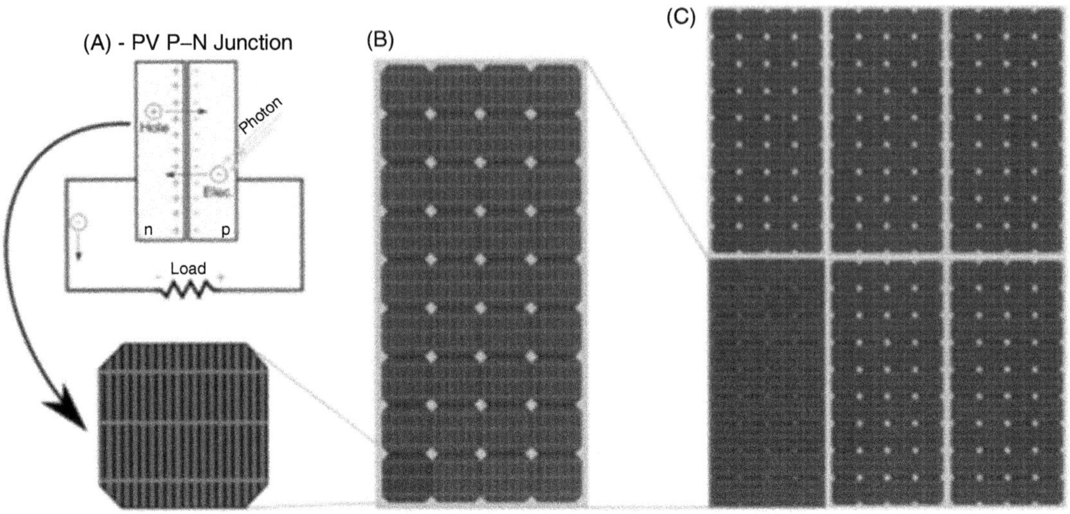

FIGURE 16.1 Form of available PV technologies (A) PN-junction and its implementation as a PV cell, (B) a PV module (containing 36 identical cells), and (C) a PV array (containing 6 modules).

and inverters) depend on primarily application requirements as well as the total cost. DC–DC converters shown in Fig. 16.2B have two main purposes: to match load characteristics and to obtain maximum power from PV module under varying solar insolation levels and/or load conditions. In addition, stand-alone inverters are sometimes used which cannot connect to the electrical grid but operate as isolated generators only to power AC loads (Fig. 16.2C and D) like refrigerators.

Energy storage (Fig. 16.2C and D) is frequently required either to store the produced electrical energy for use when sunlight is insufficient and/or to reduce/eliminate the impact of instant variations of solar irradiance, which is due to uncontrolled environmental factors. Traditionally, energy storage is done by lead acid batteries, but recently these have been replaced by lithium battery like LiIonP as they offer higher energy density and at an acceptable cost.

The grid-connected (grid-tied) inverters (Fig. 16.2E and F), which are the most common inverter systems, can supply energy not only to local loads within the household, but can also supply power to distant loads connected elsewhere on the electrical grid (as the grid acts as an infinitely large energy storage system!). This occurs when there is a surplus of generated energy. When the PV power generation runs short of supplying the local household however, power is either sourced from a local battery storage, if it exists, or from the grid.

In the absence of a grid voltage (i.e., an electrical blackout) a conventional grid-tied inverter system disconnects from the grid, which aims to protect personnel working on power lines that are assumed to not be live. In addition, such inverters also disconnect from the grid when the grid voltage level becomes outside inverter's settings, which is usually defined by acceptable voltage standards. However, high penetration of distributed renewable energy supplies like PV systems may also cause disconnections due to excessive

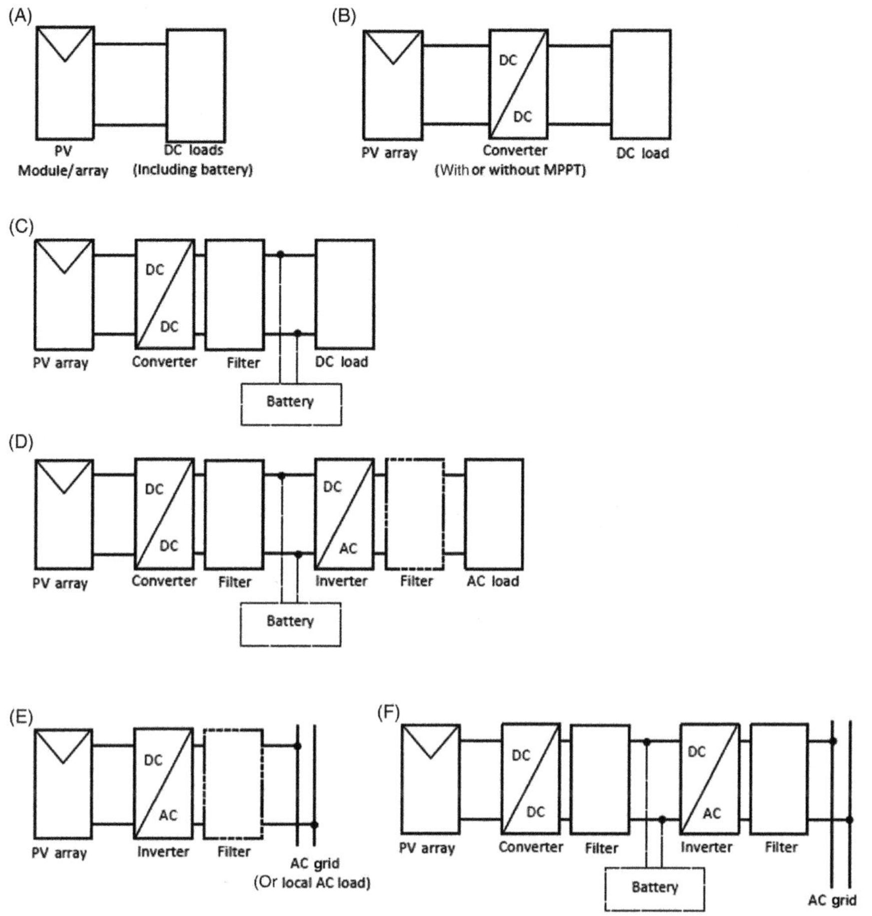

FIGURE 16.2 Block diagrams showing common PV system topologies utilized in small-scale applications, stand-alone or grid-connected (grid-tied) systems (A) direct DC connection to a load, (B) connection via a DC/DC converter, (C) with DC/DC converter and battery storage to DC load, (D) stand-alone inverter topology with a DC converter and battery storage, (E) basic grid-tied inverter topology, and (F) advanced grid-tied topology with bidirectional inverter and battery storage.

voltage rise at the point of common connection. Note that large-scale disconnection of grid-tied inverters also result further voltage fluctuations in the local distribution systems. Therefore, future grid-tied inverters are likely to accommodate advance communication capabilities to allow "localized islanding mode operation" via internet of things (IoT) in the case of blackout or voltage fluctuations in streets, campuses, or suburbs. This is likely to involve significant local battery storage capabilities at building level.

In the following four major sections of this chapter, electrical characteristics of PV cells and modules used in domestic applications will be given and the common features of the converter topologies in PV systems will be discussed. Specific attention will be given to the commonly used grid-tied PV systems as they have the biggest impact on the grid operation. In addition, the issues and failure types of PV systems, cells and modules are discussed to highlight the limitations on energy harvesting, and provide an insight into future improvements.

16.2 Electrical Characteristics of PV Cells/Modules

A solar PV cell is simply a diode that is optimized to absorb photons from the sun and convert them into electrical energy. The amount of solar power available per unit area is called the irradiance and is usually expressed in watts per square meter. The efficiency of a PV cell basically defines how effectively the irradiance (available power at the surface) is converted to a useful power at the output terminals of a PV cell that is connected to an electrical load. However, due to the unique electrical characteristics of PV cells (nonlinear current–voltage relationship), a matching circuit (converter) is commonly required to obtain the maximum power. Therefore, it is important to consider the electrical characteristics of PV cells as they have a direct impact on converter design, which may be connected to an electrical load directly or injecting power to a local AC grid.

Typical PV cells exhibit the current–voltage characteristic curve as seen in Fig. 16.3A. Note that in a converter design, five critical values of such curves are utilized namely the

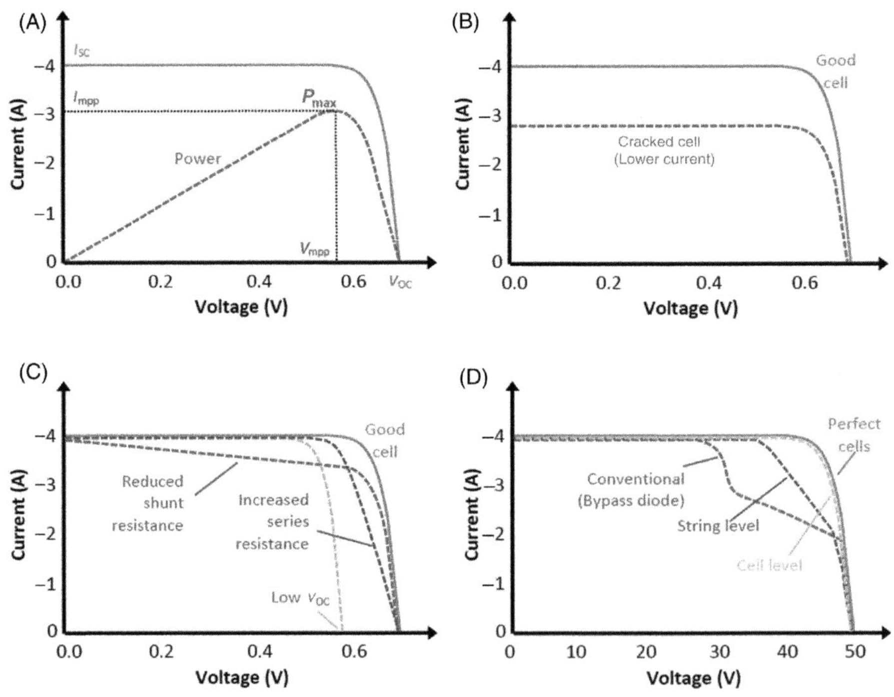

FIGURE 16.3 Typical characteristics of PV cells and modules (A) current–voltage and power curves of a healthy PV cell, (B) current–voltage curves of a cracked cell, (C) current–voltage curves of a failed cell, and (D) current–voltage curves of a module which accommodates power electronics.

open-circuit voltage (V_{OC}), the short circuit current (I_{SC}), the voltage and the current at the MPP (V_{MPP} and I_{MPP}), and the maximum power itself (P_{Max}). It is the primary aim of any converter system to operate the PV cell at its MPP in any condition for maximum utilization (higher efficiency). It is important to note that such an ideal power point tracking curve is valid for either a single uniformly illuminated cell or a set of matched cells with identical radiance.

The requirement of a converter between a PV cell and a load becomes much clearer when we consider the behavior of a cell under varying irradiance and operating temperatures, which continuously occur as a result of changes in environmental conditions. A PV panel's (module's) current–voltage characteristics under these two prevailing factors are shown in Fig. 16.4. Note that as the solar irradiance increases it is primarily the I_{MPP} of the cell that increases (Fig. 16.4A), and greatly boosting P_{Max} (Fig. 16.4C). Conversely, as

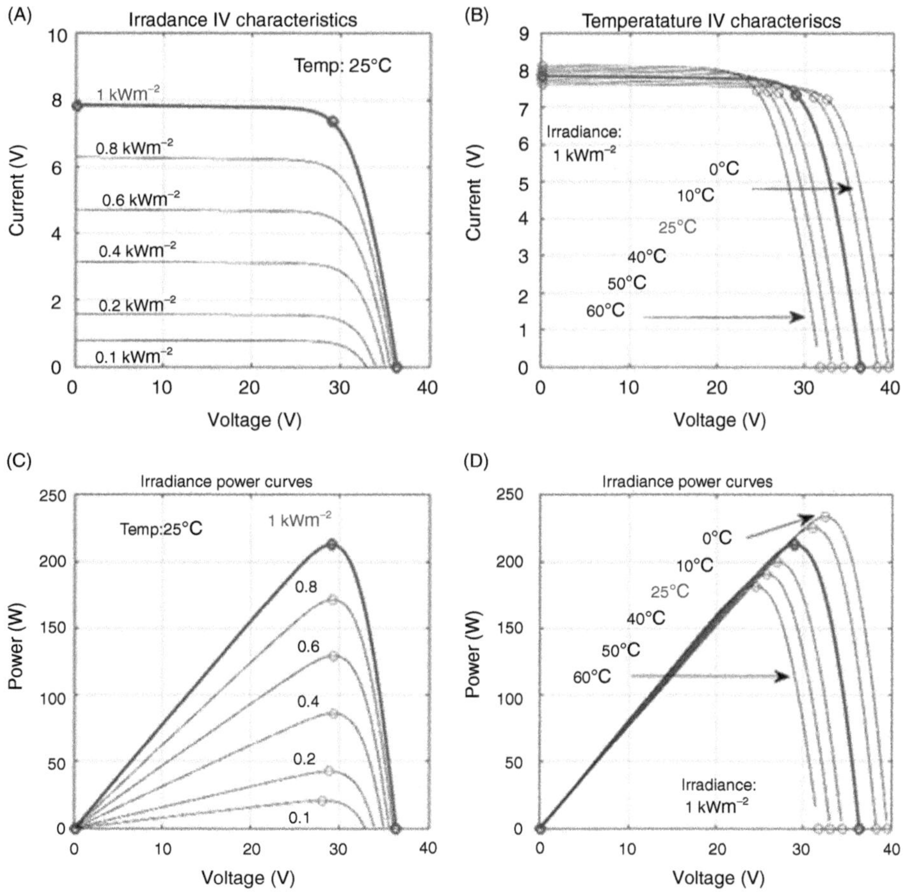

FIGURE 16.4 Typical characteristic curves of a commercial PV panel under varying solar irradiances and temperatures (A) the current–voltage curves for different solar irradiances, (B) temperatures, (C) the power curves for different solar irradiances, and (D) temperatures [1].

the temperature increases the open-circuit voltage (hence V_{MPP}) decreases (Fig. 16.4B), resulting in a significant decrease of P_{Max} (Fig. 16.4D). Ultimately, the converter aims to follow the P_{max} of a PV panel under varying irradiances and temperatures. Practically, the converter/load combination constantly re-adjusts the total load seen from the PV panel to deliver the maximum power. This load adjustment (manipulation) to follow the P_{Max} is known as maximum power point tracking (MPPT). Note that in a grid-connected inverter system the load adjustment means changing the amount of current injected into the grid through the inverter while obtaining maximum power from the PV arrays.

It should be noted here, that in practice, MPPT is achieved using various control algorithms and voltage and current feedback signals. Although the complexity and effectiveness of control algorithms may vary, the primary aim of an algorithm together with auxiliary systems is to reach to the P_{Max} value in a shortest time to accommodate the variations in irradiance and temperature.

16.3 Features of Converter Topologies in PV Systems

16.3.1 Electrical Requirements of Grid-Tied Inverters

PV systems should not only deliver maximum available power to loads for better utilization, but should also offer quality power (ideally constant magnitude and frequency sinusoidal voltage and/or current waveforms). In small-scale grid-tied PV inverters, since power electronic circuits cannot offer ideal voltage and current waveforms, this is measured by two quantities (1) the total harmonic distortion (THD) and (2) power factor (PF).

Due to the influx of the low voltage grid-tied small-scale inverters in recent years, specific standards have also been developed around the world. For example, in Australia (as of 2016), a typical residential PV system is governed by the requirements as set out by AS/NZS 4777.2 [2].

As in other standards, this standard also defines the acceptable limits on THD and PF, which are 5%, and 0.95 (leading or lagging) within 25–100% of rated power, respectively. Note that the level of PF is defined under a steady power output, grid voltage and grid frequency and translates to a voltage–current phase angle of about 18.2°.

The electrical requirements for a small grid-tied inverter defined by AS/NZS 4777.2 are summarized in Table 16.1. In addition, it defines acceptable wiring practices, which in-

Table 16.1 Typical Electrical Requirements (AS/NZS 4777.2) for Residential Grid-Tied Inverters

Grid Parameter	Requirement
THD limit	5% of rated output
Power factor	0.95–1.0 lagging/leading
Voltage range/Vrms	180–258
Frequency range/Hz	47–52
Anti-islanding	Disconnect within 2 s and reconnect after 60 s
DC injection	<0.5% of rated output current or <5 mA

clude grounding, DC injection protection and electromagnetic interference limitations. Furthermore, the standards also recommend additional protection features against so-called islanding. Therefore, grid-tied PV inverters also monitor the operating status of a grid continuously, and disconnect when the voltage or frequency exceeds certain bounds, which is known as anti-islanding. Hence the situation of having active live islands of the AC grid is avoided.

It can be noted here that the stand-alone PV inverters can have very similar electrical characteristics reference to grid-tied inverters but without an anti-islanding feature. Further features may also be present in stand-alone inverters to work seamlessly with other nano-grid supplies (such as batteries, small-scale wind turbines, and diesel generators) that may be available in local grids. It is envisaged that due to a large scale penetration of small-scale PV systems and their negative impact on power quality, the future PV systems are likely to accommodate battery systems and offer localized isolated grid operation in the case of a blackout. However, this will require faster and safer communication tools to be developed and integrated inside like PV inverters. In addition, technological changes in power electronics, like the development of wide band gap material based switching devices (SiC and GaN), will offer circuit operation at higher voltages, frequencies, and temperatures, eliminating up to 90% of the power losses. This will also allow new inverter topologies and soft switching with increased power density (highly desirable in small-scale PV systems) to be implemented.

16.3.2 Commonly Used Grid-Tied Converter Topologies

As it is illustrated in block diagram forms in Fig. 16.2, the primary aim of the converter circuits in PV systems is to deliver the maximum power to the load side (stand-alone or grid connected) for timely utilization of solar energy. Although the control aspects may vary due to the characteristics of the input power, well-known standard power electronic circuit topologies are also utilized in PV systems, include DC–DC converters (primarily for voltage gain and MPPT) and an inverter (to connect to AC grid or to local AC load, in single or three phase forms). Although such circuits are also suitable in wind generation applications, their basic features will be explained here to link to the distinct characteristics of a PV source.

Grid-connected inverters can be classified into two based on their commutation type (1) line-commutated and (2) self-commutated. The switching frequency of line-commutated inverters is controlled from the line current usually involves thyristor switches and are rarely used due to the requirements of commutation circuits, auxiliary devices, and filtering elements which all increase size and complexity and reduce efficiency.

The self-commutated inverters, however, have better and much flexible control options as they utilize metal oxide semi-conductor field effect transistors (MOSFETs) and insulated gated bipolar transistors (IGBTs) as switching devices, and can also come in two different forms (Fig. 16.5) (1) voltage-source (VSI) and (2) current-source inverters (CSI).

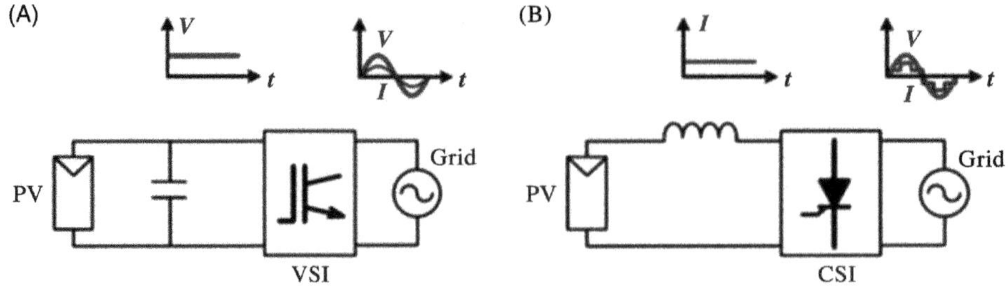

FIGURE 16.5 Single-phase self-commutated grid connected inverter topologies with typical waveforms (A) voltage-source inverters, and (B) early current-source inverters [3].

Note that the output stages of the AC circuit topologies, shown earlier in Fig. 16.2E and F, may have single phase or three phase inverters. Common single phase grid-tied PV inverters are usually current-controlled voltage-source types (Fig. 16.5A) in which instantaneous current waveform is produced using a simple pulsed width modulation (PWM) control, reference to the grid voltage, and is injected to the grid at a high PF. However, the requirement of a large and high voltage DC link capacitor (dc-link energy storage) reduces the reliability of this topology. Since three-phase voltage-source inverter has constant power flow, it reduces the size of the DC link capacitor resulting in better reliability.

The power circuit details of a single-phase voltage-sourced grid-tied PV inverter are illustrated in Fig. 16.6 [4]. This topology would only be useful for low voltage gains due to poor efficiency at high voltage ratios. If the designed PV array voltage is low with respect to the grid, a high frequency transformer in the voltage gain stage can be accommodated to improve efficiency. The capacitive energy storage element in this circuit acts as a power-decoupling buffer between the voltage gain (DC–DC) and inverter (DC–AC) stages. The DC–DC stage provides a near constant power (neglecting MPPT transients) that charges the capacitor whilst the DC–AC stages draws a sinusoidal power.

CSI on the other hand require a large series inductor (also for dc-link energy storage) to maintain constant current across the DC link (Fig. 16.7) [3]. A widely known single-phase H-bridge circuit configuration is shown in Fig. 16.7 which uses IGBTs with series diodes, or gate turn-off thyristors (GTOs). The present CSI can offer simpler control and also eliminate DC link capacitor. PWM current source inverter can supply sinusoidal current output to the AC grid at a unity PF. In addition, such grid-tied inverters do not require feedback control and are not prone to short circuit failure since the DC current never exceeds the short-circuit current of a PV panel. However, the design of the DC link inductor is vital in such inverters in order to reduce losses, weight, and cost.

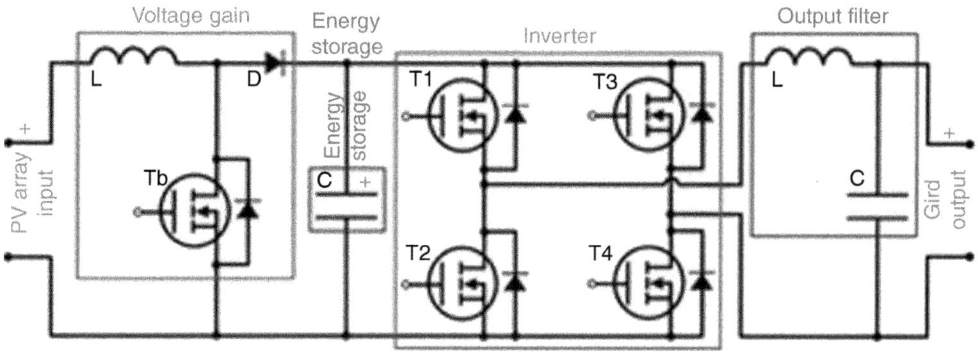

FIGURE 16.6 The details of four primary power stages of a voltage-sourced grid-tied inverter.

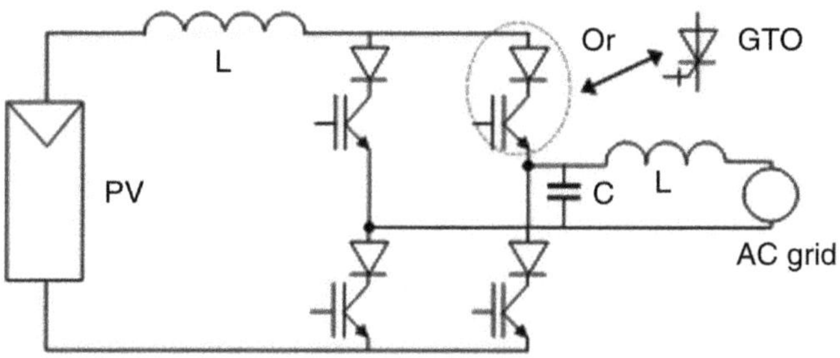

FIGURE 16.7 The details of a self-commutated current-sourced grid-tied inverter with a series inductor.

16.3.3 Emerging Converter Topologies

16.3.3.1 Cascaded Multilevel Modular Integrated Converters in Small-scale Grid-Tied PV Systems [1,4]

A hybrid of the central inverter and the micro inverter is known as a cascaded inverter, and effectively is a set of multiple H-bridges in a cascaded configuration with the grid. Cascaded multilevel modular integrated converters (MICs) (decentralized or not) have a multitude of various factors affecting their performance. Changing the number of MICs in a cascaded system has an effect on the transistor efficiency, capacitor reliability, switching frequency, fault tolerance, modularity, MPPT capability, voltage gain requirements, and ease of installation. The filter design is also affected by the cascaded nature of the system. As it can be seen in Fig. 16.8, each MIC has five primary components, which are the PV panel, boost converter, H-bridge, output filter, and controller. The PV panel, when paired with the boost converter, produces a constant DC output voltage while allowing for MPPT. The controller, which controls the boost converter, also coordinates the transistors of the H-bridge to inject the correct amount of power through the filter and into the grid to satisfy the PV, MPPT, and grid harmonic requirements. In this topology, each MIC requires

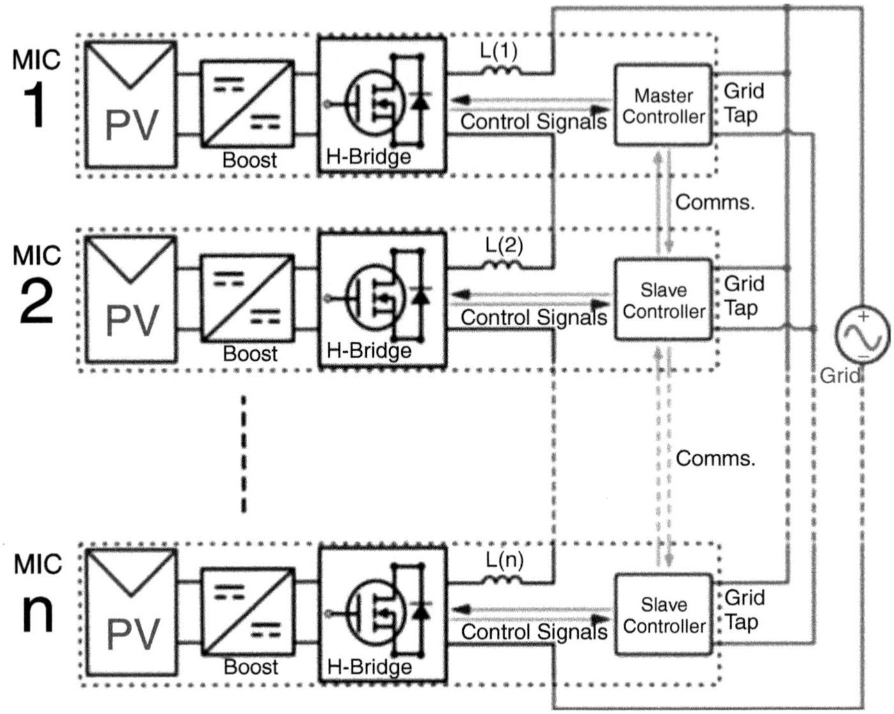

FIGURE 16.8 The decentralized cascaded multilevel modular integrated converters inverter system.

knowledge of its own DC-link voltage and current (for MPPT), its grid current (for injecting grid power) and its grid voltage (zero-crossing synchronization). Connecting the MICs in cascade requires three primary connections:

- The outputs of each MIC are connected in series (through their filters) to the grid.
- Each MIC has a separate grid connection for measurement purposes.
- Each MIC has a communication link to the neighboring MIC which may occur through power line communications (PLC), an additional wired link, or through a wireless connection (Blue tooth, Wi-fi, etc.).

16.3.3.2 Grid-Connected Current Source Inverter With Feed Forward Control [3]

A simple and low-cost single-phase current-sourced grid-connected inverter topology is given in Fig. 16.9. The topology (which has been used in wind generators previously) consists of a sinusoidally modulated boost switch (switched mode rectifier based) and line-frequency commutated unfolding circuit accommodating thyristor switches. The control algorithm adjusts the modulation index of the boost switch to control the inverter output current hence the output power. As a result of the significant losses in the DC link inductor, (due to the copper and iron) pulse area modulation can be used to reduce the size of the inductor. This modulation approach is similar to feed-forward compensation control. The topology has the potential to offer a low cost, simple, and reliable solution to reduce inductor and switch losses.

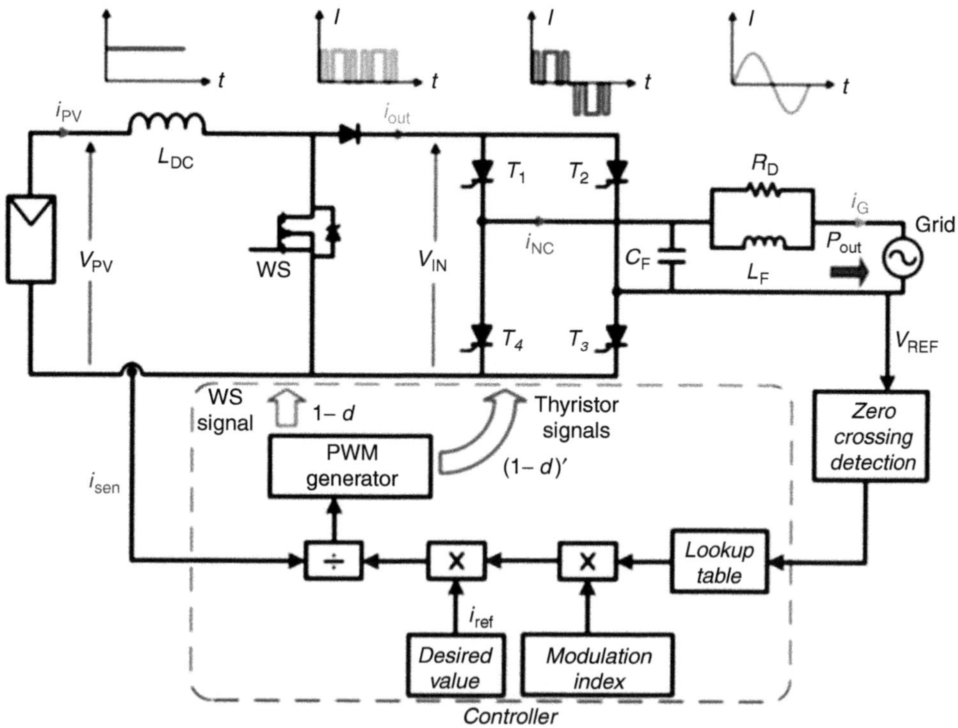

FIGURE 16.9 Grid-connected current source inverter with feed forward control and ideal current waveforms.

Although the existing PV system topologies can offer acceptable solutions at relatively higher conversion efficiency for residential applications, the developments in power electronics are likely to offer step changes while addressing the conversion efficiency and protection of PV modules at the cell level. Due to the advances, there are now opportunity for MPPT at the most localized PV cell level. This reduces power and voltage levels and also the size of both active (switches) and passive elements (inductors and capacitors) in the power circuit. The potential benefits of integrated power electronics will be robustness to shading, mismatch, ageing, cell damage, flexibility of placement, fault tolerance, and ease of installation.

16.4 Configurations of Grid-Tied PV Systems

After selecting the correct voltage and power levels for an application in a PV system, MPPT control can be done on string (array), module or cell levels. However, it is always desirable to consider localized control as it can improve energy extraction by compensating shading, cell mismatch, and aging. In Fig. 16.10, four fundamental PV inverter systems suited for MPPT in residential installations are given.

The configuration shown in Fig. 16.10A is the most common inverter system installed in the residential settings. The string inverter in this topology generates the grid-level DC voltages required by connecting PV modules in series, and it is inverted to AC by a centrally

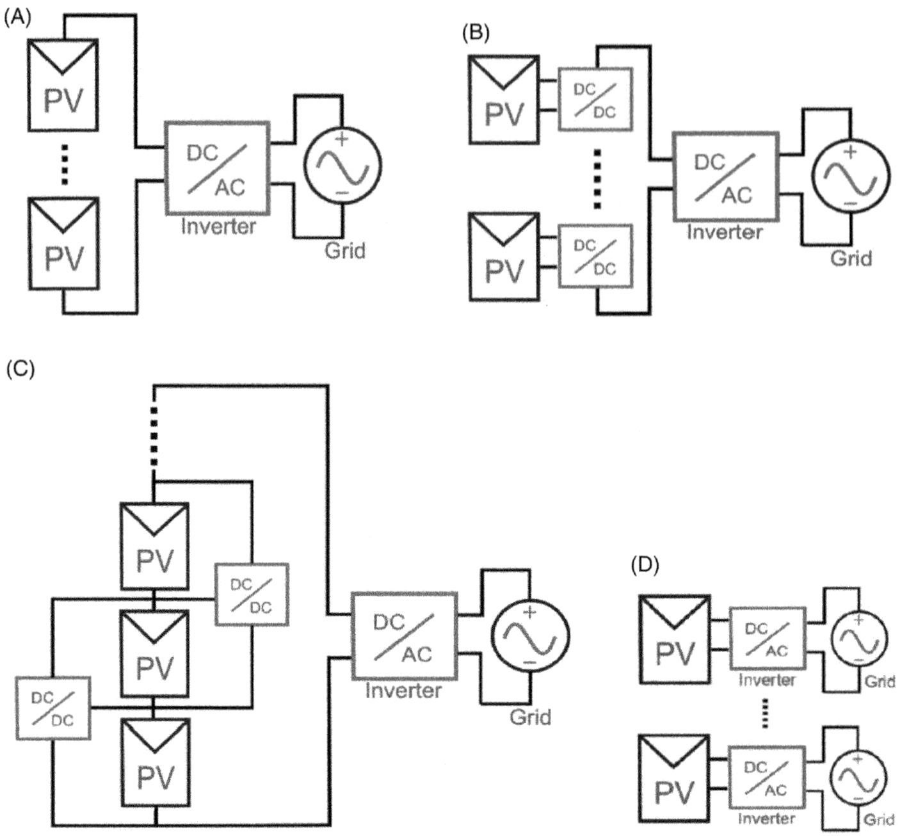

FIGURE 16.10 The four fundamental inverter systems suited for maximum power point tracking in residential installations [1] (A) String inverter, (B) Smart DC–DC optimizers, (C) Differential power processing, and (D) Micro inverters.

controlled inverter. Although it is a well-tested and simple system, the MPPT process in this arrangement has higher losses, due to the current–voltage characteristic mismatch of real PV panels preventing them to operate at their MPP.

By connecting a DC–DC converter to each PV panel as in Fig. 16.10B, it is possible to control and operate each panel independently. Although smart DC–DC optimizers do not increase the power rating of the PV panels, it does allow the true MPP of each PV panel to be tracked in a mismatched string. This is a high cost topology and has reduced conversion efficiency associated with using multiple smaller power converters, which may be offset by the increase in energy yield due to the higher overall system MPP. This topology has a potential in residential applications after the development of high temperature and low cost DC–DC converters using emerging power electronics switching devices.

The topology in Fig. 16.10C is principally the same as the smart DC–DC optimizer system, but power is instead bypassed around weaker PV panels. Note that to reduce the complexity of the previous topologies, it is necessary to remove the central DC–AC converter that ties everything to a common point. The topology in Fig. 16.10D can eliminate the central inverter by accommodating multiple miniature (PV panel-rated) inverters that operate in isolation from each other and directly connected to the AC grid. Hence, each micro inverter tracks the MPP of a PV panel and eliminates power mismatch issues in series strings. One of the main drawbacks of micro inverters (excluding the higher system cost) is the conversion efficiency, which is reduced due to the requirement to step up the low voltage of a single PV panel to the DC-link voltages required to interface the mains grid.

16.5 Issues on PV Systems and Cell and Module Level Failures

Generally, any PV system consists of three main components such as modules and arrays, central wiring point, and converter/inverter. There are internal wirings (inside a rigid and weather resistance frame) in a PV module and main wiring connecting PV panels in an array form to a central point to connect to power electronic circuits (DC/DC and/or DC/AC converters). This forms the final stage both in a stand-alone or grid-tied system. Therefore the issues and faults can also be classified around these three main components of a PV system, in addition to the integrated system-related issues.

16.5.1 Shading

In practice, to achieve higher voltage and current levels (hence powers) both in stand-alone or grid-tied applications, series and/or parallel-connected PV modules are used. In such arrangements, however, there may be power mismatches (nonuniform PV power distribution) between PV modules.

The major power mismatch in practical systems is primarily due to nonuniform shading by trees and clouds, and other factors such as dirt built up, manufacturing mismatch, aging, and angling (usually a permanent effect due to roof positions). It should be noted here that nonuniform shading conditions actually occur between PV cells of a PV module itself. However, due to the physically close proximity of the PV cells to each other in the PV module itself, it can be assumed that any of the conditions that would affect one cell would likely affect the whole module.

Although different inverter systems are adapted in practice (Fig. 16.10) to reduce or eliminate power mismatches in PV arrays while increasing reliability and reduce cost, shading still needs a special attention as it has a significant under-powering effect.

Since shaded (or under-powered) PV modules will dissipate power (due to their internal resistances) from the nonshaded/over-powered modules, causing them to heat up and create hot spots in a PV array, as a common practice, bypass diodes are connected in parallel with each PV module in an array (Fig. 16.11A). Bypass diodes simply protect system as well as reduce energy loss due to the internal resistance. Fig. 16.11B Illustrates variation of the power

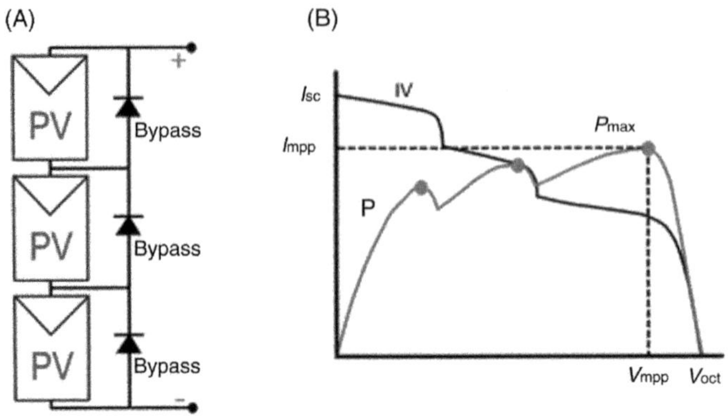

FIGURE 16.11 (A) Series connected three PV modules with bypass diodes, and (B) the typical current–voltage and power characteristics of series connected PV modules with moderate nonuniform shading [1].

curves of three series connected PV modules with moderate nonuniform shading. As it can be observed in the figure, the impact of shading is reduction in operating current and P_{max} at higher voltages. Note that shading can produce multiple power peaks that are hard to differentiate from each other with simple MPPT techniques. It was found that even a single shaded cell could significantly reduce MPP extraction by 33% or more (at module level MPPT).

As stated earlier, shading causes a serious issue in PV modules, hot spots. By blocking the flow of current and reverse biasing shaded cell, the system goes into thermal runaway or accelerated degradation. Although bypass diode can protect cell, if bypass diode fails it can generate severe heating (hot spots) and catastrophic failure. Hence, the most effective solution is to avoid the shading effect at the cell level by integrating power electronics circuits.

Note that multiple cells used in a module will not stay matched and healthy over time even when a same production batch is used and no components failed. This is due to various reasons including differential shading (clouds, leaves, and obstructions) in different sections, aging effects and direct damage for some reason. PV module defects and failures increase the risk of reliable delivery of the solar capacity, which can result unplanned replacement costs, hence increasing the system's levelized cost of energy (LCOE).

In PV systems there are a number of possible failure modes (observed characteristics of failures) both in cell and module levels, which are briefly summarized below.

16.5.2 Hot Spots

Hot spots occur due to a result of a structural defect in the solar cells or in a module or due to poor soldering joints, which can all lead to an internal short-circuit, reduced performance, partial or catastrophic failure, and burning.

16.5.3 Micro-cracks

Micro-cracks are virtually invisible microscopic level tears in the solar cells, which can occur during production of PV modules or during shipping or handling practices of panels.

They can grow over time due to thermal tension or seasonal and weather conditions, resulting reduced power output.

16.5.4 Delamination and Moisture Ingress

In a panel production, three components of panels (the glass layer, the solar cells, and the back sheet) are laminated under vacuum and designed to be weather proof. Delamination is the separation of these layers during operation, resulting in moisture penetrating into the panel, and causing internal corrosion which is visible as dark spots on the panel. Note that panels manufactured using glass substrates and frameless/thin-film PV panels can also suffer from moisture and corrosion problems.

16.5.5 Snail Trail Contamination

Snail trail is a discoloration of the panel, which usually becomes apparent after a couple of years of production, and which is usually associated with the use of defective front moralization silver paste during production of cells. This causes chemical breakdown on the front of the panel that is visible in the form of snail trails affecting panel performance.

16.5.6 Interconnects

There are four levels of interconnects in thin-film modules such as gridlines (<1 mm thick) in a cell, busbars connecting these gridlines in a single cell, cell interconnect ribbons (connecting multiple cells into a string), and string interconnect (connecting multiple strings of cells) (Fig. 16.12). Failed connection of any of these interconnects decrease the performance of a panel. Multiple ribbons and busbars may be accommodated to reduce this problem but at increased cost and slightly reduced performance.

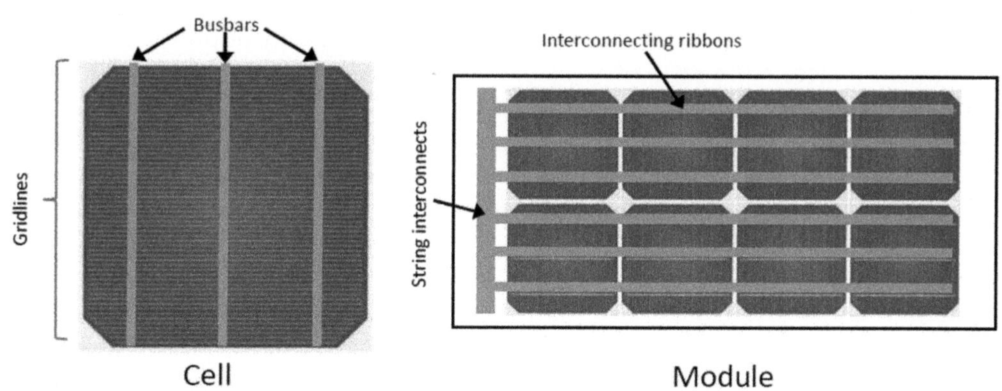

FIGURE 16.12 Four levels of interconnects in PV cells and modules.

16.5.7 Potential Induced Degradation (PID) Effect [5]

For safety, the PV arrays are earthed, which can cause a harmful potential voltage difference between the ground and the voltage generated by the panel. This effect can reduce the module's MPP and its open circuit voltage (V_{oc}) along with a reduction in shunt resistance. The consequences of this effect are an ongoing reduction in performance and accelerated ageing of the PV panel. The choice of glass, encapsulation, and diffusion barriers have all been shown to have an impact on Potential Induced Degradation (PID). It is most often associated with modules at negative potentials to ground and can often be reversed by applying positive bias to affected modules.

16.5.8 Encapsulate Discoloration

Poor quality encapsulant under UV radiation and heat may result in yellow or brown discoloration of cells. The heat absorption increases with browning and higher heat absorption enforces further browning, which results in reduced performance. Back sheet discoloration may also leads to delamination and cell corrosion.

In addition to the above cell/module level potential failures, PV modules can experience animal damage on wires (such as by squirrels, rats and cockatoos, and possums), mechanical failures (on front glass or on frame that may be due to hailstorm, poor installation, and/or wind load), main busbar and terminals failure and short circuits. In addition, electrical overstress on PV panels after direct lightning strike can cause damage such as a melted frame or shattered glass. It is also very common to observe harsh environments around coastal installations causing corroded grounding connections and galvanic reaction from incompatible metals of the PV module system construction. Furthermore, a number of potential PV inverter failures can be identified, which are usually associated with thermal management, environment, grid issues, and components. Although today's PV inverters are the weakest part of the entire PV system, their fault is easy to detect and replacement is simple.

It can be summarized here using the current–voltage characteristics that a healthy PV cell (Fig. 16.3A) will produce less power due to reduced photocurrent as a result of a failure in its structure, like cracks (Fig. 16.3B). The reduction in the output power can be visualized from the fill factor (the ratio of $V_{mpp}.I_{mpp}/V_{oc}.I_{sc}$, by comparing the maximum power to the theoretical power that would be output at both the open circuit voltage and short circuit current together) of a cell. Usually all circuit parameters are affected in a failed PV module (such as reduced open-circuit voltage or reduced shunt resistance), but increased series resistance is the most common variation (Fig. 16.3C). Although it adds complexity, cell-level integration of power electronics can recover most of lost power (Fig. 16.3D) [6].

In general, every PV module (the backbone of a PV system) experiences product failure in its life time under three categories namely infant failures, midlife-failures, and wear-out-failures. In addition to these relatively long-term failure-related power degradations, many PV modules also show a light-induced power degradation (LID) just after installation [5]. This is usually taken into account in the calculation of the rated power. Fig. 16.13

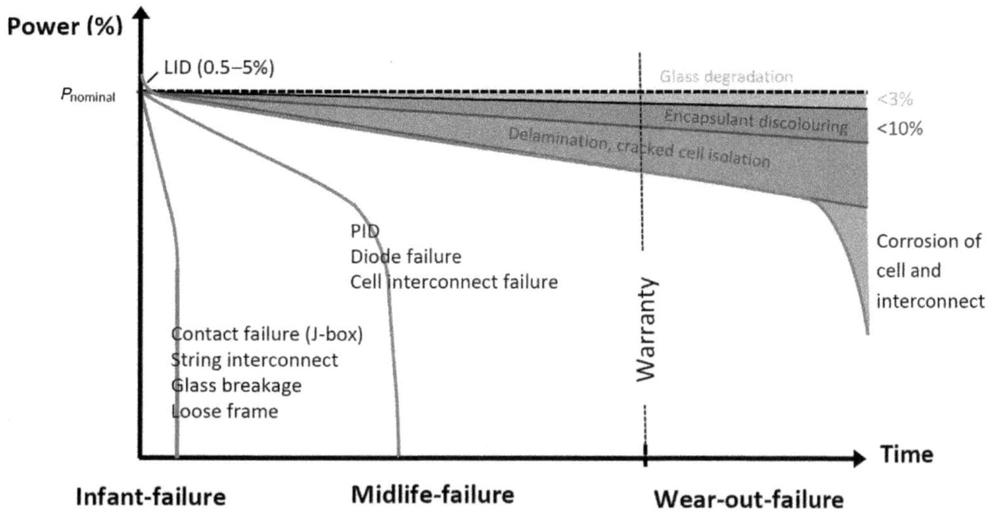

FIGURE 16.13 Three categories of failures of PV modules during its life-time [5].

shows the distribution of failure types of a PV module at the start of its working life. Note that infant-mortality failures occur in the beginning of the working life of a PV module which are primarily related to flaws in PV modules during manufacturing, and are associated with incorrect transportation and handling.

16.6 Conclusions

Since small-scale domestic applications accommodate two major components, PV modules and power electronic converters, this chapter has discussed major issues related to both. The possible PV system topologies were provided in block diagram forms to highlight their common and unique features as they are connected to a local grid or operated stand-alone.

The requirement of a converter between a PV cell and a load has been explained by utilizing the characteristics of PV cells and modules, which has also been linked to the potential degradation of power output due to faults and failures at cell and module level.

Due to the large-scale penetration of grid-tied inverters, electrical requirements of PV inverters have been discussed from the standards' viewpoint. Some insights were provided about the problems faced and future directions of the grid-tied inverters. Commonly used inverter-topologies were classified and their common features and critical components are explained. In addition, the basic application circuits of two emerging converter topologies were given.

In the final section of this chapter, the issues and faults were classified and explained. Furthermore, the chapter summarizes the distribution of failure types of a PV module at the start of its working life which aims to highlight the practical aspects of this renewable energy source.

References

[1] Scholten DM. Investigation and analysis of decentralized multilevel modular integrated converters in small-scale grid-tied PV Systems. PhD Thesis, University of Adelaide, Australia; 2017.

[2] Standards Australia. Grid connection of energy systems via inverters. Part 2: Inverter requirements, AS/NZS 4777.2; 2015.

[3] Ertasgin G. Low-cost current-source 1-ph grid-connected photovoltaic inverter. PhD Thesis, University of Adelaide, Australia; 2010.

[4] Ertugrul N, Soong WL. Analysis and control of decentralized PV cascaded multilevel modular integrated converters. IEEE Energy Conversion Congress and Exposition; 2016.

[5] Review of Failures of Photovoltaic Modules. International energy agency photovoltaic power systems programme. Report IEA-PVPS T13-01:2014; 2014.

[6] Kurtz S, Deline C, Wohlgemuth J, Marion B, Granata. Opportunities and challenges for power electronics in PV modules. ARPA E Workshop, Arlington, VA; 2011.

17

Energy and Carbon Intensities of Stored Solar Photovoltaic Energy

Charles J. Barnhart
WESTERN WASHINGTON UNIVERSITY, BELLINGHAM; INSTITUTE FOR ENERGY STUDIES, WESTERN WASHINGTON UNIVERSITY, BELLINGHAM, WA, UNITED STATES
charles.barnhart@wwu.edu

17.1 The Need for Storage

The world needs affordable, accessible, sustainable, and low-carbon energy resources [1–3]. Of the renewable resources, solar photovoltaic (PV) and wind turbines have the highest technical potential to satisfy this need, but these technologies generate electricity from variable, weather-dependent resources [4–7]. Fig. 17.1 provides a compelling visualization of 30 days of superimposed power demand time series data [black (red in the web version)] wind energy generation data [light gray (blue in the web version)] and solar insolation data [gray (yellow in the web version)]. Supply correlates poorly with demand. What will be the amount of storage needed for the operation of electrical grids incorporating increasing amounts of variable resources is a critical yet complicated question. It is complicated for two reasons: (1) the electrical grid, composed of myriad power sources and sinks, is conducted as a whole in real-time, and (2) the number of technologies and practices, their varied and evolving characteristics, and their possible implementations under differing and shifting policy landscapes presents a grossly underdetermined problem with several solutions.

Technologies and practices positioned to ensure grid-reliability include flexible conventional generation (e.g., natural gas combustion turbines and diesel generation sets), flexible renewable generation (e.g., curtailment, hydropower, and concentrated solar power with thermal storage), flexible load (e.g., demand-side management), energy storage, and resource sharing (e.g., diversity and transmission). In the future, when greenhouse gas (GHG) emissions are constrained, flexible generation will need to be achieved using low-carbon energy supplies.

Studies have made efforts to determine the amount of renewable generation an electrical grid can support by bundling these technologies and practices into an abstract resource: grid flexibility, defined as the percentage of generation and load capable of being readily dispatched or halted [5]. Less flexible grids harbor high percentages of so-called baseload

A Comprehensive Guide to Solar Energy Systems. http://dx.doi.org/10.1016/B978-0-12-811479-7.00017-8

generating plants such as nuclear, coal, and natural gas combined cycle plants. Firstly, the amount of energy storage capacity required will depend on grid flexibility. Secondly, it will depend on attributes of the renewable generation. The amount, type, mix, and degree of supply correlation affect how well supply satisfies demand. Today, storage on power grids is dominated by pumped hydroelectric storage (PHS). Table 17.1 lists worldwide storage capacity by power and energy. This chapter describes the effect storage has on the energy and carbon intensity of solar PV-generated electricity. First, key storage characteristics are listed. Second, energy return ratio results are presented. Third, carbon intensity calculations and results are presented.

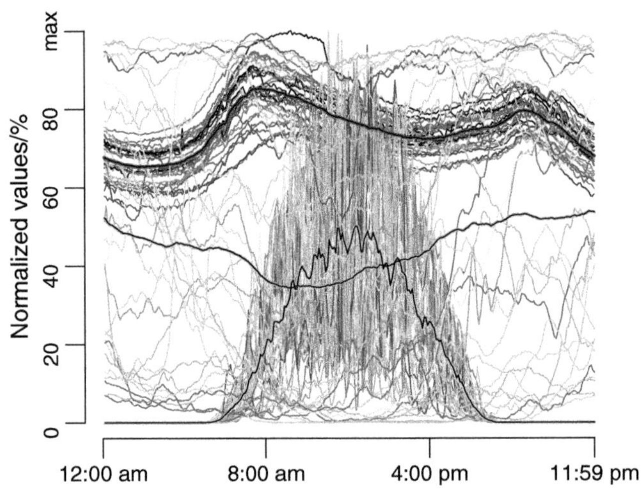

FIGURE 17.1 Wind-power generation [*light gray* (blue in the web version)], insolation [*gray* (gold in the web version)], and power demand [*black* (red in the web version)] time series data provide a compelling visualization of renewable energy's variable correlation with demand. Thirty days of data collected by the Bonneville Power Administration (Pacific Northwest United States) in April 2010 are superimposed and normalized to their maximum values. Average values are in color-highlighted black lines.

Table 17.1 Global Storage Capacity

Technology	Power (MW)	Energy (GWh)
Li-ion	~20 [8]	0.06[b]
NaS	365.3 [9]	2.191[c]
PbA	~1 800 000[a]	400[a]
Flow	3 [8]	0.024[d]
(VRB, ZnBr) CAES	400 [10] (650 [11, 12])	3.73 [10]
PHS	129 000 [8]	102 [13]

[a]Assuming total car batteries worldwide (1 billion) each 10 kg with practical power and energy densities of 180 and 40 W kg^{-1} yields 1.8T W and 0.4 TWh of capacity.

[b]Assuming 3 h storage.

[c]Assuming 6 h discharge.

[d]8 h discharge.

17.2 Key Characteristics for Storage

Energy storage incurs energetic costs and emits carbon to the atmosphere. Direct emissions of carbon are those associated with the round-trip efficiency and operation of the storage device. Indirect emissions are those resulting from the process of mining the materials and manufacturing the storage and flexible generation technologies. The energetic and carbon intensity values for energy storage technologies were obtained from life cycle assessment (LCA) and net energy analysis (NEA); studies [14–17]. Grid-scale storage requires safety, affordability, reliability, longevity, and efficiency. Technologies that satisfy these criteria include four electrochemical storage technologies—lithium-ion (Li-ion), sodium sulfur, traditional lead-acid (PbA), vanadium redox flow batteries (VRB)—and two geological storage technologies—PHS, and compressed air energy storage (CAES)—in this analysis.

Key net energy and carbon data are listed in Table 17.2. The energy intensity per unit energy storage capacity, ε_s $[(kWh_e)/(kWh_e)]$, depends on the technology's depth of discharge (*D*), its total number of charge-discharge cycles (λ), and its cradle-to-gate embodied electrical energy requirement per unit capacity of energy delivered to storage (CTG_e). Embodied energy accounts for energy expended in mining raw resources, manufacturing the device, and delivering the device to point of use. The per cycle carbon intensity [g $CO_2eq(kWh)^{-1}$] for storage technologies were calculated by adding capital ($GHG_{s,cap}$) and operational greenhouse gas ($GHG_{s,op}$) emissions per unit of electrical energy delivered per cycle.

A critical attribute of an energy storage technology is its round-trip efficiency, η. The carbon intensity of the discharged electricity is $\geq 1/\eta$ times the carbon intensity of the input electricity. Using storage increases the carbon intensity of delivered electricity by a

Table 17.2 Data Used in Net Energy Analysis of Storage Technologies

Technology	CAES	Li-ion	NaS	PbA	PHS	VRB
$GHG_{s,cap}$ (kg (MW h)$^{-1}$)	19 400	600 960	687 500	153 850	35 700	161 400
$GHG_{s,op}$ (kg (MW h)$^{-1}$)	288	0	0	0	1.8	3.3
Discharge depth (*D*)	1	0.8	0.8	0.8	1	1
Cycles (λ)	25 000	6 000	4 700	700	25 000	10 000
Efficiency (η)	0.7 (1.36)	0.9	0.75	0.9	0.85	0.75
CT G_e	22	136	145	96	30	208
Carbon multiplier (χ)	0.735[a]	1.111	1.333	1.111	1.764	1.333
Energy intensity (ε_s)	0.00088	0.028	0.039	0.17	0.0012	0.072

Definitions are in the text. Detailed analysis and references in Supplementary Materials.

[a]CAES operation delivers more electricity than enters storage by combusting natural gas.

Sources: Denholm P, Kulcinski GL. Life cycle energy requirements and greenhouse gas emissions from large scale energy storage systems. Energy Convers Management 2004;45:2153–72, Sullivan J.L, Gaines L. A review of battery life-cycle analysis: state of knowledge and critical needs ANL/ESD/10-7. Technical report. Oak Ridge, TN: Argonne National Laboratory; 2010.

factor, χ, as listed in Table 17.2, where χ is a carbon intensity multiplier. If storage is 90% efficient, the carbon intensity of the delivered electricity increases by 11%, $\chi = 1.11$. Manufacturing storage also incurs its own energetic and carbon costs.

Despite higher energy and carbon intensities when compared to PHS, electrochemical storage technologies present one clear advantage: energy density. Batteries are able to store several hundred times the amount of energy per unit mass and volume than PHS (Fig. 17.2). Additionally, batteries do not require geological features, that is steep topography, that PHS requires and therefore can be deployed anywhere including city centers, residences, and commercial buildings (Fig. 17.3).

17.3 Net Energy Analysis of Storing and Curtailing Solar PV Resources

Curtailing renewable resources results is an immediate and obvious forfeiture of energy. However, flexible grid technologies can also consume significant amounts of energy in their manufacture and operation. These embodied energy costs are not as immediately apparent, but they are an energy sink from a societal perspective.

In this section, I compare the energetic costs of electrical energy storage (EES) to the energetic costs of curtailment. In lieu of storage or other means of grid flexibility, variable resources are curtailed during periods of oversupply or of strong market disincentives [19, 20]. Consequently, electricity is squandered, capacity factors are reduced and revenue

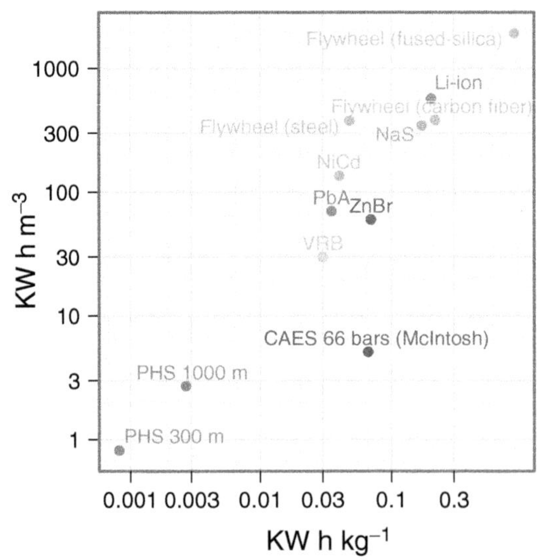

FIGURE 17.2 A plot comparing volumetric and specific energy densities for energy storage technologies. Technologies considered for large-scale energy storage have labels in color (data obtained for PHS and CAES from calculations, battery data [9] and flywheel data [18].

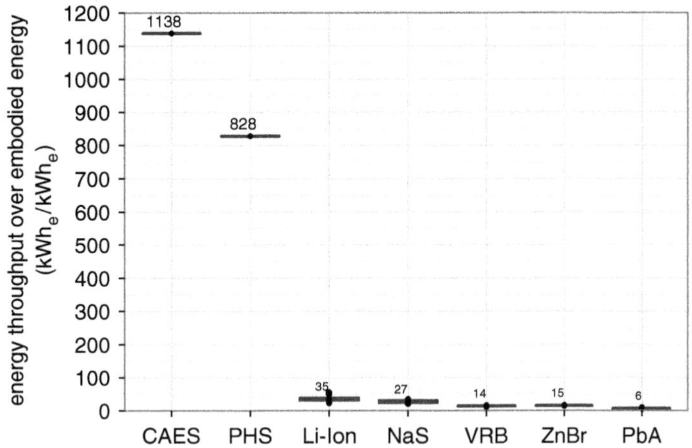

FIGURE 17.3 The total energy stored over the life a storage device divided by the embodied energy required manufacture the device [(kWhe)/(kWhe)] provides a comparative metric for comparing societal energy costs. Higher values are better.

for generation asset owners in certain markets is lost. Worldwide, curtailment rates are projected to increase as wind and solar comprise a larger fraction of the generation mix [5, 19]. We ask whether storage provides societal net energy gains over curtailment. EES has significant value not quantified or analyzed in this study, including electricity market economics [21], insuring reliable power supplies to critical infrastructure [22], ancillary benefits to power grid operation [8], and application in disaster relief and war zone scenarios.

Storage affects net energy ratios as shown by mathematical models developed in previous work [17, 23]. This framework accommodates any type of generation or storage technology. Using LCA data for generation and storage technologies, we calculate which storage and generation technologies result in a net energy gain over curtailment. We present our data and results in terms of energy return on investment (EROI); the amount of electrical energy returned per *unit* of electrical energy invested. Fig. 17.4 shows calculated grid EROI values, $EROI_{grid}$, for solar PV used with storage technologies (colored lines) as a function of φ. The solid black line bisecting the plots indicates the EROI value due to curtailment, spanning a range from original resource EROI (8 for solar PV) to zero. The (green in the web version) light gray region to the right of this line indicates combinations of EROI, $ESOI_e$, and φ in which storage yields better energy returns than curtailment, $EROI_{grid} > EROI_{curt}$. To the left, in (blue in the web version) dark gray, $EROI_{grid} < EROI_{curt}$, storage implementation is more energetically costly than simply curtailing the resource.

Several interesting results emerged that are shown in Fig. 17.4. First, storage technologies with low $ESOI_e$ values, such as PbA and ZnBr, reduce the grid EROI down much more severely than technologies with high $ESOI_e$ values, such as PHS, CAES, and Li-ion. Second, all battery technologies except for PbA paired with solar PV yield grid EROI values that are greater than curtailment alone for reasonable values of φ. However these grid EROI values are below the average US power grid values ~20.

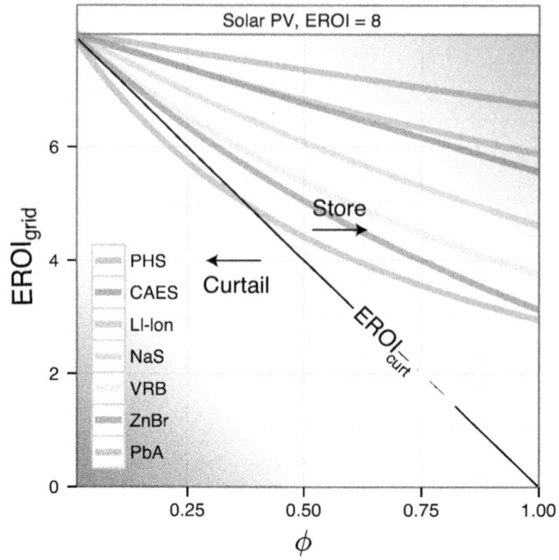

FIGURE 17.4 Grid $EROI_{grid}$ values as a function of storage or curtailment fraction, ϕ, and EES technology paired with solar PV. All technologies except for PbA perform better than curtailment on a societal energy cost basis.

Ideally, storage technologies that support generation resources should not diminish energy return ratios below curtailment energy return ratios for reasonable values of φ. This means that storage technologies with high round-trip efficiencies and long cycle life values such as PHS and Li-ion are much more favorable for storing electricity generated from solar PV than short-lived batteries for example traditional lead-acid batteries.

Curtailment of solar PV resources during times of excess generation is a viable form of grid flexibility. Curtailment of solar PV yield carbon and energy intensities that are lower than respective pairings with low cost PbA electrochemical storage technologies. Although curtailment appears to be an immediate waste of a resource, the life cycle energy costs of storage should be considered. Avoiding curtailment may not lead to the most environmentally sound decisions. Curtailment is not the only option, nor is it ideal. Useful applications for excess electricity occur beyond the power grid. Excess electricity could be used for thermal storage, producing heat or ice for later use. Additionally, electricity could be used to pump or desalinate water, smelt metal ores, or manufacture goods. The energy is "stored", that is embodied elsewhere in the economy.

17.4 The Carbon Footprint of Storing Solar PV

Energy storage emits carbon to the atmosphere. Direct emissions of carbon are those associated with the round-trip efficiency and operation of the storage device. Indirect emissions are those resulting from the process of mining the materials and manufacturing the storage and flexible generation technologies. The carbon intensity values for energy storage technologies were obtained from LCA and NEA studies [14–17]. Carbon intensity values for the average US power grid emissions and subgrid emissions were obtained from the US Department of Energy [24]. Table 17.3 lists the life carbon intensity values for wind, PV, and gas.

The per cycle carbon intensity [g CO_2eq(kWh)$^{-1}$] for storage technologies were calculated by adding capital and operational GHG emissions.

$$\mathrm{GHG}_s = \frac{\mathrm{GHG}_{s,\mathrm{cap}}}{\lambda_D} + \mathrm{GHG}_{s,\mathrm{op}} \tag{17.1}$$

The storage technology's depth of discharge (*D*), modulates per cycle capital ($\mathrm{GHG}_{s,\mathrm{cap}}$) emissions from storage meaning that a shallow depth of discharge requires larger batteries (with associated manufacturing costs) to provide equivalent storage capacities. Values used in these calculations can be found in Table 17.2.

The life cycle carbon intensity of electricity delivered to the power grid from generation resources via energy storage technologies was calculated by summing per cycle storage carbon intensities with life cycle generation carbon intensities.

$$\mathrm{GHG}_g = \mathrm{GHG}_s + \mathrm{GHG}_r / \eta \tag{17.2}$$

Table 17.3 Generation Technology Life Cycle

Resource	Reference and Notes	[(kg CO_2eq)kWh]$^{-1}$				
		Min	25th%	Median	75th%	Max
Wind	[25] On Shore (Harmonized) 107 estimates from 44 studies	22	50	91	119	333
PV	[26] Crystalline Silicon PV irradiation of 1700 (kW h) m^{-2} a^{-1} 41 estimates from 13 studies	5	20	22	25	38
NGCC	[27] 51 estimates, 42 studies Capital emissions: 1 g (kW h)$^{-1}$	1.4	2	2.2	2.4	3
NGCT		1.2	1.3	1.5	1.8	1.9
NGCC	[27]	–	15	21	36	–

PV, Photovoltaic; *NGCC, Natural Gas Combined Cycle; NGCT, Natural Gas Combustion Turbine.*

Fig. 17.5 shows carbon intensity values in terms of kg CO_2eq per electricity delivered to the power grid from various solar-storage technology pathways. For reference, the average carbon intensity values for the US power grid (518 kg $(MWh)^{-1}$) are shown. Additionally, carbon intensity values direct solar and grid-storage pairings are shown. All solar-storage pairings emit less carbon than the US power grid average. The best performing storage technologies are PHS, VRB, and Li-ion. The worst performing technologies are CAES, which emits carbon via natural gas combustion upon discharge and PbA, which has a short cycle life. For power grids with low-carbon intensity resources such as the Northwest Western Electricity Coordinating Council, solar PV should be stored using low-carbon storage technologies such as Li-ion and PHS.

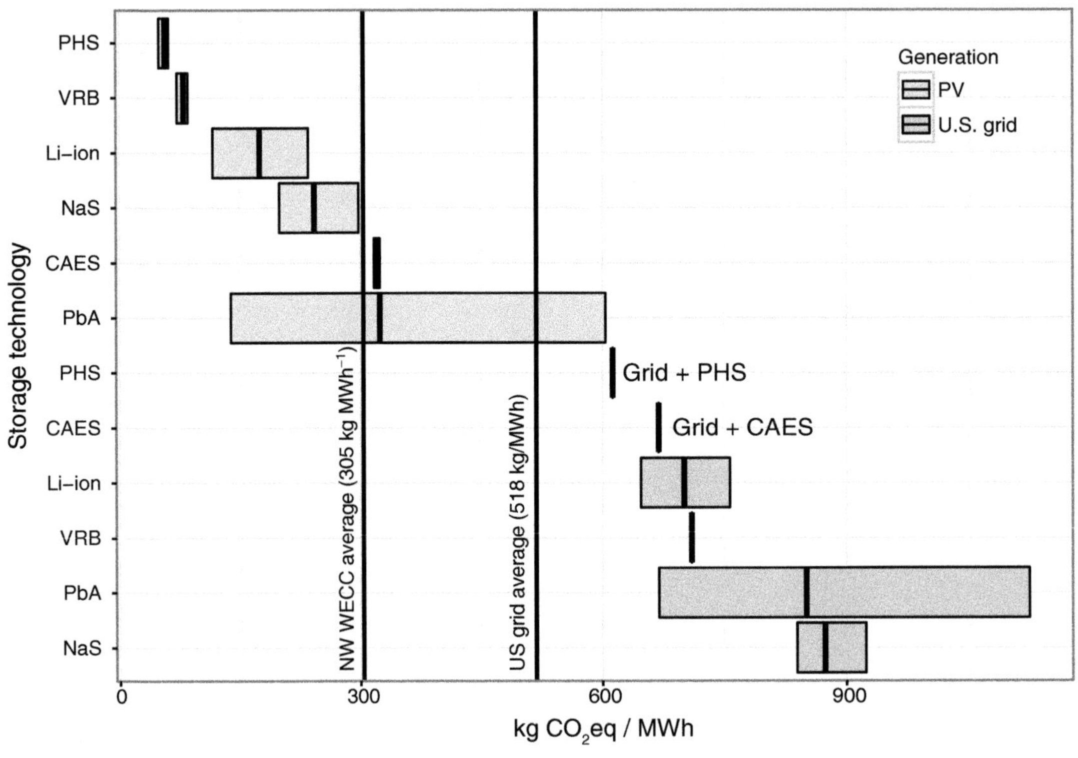

FIGURE 17.5 The carbon intensity of electricity provided from solar PV charged storage technologies. For comparison, the carbon intensity of electricity provided from storage technologies charged by a hypothetical US average power grid is displayed on the right half of the bar chart. Vertical lines represent the average US power grid carbon intensity and the intensity of a particularly low-carbon regional grid, the Northwest Western Electricity Coordinating Council (NW WECC).

17.5 Conclusions

Energy storage promises many benefits for electrical power grids and societal energy use in general. Our analysis shows how to calculate and compare their energy and carbon footprints. In conclusion, the analyses presented in this chapter reveal the following insights.

1. Flexible power generation and energy storage come with a cost. Energy delivered from storage has greater carbon and energetic intensities than energy delivered directly from power generation technologies, and depending on the technology, the energy, and carbon penalties for storage can be large.
2. The carbon intensity of solar PV plus most storage technologies including PHS, Li-ion, and VRB are far lower than those of the US grid. CAES and PbA technologies yield higher carbon intensities than the US Grid. All storage technologies when paired with solar PV trade higher energy intensities for lower carbon intensities.
3. Not all storage technologies are created equal. PHS performs best and traditional PbA batteries perform worst. CAES trades low energy intensity for high carbon emissions associated with combustion of natural gas. Li-ion and VRB perform best among electrochemical storage solutions with Li-ion providing the lowest energy intensity and VRB the lowest carbon intensity. Traditional PbA batteries perform poorly by these metrics. Although they have low energy requirements for manufacture, their low number of charge-discharge cycles leads to frequent replacement and a high-energy intensity of 0.17.
4. The curtailment of solar resources provides flexibility with higher carbon and energy costs in comparison to the implementation of energy storage technologies except for PbA. This is due to the fact that solar PV panels are energetically intensive to produce. Curtailing these resources forfeits energy that incurred high energetic costs. These costs, in the case of PV, are greater than the cost of incorporating storage, especially storage with low energy intensity values such as PHS and Li-ion.

Energy storage and curtailment can provide the flexibility the power grid will require as the fraction of variable resource supply increases. This chapter shows the benefits of using systems-level energy intensity and carbon intensity analysis to compare performance of flexible options for solar PV [28]. Policy makers and consumers that consider the effects of deploying storage with solar PV can better identify environmentally sound solutions.

References

[1] Chu S, Majumdar A: Opportunities and challenges for a sustainable energy future, *Nature* 488:294–303, 2012.

[2] Conti J, Holtberg P, Doman LE, Smith KA, Sullivan JO, Vincent KR, et al. International energy outlook 2011, DOE/EIA-0484(2011). Technical report. Washington, D.C.: US Energy Information Administration; 2011.

[3] Kojm C: *Global trends 2030: alternative worlds*, National Intelligence Council, 2012, p. 160.

[4] Lew D, Piwko D, Miller N, Jordan G, Clark K, Freeman L. How do high levels of wind and solar impact the grid? The western wind and solar integration study. Technical Report December. National Renewable Energy Laboratory; 2010.

[5] Denholm P, Margolis RM: Evaluating the limits of solar photovoltaics (PV) in traditional electric power systems, *Energy Policy* 35:2852–2861, 2007.

[6] Marvel K, Kravitz B, Caldeira K: Geophysical limits to global wind power, *Nat Clim Chang* 3:118–121, 2012.

[7] Hand M, Baldwin S, DeMeo E, Reilly J, Mai T, Arent D, et al. Renewable electricity futures study. NREL TP-6A20-52409; 2012.

[8] Rastler D. Electricity energy storage technology options: a white paper primer on applications, costs, and benefits. Technical report. Palo Alto: Electric Power Research Institute; 2010.

[9] Reddy T, Linden D. Linden's handbook of batteries; 2010.

[10] Succar S: Compressed air energy storage. In Barnes FS, Levine JG, editors: *Large energy storage systems handbook*, Boca Raton, FL, 2011, CRC Press, pp 111–153, chapter 5.

[11] Sandia National Laboratory. American Recovery and Reinvestment Act: energy storage demonstrations. Technical report. Sandia National Laboratory Energy Storage Systems; 2011.

[12] RWE. Adele—adiabatic compressed air energy storage for electricity supply. Technical report. Essen/Köln: RWE Power AG; 2010.

[13] EIA. Electric power monthly April 2012. Technical Report April. US Energy Information Adminsitration; 2012.

[14] Denholm P, Kulcinski GL: Life cycle energy requirements and greenhouse gas emissions from large scale energy storage systems, *Energy Convers Manag* 45:2153–2172, 2004.

[15] Rydh C, Sandén B: Energy analysis of batteries in photovoltaic systems. Part I: performance and energy requirements, *Energy Convers Manag* 46:1957–1979, 2005.

[16] Sullivan JL, Gaines L. A review of battery life-cycle analysis: state of knowledge and critical needs ANL/ESD/10-7. Technical report. Oak Ridge, TN: Argonne National Laboratory; 2010.

[17] Barnhart CJ, Benson S: On the importance of reducing the energetic and material demands of electrical energy storage, *Energy Environ Sci* 6:1083–1092, 2013.

[18] Semadeni M: Storage of energy, overview, *Encycl Energy* 5:719–738, 2004.

[19] Wiser R, Bolinger M. 2011 wind technologies market report. Technical Report August. US Department of Energy; 2012.

[20] Lannoye E, Flynn D, O'Malley M: Evaluation of power system flexibility, *IEEE Trans Power Syst* 27:922–931, 2012.

[21] Budischak C, Sewell D, Thomson H, Mach L, Veron DE, Kempton W: Cost-minimized combinations of wind power, solar power and electrochemical storage, powering the grid up to 99.9% of the time, *J Power Sources* 225:60–74, 2013.

[22] Armand M, Tarascon JM: Building better batteries, *Nature* 451:652–657, 2008.

[23] Barnhart CJ, Dale M, Brandt AR, Benson SM: The energetic implications of curtailing versus storing solar- and wind-generated electricity, *Energy Environ Sci*, 2013.

[24] EGRID. EGRID 2013 data; 2014.

[25] Dolan SL, Heath GA: Life cycle greenhouse gas emissions of utility-scale wind power systematic review and harmonization, *J Ind Ecol* 16:S136–S154, 2012.

[26] Hsu DD, Donoughue PO, Fthenakis V, Heath GA, Kim HC, Sawyer P, et al: Life cycle greenhouse gas emissions of crystalline silicon photovoltaic electricity generation, *J Ind Ecol* 16(s1): 122–135, 2012.

[27] O'Donoughue PR, Heath GA, Dolan SL, Vorum M: Life cycle greenhouse gas emissions of electricity generated from conventionally systematic review and harmonization, *J Ind Ecol* 18:125–144, 2014.

[28] Carbajales-Dale M, Barnhart CJ, Brandt AR, Benson SM: A better currency for investing in a sustainable future, *Nat Clim Chang* 4:524–527, 2014.

18

Thin Film Photovoltaics

Senthilarasu Sundaram*, Katie Shanks*, Hari Upadhyaya**

*ENVIRONMENT AND SUSTAINABILITY INSTITUTE (ESI), UNIVERSITY OF EXETER, PENRYN, UNITED KINGDOM; **WOLFSON CENTRE FOR MATERIALS PROCESSING, INSTITUTE OF MATERIALS AND MANUFACTURING, DEPARTMENT OF MECHANICAL, AEROSPACE AND CIVIL ENGINEERING, BRUNEL UNIVERSITY, LONDON, UXBRIDGE, UNITED KINGDOM

s.sundaram@exeter.ac.uk, hari.upadhyaya@brunel.ac.uk

18.1 Introduction

At present, crystalline silicon (c-Si) dominates more than 90% of the market with module efficiencies of 15%–21% and a record lab cell efficiency of 26.6% under 1000 W m^{-2} [1]. However, the frail nature of the wafers and the bulky nonflexible modules limits the first generation of silicon technology. Thin film technology has the answers and potential to eliminate many existing bottlenecks of c-Si photovoltaic (PV) programs experienced at different levels from module production to its applications in terrestrial, space and building integration photovoltaics (BIPV). Not only this, but thin film technology lends itself more easily to improved aesthetics, color, flexibility, and light weight options. Thin film PV modules can achieve minimum material usage and be manufactured on a large range of substrates. Some of the advantages of thin film technologies are:

- The high absorption coefficient ($\sim 10^5$ cm^{-1}) of the absorber materials is about 100 times higher than c-Si, thus about 1–2 μm of material thickness is sufficient to harness more than 90% of the incident solar light. This helps in reducing the material mass significantly to make modules cost effective.
- The estimated energy payback time of thin film PV is considerably lower than that of c-Si PV. Estimations suggests that CdTe has the lowest payback time among all PV technologies. With recent improved cell and modules efficiencies of 17%, the payback time could be as low as 6 months.
- The formation of heterojunction cells and better device engineering for reduction of photon absorption losses and enhanced collection of photogenerated carriers are possible.
- Large-area deposition (on the order of m^2) along with monolithic integration (interconnection of cells during processing of devices) is possible which minimizes area losses, handling, and packaging efforts.
- Roll-to-roll manufacturing of flexible solar modules is possible, further reducing the energy payback time significantly.

A Comprehensive Guide to Solar Energy Systems. http://dx.doi.org/10.1016/B978-0-12-811479-7.00018-X

- Flexible and lightweight PV facilitates several attractive applications.

A summary of the thin film properties is given in Table 18.1.

However, the efficiency and the robustness of thin film PV technologies have to match or exceed the current c-Si technology to be competitive in the market share. Once optimized, the manufacturing methods can be scaled up and provide factors of magnitude cheaper processing cost and low energy payback time in comparison to c-Si. Currently, amorphous silicon (a-Si), cadmium telluride (CdTe), and copper-indium gallium di-selenide (CIGS) are considered the mainstream thin-film technologies. The world record efficiency of 23.3% has been obtained by NREL for CIGS cells under 14.7× concentration and of 22.6% without concentration by ZSW [2]. CdTe technology is not far behind with a record research-cell efficiency of 22.1% obtained by First Solar [2] and for CIGS and CdTe this substantial progress has occurred over a short span of time [2]. For amorphous silicon the stabilized record cell efficiency is behind at 14% by AIST [2]. So, although a-Si had promising manufacturing targets a decade ago, it lost out to competition due to its intrinsic issues of light dependent degradation referred to as Staebler Wronsky Effect (SWE). With the cost of c-Si modules available well below $0.4 per Wp and with life time guarantees of 25 years, a-Si technology, although a robust option, cannot compete without efficiencies levels matching around 20%.

Recent developments on solid state perovskite based solar cells (a kind of solid state analogue of dye-sensitised solar cells) are equally promising. In year 2012–13 alone the efficiency figures acquired an unprecedented leap from near 4% [3] to 17% and are now currently 22.7% [2,4]. These technologies with their low-temperature and cheaper processing cost on flexible substrates [5,6] make them potentially very attractive as a cost effective option for high volume manufacturing. However, there are issues of instability and intrinsic issues that need to be addressed. The material, methyl ammonium lead iodide (MAPI), or $CH_3NH_4PbI_3$, is labile and can be attacked by moisture in the atmosphere as well as undergo a phase change at 93°C. These are considered to be the major scientific and engineering challenges. Worldwide attempts are underway to overcome these challenges along with the replacement of Pb in the materials by Sn (Tin) or other benign elements to overcome the environmental concerns.

Table 18.1 Summary of Thin Film Properties

	c-Si	a-Si (stabilised)	CdTe	CIGS	Perovskite
Absorption coefficient/cm^{-1}	$\sim 1 \times 10^4$	1.7×10^6	1.1×10^6	$>1 \times 10^5$	1.5×10^4
Direct band gap/eV	3.4	1.75	1.44	1–1.6	1.55
Sufficient thickness/μm	170–200	1	3–5	1–2	0.6–2
Record cell efficiency/%	26.6	14	22.1	22.6	22.7

Compound thin film solar cells viz. CdTe, CIGS, MAPI etc. also provide advantage of development of ternary, quaternary variants of the materials for the manipulation of the band gap, which essentially can be used for development of tandem or multijunction solar cells. This offers the possibility of significant absolute jump in efficiency with the cascade of absorbers utilizing the wider part of the spectrum of light. The possibilities of developing CIGS/Perovskite or c-Si/Perovskite is also being attempted by several groups, which provides a theoretical efficiency limit over 40%, where practical efficiencies over 30% is achievable with a marginal addition of processing costs.

18.2 Thin Film Cell Configurations

18.2.1 Amorphous Silicon

The intrinsic (i) form of a-Si thin film can be doped as p-type or n-type to form a p-n junction, however, initial p-n junction device trials could not result in a solar cell action. This was because of significant recombination losses due to the presence of a large number of surface defects as dangling bonds in the material. The passivation of these defects are possible and hydrogen doped amorphous silicon, a-Si (H), could achieve significant efficiencies in a p-i-n junction configuration as shown in Fig. 18.1. Thus, this was the choice of PV for industries in the last millennia.

The total typical thickness of the p (25 nm)-i (500 nm)-n (30 nm) stack is within 1 micron and thus cannot absorb all the incident photons causing significant transmission losses. Therefore, there was a need to develop double and triple junction cells to achieve high efficiency numbers, along with strategies to apply texturing (roughness) of the substrates that enhances light trapping in this "superstrate" configuration. In the "superstrate" configuration, the a-Si cell is grown in the p-i-n sequence onto a transparent substrate. While the "substrate" configuration can be grown on any type of substrates that could be rigid glass or flexible metal or polymer foil. It bears a reverse, n-i-p, configuration and the light enters through the last grown p-layer.

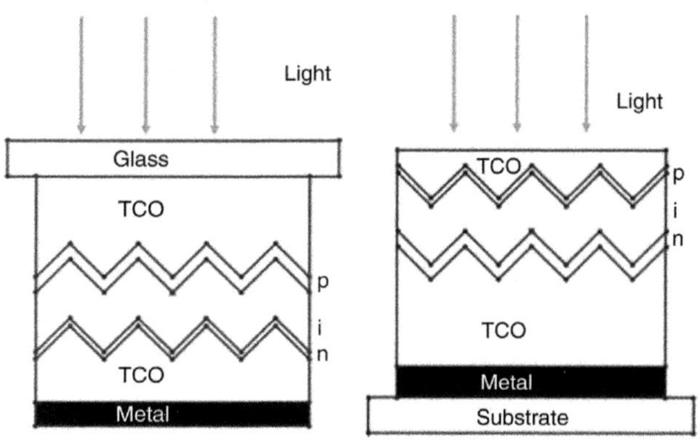

FIGURE 18.1 Schematic presentation of a-Si solar cell in "superstrate" (p-i-n) configuration (*left*) and "substrate" (n-i-p) configuration (*right*).

Amorphous silicon cells have also been combined with nanocrystalline silicon junctions [1] and the cells of other materials. Hetero-structures between a-Si:H layer and c-Si wafer referred to as HIT (Hetero-junction with Intrinsic Thin film layer) cells have also been developed and currently hold the record efficiency for crystalline Si. Panasonic manufactured a cell of "practical size" (101 cm^2) with an efficiency of 26.7% [7].

18.2.2 Cadmium Telluride Solar Cells

CdTe thin film solar cell structure comprises of a p-type CdTe absorber layer and n-type CdS based window layer forming a heterojunction, which has an intermixed interface region. Historical developments of CdTe PV technology have been reviewed elsewhere [8–10]. CdTe thin film absorbers possess good electronic property favorable for solar cells action. It has an ideal direct band gap ~1.5 eV which matches well with the maxima of the Sun's spectrum offering maximum theoretical efficiency ~30% limit. The most attractive features of CdTe compound are its chemical simplicity and the robust stability. CdTe is not only stable for terrestrial applications but it has also been demonstrated that CdTe has excellent stability under high energy-photon and electron irradiation for space applications, superior to Si, GaAs, CIGS etc. [11].

So far, the highest efficiencies have been achieved in "superstrate" configuration for CdTe. Fig. 18.2 gives the schematics of CdTe solar cell grown on TCO coated glass substrate in a superstrate configuration. The substrate configuration has however been reconsidered and the development of a novel doping method allowed solar cell efficiencies close to 14% [12]. CdTe layers may be grown by a variety of vacuum and nonvacuum methods classified into high temperature and low temperature processes and resulting typical thicknesses between 2 and 6 μm [8,10,13].

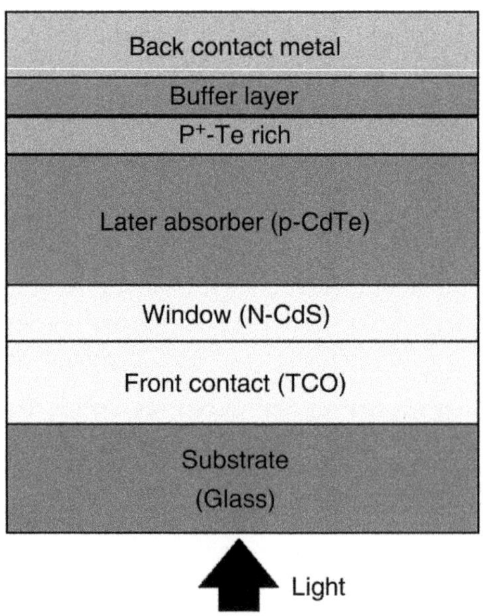

FIGURE 18.2 Schematic presentation of CdTe/CdS solar cell in "superstrate" configuration showing different layers.

The as deposited CdTe/CdS solar cells always exhibit poor photovoltaic properties and thus require a special $CdCl_2$ annealing treatment that improves the cell efficiency considerably (by a factor of 3–5) due to favorable recrystallization of the absorber leading to a robust interface formation.

First Solar (USA) leads the production of CdTe modules (2.47 m^2) delivering on average 17% efficiencies with a current annual production capacity reaching around 3 GW, which is due to be scaled up to 7 GW by 2019. Due to the ease and speed of its production CdTe offers the least payback time <1 year among all competing technologies including c-Si, thus making it one of the fastest growing technologies.

18.2.3 CIGS Solar Cells

Copper Indium Gallium di-Selenide or $Cu(InGa)Se_2$ or CIGS is one of the best absorber materials available owing to its high absorption coefficient and wide spectral response. By adjusting the In and Ga ratio, we can adjust the band gap of the material from $CuInSe_2$ (1.1 eV) to $CuGaSe_2$ (1.68 eV). For CIGS, the substrate configuration gives the highest efficiency owing to favorable process conditions (Fig. 18.3). This configuration gives this technology an added advantage of utilizing a range of substrates for the deposition of thin films from rigid glass to flexible metal or polymeric foils. The downside of this technology is the processing of the absorber at relatively higher temperatures >450°C. Also, the introduction of alkali elements viz. Na, K into the thin film absorber stack provides an almost 3–4 folds increase in the efficiency which has been a "holy grail" for sometime in the thin film field. *Superstrate* structures were investigated in early 80s but efficiency was below

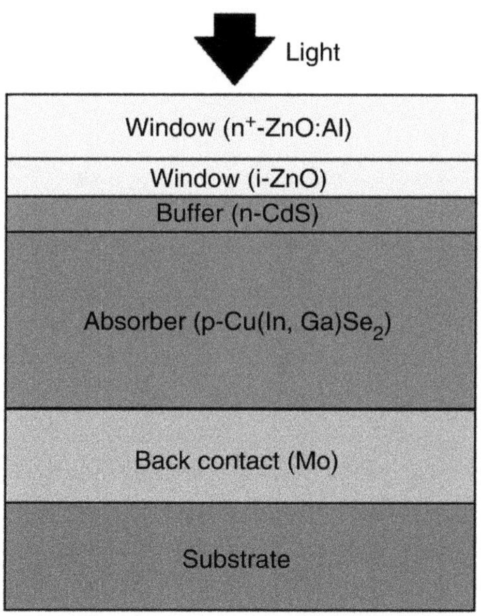

FIGURE 18.3 Schematic presentation of CIGS solar cell in *substrate* configuration.

5%. However, recent efforts have improved the efficiency to ~13%. This has been possible with the introduction of undoped ZnO instead of CdS buffer layer and coevaporation of Na_xSe during the CIGS deposition. Further, bifacial CIGS solar cells with both front and rear transparent conducting contacts were also investigated [14].

Molybdenum (Mo), grown by sputtering or e-beam evaporation is the most commonly used electrical back contact material for CIGS solar cells. When CIGS is grown on Mo an interface layer of $MoSe_2$ is automatically formed which helps in ohmic transport between CIGS and Mo. Recently, alternative back contact materials have been explored but industrial production is still based on Mo layers. High efficiency cells have p-type $Cu(In,Ga)Se_2$ bulk while a defect-chalcopyrite $Cu(In,Ga)_3Se_5$ phase in the form of thin layer segregates at the top surface which is n-type especially when doped by cation atoms diffusing from the buffer layer. Several semiconductor compounds with n-type conductivity and band gaps between 2.0 and 3.4 eV have been applied as a buffer to form a heterojunction in CIGS solar cells. However, CdS remains the most widely investigated buffer layer. *T*COs with band gaps of above 3 eV are the most appropriate for front electrical contact due to their optical transparency (greater than 85%) and reasonably good electrical conductivity. Today, CIGS solar cells employ either ITO or, more frequently, RF-sputtered Al-doped ZnO. One of the breakthroughs in CIGS PV technology was the introduction of potassium [15]. These observations inspired the development of a post deposition treatment (PDT) which led to increased efficiencies of >20% [16–18].

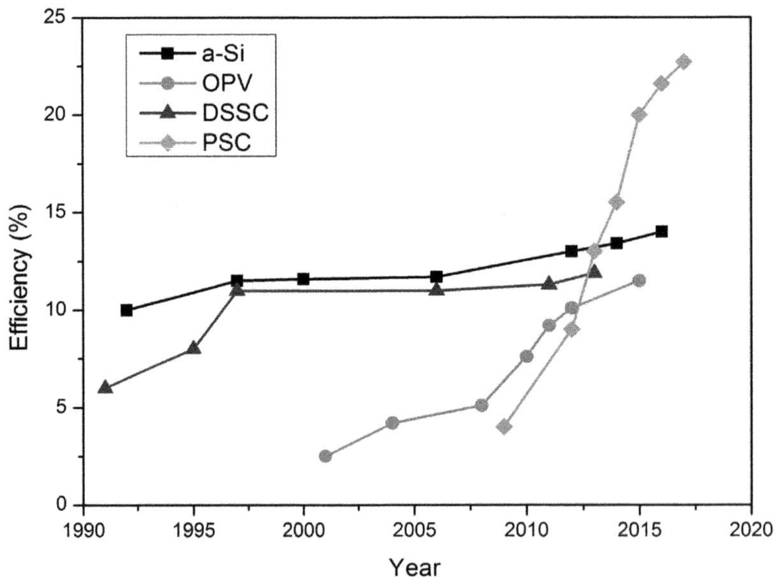

FIGURE 18.4 Comparing the rate of increase in perovskite solar cell efficiencies with the other thin-film PV technologies [19].

Solar Frontier (Japan) leads the production on CIS technology with annual production capacity over 1 GW and has plans to enter BIPV market with the use of aluminium in place of glass to reduce the weight of the modules.

18.2.4 Perovskite

Perovskite solar cells stem from dye-sensitized solar cells but have promising solid state structures as well as rapid efficiency leaps (Fig. 18.4), which have now reached 22.7% [2].

Most commonly, $CH_3NH_3PbI_3$ based organic–inorganic perovskite materials have been used in these types of solar cells due to their high charge carrier mobility. High mobility is important because, together with high charge carrier lifetimes, it means that the light-generated electrons and holes can move large enough distances to be extracted as current, instead of losing their energy as heat within the cell [19]. Moreover, the perovskite solar cells can be deposited by low-temperature methods such as solution process viz. spin coating, spray deposition, and thermal evaporation methods.

18.3 Deposition and Growth Techniques

Room temperature deposition allows the use of a variety of substrates such as glass, metal, and plastics but often results in inferior module efficiencies and quality. Two most commonly used methods for a-Si are plasma enhanced chemical vapor deposition (PECVD) and glow discharge CVD. For high deposition rates the deposition technologies based on very high frequency (VHF), microwave and high-pressure plasma are currently being pursued. Alternative deposition methods using hot wire CVD (HWCVD) technique, electron cyclotron resonance reactor (ECR) and also the combination of HWCVD and PECVD are also being carried out to increase the deposition rate [1,10,20–22]. Vacuum Evaporation is a simple deposition method allowing low temperatures for CdTe and CIGS cells. Vacuum evaporation method involves simultaneous evaporation of the constituent elements from multiple sources in single or sequential processes during the whole absorber deposition process [23]. There is a substantial interest in testing different deposition techniques for CIGS which could improve the technology greatly. Some methods include: nanosized precursor particles and electro-deposition from a chemical bath. A hybrid approach that uses additional vacuum deposition on electrodeposited precursor layers has also been investigated [24]. Closed Space Sublimation (CSS) and vapor transport (VT) are the prominent and industrially used processes for CdTe deposition owing to its very high rate ($2–5\ \mu\ min^{-1}$) of deposition. First Solar is the most successful CdTe company to date with an annual production capacity of approximately 3 GWp for modules on $60 \times 120\ cm^2$ glass substrates. GE global research has achieved a ground-breaking efficiency of 19.6% on glass substrates. Early 2014 First Solar communicated 20.4% solar cell efficiency and a very remarkable full scale module efficiency of 17%. Details of other alternative methods such as screen-printing, spray pyrolysis, MOCVD, CVD and atomic layer deposition (ALD) and electro-deposition (ED) are also possible [11,13,23,24]. For flexible substrates such as polymers low temperature methods like sputtering, HVE, and electro-deposition (ED) are suitable.

18.4 Flexible Cell Formations

Probably the ultimate advantage of thin-film technology is the application of roll-to-roll manufacturing for production of monolithically interconnected solar modules for low capex, lightweight, flexible modules leading to low energy payback time because of high throughput processing and low cost of overall system. Out of all the thin film technologies, CIGS cells have shown the most promise and progress for flexible products, owing to its high efficiency achieved under substrate configuration. A large number of activities on highly efficient, stable and flexible thin film modules based on CIGS has recently drawn much interest for flexible solar cells on metal and plastic foils. Apart from the expected high efficiency and long term stability for terrestrial applications, flexible CIGS has excellent potential for space application because of their space-radiation-tolerant properties which are 2–4 times superior to the conventional Si and GaAs cells.

High record efficiencies of flexible CIGS solar cells are 17.7% on stainless steel, 18.6% on enamelled steel and 20.4% on polymer foil [16–18].

18.5 Challenges

The main challenge for most thin film technologies is their stability. Amorphous silicon cell efficiency falls instantly at module level with current commercial PV module efficiencies ranging around 4%–8% [25]. The SWE [26] causes a-Si cells to suffer light induced degradation and the short orders in the amorphous material and dangling bonds also reduce efficiency [10]. An important issue in CdTe solar cell technology is the formation of efficient and stable ohmic contact on p-CdTe layer and avoiding Schottky contacts forming. Another key issue for both CdTe and CIGS, is the toxicity of the materials involved. CdTe technology relies on the toxic cadmium Cd, but there is some manoeuvrability in CIGS technology in the elimination of very thin (typically ~50 nm) CdS buffer layer.

Some key points in favor of CdTe in terms of environmental impact however are:

- Cd is produced as a by-product of Zn and can either be put to beneficial uses or discharged to the environment posing another risk.
- CdTe in PV is much safer than other current Cd uses.
- CdTe PV uses Cd 2500 times more efficiently than Ni-Cd batteries.
- Occupational health risks are well managed.
- Absolutely no emission during PV operation.
- A risk from fire emission is minimal.
- CdTe technology and modules are safe and don't pose significant risks.

Similarly for perovskites, critical issues involve the stability of the organic spiro OMeTAD layer; and the need for Pb-free compounds in the cells, to cover the environmental concerns [27]. In addition to improving material stability, the prevention of water, air and high energy photons getting into the device and destroying the perovskite are current technology aims. Currently, relatively low temperatures of around 95°C can cause degradation.

18.6 Conclusions

Thin film PV has clearly demonstrated an excellent potential for cost effective generation of solar electricity, especially using CdTe technology. It is anticipated that a mix of c-Si and thin film PV technologies will cater the market needs in near to mid-term future, followed by the dominance of thin film and other PV technologies in long term. The a-Si technology has suffered a setback mainly because of the stagnancy and the low efficiency figures arising from the intrinsic light dependent degradation issues. Thin film PV industries are growing fast however, there are several issues such as reducing the gap between lab efficiency and larger area industrial production efficiency. This is achievable with the design of better equipment with in situ diagnostics. Nonavailability of standard deposition system for thin film PV has been a problem so effort is needed to develop large area equipment suitable for thin film PV. For lower cost, high throughput and yield efforts are needed that require further simplification and increased robustness of the process and device structures.

Further improvements in the stability of device structure and still higher efficiencies (greater than 25%) along with even thinner layers would see the way forward for mass thin film productions.

The cost reduction of modules aimed below \$0.3/W by CdTe and c-Si technologies have brought about a revolution but the Balance of System (BOS) costs viz. battery storage, inverter and electronic component costs which determine the overall cost of electricity, require drastic improvements for further success of solar technologies.

References

[1] Meillaud F, Boccard M, Bugnon G, et al: Recent advances and remaining challenges in thin-film silicon photovoltaic technology, *Mater Today* 18(7):378–384, 2015.

[2] National Renewable Energy Laboratory (NREL) – National Center for Photovoltaics: 'Research Cell Record Efficiency Chart'. Available from: https://www.nrel.gov/pv/assets/images/efficiency-chart.png.

[3] Kojima A, Teshima K, Shirai Y, Miyasaka T: Organometal halide perovskites as visible-light sensitizers for photovoltaic cells, *J Am Chem Soc* 131(17):6050–6051, 2009.

[4] Zhou Y, Zhu K: Perovskite solar cells shine in the "Valley of the Sun", *ACS Energy Lett* 1(1):64–67, 2016.

[5] Kumar MH, Yantara N, Dharani S, et al: Flexible, low-temperature, solution processed ZnO-based perovskite solid state solar cells, *Chem Commun* 49(94):11089, 2013.

[6] Docampo P, Ball JM, Darwich M, Eperon GE, Snaith HJ: Efficient organometal trihalide perovskite planar-heterojunction solar cells on flexible polymer substrates, *Nat Commun*:4, 2013.

[7] Wang HP, Lin TY, Hsu CW, et al: Realizing high-efficiency omnidirectional n-type Si solar cells via the hierarchical architecture concept with radial junctions, *ACS Nano* 7(10):9325–9335, 2013.

[8] Luque A, Hegedus S: *Handbook of photovoltaic science*, John Wiley & Sons Ltd, 2003.

[9] Luque A, Hegedus S. Handbook of Photovoltaic Science and Engineering; 2011.

[10] Lee TD, Ebong A: Thin film solar technologies- a review, 2015 12th Int Conf High-Capacity Opt Networks Enabling/Emerging Technol:33–42, 2015.

[11] Bätzner DL, Romeo A, Terheggen M, Döbeli M, Zogg H, Tiwari AN: Stability aspects in CdTe/CdS solar cells, Thin Solid Films 451–452:536–543, 2004.

[12] Kranz L, Gretener C, Perrenoud J, et al: Tailoring impurity distribution in polycrystalline cdte solar cells for enhanced minority carrier lifetime, Adv Energy Mater 4(7), 2014.

[13] Zeng G, Zhang J, Li B, Wu L, Li W, Feng L: Effect of deposition temperature on the properties of CdTe thin films prepared by close-spaced sublimation, J Electron Mater 44(8):2786–2791, 2015.

[14] Nakada T, Hirabayashi Y, Tokado T, Ohmori D, Mise T: Novel device structure for Cu(In,Ga)Se2 thin film solar cells using transparent conducting oxide back and front contacts, *Solar Energy* 77(6):739–747, 2004.

[15] Wuerz R, Eicke A, Kessler F, Paetel S, Efimenko S, Schlegel C: CIGS thin-film solar cells and modules on enamelled steel substrates, *Solar Energy Mater Sol Cells* 100:132–137, 2012.

[16] Chirilă A, Reinhard P, Pianezzi F, et al: Potassium-induced surface modification of Cu(In,Ga)Se2 thin films for high-efficiency solar cells, *Nat Mater* 12(12):1107–1111, 2013.

[17] Jackson P, Hariskos D, Wuerz R, Wischmann W, Powalla M: Compositional investigation of potassium doped Cu(In,Ga)Se2 solar cells with efficiencies up to 20.8%, *Phys Status Solidi – Rapid Res Lett* 8(3):219–222, 2014.

[18] Feurer T, Reinhard P, Avancini E, et al: Progress in thin film CIGS photovoltaics – Research and development, manufacturing, and applications, *Prog Photovoltaics Res Appl* 25(7):645–667, 2017.

[19] Hodes G. Perovskite-Based Solar Cells; 2013, 342, (October), p. 317–319.

[20] Green MA: Thin-film solar cells: review of materials, technologies and commercial status, *J Mater Sci Mater Electron* 18:15–19, 2007.

[21] Carabe J, Gandia JJ: Thin-film-silicon solar cells, *Opto-Electron Rev* 12(1):1–6, 2004.

[22] Kamalisarvestani M, Saidur R, Mekhilef S, Javadi FS: Performance, materials and coating technologies of thermochromic thin films on smart windows, *Renew Sustain Energy Rev* 26:353–364, 2013.

[23] Perrenoud J, Schaffner B, Buecheler S, Tiwari AN: Fabrication of flexible CdTe solar modules with monolithic cell interconnection, *Solar Energy Mater Sol Cells* 95(Suppl. 1):S8–S12, 2011.

[24] Compaan AD, Gupta A, Lee S, Wang S, Drayton J: High efficiency, magnetron sputtered CdS/CdTe solar cells, *Solar Energy* 77(6):815–822, 2004.

[25] Sharma S, Jain KK, Sharma A: Solar Cells: in research and applications—A review, *Mater Sci Appl* 6(12):1145–1155, 2015.

[26] Staebler DL, Wronski CR: Reversible conductivity changes in discharge-produced amorphous Si, *Appl Phys Lett* 31(4):292–294, 1977.

[27] Yang S, Fu W, Zhang Z, Chen H, Li C-Z: *Recent advances in perovskite solar cells: efficiency, stability and lead-free perovskite* 5(23):11462–11482, 2017.

PART 5

Environmental Impacts of Solar Energy

19

Solar Panels in the Landscape

Beatrice Dower
MVGLA, COMRIE, UNITED KINGDOM
beatrice.dower@mvgla.com

19.1 Introduction

People attach value to the landscape around them, whether it provides a functional role as agricultural land, recreational open space or the setting for their local community; or invokes an aesthetic response to scenic views, dramatic topography, or a familiar place [1]. These values, either openly acknowledged or unexpressed, are tested when change is proposed. People can react strongly to change when landscape they value is threatened with development, particularly if the landscape in question is local to them. Solar photovoltaic (PV) panel installations are becoming increasingly common in response to the need to generate renewable energy as part of the measures being taken to combat climate change. Installations can range from a few panels on a house roof, to large areas taken over by arrays. In this chapter we will discuss how PV solar panels affect the landscapes around us, and how the design of installations can be modified to provide "best fit" with the landscape.

19.1.1 What Is Landscape?

The landscape is the environment around us that forms the setting for our daily lives. The landscape we experience has character—made up of the topography, vegetation, land use, and elements than give the landscape structure and pattern, such as field boundaries, trees, and roads. The European Landscape Convention defines landscape as "*an area, as perceived by people, whose character is the result of the action and interaction of natural and/or human factors*" [2]. This definition takes in the perception of the area by the viewer as well as the physical elements that form it. The perception of an area can include a response to how the landscape is used, or could be used, and the values that the viewer places on aspects of the landscape. The European Landscape Convention also gives value to the landscape as a resource in its own right and reminds us that it is not just scenic areas that have value, but the degraded landscapes around our cities are important in their own way, as are the remote places that we will never visit but we know are there.

Opinions on land use vary widely; some prefer untouched vegetation, grassland, or forest; for others the green fields and hedges of a productive verdant area are seen as the "ideal countryside," while yet others may see a landscape with a more intensive or

A Comprehensive Guide to Solar Energy Systems. http://dx.doi.org/10.1016/B978-0-12-811479-7.00019-1

FIGURE 19.1 An ideal landscape? Looking over the Coquet Valley in Northumberland, UK. *Photo: B. Dower (all rights reserved).*

modern function as the ideal (Fig. 19.1). Understanding the landscape, how it is made up and how it is valued, is important to understand how developments will affect perceptions and character of the landscape that surrounds us.

Paradoxically, the one constant in our landscape is that it will change. Landscape change is an inevitable process, driven by climate, ecology, and human activity. Though it is not possible to stop change, it is sometimes possible to guide changes in our landscape for the purposes of improving it or maintaining valuable aspects. Development is a noticeable force for change, and solar installations, using increasingly efficient technology, are becoming larger with the drive for renewable energy production. Planning and siting of solar farms can have wide-ranging effects on the character of the landscape as installations take up extensive areas of land. Since the first solar park of 1 MW in California in 1982, many countries have built a substantial number of installations that make a meaningful contribution to their renewable energy production.

19.2 Solar Installation Types

The two main technologies for solar power generation are PVs and thermal or concentrated solar power. In this chapter we concentrate on PV technologies. PV solar panels are rectangular, usually in the order of 1.3–1.7 m^2, with an average size of 1.65×1 m, used singly or in sets or arrays. Solar panels can be used in different situations, depending on the scale of the installation and the end use for the energy created.

19.2.1 Building-Mounted Panels

Solar panels can be mounted on buildings and are usually retrofitted to existing buildings. The building provides the support structure, ideally on a south-facing roof slope that provides both aspect and tilt. The panels are fixed to the existing roof or mounted on low supports to gain tilt on flat roofs. Cabling can then be internal to the building.

There has been a significant uptake of domestic solar schemes by private homeowners in the UK, such that roof mounted solar panels are now a common sight. Several local authorities in the UK have invested heavily in installing solar panels to local authority-owned properties throughout the country. In addition, businesses have sought the benefits of renewable and micro-energy generation subsidies (such as the recent "feed in tariff" subsidy) by installing solar panels on farm buildings, industrial buildings, office buildings, etc. Public buildings such as schools, hospitals, and council buildings have also seen an increasing number of solar installations (Fig. 19.2).

As a result, solar panels are a common sight on buildings throughout the UK, even as far as northern Scotland, where although sunlight duration and intensity may be lower than southern England and Europe, there is sufficient light to generate power using the latest and most efficient technology (Fig. 19.3).

Retrofitted solar panels are classed as permitted development (not requiring planning consent) in the UK, provided that the building is not listed, is not in a designated area, and the panels are of a standard design. For listed buildings valued for their architectural merit, solar panels would need to be carefully sited (e.g., located on the rear

FIGURE 19.2 Solar panels retrofitted onto existing buildings, Otterburn First School, Northumberland UK.
Photo: B. Dower (all rights reserved).

FIGURE 19.3 Solar panels on a house on the northern coast of Scotland, with the Atlantic Ocean visible to the right.
Photo: B. Dower (all rights reserved).

facing roof slopes) to be given consent. Similarly, solar panels on buildings in conservation areas are unlikely to be consented if they are visible from main streets or key vistas through the settlement.

As solar installations on buildings are permitted development (with the exceptions set out above), there is little control over how solar panels are placed, which can lead to widely varied results in terms of visual composition. Solar panels are least intrusive if they clearly relate to the geometry of the roof. As solar panels are standardized as rectangular panels, they tend to be most satisfactory on rectangular roofs. The placing of rectangular panels can be difficult for roofs interrupted by dormer windows and can greatly reduce the visual clarity and appearance of the building (Fig. 19.4).

Wind lift is an issue for building mounted panels. To avoid panels being damaged in high winds, they should be fixed close to the surface and set down below the roof ridge and with clearance from the edges of the roof [3]. The method of fixing can further limit the design of the installation, the placement, and type of fixings on a slate or tiled roof must relate to both the surface material and the positions of rafters below. These aspects can limit the scale of the installation and affect the arrangement of panels in relation to the geometry of the roof.

Modern farm or industrial buildings are generally well suited to taking solar panels, as they are usually of simple rectangular form, often of a gentler pitch than domestic buildings and can be large in extent. Farmers seeking to diversify can make good use of shed roof space with solar panels, without disruption to the workings of the farm (Fig. 19.5).

FIGURE 19.4 Solar panels ruin the appearance of this vernacular style house in northern Scotland. The better sited panels are those in the garden behind. *Photo: B. Dower (all rights reserved).*

In recent years there has been an increasing drive toward designing solar panels as an integral part of new developments. Planning applications for public and domestic buildings with solar panels as part of the new-build proposals are seen favorably, and in line with government support for sustainable development [4]. Solar panels can be integrated into buildings at the design stage, often with better results than retrofitting.

Large-scale solar panel installations are possible on structures such as stations and factory roofs. London's new Blackfriars Station spans the River Thames and has 4400 solar panels designed into the roof structure, which generate approximately 1 MW of energy. Another example is the high-speed rail line from Paris to Amsterdam, which passes an ancient woodland near Antwerp, Belgium. A 2 mile long cover shelter was used to avoid the need to fell the trees that are within topple-distance and has been fitted with 16,000 solar panels on its roof (Fig. 19.6).

Small-scale solar panels on structures such as roadside lighting or sign posts are also a familiar site in the UK. Panels of less than 0.5 m^2 provide sufficient power to light up warning signs, particularly in rural locations where other sources of power are more difficult to provide.

FIGURE 19.5 Solar panels on farm buildings, Dorset UK. *Photo: B. Dower (all rights reserved).*

FIGURE 19.6 Sketch of train cover tunnel. *Source: B. Dower (all rights reserved).*

19.2.2 Integrated Materials

Recent developments in building materials have increased the availability of building-integrated solar generation, with, for example, tiles or slates for roofing that have PV cells within them [5]. By integrating the solar units with the fabric of the roof, the design of the building is not compromised by panels added to the surface. The cost of solar integrated materials is, however, still high.

19.2.3 Free-Standing Solar Farms

While most domestic solar installations are mounted on buildings, utility-scale installations are free-standing or ground-mounted. Arrays of panels mounted on frame structures have both advantages and disadvantages. A key advantage is that the scale of the development is not limited by the size of a roof or building structure and can be extended over many hectares. The main disadvantage is the removal of the land from otherwise potentially more productive agriculture.

Free-standing solar farms can vary in size from a few panels or a few hectares, to vast areas of panels, such as the recently constructed Tengger Desert Solar Park in India which is spread over 43 km^2 and generates 1500 MW. Most UK solar farms are in the order of 8–12 ha (20–30 acres). The largest solar farm in the UK is currently Shotwick Solar Farm, Wales, which covers 90 ha and generates 72 MW, but now there are proposals for schemes of over 100 ha.

Large solar farms made up of arrays of panels may stretch for considerable distances. In the UK, arrays of panels are typically designed to fit within the existing field pattern, to preserve the landscape pattern of hedgerows or walls [6]. This is done for reasons of preserving landscape character, long-established field boundaries that can be important ecological resources in the landscape, and for purposes of screening from nearby locations. As arrays are usually in south-facing rows, this can lead to a diagonal orientation with respect to rectilinear field boundaries. Space must be left around the edges of the fields to avoid shadow by hedgerows or trees and to allow turning access for maintenance vehicles. As a result, large solar farms in the UK are typically made up of several fields of arrays rather than long uninterrupted lines that are possible with extensive unenclosed land in other parts of the world.

Panel arrays are arranged in rows to avoid shadowing and allow maintenance access, and are elevated off the ground by 1–2.5 m. As a result, there is light that will reach the ground between the arrays, and although the panels shade the ground below them, there is sufficient light for vegetation to grow. This vegetation ground layer can be maintained as a wildlife resource or can be grazed by sheep, which have the benefit of a secondary agricultural use for the land. Cattle are not suitable for grazing among solar panels, as the larger animals could damage the panels or infrastructure and would need greater clearance off the ground.

The mounting frames hold the panels at an inclined angle that can be calculated for optimum solar yield for a given latitude [7]. Panels are usually set at a fixed orientation and tilt, though some developments are mounted on frames that allow for adjustments in tilt and orientation, for tracking the sun to gain maximum yield. These solar tracker arrays have higher maintenance costs, require greater land take to avoid shadowing each other, and are generally not seen in new installations.

Mounting frame ground anchors can be piled or poured concrete foundations, though the former is favored for speed of construction and ease of decommissioning and may form part of conditions for consent (Fig. 19.7). In areas of archaeological interest, premolded concrete blocks or "shoes" may be required as they minimize ground disturbance and are easy to remove [8].

In the UK, solar farms are considered to be temporary structures and planning conditions can be used to ensure that the installations are removed when no longer in use and the land restored to its previous use. However, in the lifespan of the installation (usually around 20 years), it is likely that technologies will advance and "repowering" of old sites may be an alternative to full decommissioning.

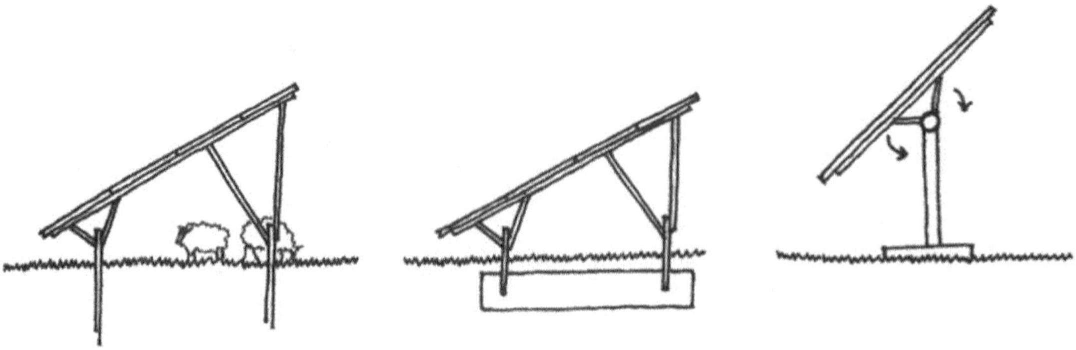

FIGURE 19.7 Sketch of ground mounted PV panels, with piled or concrete shoe foundations and solar tracker panel mount. *Source: B. Dower (all rights reserved).*

19.2.4 Floating Solar Farms

Solar panels can be mounted on pontoons floating on open bodies of water, making use of not only open flat space, but also benefiting from the cooling effect of water below the arrays. Solar pontoons have not been built on the sea, but the calmer waters of reservoirs can be suitable for extensive arrays. Solar power generation is an interesting secondary use for reservoirs, particularly on drinking water reservoirs where recreational access is not encouraged. The use of the water surface on a reservoir may also be preferable to using agricultural land or disturbing terrestrial ecosystems. The visibility of the panels mounted on low pontoons at a distance from the shore is limited, and the recreational experience of walking along the shore is not affected greatly.

Europe's largest floating solar farm opened in 2016, on the Queen Elizabeth II reservoir in Surrey, UK. It is made up of pontoons that cover 57,500 m^2 and provides a 6.3 MW private electricity supply for the water plant at the reservoir. The largest scheme proposed is on the Yamakura reservoir (Chiba Prefecture Japan), a proposal for 180,000 m^2 of panels, while the largest scheme in operation is the 2017 floating plant in Huinan (Anhui Province China), generating 40 MW on waters flooding a former coal mining area [9].

Floating pontoons do not cover the water surface completely and are tethered together for flexibility with water movement, so that there is some light penetration and wind to aerate the water below. The environment below the floating raft is therefore shaded in part and provides a variation in habitat within an open reservoir, which can be beneficial to the ecosystem within the reservoir.

19.3 Key Visual Elements

The main visible elements of free-standing solar farms are the panels, which are typically dark blue to black and finished with a nonreflective coating to maximize the absorption of light. Early models had aluminum frames and grid lines between the cells, and more recent models have no frame or grid and are plain black, making them simpler in appearance. Fixtures and free-standing panel supporting frames can be galvanized or painted. The cabling is undergrounded as close to the panel as possible and is not normally a noticeable element.

When seen from the reverse side (from the north in the northern hemisphere), the shaded undersides of solar panels are visible. When seen from the front or sides (south, east, or west in the northern hemisphere), the solar panels can appear black or dark blue from a distance, although they appear pale at an angle when there is some reflection (glare is the term for reflection of ambient brightness, glint is reflection of direct sunlight). Set on buildings, these colors can mimic smooth dark or gray roofing materials that can also appear pale when wet or with a slight reflection of the sky. However, black, dark blues, and grays are unfamiliar colors for large-scale areas of the landscape, particularly in verdant countries. As a result, solar farms with large numbers of freestanding panels will stand out in landscapes with fields of greens, yellows, or browns. Colors in the landscape change with the seasons more than the variation in grays of solar panels in different light conditions. Grazing fields will remain green (albeit different greens through the seasons), but arable fields vary from brown/black/red when ploughed to fresh or rich greens during growth and yellows or ochres during harvest. The flowering of crops such as oil seed rape or flax can also create bright flushes of color in the landscape.

The color of a solar farm can therefore mark it out as different from its surroundings when seen from a distance. There are comparisons that can be drawn to large industrial buildings, or in an agricultural setting to glasshouses and polytunnels, or the increasing use of plastic agricultural netting to protect young crops, which creates whole fields covered in white fabrics. These are noticeable in late spring, but are not in place all year round (Fig. 19.8).

FIGURE 19.8 Crop netting in Perthshire, UK. *Photo: B. Dower (all rights reserved).*

19.3.1 How People Experience Solar Farms

People, in their daily lives, move around the landscape to commute, to make other journeys for work, recreation or to get from one place to another. They also look around while traveling and may stop to appreciate a particular view. Most people therefore experience solar farms in passing from roads or paths nearby. The viewing experience is different from residential properties or settlements nearby, with a small number of people seeing an installation repeatedly, rather than many people seeing the installation in passing along a busy road.

From a distance, a solar farm appears as a metallic surface, the color and reflectivity of which depends on the relative direction of view and the light conditions. It may not be readily perceptible as a solar farm if the lines of panels cannot be distinguished and it can be mistaken for glasshouses or large industrial building roofs. Only at relatively short distances can the arrays be seen separately, and the mounting structures are only visible from the reverse side. Other infrastructure of the solar farm, such as transformers, substation, worker welfare facilities, security fencing and pole-mounted CCTV equipment are also visible when close to the installation (Fig. 19.9).

19.3.2 Impressions of a Solar Farm

For many, commercial scale solar farms are still relatively new and there is an element of curiosity when viewing them, but for other people the panels will be an intrusion in the

FIGURE 19.9 Solar farms include transformers, CCTV cameras and fences as well as panels. Dorset, UK. *Photo: B. Dower (all rights reserved).*

landscape. People's perception of solar farms is affected by their predisposition toward solar farms as a means of generating renewable energy.

Often the reaction to a solar farm is a reaction to the change to the site from its previous state, rather than to the solar panels per se. A negative response can arise when the solar farm is a change from the green field that the viewer remembers. This is in part because green arable or grazing fields are seen as better—more natural, healthier, less industrialized—land uses than solar farms. However, reactions can be different if the site was previously a brownfield or derelict site, such that the solar farm is seen as a better land use than the site in its former state. This may also be the case with different proposals on the same site. For example, a solar panel farm was proposed on an arable field in the south of England and gained local support as it had previously been proposed as a windfarm site—the solar panels were seen as a better option than turbines!

This highlights a key difference between solar and wind as renewable energy technologies. Wind turbines, by their very nature, must reach upwards to catch the wind resource and are therefore tall structures. Wind turbines of 150–200 m to blade tip are becoming increasingly common with technological advances and the pressure to increase yields in an economy with decreasing subsidies. These are inevitably very visible structures, with even "small" turbines such as those of Novar Windfarm in Scotland (55.5 m to blade tip) being visible in excess of 40 km in good viewing conditions, and larger turbines potentially being visible from up to 50–60 km away [1]. Solar panels, in contrast, are low structures, with the panels often only up to approximately 2.5 m off the ground. As low-level structures, they do not have long-distance visibility in flat landscapes, and elsewhere effective screening and appropriate use of the topography can enable the zone of visual influence to be very contained, to within the site or within close proximity of the site only. In such cases, the public can go by without knowing that the installation exists—quite a contrast to passing a windfarm.

Another aspect of public reaction to solar panel farms is that solar panels are becoming increasingly common on domestic buildings, and many people seeing installations can relate them to solar panels on their own house or on the house of someone they know. This brings an aspect of familiarity to solar panels and a sharing of renewable energy goals, which allows some people to view larger installations in a more favorable light. Windfarms too are becoming more familiar, but being visible over greater distances, may have a greater bearing on the character of landscapes and views, which exercises people's opinions on landscape change [1].

19.4 Environmental Issues in Planning

Environmental impact assessment (EIA) of development in the landscape considers the potential effects of the installation on the proposed area and is a wide ranging assessment covering many aspects of the environment, such as effects on landscape, visual amenity, ecology, ornithology, archaeology and cultural heritage, noise, and socioeconomic effects. The predicted effects of a large-scale solar panel installation will determine whether it is given planning consent and what conditions or mitigation measures are required to reduce or avoid the environmental effects that have been identified by the EIA.

Large-scale solar farms are considered to have negative effects on the rural environment, with visual effects being the aspect that most people will respond to. In the UK at least, the most common reasons for refusal of planning permission are related to visual and landscape matters and the use of agricultural land.

9.4.1 Landscape and Visual Effects

Assessment of effects on landscape character and on visual amenity are normally required in the UK for large developments such as commercial scale solar farms, and are undertaken following guidance set out in the *Guidelines for Landscape and Visual Impact Assessment* [10]. Landscape and visual impact assessments should consider the sensitivity of the receptor and the magnitude of change as a result of the development. Evaluation of the sensitivity of the receptor includes identification of the susceptibility of the landscape or view to change and the value placed on the landscape and view as a resource. Identifying the magnitude of change includes consideration of the scale of the change, the duration and permanence of the change, and the geographical area over which the change will be experienced. These aspects are brought together to identify the level of effect and whether it is significant in planning terms.

An assessment of the potential visual effects of a proposed solar farm should consider the appearance of the installation (including panels and infrastructure) from local routes, including consideration of the effect it will have on the experience of the routes for people passing by. Views from particularly sensitive locations, such as visitor attractions, places where people congregate, or places where the view of the landscape is important, may influence the design and location of the installation, including how and whether screening is possible. A visual impact assessment should consider potential views from these locations, as well as effects on views from residential properties or settlements nearby.

A solar farm can also change the character of the site and its immediate surrounding landscape, introducing a modern technological layer to the pattern and land use of the area. The orientation of the arrays with respect to the existing pattern of the landscape can have effects on landscape character, where arrays cross rectilinear fields diagonally. The effects of the installation and its infrastructure on landscape character should be considered as part of an EIA.

UK guidance on solar farms, prepared for counties such as Cornwall and Devon that have seen increasing pressure for solar farm developments [6,11], encourages careful siting and screening of solar installations to address visual effects, with particular emphasis on retention of existing hedgerows and trees that give structure and pattern to the landscape. Further mitigation through additional screening planting of hedges and trees is also encouraged, and maintenance regimes can be designed to allow hedgerows to grow taller than normal, for screening purposes. In undulating landscapes, it is often views from a distance (although generally no more than 5–10 km) from elevated vantage points with north-facing views over the site that will have visibility of solar farms that cannot be mitigated by local screening.

9.4.2 Effects on Land Use

The area that a solar farm occupies is an important consideration in most locations. As the surface area of the installation determines the yield, utility-scale installations require large areas of land. Solar installations take up considerably larger areas of land in comparison to windfarms. A windfarm of 25 MW may cause approximately 12 ha of habitat loss under the turbines, tracks, and compounds, with the land between the turbines (spaced for safety and to avoid turbulence [1]) remaining intact and available for continued land use. A solar farm generating 25 MW will take up approximately 50 ha, with a larger land-take in higher latitudes. The loss of previous function for this land may be significant in determining whether it gains consent.

The use of agricultural land for free-standing solar installations is a factor that planning authorities need to consider, and guidance contained in the UK *National Planning Policy Framework* [12] encourages the effective use of land by focusing large-scale solar farms on previously developed and nonagricultural land, provided that it is not of high environmental value. If a proposal involves "greenfield" land, it should be demonstrated that the proposed use of the agricultural land is necessary (i.e., no nonagricultural land is available); that poorer quality land has been used in preference to higher quality land; and that there are provisions for continued agricultural use where applicable and/or biodiversity improvements around arrays (Fig. 19.10).

In some countries, large areas of nonagricultural land are under pressure for development, with barren or desert landscapes preferred as there is little settlement and little vegetation to disturb. While these landscapes are perhaps less affected than smaller scale verdant landscapes, there are nonetheless impacts on the ecology and environment, ranging from altering the soil ecosystems below the panels, to interrupting migration of species across the area.

9.4.3 Other Environmental Issues

As set out above, landscape and visual matters are not the only issues that need to be considered for large solar installations. Although each area of environmental effect is a subject in its own right, we can briefly explore the main challenges.

The introduction of panels alters the ecology of a solar farm site, by introducing different habitats with varied shading from light to dark, varied water regimes with dry areas below the panels that receive little or no rainfall and changes to the pattern of the landscape with corridors of vegetation along the arrays. The partial shade below the panels can benefit species that seek refuge from the sun and different, shade tolerant plants can grow below the panels rather than between the arrays. Research is still needed as to how these localized environmental changes affect the ecology in different settings, with great contrasts between these localized effects for solar farms set in temperate or desert climates.

FIGURE 19.10 Sheep grazing below solar panels, Dorset UK. *Photo: K. Harris (all rights reserved).*

Whether these changes are beneficial or detrimental will also vary with location, there may be significant threats to local populations of rare or endangered species, or in contrast, there may be opportunities to grow food crops that cannot tolerate open sunlight. The likely effects on the ecology of the site should be investigated as far as possible during the design process and appropriate mitigation measures included for ecological benefits. Monitoring of the site after construction is also important to increase our understanding of how ecosystems react.

The effects of solar panels on birds is an area of ongoing research and is of concern where extensive arrays of panels can be mistaken for open water by migrating birds. Bird deaths due to impact with solar panels are a real problem for some species where resting places on long journeys are critical to survival. Understanding the routes and vulnerabilities of passing bird populations is important to reduce the potential consequences for birds.

Solar farms can affect sites of cultural or archaeological significance, both directly through construction and indirectly by affecting the setting of the asset. Modern panels can be seen as an intrusion upon the historic setting of cultural heritage features, particularly where they affect key views to or from the asset or affect the ability of the viewer to understand the location of the asset for its historical purpose. Examples may be where a solar farm would intrude on a vista in a historic designed landscape or affect the appreciation of the relationship between a fort located on a vantage point and the landscape the occupants once surveyed. Physical disturbance to the site during construction can damage cultural heritage features, and appropriate design mitigation should be used to avoid this. Construction works for any type of development can also reveal hidden archaeological features; such findings may have consequences for further construction, and sites with particular local sensitivity should have an archaeological watching brief in place during construction.

Operational solar farms do not generate noise, but noise may be an issue during construction for properties or public routes nearby, particularly during the sinking of piles if they are used. However, construction effects are generally short-lived and can be reduced by the use of other anchoring techniques or controlled to a degree by conditions such as restricting working hours.

Views of solar farms from above come into play when considering aircraft safety, and solar sites close to airports risk glint and glare for pilots on approach routes. It is noticeable however, that numerous airfields have solar panel installations on buildings or land as a source of renewable energy to offset high airport energy costs. Research carried out by the Solar Trade Association has suggested that glint and glare are less of a concern than could perhaps be expected, with antireflective coating on panels reducing glare to similar levels to black asphalt surfaces or water bodies [13].

19.5 Offset Mitigation

Some of the environmental effects of solar panel farms can be mitigated to reduce or avoid adverse effects. Mitigation can also be used to offset or compensate for environmental effects.

Mitigation to offset ecological impacts with alternative biodiversity benefits takes various forms in the UK, including maintaining or planting native species in the screening hedges, which provide food sources and habitats for wildlife. Additional measures often include planting wildflowers around site edges or among the panels (Fig. 19.11), and small-scale interventions such as supporting the installation of bat, bird or owl nesting boxes can have local benefits, particularly where these species have been identified in survey work.

Another form of mitigation for the loss of the former primary agricultural use of the land is to introduce alternative forms of agriculture that can carry on below the panels and be an extra source of income from the land. These include grazing sheep, which can pass under the panels with ease and can feed on the grassland between the arrays. The panels will provide shade and shelter for the sheep, and are not damaged by the sheep, although there can be a concern about sheep chewing on cabling left within reach. Free range poultry rearing is also potentially a good use of the land below the panels, poultry housing can be fitted below the panels, the darker drier areas below the panels provide dust baths for the birds, and perimeter security fencing will keep predators away.

Farm diversification is important for continued viability of both small and large farms, and the continuation of other forms of agriculture below the solar installation is seen as a

FIGURE 19.11 Wild flowers planted round a solar panel farm, Dorset, UK. *Photo: B. Dower (all rights reserved).*

benefit that can help to mitigate against the loss of arable land. However, this is not always considered sufficient to compensate for the loss of the best and most versatile agricultural land, and sites of lower agricultural quality are favored.

19.6 Concluding Remarks

As has been explored in this chapter, solar installations can have environmental effects that will vary with site and location of the panels, ranging from significant impacts of ground-mounted panels on areas of high ecological, cultural heritage, scenic, or agricultural quality; to minimal impacts for panels located on industrial building roofs. Small-scale and domestic solar installations are permitted developments in the UK, but larger installations require an assessment of potential environmental effects. While effects can be readily perceptible or subtle, localized or spread across kilometers of view, others may not become apparent until several years after construction, with continued research and monitoring of environmental change. The process of developing a solar installation needs to include the identification of potential significant environmental effects, and to address them through siting, design, and mitigation. For landscape and visual effects of solar farms, the most effective form of mitigation is retention of hedged field boundaries where possible. Hedges have multiple benefits for solar farms, providing screening (grown to a desired height but not overshadowing panels), reinforcing landscape pattern and structure, and providing opportunities for ecological benefits. The loss of agricultural potential of higher quality land is a concern, although offset mitigation can include alternative forms of agriculture than can coexist with the panels.

The author wishes to thank Rebecca Knight of LUC and Marc van Grieken of MVGLA for contributions and review.

References

[1] Wind turbines and landscape. In: Letcher T, edtior. Wind energy engineering, a handbook for onshore and offshore wind turbines. London: Academic Press; 2017.

[2] Council of Europe: *European landscape convention*, Council of Europe, 2000.

[3] UK planning guidance website; 2017. Available from: www.planningportal.co.uk/info/200130/common_projects/51/solar_panels/2.

[4] Planning practice guidance for renewable and low carbon energy, online guidance website; 2015. Available from: https://www.gov.uk/guidance/renewable-and-low-carbon-energy.

[5] Solar PV slates website; 2017. Available from: http://www.gb-sol.co.uk/products/pvslates/default.htm, or similar websites.

[6] LUC. Devon landscape policy group advice note No. 2: Accommodating wind and solar PV developments in Devon's landscape; 2013.

[7] Solar angle calculator website; 2017. Available from: http://solarelectricityhandbook.com/solar-angle-calculator.html, or similar websites.

[8] BRE National Solar Centre. Planning guidance for the development of large scale ground mounted solar PV systems; 2013.

[9] South China Morning Post. China flips the switch on world's biggest floating solar farm; 2017.

[10] Landscape Institute and the Institute of Environmental Management & Assessment: *Guidelines for landscape and visual impact assessment*, 3rd ed., London and New York, 2013, Routledge.

[11] Cornwall Council. Cornwall renewable energy planning advice; 2016.

[12] UK Government Department for Communities and Local Government. National planning policy framework; 2012.

[13] Solar Trade Association. Impact of solar PV on aviation and airports; 2016.

20

Solar Energy Development and the Biosphere

Michelle Murphy-Mariscal*, Steven M. Grodsky**,†, Rebecca R. Hernandez**,†

*MT. SAN JACINTO COLLEGE, MENIFEE, CA, UNITED STATES; **UNIVERSITY OF CALIFORNIA, DAVIS, CA, UNITED STATES; †WILD ENERGY INITIATIVE, JOHN MUIR INSTITUTE OF THE ENVIRONMENT, DAVIS, CA, UNITED STATES

rrhernandez@ucdavis.edu

20.1 Introduction

As of April 2017, atmospheric concentration of CO_2 has reached an unprecedented mark of 410 parts per million. Despite support for renewable energy development as a means to combat greenhouse gas emissions to mitigate climate change impacts and reduce reliance on finite energy resources, rapid renewable energy deployment is complicated by environmental trade-offs. Potential environmental impacts of renewable energy development include, but are not limited to, habitat fragmentation, degradation or disruption of valuable ecosystem services, biodiversity loss, and increasing land scarcity [1,2]. These ecological impacts may be overlooked as minor when compared to those of global climate change, which threatens biodiversity on a global scale; however, cumulative disturbances associated with renewable energy development are complex, difficult to mitigate, and poorly understood [3].

The development of solar energy is unique in that adverse environmental impacts and associated costs can be avoided with appropriate siting and decision-making. Increased awareness of these potential tradeoffs is the first step towards achieving greater sustainability in solar energy design and enterprise. Here, we discuss: (1) potential impacts from construction, operation, and decommissioning of solar energy facilities, focusing particularly on ground-mounted, utility-scale solar energy (USSE, > 1 MW_{DC}) USSE installations; (2) potential environmental effects over the lifetime of solar energy installations; and (3) potential ecological responses of wildlife and other biosphere attributes with options for mitigating or reducing those impacts.

A Comprehensive Guide to Solar Energy Systems. http://dx.doi.org/10.1016/B978-0-12-811479-7.00020-8

20.2 Solar Energy Effectors and Potential Effects on the Environment

Effectors may be temporally categorized over the lifetime of a photovoltaic (PV) or concentrating solar power (CSP) installation, from construction through decommissioning (Fig. 20.1) and may have one or more potential effects on the environment with multiple potential ecological responses. Additionally, the technology, size, and location of solar energy infrastructure may impact biota and the environment in different ways. For example, integrated solar energy is that which has zero land-use and land-cover change impacts beyond those associated with raw materials acquisition and manufacturing. Thus, it has minimal to zero adverse effects on the biosphere (beyond life-cycle emissions), resources (e.g., cultural), and legal entitlements (e.g., religious rights of indigenous communities)

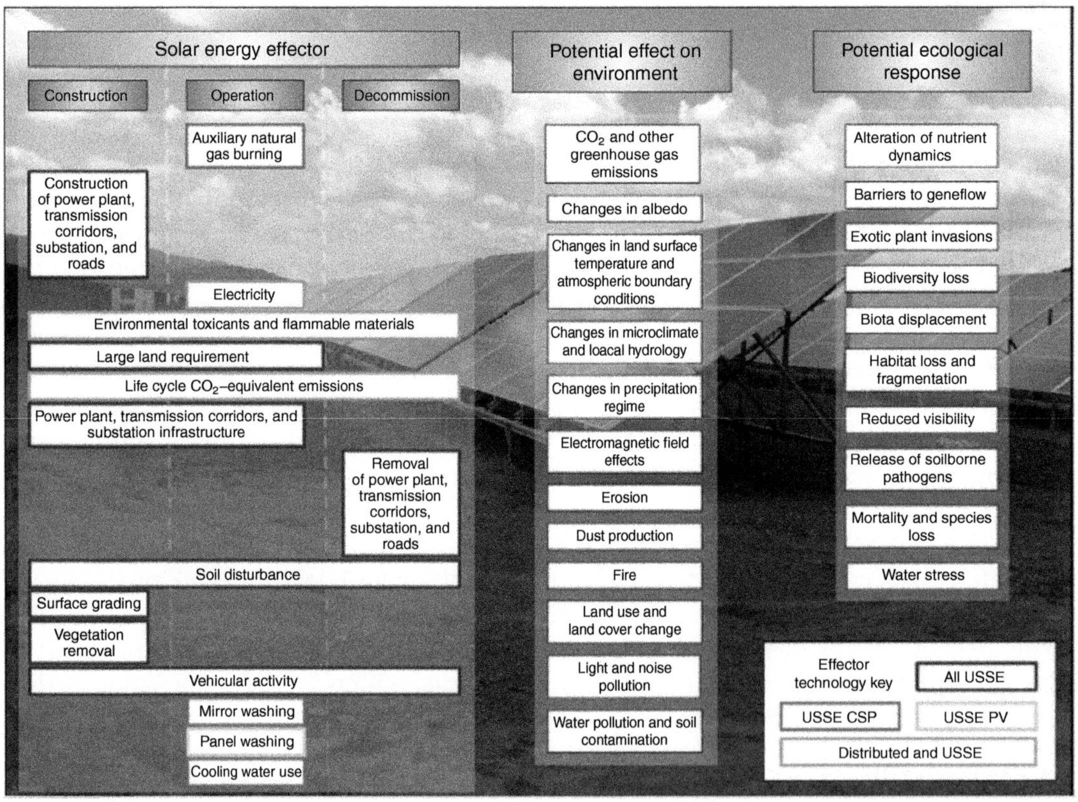

FIGURE 20.1 Solar energy effectors for utility-scale solar energy technologies (ALL USSE), including concentrating solar power (USSE CSP) and photovoltaics (USSE PV), and for both utility-scale and distributed schemes (Distributed and USSE). *Photo credit: Rebecca R. Hernandez. From Hernandez RR, Easter SB, Murphy-Mariscal ML, Maestre FT, Tavassoli M, Allen EB, et al. Environmental impacts of utility-scale solar energy. Renew Sust Energy Rev 2014;29:766–79.*

FIGURE 20.2 Aerial view of 4.1 MW of integrated solar energy at the UC Davis West Village. West Village is the largest net zero energy community in the United States, combining energy efficient technology with on-site energy production via rooftop and vertical photovoltaic installations (Davis, California). *Photo courtesy of UC Davis.*

[3a]. Integrated solar energy is cohesively constructed into elements of the built environment in urban and suburban areas (e.g., commercial and residential building rooftops, parking garages, and carports) relatively close to consumers (Fig. 20.2). Although geographically diffuse, integrated solar energy offers high levels of solar energy potential [4]; it has been estimated that 20%–27% of all residential rooftop space and 60%–65% of commercial rooftops in the United States are conducive to photovoltaic and solar thermal systems [5]. In contrast, displacive solar energy is that which incurs additional land-use or land-cover change and therefore reduces biophysical capacity or facilitates the loss of other resources of value (e.g., cultural) across the Earth's surface. These installations are typically ground-mounted and large in capacity (e.g., utility-scale solar energy [USSE]) They are often geographically far from demand loads and preexisting transmission, and have large land area requirements (i.e., installed capacity increases concomitantly with land area).

20.2.1 Land Requirements

To meet projected 2040 energy consumption demands, it is estimated that approximately 800 000 km^2 of additional land (with spacing), an area two times that of the state of

California, would be affected by carbon-intensive and renewable energy development [6]. Ground-mounted solar energy requires relatively large expanses of land to support power plant infrastructure, mirrors and towers (e.g., CSP), and panels (e.g., PV), and therefore, such installations are often sited far from urban population centers where most electricity is consumed. This may necessitate additional transmission infrastructure (i.e., power line corridors, roads, and substations) to transport electricity, expanding impacts beyond the immediate footprint of facilities themselves. Hernandez and colleagues [7] found that PV and CSP ground-mounted USSE installations in California have a land-use efficiency of 35.1 and 33.9 Wm^{-2}, respectively, based on the nominal (i.e., nameplate) capacity. Land use efficiency can vary significantly from these averages. For example, Zichella and Hladik [8] reported that a currently operational, 354 MW USSE facility in the California desert, United States, occupies over 645 ha, equal to 54.9 Wm^{-2} and is a more efficient installation based on nameplate capacity. In California, it was also found that installations on public land required significantly more land per installed MW of capacity than those sited on private land (10 more Wm^{-2}), demonstrating flexibility in USSE design perhaps driven by differences in the price of land [7]. This agrees with land estimates comparing PV with coal life-cycles by Fthenakis and Kim who showed that coal with surface mining in the U.S. uses more land than PV installed in the Southwest [63]. Also, they found that integrated PV and CSP technologies incorporated into the built environment had the lowest land-use intensity across all sources of electricity, underscoring the potential to avoid negative impacts on the biosphere through appropriate siting decisions.

20.2.2 Land-Use and Land-Cover Change

In light of the land area requirements of displacive solar energy installations, land-use and land-cover change is a significant conservation concern. Relative to other energy production systems, renewable sources of energy may occupy a relatively small percentage of new land area being affected by energy development in the United States under the assumption that PV and CSP would comprise only 0.5% and 0.04% of United States energy production from 2012 to 2040 [6]. However, if larger percentages of displacive installations are realized, it would scale accordingly. Additionally, displacive facilities may be disproportionately sited in areas where high biological endemism (species with very limited distributions that are often highly adapted to their environments), fragile habitats, and high solar resources co-occur, such as the Mojave Desert in the southwestern United States. Indeed, deserts and xeric shrubland habitat types in the United States are expected to experience the greatest land-cover change impacts due to PV and CSP siting by 2030 [9]. Hernandez and colleagues [7] found that the plurality of USSE developments in California are sited in shrublands and scrublands, the land cover type with the highest inherent biodiversity of those included in the study, necessitating 375 km^2 of additional land. Additional conflicts may arise when solar facilities are sited in areas where land has agricultural value; for example, 118 km^2 of land categorized as cultivated cropland has been converted to or is earmarked for USSE development in California [7]. Lastly, USSE may also be disproportionately sited in areas with biophysical capacity to support ecosystem services, including carbon sequestration and storage. De Marco and colleagues [10] found that 42 of the 82 permitting requests for new USSE sites in Leece, Italy (238.4 km^2) were in ecologically unsuitable areas, comprising 18 563 ha of land-cover change, including in places with century-old olive groves notable for their cultural value and that provide the largest contribution to carbon sequestration, relative to other land-cover types evaluated (>1.5 million tons of CO_2).

20.2.3 Surface Grading and Vegetation Removal

During the construction phase of a solar energy power plant, preparation of the facility site may include grading and scraping, which removes all aboveground biomass [9]. Grading reduces wildfire risks on-site and prevents the panels from being shaded by vegetation [10]; however, from an ecological perspective, these activities constitute a loss of habitat within the footprint of the facility and degradation of surrounding land, which may result in mortality of wildlife or species displacement. Ecosystems with limited resources (e.g., precipitation, nutrients) may be slow to recover from disturbance, either from the construction of the USSE facility itself or its decommissioning, making restoration an inviable and/or costly option. For example, natural recovery times for desert plant communities to return to predisturbance species composition is 215 years based on a meta-analysis of 31 individual studies [10a]. If topsoil has been removed from the site, this recovery time may be longer and thus restoration potential may be diminished depending on the size and intensity of the disturbance [11,12]. Additionally, more carbon is sequestered in soils than in the atmosphere and terrestrial vegetation combined [13]. Therefore, soil disturbance resulting from site development may release a significant amount of stored organic (and possibly inorganic) carbon, potentially offsetting benefits of establishing the renewable energy source (in terms of reducing greenhouse gas emissions). Significant soil processes are negatively impacted by disturbance, including nutrient cycling and water holding capacity [14]; soil biota that contribute to these processes, such as biological soil crusts, may take 20–1000+ years to recover in aridland environments [12], necessitating costly active restoration techniques that require salvaged material to expedite recovery [14a]. Disturbed soils are more prone to wind erosion, thus potentially impacting human health (e.g., valley fever), reducing fertility of biological soil crusts and vegetation through reduced photosynthesis, and contributing to sedimentation in surface water [15].

20.2.4 Hydrologic Changes and Water Degradation

Construction activities may impact surface-water flow pathways and water quality, especially when projects are sited on bajadas, individual alluvial fans, floodplains, or near washes. Flood control structures may be constructed on-site to intentionally divert water around facility footprints in an effort to reduce soil erosion near facility infrastructure. Modifications to surface-water flow may alter geomorphological processes and downstream aquatic ecosystems and habitats by altering transport of organic matter, nutrients, minerals, and sediments [16].

Large concentrating solar power facilities require large quantities of water for operation, which may stress water resources, especially in arid environments where water scarcity contrasts abundant sunlight levels optimal for solar energy production. Wet-cooled CSP facilities, which need steam to generate electricity and water for cooling, may use between 5551 and 17 886 $m^3\ a^{-1}\ MW^{-1}$ (4.5–14.5 ac-ft of water per year per MW of electricity produced), where "a" refers to annum [15]; this amount equals or exceeds nuclear and coal power plants [17]. Alternatively, dry-cooled CSP technology requires between 247 and 1234 $m^3\ a^{-1}\ MW^{-1}$ (0.2–1.0 ac-ft $year^{-1}\ MW^{-1}$) of electricity produced [15]. Additionally, CSP installations require water for mirror washing, potentially equating to 617 $m^3\ a^{-1}\ MW^{-1}$ (0.5 ac-ft $year^{-1}\ MW^{-1}$) or more [15]. It is noted that PV in the southwestern United States uses marginal amounts of water for panel washing, and therefore, has a clear advantage to any thermoelectric power generation in arid areas.

20.2.5 Changes in Land-Surface Temperature, Albedo, and Microclimate

Photovoltaic panels have low reflectivity, owing to a large proportion of solar radiation that reaches them being converted to electricity and heat [18]. There is growing concern that PV installations may cause a "heat island" effect similar to those that occur in urban areas, especially in desert environments, whereby mean air temperatures surrounding the installation increase due to a decrease in albedo. The effective albedo of a PV panel (current upper maximum >35.2%) is the sum of its reflectivity (e.g., 0.06–0.1) and solar conversion efficiency (e.g., 0.12–0.252 [e.g., SunPower panels for upper limit] [19]). Ultimately, the variation in the albedo of natural and built environments where solar energy installations are sited and the variation in the effective albedo of PV can lead to different results; thus additional research in diverse environments is needed to determine generalized patterns of altered land-surface temperature by solar energy. A modeling study by Taha [19] found that a large scale PV installation in Los Angeles—characterized by common roofing materials, concrete, and asphalt—would reduce the urban heat island effect and cool cities up to 0.2°C. Another modeling study in the desert found that local night-time temperatures were 3–4°C higher in solar facilities than nearby control areas [20]. Furthermore, panels and mirrors may create an insulation effect due to physical shading and airflow alterations. This was demonstrated at a CSP plant in China, where soil temperatures were reported as being 0.5–4°C lower in spring and summer and higher by the same range in winter relative to control sites [21]. However, Nemet [22] found that the effect of albedo change due to widespread deployment of PV globally would be negligible in comparison to the benefits of reducing greenhouse gas emissions from the same deployment intensity.

20.3 Ecological Impacts and Responses

Ecological responses to disturbance and development are well studied, although few studies have quantified effects of displacive installations on the biota, habitats, and ecosystems occurring within or near their footprints. Displacive and large (e.g., USSE) installations may negatively or positively affect a diversity of biota (Table 20.1). Effects likely depend on the

Table 20.1 Known or Expected Impacts of USSE on a Subset of Taxa

		Habitat Fragmentation	Panels and Mirrors	Fences	Air-Cooled Condenser (CSP Only)	High Energy Flux Field (CSP Only)
Birds	Passerines and insectivorous birds	–	–	o	–	–
	Raptors	o	–	o	o	–
	Ravens	+	o	+	o	+
	Waterbirds	o	–	o	o	o
Mammals	Bats	–	o	–	–	o
	Bighorn sheep	–	o	–	o	o
	Coyotes	–	o	–	o	o
	Kit foxes	–	+	–	o	o
Reptiles	Desert tortoise	–	o	–	o	o
Insects	Flying insects	–	–	o	–	–
Plants	Native annuals	–	o	–	o	o
	Native perennials	–	–	–	o	o
	Invasive plants	o	o	+	o	o
Total type disturbance known effect	Negative	14	10	10	6	5
	Positive	1	2	2	0	0

Impacts are listed as positive (+), negative (−), or neutral (o) based on experience and judgment of the authors and the literature.
Source: From Moore-O'Leary R, Hernandez R, Johnston, DS, et al. Sustainability of utility-scale solar energy—critical ecological concepts. Front Ecol Environ 2017. doi:10.1002/fee.1517.

design, technology, size, siting, and land-use efficiency of each facility. At the individual species scale, disturbance may elicit behavioral responses (e.g., avoidance of noise and light), reduce resource acquisition opportunities, and alter social dynamics, each of which concurrently occur with physiological responses (e.g., increased heart rate). These responses may result in energy and nutritional expenditures, which lead to reduced vitality, reduced fecundity, and increased mortality in wildlife species [23], although the effects often are species- and habitat-specific. For example, desert tortoises (*Gopherus agassizii*) translocated to adjacent habitats outside of a solar facility footprint prior to construction activities have been shown to experience higher body temperatures and increases in energy expenditure during the first year following displacement; however, negative effects on tortoise growth and body condition were not documented [24]. Displacive installations may also lead to ecological effects spanning beyond individual taxa, affecting species–species and species–process interactions (e.g., trophic interactions) in ecosystems [25,26]. In addition to direct impacts experienced on-site, wildlife communities and habitats may be affected outside of facility footprints. For example, wildlife abundance and composition downstream of a large power plant may be modified due to altered magnitudes of stream surface flow, timing, duration, and velocity [16]. Wildlife responses may vary temporally, including temporary movement of individuals away from disturbance during construction activities and permanent displacement of individuals due to habitat loss. Such has been reported for bird densities and diversities, which are lower within USSE development footprints than surrounding areas [27–29].

However, not all wildlife responses to USSE development are negative; some species, especially generalist species that do not require specialized habitats or diets, may benefit from human development and disturbance. For example, common raven (*Corvus corax*) abundance has been positively correlated with development in the desert region of California [30], potentially due to subsidies of anthropogenic resources (e.g., food, nest and roost sites, water; [31]). While this translates to increased fecundity for the subsidized predator [31], it is often to the detriment of native prey (i.e., desert tortoise; [32]). Similarly, mammalian scavengers, such as coyotes (*Canis latrans*), may be attracted to solar energy facilities by availability of unmanaged refuse [33] and carcasses of birds that succumbed to operation-related injuries (e.g., collision with infrastructure) [29]. Furthermore, during operation of a displacive solar energy facility, wildlife and plants may acclimate to development. Wildlife species, such as invertebrates and small reptiles, may recolonize installations sited in natural or other valuable environments where vegetation is allowed to reestablish in between panels, which may, in turn, attract larger predators. At a PV facility in South Africa, Visser [29] found that raptors and terrestrial birds utilized the installation for foraging and hunting, flocking birds used the evaporation pond as a drinking site, and several species of birds nested on the mountings directly beneath the panels or on the ground.

20.3.1 Habitat Fragmentation

Perhaps the least debated ecological impact of displacive solar energy is habitat loss and concurrent habitat fragmentation resulting from its development. This impact is of paramount concern because habitat fragmentation is among the leading causes of global biodiversity decline [34]. Habitat fragmentation occurs when once contiguous tracts of natural landscape are disturbed or converted, resulting in spatially distinct patches of remnant habitat [35]. Among other impacts, long-term ecological studies have demonstrated that habitat fragmentation results in decreased species richness [36], impaired ecosystem function, increased edge effects, and isolation of resident populations or communities from adjacent patches [37]. Hernandez and colleagues [38] found that of the USSE installations planned and under construction in the state of California, over 73% of PV and 90% of CSP installations were sited less than 10 km away from the nearest protected area, thereby increasing edge effects and undermining the effectiveness of those protected areas as corridors for wildlife movement. While wide-ranging wildlife species may have the ability to circumvent USSE infrastructure during seasonal migration or movement associated with resource acquisition and mating, displacive solar energy projects may prohibit movement of less-mobile wildlife species and plant propagates, thus increasing gene flow disruption between populations [2].

20.3.2 Roads, Transmission Lines, and Fences

The roads, transmission lines, and fencing that radiate from and surround large and displacive facilities contribute to habitat fragmentation and degradation and may cause a considerable amount of negative ecological impacts. The effects of both paved and dirt roads on wildlife have been well documented [39], including direct mortality from vehicle collision, modified behavior (e.g., avoidance), and edge effects (e.g., altered microclimate, increased predation risk and invasion of exotic species). Larger, motile wildlife may easily traverse roadways; however, their risk of collision increases with traffic volume. In contrast, roadways may be insurmountable linear barriers to less-motile species, potentially leading to inbreeding and greater vulnerability to catastrophic events, such as wildfire. Additionally, roads impact species spatial distribution and habitat use, as demonstrated by the decreasing density of desert tortoises with increased proximity to roadways [40]. Invasive plant species often colonize disturbed areas and thus benefit from disturbance associated with the construction of roadways [41]. Propagules of exotic species may be carried by vehicles and construction equipment along roadways [11], aiding in their invasion and spread across the landscape [42]. In contrast, road edges may enhance the vigor of some perennial shrubs and the germination of some annual species, which benefit from water runoff from impervious surfaces and support greater densities of herbivorous arthropods than sites further away from roadways [11]. However, wildlife may be attracted to road edges by the availability of forage, thus increasing their risk of collision.

Transmission and distribution lines are essential for transporting electricity generated from any type of power generation facility. Similar to roadways, the construction of transmission corridors may degrade surrounding habitats; furthermore, maintenance of transmission corridors (e.g., vegetation removal to decrease fire risk) is a continual source of disturbance. Because of these factors, the ecological impacts of transmission infrastructure include their potential to become linear barriers to wildlife movement (e.g., species may avoid the degraded or altered habitat within the corridor), edge effects, and altered community compositions. For example, in Australia, the community composition and abundance of small mammals was shown to differ between transmission corridors and adjacent forested habitat [43,44], with introduced and grassland species being favored over native, forest species. Bird diversity may be lower in corridors than surrounding forested habitat in the United States, with generalist forest species and shrubland birds dominating transmission corridors [45]. However, mid-seral vegetation management that retains structural complexity of vegetation in the corridor (as opposed to complete and frequent vegetation removal) may promote biodiversity and maintain connectivity for forest species [44], highlighting the need for site- and habitat-specific management within transmission corridors to reach conservation goals.

In addition to indirect ecological effects, overhead transmission lines may pose direct collision and electrocution risks to birds. On the basis of known fatality rates, an estimated 10^9 (1 billion) bird strikes may occur annually in the United States alone [46]. Weak fliers (based on wing morphology and wing loading [i.e., ratio of weight to wing area]), were found to have high probabilities for powerline collision in Spain; birds of prey, ravens, and thermal soarers also were among electrocution victims [47]. Several studies have identified powerline electrocution as a conservation problem for several species of rare and endangered raptors worldwide, including California condor (*Gymnogyps californianus*; [48]), Spanish imperial eagle (*Aquila adalberti*; [49]), Bonelli's eagle (*Aquila fasciata*; [50]), and the Eurasian eagle owl (*Bubo bubo*; [51]). Guidelines for reducing electrocution risks, such as minimum conductor spacing, may help mitigate some avian mortality [52]. Lastly, steel towers and power poles provide hunting perches for opportunistic predatory birds, which may increase predation risk for slow or sedentary wildlife (e.g., ravens predating desert tortoise; [11]). These effects are not unique to USSE and no studies to date have studied ecological impacts from transmission lines specifically associated with solar energy power plants.

In addition to fragmenting habitat, fence lines surrounding USSE developments for security may act as dispersal barriers to some species of wildlife. Bats and most birds can fly over fences, with a few exceptions (e.g., roadrunners), and insects and small bodied animals (e.g., lizards, snakes, and rodents) may travel unimpeded through some fences. However, larger bodied animals (e.g., kangaroo rats—*Dipodomys* spp.) and animals with small home ranges (e.g., desert tortoises) may be excluded. This may prevent gene flow between individuals located on either side of the fence line. Promisingly, fences may be engineered to accommodate the needs of some species (e.g., kit foxes in the San Joaquin Valley of California, United States; [53]).

20.3.3 Panels and Mirrors

Large expanses of PV panels and mirrors may be perceived by flying species as flat-water bodies [54]. This phenomenon, known as "the lake effect", occurs when flying species mistake flat surfaces of mirrors and modules for water. Some species may suffer impact trauma from collision as they attempt to land whereas others (e.g., waterfowl) may strand themselves because they are unable to easily take off from a terrestrial surface. Both scenarios increase risk of mortality or injury leading to starvation or predation [55]. Non-fatal collisions of large-bodied birds with panels were documented at PV facilities in South Africa [29] and southern California [55], and impact trauma was the leading cause of avian death documented at a PV and parabolic trough facility in the Mojave Desert, United States [55]. Additionally, the presence of ponds at PV facilities may serve as an attractant to waterbirds and flocking birds [29,55]. No positive effects of panels or mirrors are documented for waterbirds or flocking birds. Polarized light from PV panels and mirrors can attract insects [56], which, in turn, may attract insectivorous raptors (e.g., kestrels—*Falco* spp.) and insect gleaning bats that might utilize PV fields and evaporation ponds for foraging [29]. Subadult bats have been observed attempting to drink off of panels [57], suggesting that they are attracted to and confused by the panels; it is not known if these wasted attempts cause detrimental energy expenditures. If vegetation is allowed to regrow between panels, terrestrial foraging birds may utilize those areas for shade and shelter. Birds may also utilize the underside of panels or the ground beneath panels as nesting sites [29]. Nesting success may depend upon the presence of predators within the facility footprint. Small carnivores (e.g., kit fox, *Vulpes macrotis*) may be able to establish natal dens within PV arrays [58].

PV panels and mirrors may have a negative impact on both annual and perennial native plant species, which are well adapted to their local, unshaded environments. For example, desert plants tolerate high temperatures and solar radiation levels and low precipitation; however, plants within solar arrays experienced altered microclimates, including 11°C cooler temperatures from panel shading and increased water from water runoff at the edges of panels [59]. These altered conditions may be beneficial to generalist, invasive annual plant species. Meanwhile, altered microhabitat conditions in solar facilities may decrease seed production, density, species richness, and community abundance of native annual species [59]. In aridlands, decreased plant cover and biomass is associated with decreases in diversity and abundance of small reptiles and other wildlife species [11].

20.3.4 Air-Cooled Condensers and High-Energy Flux

Impacts from air-cooled condensers and high-energy flux are unique to CSP power plants. Bats may collide with the fans of air-cooled condensers while foraging or in their search to locate roosting sites, although acoustic deterrents may mitigate this impact. Insects may collide with fans and are expected to be negatively impacted. Passerines are not expected to be negatively impacted by air-cooled condensers [60]. More research is needed to better understand potential effects of air-cooled condensers on wildlife.

Solar flux is created by the high intensity concentration of light reflected off mirrors, creating temperatures exceeding 800°C. Insects are attracted to the flux field as a source of polarized light [56,61], resulting in potential incineration of flying insects. Mortality from solar flux has been documented for both dragonflies and butterflies [55]. Attracted to their insect food source, insectivorous birds experience singeing of flight feathers when foraging near flux towers, resulting in mortality. Minor singeing causes impairment of flight, which may lead to inability to forage and evade predators, while severe singeing may cause loss of flight leading to impact trauma and mortality from collision with mirrors [54,55,62]. Scavenger species (e.g., corvids, small carnivores) may benefit from bird fatalities at USSE facilities (i.e., from flux and impact; [29]). Bat carcasses have been retrieved from CSP facilities. While the cause of death remains unknown, bats may be lured into flux fields while foraging, although neutral effects are expected due to bat activity being concentrated after sunset. Flux fields and air-cooled condensers are not expected to negatively impact annual or perennial plants, ungulates, small mammals, carnivores, or reptiles, although no research on any of these taxa has been conducted.

20.4 Summary

Globally, solar energy can provide great environmental benefits, not the least of which is reduced greenhouse gas emissions when substituted for carbon-intensive sources of energy. Indeed, integrated solar energy and other appropriate siting decisions (e.g., reclamation of contaminated land) provide additional benefits associated with land sparing. These benefits should be conjunctively considered in contrast to the environmental costs of solar energy development in places with high biophysical capacity, including natural aridland environments. Displacive USSE development requires land and, to date, rapid deployment of USSE facilities associated with power purchase agreements have emphasized displacive environments. Ecological impacts of displacive USSE development on the biosphere likely are exacerbated when solar facilities are sited in ecosystems with low rates of recovery from disturbances like sensitive areas within the Mojave Desert. Ecological effects of USSE may span the lifetime of a solar facility, from construction to decommissioning. Specifically, siting, site preparation, construction, operations and maintenance, and decommissioning of displacive facilities all may affect ecosystem integrity. Alterations to geohydrology and microclimate from USSE infrastructure may disrupt the physical, chemical, and biological properties of soils, which, in turn, can affect plants, animals, and ultimately "bottom-up" ecosystem processes and interactions. At the landscape-level, new solar energy development beyond the built environment can disturb and fragment habitat. In terms of wildlife response to disturbance, most often sensitive, specialist species are negatively affected, while generalist species typically benefit. Further, invasive plant species often thrive on disturbance and may outcompete native plant species not adapted to disturbance following environmental perturbations. Habitat fragmentation from solar energy infrastructure, including roads, may reduce animal movement and dispersal capacity near solar facilities, which may, in turn, lead to decreased gene flow among subpopulations. Plants and animals may be affected by displacive development directly (e.g., mortality) and indirectly (e.g., displacement). In general, studies on direct or indirect effects of infrastructure associated with solar energy on biota are few, but current research efforts will soon lead to an influx of literature on this subject. However, studies have shown that displacive solar energy projects may cause mortality and extirpation of some species. Assessment of the true sustainability of solar energy hinges on understanding both environmental benefits and costs to the biosphere. Engineering focused on capturing the full potential of integrated solar and the design of solar energy to support positive technological and ecological outcomes simultaneously will contribute to conservation of the biosphere and greater sustainability for humans.

References

[1] Lovich JE, Ennen JR: Wildlife conservation and solar energy development in the desert southwest, United States, *BioScience* 61:982–992, 2011.

[2] Hernandez RR, Easter SB, Murphy-Mariscal ML, Maestre FT, Tavassoli M, Allen EB, et al: Environmental impacts of utility-scale solar energy, *Renew Sust Energy Rev* 29:766–779, 2014.

[3] Scholes R, Settele J, Betts R, Bunn S, Leadley P, Nepstad D, et al: Climate change (2014) terrestrial and inlandwater systems. In Barros V, Field C, editors: *Final draft report: impacts, adaptation and vulnerability. Contribution of working group II to the fifth assessment report of the intergovernmental panel on climate change*, Cambridge, 2014, Cambridge University Press, pp 153.

[3a] Mulvaney D: Identifying the roots of Green Civil War over utility-scale solar energy projects on public lands across the American Southwest, *J Land Use Sci* 12(6):493–515, 2017.

[4] Hernandez RR, Hoffacker MK, Field CB: Efficient use of land to meet sustainable energy needs, *Nat Clim Change* 5:353–358, 2015.

[5] Paidipati J, Frantzis L, Sawyer H, Kurrasch A: *Rooftop photovoltaics market penetration scenarios*, National Renewable Energy Laboratory, 2008.

[6] Trainor AM, McDonald RI, Fargione J: Energy sprawl is the largest driver of land use change in United States, *PLoS ONE* 11(9):e0162269, 2016.

[7] Hernandez RR, Hoffacker MK, Field CB: Land-use efficiency of big solar, *Environ Sci Technol* 48:1315–1323, 2014.

[8] Zichella C, Hladik J: Siting: finding a home for renewable energy and transmission, *Elect J* 26(8):125–138, 2013.

[9] McDonald RI, Fargione J, Kiesecker J, Miller WM, Powell J: Energy sprawl or energy efficiency: climate policy impacts on natural habitat for the United States of America, *PLoS ONE* 4(8):e6802, 2009.

[10] De Marco A, Petrosillo I, Semeraro T, Pasimeni MR, Aretano R, Zurlini G: The contribution of utility-scale solar energy to the global climate regulation and its effects on local ecosystem services, *Global Ecol Conservation* 2:324–337, 2014.

[10a] Abella D: Disturbance and plant succession in the Mojave and Sonoran Deserts of the American Southwest, *Int J Environ Res Public Health* 7:1248–1284, 2010.

[11] Lovich JE, Bainbridge D: Anthropogenic degradation of the southern California desert ecosystem and prospects for natural recovery and restoration, *Environ Manage* 24:309–326, 1999.

[12] Webb R, Belnap J, Thomas KA: Natural recovery from severe disturbance in the Mojave desert. In Webb R, Fenstermaker L, Heaton J, Hughson D, McDonald E, Miller D, editors: *The Mojave desert: ecosystem processes and sustainability*, Reno, 2009, University of Nevada Press, pp 343–377.

[13] Swift RS: Sequestration of carbon by soil, *Soil Sci* 166:858–871, 2001.

[14] Miller DM, Hughson DL, Webb RH: Recovery, restoration, and ecosystem monitoring. In Webb R, Fenstermaker L, Heaton J, Hughson D, McDonald E, Miller D, editors: *The Mojave desert: ecosystem processes and sustainability*, Reno, 2009, University of Nevada Press, pp 339–342.

[14a] Chiquoine LP, Abella SR, Bowker MA: Rapidly restoring biological soil crusts and ecosystem functions in a severely disturbed desert ecosystem, *Ecol Appl* 26(4):1260–1272, 2016.

[15] US-BLM, US-DOE. Final programmatic environmental impact statement for solar energy development in six southwestern states (Arizona, California, Colorado, Nevada, New Mexico and Utah) (FES 12-24; DOE/EIS-0403). 2012. Available from: https://energy.gov/sites/prod/files/EIS-0403-FEIS-Volume1-2012_0.pdf.

[16] Grippo M, Hayse JW, O'Connor BL: Solar energy development and aquatic ecosystems in the southwestern United States: potential impacts, mitigation, and research needs, *Environ Manage* 55:244–256, 2015.

[17] Pizzo S: When saving the environment hurts the environment: balancing solar energy development with land and wildlife conservation in a warming climate, *Colo J Int Environ Policy* 22:123, 2011.

[18] Turney D, Fthenakis V: Environmental impacts from the installation and operation of large-scale solar power plants, *Renew Sust Energy Rev* 15(6):3261–3270, 2011.

[19] Taha H: The potenital for air-temperature impact from large-scale deployment of solar photovoltaic arrays in urban areas, *Sol Energy* 91:358–367, 2012.

[20] Barron-Gafford GA, Minor RL, Allen NA, Cronin AD, Brooks AE, Pavao-Zuckerman MA: The photovoltaic heat island effect: larger solar power plants increase local temperatures, *Sci Rep* 6:35070, 2016.

[21] Gasparatos A, Doll CN, Esteban M, Ahmed A, Olang TA: Renewable energy and biodiversity: implications for transitioning to a green economy, *Renew Sust Energy Rev* 70:161–184, 2017.

[22] Nemet GF: Net radiative forcing from widespread deployment of photovoltaics, *Environ Sci Technol* 43:2173–2178, 2009.

[23] Johnson CJ, St-Laurent MH: Unifying framework for understanding impacts of human developments on wildlife. In Naugle DE, editor: *Energy development and wildlife conservation in western North America*, Washington, DC, 2011, Island Press, pp 27–54.

[24] Brand LA, Farnsworth ML, Meyers J, Dickson BG, Grouios C, Scheib AF, et al: Mitigation-driven translocation effects on temperature, condition, growth, and mortality of Mojave desert tortoise (*Gopherus agassizii*) in the face of solar energy development, *Biol Conserv* 200:104–111, 2016.

[25] Grodsky SM, Moore-O'Leary KA, Hernandez RR: From butterflies to bighorns: multi-dimensional species-species and species-process interactions may inform sustainable solar energy development in desert ecosystems. In Reynolds RL, editor: In *Proceedings of the 31st annual desert symposium*, Berkeley, CA, 2017, California State University Desert Studies Center, April 14–15, 2017.

[26] Moore-O'Leary RR, Hernandez DS, Johnston, et al: Sustainability of utility-scale solar energy—critical ecological concepts, *Front Ecol Environ*, 2017doi: 10.1002/fee.1517.

[27] DeVault TL, Seamans TW, Schmidt JA, Belant JL, Blackwell BF, Mooers N, et al: *Bird use of solar photovoltaic installations at US airports: implications for aviation safety*, USDA National Wildlife Research Center—Staff Publications, 2014, Paper 1418.

[28] Harvey HT and Associates. 2014. California Valley Solar Ranch Project avian and bat protection plan: Quarterly post-construction fatality report, 16 November 2013–15 February 2014. Unpublished report to HPR II, PLC, California Valley Solar Ranch, San Luis Obispo.

[29] Visser E. 2016. The impact of South Africa's largest photovoltaic solar energy facility on birds in the Northern Cape, South Africa. Thesis, University of Cape Town, South Africa.

[30] Duerr AE, Miller TA, Cornell Duerr KL, Lanzone MJ, Fesnock A, Katzner TE: Landscape-scale distribution and density of raptor populations wintering in anthropogenic-dominated desert landscapes, *Biodivers Conserv* 24:2365–2381, 2015.

[31] Kristan WB, Boarman WI: Effects of anthropogenic developments on common raven nesting biology in the west Mojave desert, *Ecol Appl* 17:1703–1713, 2007.

[32] Kristan WB, Boarman WI: Spatial pattern of risk of common raven predation on desert tortoises, *Ecology* 84:2432–2443, 2003.

[33] Fedriani JM, Fuller TK, Sauvajot RM: Does availability of anthropogenic food enhance densities of omnivorous mammals? An example with coyotes in southern California, *Ecography* 24:563–566, 2001.

[34] Wilcove DS, Rothstein D, Dubow J, Phillips A, Losos E: Quantifying threats to imperiled species in the United States: assessing the relative importance of habitat destruction, alien species, pollution, overexploitation, and disease, *BioScience* 48(8):607–615, 1998.

[35] Wilson MC, Chen X, Corlett RT, Didham RK, Ding P, Holt RD, et al: Habitat fragmentation and biodiversity conservation: key findings and future challenges, *Landscape Ecol* 31(2):219–227, 2016.

[36] Diamond JM: The island dilemma: lessons of modern biogeographic studies for the design of natural reserves, *Biol Conserv* 7:129–146, 1975.

[37] Haddad NM, Brudvig LA, Clobert J, Davies KF, Gonzalez A, Holt RD, et al: Habitat fragmentation and its lasting impact on Earth's ecosystems, *Sci Adv* 1:e1500052, 2015.

[38] Hernandez RR, Hoffacker MK, Murphy-Mariscal ML, Wu GC, Allen MF: Solar energy development impacts on land cover change and protected areas, *PNAS* 112(44):13579–13584, 2015.

[39] Forman RTT, Alexander LE: Roads and their major ecological effects, *Annu Rev Ecol Syst* 29:207–231, 1998.

[40] Von Seckendorff Hoff K, Marlow RW: Impacts of vehicle road traffic on desert tortoise populations with consideration of conservation of tortoise habitat in southern Nevada, *Chelonian Conserv Biol* 4:449–456, 2002.

[41] Zink TA, Allen MF, Heindl-Tenhunen B, Allen EB: The effect of a disturbance corridor on an ecological reserve, *Restor Ecol* 3:304–310, 1995.

[42] Gelbard JL, Belnap J: Roads as conduits for exotic plant invasions in a semiarid landscape, *Conserv Biol* 17:420–432, 2003.

[43] Goosem M, Marsh H: Fragmentation of a small-mammal community by a powerline corridor through a tropical rainforest, *Wildl Res* 24:613–629, 1997.

[44] Clarke DJ, Pearce KA, White JG: Powerline corridors: degraded ecosystems or wildlife havens? *Wildl Res* 33(8):615–626, 2007.

[45] Andersen MC, Watts JM, Freilich JE, Yool SR, Wakefield GI, McCauley JF, et al: Regression-tree modeling of desert tortoise habitat in the central Mojave desert, *Ecol Appl* 10:890–900, 2000.

[46] Hunting K: *A roadmap for PIER research on avian collisions with power lines in California*, Sacramento, 2002, California Energy Commission, Public Interest Energy Research (PIER) Program, Technical Report P500-02-071F.

[47] Janss GFE: Avian mortality from power lines: a morphologic approach of a species-specific mortality, *Biol Conserv* 95:353–359, 2000.

[48] Rideout BA, Stalis I, Papendick R, Pessier A, Puschner B, Finkelstein ME, et al: Patterns of mortality in free-ranging California condors (*Gymnogyps californianus*), *J Wildl Dis* 48(1):95–112, 2012.

[49] Lopez-Lopez P, Ferrer M, Madero A, Casado E, McGrady M: Solving man-induced large-scale conservation problems: the Spanish imperial eagle and power lines, *PLoS One* 6(3):e17196, 2011.

[50] Rollan A, Real J, Bosch R, Tintó A, Hernández-Matías A: Modelling therisk of collision with power lines in Bonelli's Eagle Hieraaetus fasciatus and its conservation implications, *Bird Conserv Int* 20:279–294, 2010.

[51] Martinez JA, Martinez JE, Manosa S, Zuberogoitia I, Calvo JF: How to manage human-induced mortality in the Eagle Owl Bubo bubo, *Bird Conserv Int* 16(3):265–278, 2006.

[52] APLIC Avian Power Line Interaction Committee: *Reducing avian collisions with power lines: the state of the art in 2012*, Washington, DC, 2012, Edison Electric Institute and APLIC.

[53] Cypher BL, Van Horn Job CL. Permeable fence and wall designs that facilitate passage by endangered San Joaquin kit foxes. Stanislaus, CA, 2009.

[54] Walston LJ, Rollins KE, LaGory KE, et al: A preliminary assessment of avian mortality at utility-scale solar energy facilities in the United States, *Renew Energy* 92:405–414, 2016.

[55] Kagan RA, Viner TC, Trail PW, Espinoza EO: *Avian mortality at solar energy facilities in southern California: a preliminary analysis*, Ashland, OR, 2014, National Fish and Wildlife Forensics Laboratory.

[56] Horváth G, Blaho M, Egri A, Kriska G, Seres I, Robertson B: Reducing the maladaptive attractiveness of solar panels to polarotactic insects, *Conserv Biol* 24(6):1644–1653, 2010.

[57] Greif S, Siemers BM: Innate Recognition of water bodies in echolocating bats, *Nature Comm* 1:107, 2010.

[58] Grodsky SM, pers. comm.

[59] Tanner K, Moore K, Pavlik B: Measuring impacts of solar development on desert plants, *Fremontia* 42, 2014.

[60] McCrary MD, McKernan RL, Schreiber RW, Wagner WD, Sciarrotta TC: Avian mortality at a solar energy power plant, *J Field Ornithology* 1:135–141, 1986.

[61] Kriska G, Horváth G, Andrikovics S: Why do mayflies lay their eggs en masse on dry asphalt roads? Water-imitating polarized light reflected from asphalt attracts Ephemeroptera, *J Exp Biol* 201:2273–2286, 1998.

[62] Walston LJJ, Rollins KE, Smith KP, LaGory KE, Sinclair K, Turchi C, et al: A review of avian monitoring and mitigation at existing utility scale solar energy facilities, *US Department of Energy*, 2015.

[63] Fthenakis VM, Kim HC: Land Use and Electricity Generation: A Life Cycle Analysis, *Renewable and Sustainable Energy Reviews* 13:1465–1474, 2009.

21

Energy Return on Energy Invested (EROI) and Energy Payback Time (EPBT) for PVs

Ajay Gupta
EROI ENERGY ADVISORS INC., BRAMPTON, ON, CANADA
akg78002@me.com

21.1 Introduction

Energy return on investment (EROI) or as it sometimes called, energy return on energy invested (ERoEI), is a tool for analyzing and comparing different types of fuels. EROI refers to the ratio of the usable energy returned during a systems lifetime, to all the invested energy needed to make this energy usable. It is related to net energy analysis (NEA), which calculates energy output minus energy inputs of a system, and also to life cycle analysis (LCA), which describes the total energy inputs for a system. As EROI is a ratio, it is a property without units. For example, EROI can be calculated using joules per joules or barrels of oil output per barrels of energy equivalent input. As such, EROI analysis has certain advantages over other tools when comparing different fuels and their impact on organisms or society.

Dale et al. [1] offer a short list of historic conceptualizations of EROI before it was popularized by systems ecologist Charles Hall. It was Hall who first coined the term "Energy Return on Investment" in the 1970s with a focus on migrating fish [2], and during the 1980s, along with others, expanded the concept to energy sources fuelling the US economy, such as oil [3]. The concept of comparing energy outputs with inputs, expressed as "net energy" has been in the anthropological [4], economic [5,6], and ecological [7] literature for sometime; however, EROI offers some additional insight into fuels. EROI analysis can be applied to the fuel of a variety of systems, including biological organisms. Hall argues that it can even be the "master driver" of evolution, as energy can be viewed as the master resource for evolution and EROI the means of obtaining Darwinian fitness [8]. Although it is a physical concept, EROI analysis can also have economic implications in terms of rates of growth [9], finance, and the economics of fuels [10]. The higher the EROI of a fuel technology, the more valuable it is in terms of producing an economically useful energy output [11]. A higher EROI allows more net energy to be available to the economy; and all economic activity relies, to some degree, on energy use.

A Comprehensive Guide to Solar Energy Systems. http://dx.doi.org/10.1016/B978-0-12-811479-7.00021-X

Net energy alone can be misleadingly large when evaluating an abundant energy resource of poor quality, such as ethanol from corn, or shale oil for which the energy cost of fuel production is slightly less than the energy gained from burning it. EROI analysis when used with NEA can help to assess the quality of fuels and this gives it the added ability of ranking different fuels within given system boundaries over time. By evaluating the inputs for the extraction of a resource and the value of the output in the same units, EROI can interpret the difference between the effects of technology (which would increase EROI) and those of depletion (which would decrease EROI) over time. As it pertains to fuels for a society, such as solar PV electricity, EROI is the ratio of the amount of energy delivered to society as a useful energy carrier by a chain of processes exploiting a primary energy source, to the total energy invested in exploring, extracting, processing, and delivering that energy [12,13]. A primary energy source is a natural form of energy (solar radiation, fossil fuels, waterfalls, etc.) that can be used to create energy carriers (electricity, gasoline, steam, etc.), which are used to fuel work in society [14]. EROI is represented in its simplest form by the following equation:

$$\mathrm{EROI} = E_{\mathrm{output}} / E_{\mathrm{input}} \tag{21.1}$$

Once the EROI reaches a ratio of unity, 1:1, or lower, in other words when the inputs of energy are equal to or greater than the outputs of energy, it is no longer considered useful to society, unless it is somehow being used to produce a higher quality fuel. Above 1:1, however, the impact is nonlinear and as illustrated in Fig. 21.1, shifts in high EROI values may have little impact on society, whereas those in low EROI values, especially below 5:1, may have far greater impacts.

Historically, it seems that EROI for oil and gas have been decreasing and that for coal has had its ups and downs. Assuming financial costs are related to energy costs, it has been estimated that the EROI of global fossil fuels from 1800 to 1920 was 30–40:1, but increased to 60:1 for oil and coal and to more than 100:1 for gas during the 1960s. All declined during the 1970s and then increased again into the 1990s [15]. Oil and gas subsequently declined irregularly to 10:1 and 20–40:1 respectively, while coal increased, also irregularly, in more recent years [16]. Current ranges for such fuels are illustrated in Fig. 21.2 and those for modern electric power generation, including PV, are in Fig. 21.3.

It is important to note that the values in Figs. 21.2 and 21.3 should not necessarily be taken at face value. This is because of differences in the methods, data, and boundaries of studies used to generate them [16]. Also, there is inherently a significant amount of assumptions necessary to generate global values of EROI for fuels. Ideally, EROI is most informative when given specific and well-defined geographic, temporal, and technological boundaries. It can also be very useful to follow the trend over time. In this regard, US oil/gas and oil in general have the only robust body of EROI information currently available. Unfortunately, data, research, and personnel for other EROI analyses have been scarce and the body of work has been small until 2011 [17]. Since then, however, there has been a growing interest in EROI and more analyses are available, in part due to a growing body of LCA publications in Europe. This is especially true for solar PV.

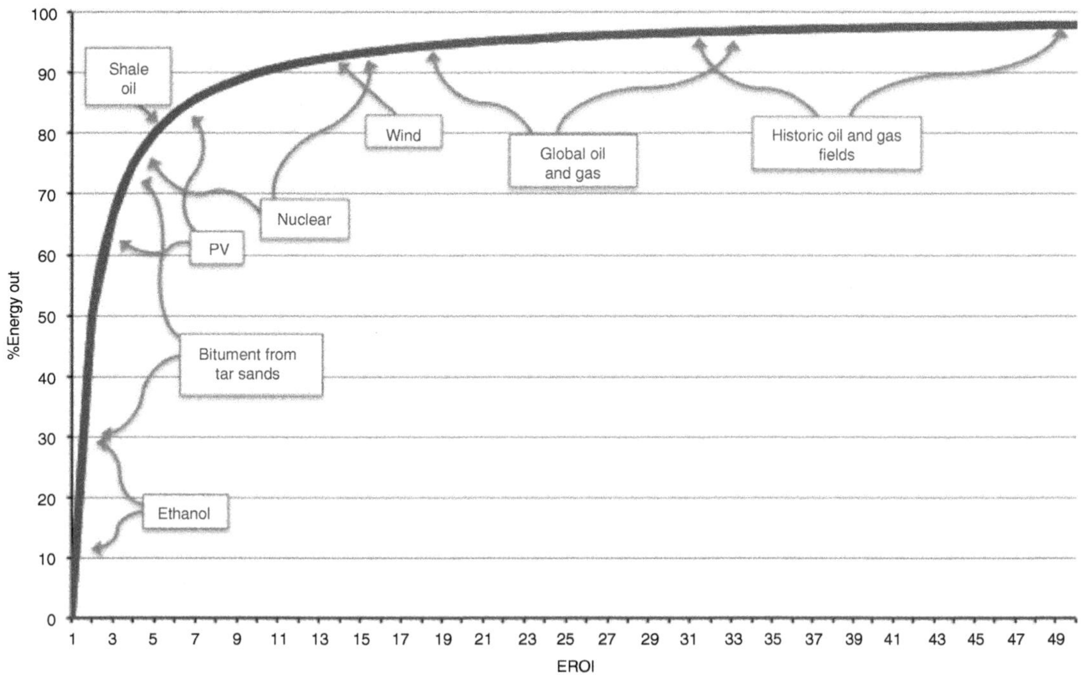

FIGURE 21.1 Energy output versus EROI for various energy sources. *Adapted from Hall CAS. Energy return on investment: a unifying principle for biology, economics, and sustainability. Springer Nature, Lecture Notes in Energy, vol. 36. Cham: Springer International Publishing AG; 2017 and concept from Euan Mearns. As the EROI approaches 1:1 the ratio of energy gained to energy consumed in the process decreases exponentially [13]. Whereas high EROI fuels deliver a greater proportion of their energy to society, low EROI fuels deliver a much smaller proportion of their energy to society.*

21.2 Methods of EROI Analysis

21.2.1 Introduction to Methods of EROI Analysis

To get the best information from a solar PV EROI analysis, one should strive for a well-defined system within a well-defined set of boundaries and good data. Unfortunately, much of the questions surrounding PV technologies and deployment are from an industry-wide or nationwide point of view. These questions have always been more difficult to answer with EROI analysis, as they inherently require more data and usually more assumptions to obtain a complete picture. EROI analysis is not a precision science, however, with attention to assumptions and uncertainties involved, reasonably accurate and repeatable results can be obtained to answer important and specific questions.

The first formal protocol for developing EROI analysis was published by Murphy and Hall [14], which provides a general overview on how to proceed toward a "standard" EROI calculation. The standard EROI calculation has its output boundary set at the wellhead, farm gate, buss bar, or other source. For example, when calculating the standard EROI for crude oil, the boundary is at the wellhead, or where the oil comes out of the ground. The subsequent activities of storing, transporting, and refining the crude oil are not included

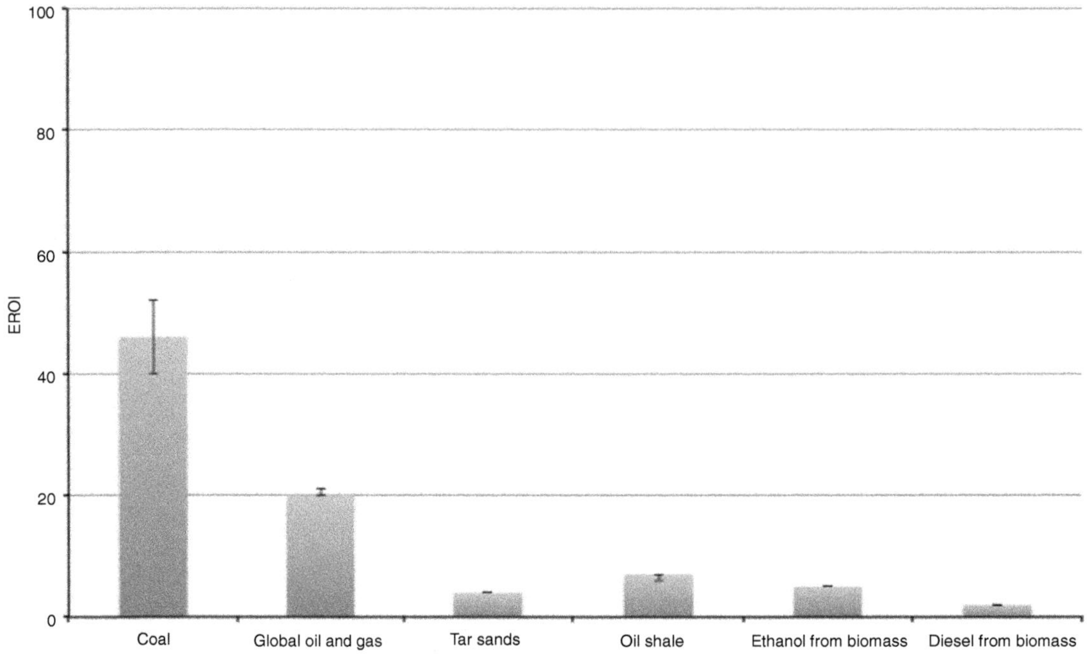

FIGURE 21.2 Mean EROI (and standard error) for thermal fuels. *Adapted from Hall CAS. Energy return on investment: a unifying principle for biology, economics, and sustainability. Springer Nature, Lecture Notes in Energy, vol. 36. Cham: Springer International Publishing AG; 2017; Lambert J, Hall CAS, Balogh S, Gupta A, Arnold M. Energy, EROI, and quality of life. Energy Policy 2014;64:153–67. For explanation and detailed references please see Hall CAS, Lambert JG, Balogh SB. EROI of different fuels and the implications for society. Energy Policy 2014;64:141–52.*

in the standard analysis and constitute what can be called "extended boundary analysis." This allows for a comparison among different fuels and resources as well as a procedure for undertaking other approaches an analyst might want to control. Among the protocols outlined by Murphy and Hall are those of correcting for energy quality, for example, the difference between thermal energy versus electricity for coal. A very basic procedure for an EROI analysis of any fuel or system as adapted from Hall [18] and Murphy and Hall [14] is as follows:

1. *State objectives.*
2. *Create a flow diagram and identify system boundaries*—all direct, indirect, and embodied energy inputs and outputs should be included, as well as the boundaries used for analysis or sensitivity analysis.
3. *Quantify all appropriate inputs and outputs within system boundaries.*
4. *Identify and convert financial flows if necessary*—in the case where direct flow measurements are not available and financial information must be used, the data needs to be converted to energy units using an energy intensity value.
5. *Make the calculation*—at minimum a standard EROI value should be calculated.
6. *Choose method of energy quality adjustments as part of sensitivity analysis.*

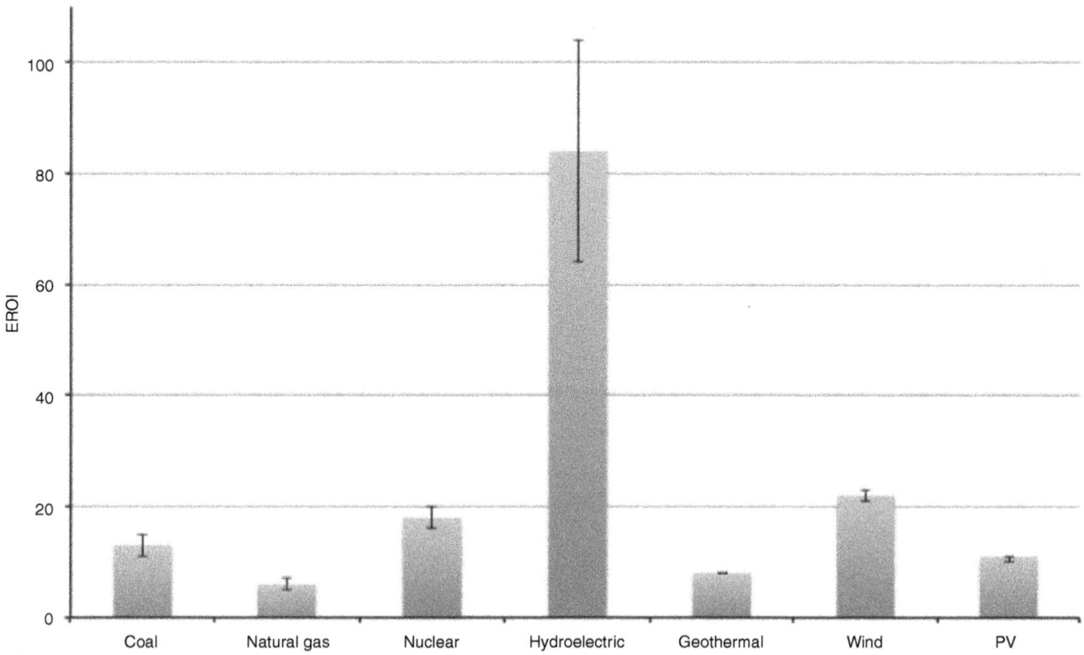

FIGURE 21.3 Mean EROI (and standard error) for electric generation systems. *Adapted from Hall CAS. Energy return on investment: a unifying principle for biology, economics, and sustainability. Springer Nature, Lecture Notes in Energy, vol. 36. Cham: Springer International Publishing AG; 2017; Lambert J, Hall CAS, Balogh S, Gupta A, Arnold M. Energy, EROI, and quality of life. Energy Policy 2014;64:153–67. For explanation and detailed references please see Hall CAS, Lambert JG, Balogh SB. EROI of different fuels and the implications for society. Energy Policy 2014;64:141–52.*

Despite the protocols offered by Murphy et al. [19], given the relatively novel nature of EROI, and the state of energy data in the world, many issues arose from differing calculations in EROI analyses over time. Although output data is generally published in technical or industrial reports, for PV the nameplate or total capacity of a system is not always realized and intermittency is an unpredictable variable. Also, the system lifetimes can vary from region to region as well. Deriving energy inputs is even more difficult and less certain. As most EROI analyses for fuels involve large industries and resources, direct measurements are usually not available as an option. Also, it can be difficult to get the desired energy cost data for calculating EROI because they can be privately owned by corporations that tend not to make data of any kind available to the public and few nations maintain industry-level energy data over time. There are also two different types of inputs—direct and indirect energy costs. Direct energy is used on site, such as electricity used to dope or cut wafers of solar cells. Indirect energy is used off site to derive materials or services used later on site or other places in the development process. Indirect energy costs can add a significant amount of ambiguity to methods not just because of data availability, but also because of the problem of defining boundaries of analysis. For example, calculating the indirect cost of the manufacturing of an aluminum frame for a PV array can be straightforward, but what about including the energy costs of business services, labor, and taxes?

While it may seem straightforward to decide what the boundaries of a specific calculation should be, in terms of the protocol for all EROI analysis, it is quite difficult. At some point the question becomes less scientific and more an issue of philosophy. The literature on EROI analysis has yet to reconcile the issue of differing perspectives on boundaries of analysis. According to Hall [8], for now, the general method in dealing with this uncertainty in methods is through sensitivity analysis—report the results for systems using different assumptions about the data or philosophy and leave the final choice with the reader.

21.2.2 Energy Payback Times

As it is a renewable resource, EROI for PV is not calculated using the same method as for finite resources. In general, the energy cost for renewables is a very large capital cost per unit output, especially given backup systems such as batteries. As a result the input for the EROI equation is mostly upfront, while returns are realized throughout the lifetime of the system. Historically, there are very few attempts at studies that perform "bottom-up" analysis. Alternatively, we can calculate EROI by dividing the lifetime ($T_{life(yr)}$) of a system by its "Energy Pay Back Time," ($EPBT_{(yr)}$).

$$EROI = T_{life(yr)} / EPBT_{(yr)} \tag{21.2}$$

EPBT is the time it takes for the system to generate the amount of energy equivalent to the primary energy or kWh equivalent that was used to produce the system itself. PV EPBT can vary depending on the technology, location of production and installation, material requirements, and operating efficiency. Factors that can lower the EPBT include lower ore grades of rare metals used in production caused by depletion or competing industries (Chapter 25), lower than projected lifetimes and efficiencies, problems with energy storage, and intermittence. The following is an example of a common EPBT calculation [20]:

$$EPBT = (E_{mat} + E_{manu} + E_{trans} + E_{inst} + E_{eol}) / (E_{agen} - E_{aoper}) \tag{21.3}$$

where E_{mat} is the primary energy demand to produce materials for the system; E_{manu} is the primary energy demand to manufacture the system; E_{trans} is the primary energy demand to transport materials used during the lifecycle; E_{inst} is the primary energy demand to install the system; E_{eol} is the primary energy demand for end-of-life management; E_{agen} is the annual electricity generation in primary energy terms; and E_{aoper} is the annual energy demand for operation and maintenance in primary energy terms. The annual electricity generation (E_{agen}) is defined as primary energy based on the efficiency of electricity transformation on the demand side.

Eq. (21.3) employs a method and boundaries derived from the LCA of a given system. Raugei et al. [21,22] and Hall [8] summarized the differences and similarities between this method and others, discussed in the following section, stressing the importance of boundaries and physical definitions. Variables that affect the energy cost of PV systems include cell material, encapsulation matter, cell production process, capital equipment, frames, and body of system (BOS). The output during the lifecycle must be estimated. Variables for output include solar irradiation, conversion efficiency including lifetimes, performance ratios, and electricity generation efficiency.

21.2.3 Overlapping Energy Input Accounting Methods

In general, there are three different scopes used for collecting energy cost data for an energy system in the literature [8]. The first is using national energy accounts for direct energy used. Some countries maintain records of energy used by various industries including those of their energy industry. This approach takes serious investment in time, a good library and Internet services, and ideally the assistance of professional experts in the varying fields being examined. The second is using national accounts for capital expenditures and other indirect uses. This includes energy used offsite to produce and maintain the capital equipment used to make PV modules. The indirect energy costs of some materials and processes are usually known, for example, forming concrete or aluminum. Other forms of capital must be estimated by converting known financial values into quantities of energy. The indirect energy cost of dollars spent on chemicals, steel products, and other relevant capital can be calculated as the dollar cost times the *energy intensity* of the formation of that capital. The energy intensity is measured by the quantity of energy used to produce a dollar worth of output in the industrial sector of the economy (joule $(\$)^{-1}$) [19]. Using dollar values is not an ideal method for capturing energy costs of capital inputs because it carries the errors in generating financial variables; however, the advantage in using energy intensities to estimate energy costs is that financial data is more readily available than energy data. These two procedures have a critical issue of determining a set of boundaries for analyses. Note that there is no emphasis on differentiating between expressing energy inputs as primary energy or energy carriers, although a quality correction is required. This correction is typically applied by multiplying higher quality electricity values by 3 to equate thermal values [8,19].

The third procedure involves using the values for energy inputs provided in LCAs and deriving the calculation for measuring cumulative energy demand (CED). The LCA is a standardized method for analyzing various aspects involved with the development and lifetime of a product [23]. The methodology of LCA calculates CED, which describes the total primary energy extracted from the environment to deliver, support and retire a given system. The CED is informed by data provided in life cycle inventory (LCI) Databases, manufacturer's technical specifications, and indirect estimates. LCI data includes direct measurements, expert assessments, company data surveys, and theoretical calculations. It is common to borrow data from other studies to cover parts of LCA supply chains. As it is standardized, and therefore, has clear definitions and boundaries of analysis, in theory the CED method provides the most detailed information on the energy costs of PV systems available to EPBT and EROI analyses today, although it is not always considered comprehensive.

21.2.3.1 Confusion in PV EROI Results Caused by Inconsistencies in Objectives and Energy Input Accounting

Given that the concept of EROI analysis is relatively young; covers a broad range of disciplines; and attempts to answer very large questions of social importance, it is no surprise that there are some disagreements on EROI terminology. As EROI analysis has only recently become more popular, especially in Europe, communication and standardizations are more important than ever. Issues of compatibility and broad conclusions are often argued among an interdisciplinary body of analysts interested in the subject of particular fuels. Specifically, issues tend to relate to poor or evolving definitions, and boundaries of analyses and how that relates to their compatibility.

More explicit terms are necessary to move the field of EROI analysis into more complex analysis. King [24] argues this point and introduces some explicit terminology. The first distinction King makes is between "EROI" and "power return on investment." Power is the ability to do work over time and King argues that if the units in the equation are that of (energy per year output)/(energy per year input) then that constitutes "power return on investment." According to Hall [8], as EROI is a ratio and the units cancel out, the distinction is most important when discussing rapidly shrinking or growing technologies. The second distinction made by King involves cases where energy output includes that invested in getting the fuel or not. For example, if natural gas is used to pressurize oil fields for extraction, is that gas subtracted from the output? Although this distinction may not seem relevant to PV EROI at the moment, it is possible that it can become an issue in the future and it might be when comparing PV to other fuels for electricity.

Specific to PV EROI, confusion from results has been mainly due to a lack of clear objectives, gaps in data availability, and inconsistencies in energy input accounting or issues of analysis boundaries. Original studies into PV energy costs related an EROI in the range of 3–10:1 with an average of about 6.5:1 [17,25,26]. Since then, however, a growing number of calculations are appearing well below this range. Prieto and Hall [27] estimated an EROI of 2.45:1 for Spanish PV; Palmer [28] estimated a similar EROI for rooftop systems with battery back up in Australia; Weissbach et al. [29] for PV in Germany; and recently Ferroni and Hopkirk [30] in Switzerland and Germany. After some clarifications and arguments about methods, objectives, and definitions [21,31,32], Leccisi et al. [33] and Raugei and Leccisi [34] estimated an EROI of 9:1 for the same system Ferroni and Hopkirk claimed had an EROI below unity in their study. How can two EROI analyses of the same systems reach such drastically differing conclusions?

Sorting through all of the data, calculations and sources of information is outside the scope of this chapter, but a summary is available in Hall [8]. Instead, what follows are lessons learned from exchanges in subsequent PV EROI, publications. First, correcting for energy quality was inconsistently defined. In a simplified example, for the case of measuring EROI for crude oil, the output is measured at the wellhead in barrels whereas for PV EROI electricity is measured as output in kilowatt-hours. Important to note is that crude oil in this case is a primary energy source and electricity from PV is an energy carrier.

To properly compare the two EROIs the crude oil must first also be converted into electricity, or the electricity must be expressed as primary energy equivalent. Another lesson from the exchanges is that it is important to define energy quality of energy input data in EPBT calculations. Energy invested is the readily available energy diverted from other possible societal uses [19] and as such it is provided as energy carriers, however, they can be measured in terms of thermal energy and electricity. Therefore, a quality correction must be made to provide comparability. As previously discussed, in some cases the electricity value is multiplied by 3 to provide this correction. The CED, however, describes the total primary energy that must be harvested from the varying environmental sources to produce a given amount of usable energy carrier throughout the lifecycle [35], and thus arguably already provides the energy mix in comparable units. Discounting the quality of energy when accounting can lead to incomparable analyses. Therefore, explicitly stating objectives of evaluating economical and effective use of available energy carriers versus efficient use of primary energy resources and using consistent methods in measuring these differing flows is paramount to comparable EPBT and EROI studies for PV and other fuels.

21.2.4 Pathways to PV Net Energy Analysis Using CED

As can be inferred from the methodological issues surrounding the EROI of PV systems, the more precise is the question being asked, the more precise will be the answer the EROI analysis can provide. This section is adapted from Raugei et al. [22] and summarizes different pathways for PV NEA and EROI analysis using two families of energy metrics. The following definitions apply to the equations in this section and are reported per square meter of the PV system over the system lifetime:

Irr	total solar irradiation over system lifetime (MJ m^{-2})
η_{PV}	PV module energy harvesting efficiency (MJ MJ^{-1})
PE	primary energy directly harvested over system lifetime (MJ m^{-2})
	(Irr)($-\eta_{PV}$)
Inv	energy investment to build, operate and dismantle PV system, in terms of its primary energy demand (M m^{-2})
PR	performance ratio (MJ MJ^{-1})
Out	can be accounted for in direct energy units such as electricity (Out_{el}), or in terms of its equivalent primary energy ($Out_{PE\text{-}eq}$)

21.2.4.1 $EROI_{el}$: Energy Output Expressed in Terms of Direct Energy

When expressing energy output in terms of direct energy, which in the case of PV is always electricity, the EROI of PV electricity may be calculated as:

$$EROI_{el} = Out_{el} / Inv \tag{21.4}$$

This relation expresses the energy delivered to society, in units of electricity, per one unit of the sum of the energy carriers diverted from other societal uses (excluding energy delivered to society), in terms of their total primary energy demand. There is no distinction between renewable and nonrenewable energy inputs. The purpose of this metric is to evaluate and compare its economical and effective use of available energy carriers from a short-term perspective.

21.2.4.2 $EROI_{PE\text{-}eq}$: Energy Output Expressed in Terms of Equivalent Primary Energy

When expressing energy output in terms of equivalent primary energy, the EROI of PV may be calculated as:

$$\mathrm{EROI}_{\mathrm{PE-eq}} = \mathrm{Out}_{\mathrm{PE-eq}} / \mathrm{Inv} = (\mathrm{Out}_{\mathrm{el}} / \eta_G) / \mathrm{Inv} = \mathrm{EROI}_{\mathrm{el}} / \eta_G \tag{21.5}$$

where η_G is the lifecycle energy efficiency of the electricity grid (G) of the country or region where the PV system being analyzed is deployed. It is calculated as the ratio of the yearly electricity output of the entire grid to the total primary energy harvested from the environment for the operation of the grid in the same year: $\eta_G = 1/\mathrm{CED}_G$.

This relation expresses the energy delivered to society, in units of equivalent primary energy, per one unit of the sum of the energy carriers diverted from other societal uses (excluding energy delivered to society), in terms of their total primary energy demand. There is no distinction between renewable and nonrenewable energy inputs. The purpose of this metric is to evaluate and compare its economical and effective use of available energy carriers from a short-term perspective. It is important to note that as the grid performance is intrinsic to this metric, observed change in $\mathrm{EROI}_{\mathrm{PE\text{-}eq}}$ may depend not only on the technological system being evaluated, but also on a change in the average life-cycle efficiency of the grid (η_G).

21.2.4.3 The Cumulative Energy Demand (CED) Metric

CED per unit output may be operationally defined as:

$$\mathrm{CED} = (\mathrm{PE} + \mathrm{Inv}) / \mathrm{Out} \tag{21.6}$$

where both PE and Inv are expressed in terms of primary energy, while Out is expressed in direct energy units of the delivered energy carrier, which for PV is electricity.

This relation expresses the total primary energy harvested from nature per unit of energy delivered to society as electricity. It is recommended to distinguish between renewable and nonrenewable energy. The purpose of this metric is to evaluate the efficient use of primary energy resources from a long-term perspective.

21.2.4.4 The Nonrenewable Cumulative Energy Demand (nr-CED) Metric

LCAs keep track of all renewable and nonrenewable energy flows separately, which leads to the possibility of calculating Nonrenewable Cumulative Energy Demand (nr-CED) per unit output. It can be operationally defined as:

$$nr - CED = (PE_{nr} + Inv_{nr}) / Out \tag{21.7}$$

where PE_{nr} is the nonrenewable share of the primary energy directly harvested over system lifetime; and Inv_{nr} is the nonrenewable share of the energy investment in terms of its nonrenewable primary energy demand.

This relation expresses the nonrenewable primary energy harvested from nature per unit of energy delivered to society as electricity. It is recommended to distinguish between renewable and nonrenewable energy. The purpose of this metric is to evaluate the sustainability and efficient use of nonrenewable primary energy resources from a long-term perspective.

As previously described, using CED in accounting for energy inputs to a PV system intrinsically handles the issue of energy quality corrections. Also, CED can analyze the impact of an energy carrier on the stocks of primary energy sources and thus evaluate efficiency in that regard. As it can also differentiate between renewable and nonrenewable primary energy inputs, this method also provides information on the sustainability of a technology within a given primary energy source mix over the long-term. Unfortunately, it becomes obvious that there is no one method for analysis that can answer every question of PV system efficiency in society, and there is no *one* EROI for all PV systems. Every system must be analyzed within its own set of boundaries and carefully stated objectives. Then, perhaps the analyses can be harmonized given certain assumptions and comparisons can be made.

21.3 Results of EROI Analysis of PV Systems, Harmonization and Trends Over Time

21.3.1 Results of a UK Case Study Comparing PV and Nonrenewable EROIs

In 2016 Raugei and Leccisi published a study that performed an NEA through EROI analysis of the United Kingdom and its full range of electricity generation technologies, including PV, using the methods described previously [34]. To their knowledge, this was the first such national-level analysis at the time and is relevant to the body of information concerning the EROI of PV systems. Although the vast majority of PV in the United Kingdom is currently in the form of mono-Si and poly-Si, Raugei and Leccisi also included CdTe in their study. The conclusions offer a comparison between those technologies and wind (on-shore and off-shore), hydro, biomass, nuclear, gas (including combined cycle), oil, and coal in terms of their impacts on UK economic growth. The comparisons calculated as $EROI_{el}$ are illustrated in Fig. 21.4 along with the value for the UK electric grid as a whole.

The range of resulting $EROI_{el}$ of the different technologies comprising UK electricity is fairly large due to the very low value of biomass and the very high value and potential of hydropower. Gas-fired, nuclear, and wind energy can all be classified as well performing,

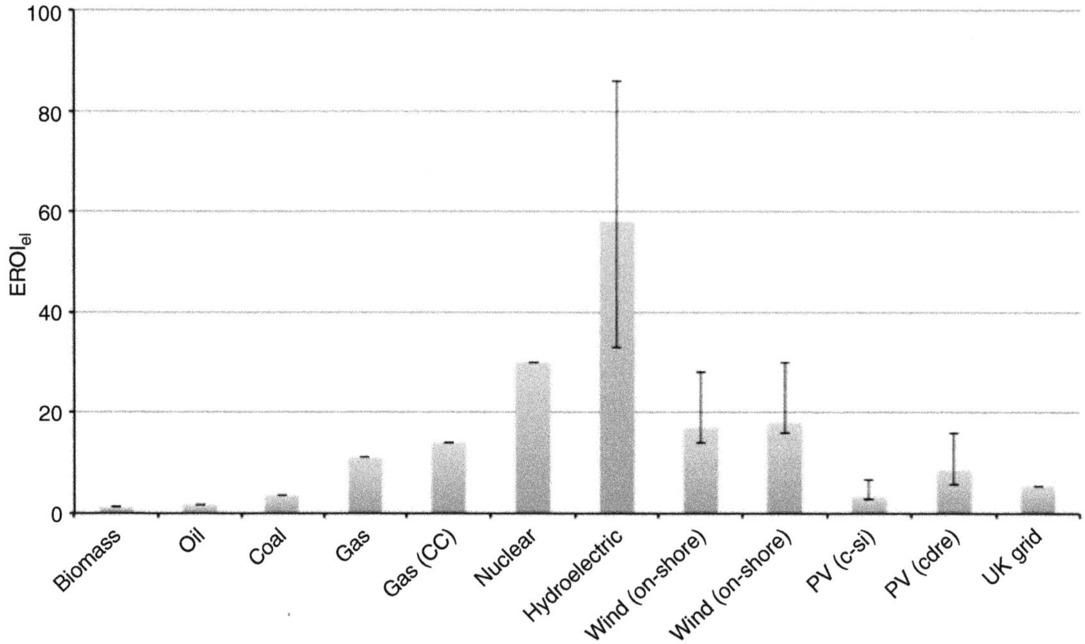

FIGURE 21.4 EROI$_{el}$ of all electricity generation technologies in the United Kingdom. *Adapted from Raugei M, Leccisi E. A comprehensive assessment of the energy performance of the full range of electricity generation technologies deployed in the United Kingdom. Energy Policy 2016;90:46–59. Values for hydro, wind, and PV technologies include variability on capacity factors and uncertainty in system lifetimes.*

followed by PV from CdTe which is above the grid mix average and finally PV from c-Si and coal. Figure 21.5 illustrates comparisons of these same technologies in terms of their EROI$_{PE\text{-}eq}$ in relation to the percent energy output contributed to the UK economy.

The ranking of the fuels are the same and hydropower is actually off the chart. It can be inferred from Fig. 21.5 that biomass and oil-fired electricity contribute very little net energy to the UK economy, whereas coal and PV from c-Si contribute significantly more, although still below the average for the UK grid as a whole. Most important is the illustration of how similar PV from CdTe, gas, wind, nuclear, and hydropower are in terms of their respective contributions to the UK economy in this regard.

This kind of analysis has much to offer policy-makers in assessing options moving toward sustainability, economic growth, and low-carbon futures. This particular study illustrates that fuels such as coal and oil, traditionally thought to be high EROI fuels, are not the powerhouses they seem when producing electricity, at least not in the UK's electricity generating mix. This raises certain issues such as the feasibility of "clean coal" through carbon capture sequestration (CCS) in the United Kingdom. Brand-Correa et al. [36] also conducted a nationwide EROI analysis for the United Kingdom using an Input-Output method. Interestingly, even though they used a completely different approach, which included trade data and relied heavily on energy intensities to measure energy flows, some

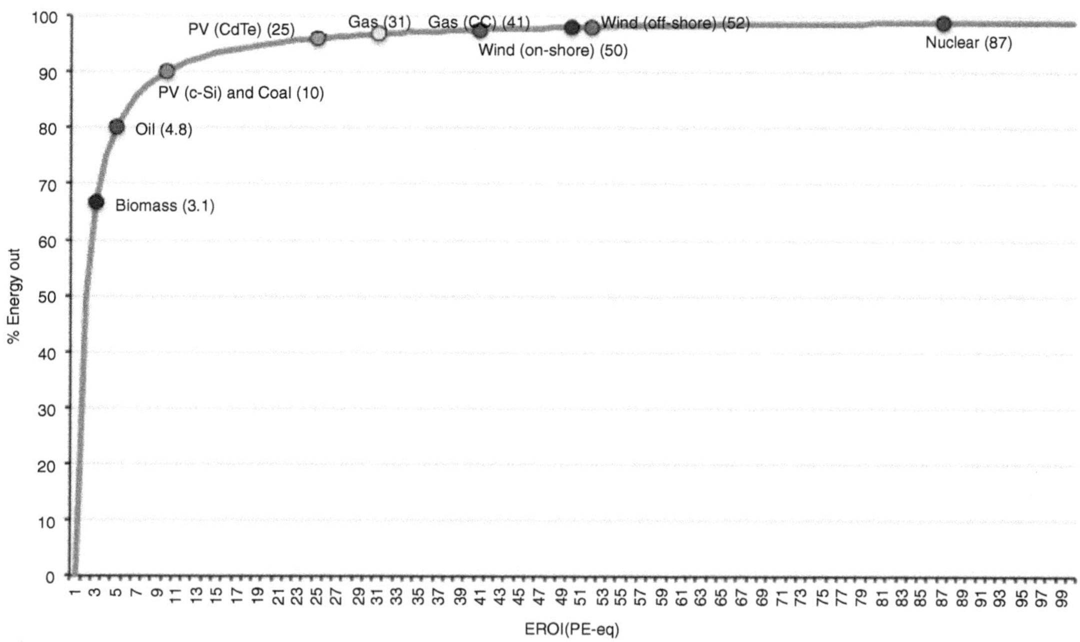

FIGURE 21.5 $EROI_{PE\text{-}eq}$ of all electricity generation technologies in the UK in relation to their percent energy output. *Adapted from Raugei M, Leccisi E. A comprehensive assessment of the energy performance of the full range of electricity generation technologies deployed in the United Kingdom. Energy Policy 2016;90:46–59. The $EROI_{PE\text{-}eq}$ value for hydropower is calculated at 170 and is off the chart.*

results were similar to those of Raugei and Leccici [34]. These are two examples of how EROI analysis can be used to generate national-level information about a specific set of fuels and how they compare to one another concerning PV. EROI analysis can also be very useful in determining certain aspects of the PV industry itself.

21.3.2 Results From Harmonizing EROI and EPBT Analyses and Trends in the Industry

To illustrate and evaluate large trends in the PV industry, such as overall trends in CEDs and EPBTs and eventually EROIs, the individual analyses must first be harmonized so that boundaries and methods are consistent and results are comparable. This moves the scope away from evaluating definitive objectives using precision data for a particular PV system under specific boundaries, toward analyzing a broader trend in technology as a whole over time. The results are more robust with the more studies there are available to aggregate into a harmonized body. Variables that are typically controlled include system life expectancy, irradiation, performance ratios, and degradation. There are some recommended sets of values for these variables available in the literature [22,35]. Other areas of system harmonization include the BOS including aluminum frames, use of energy intensities if applicable, and inconsistent factors considered extensions in analysis, such as labor or debt.

There is a growing body of work on the CEDs and EPBTs of various system types across varying geological boundaries. In a recent metaassessment Koppelaar [37] harmonized several studies for annual solar radiation, supply chain technologies, lifecycle boundaries, and technological configurations of mono-Si and poly-Si systems. Koppelaar identified quality variation and study shortcomings and examined the ability to reproduce existing results. The following parameters were used for harmonization: EROI was calculated for studies that only derived EPBT; energy input values missing from the solar module production chain were complimented with mean values; studies lacking energy input values for BOS, installation, transportation, operation, maintenance, and decommissioning were complimented with mean values; energy input values for batteries, auxiliary services, power lines operation, restructuring, labor, and capital investment cost based on energy intensities were removed from studies; electricity output was recalculated using a 1700 $kW\,h\,m^{-2}\,yr^{-1}$ (where yr refers to annum) radiation value; performance ratio of 0.8% with a degradation rate of 0.7% yr^{-1} were applied; solar module packing factors to adjust for noncell module area were assumed at 0.94 for poly-Si and 0.8 for mono-Si modules; and a plant lifetime of 25 years. The results of the harmonization on EROIs are illustrated in Fig. 21.6 in terms of % Energy Output related to EROI.

In general, the harmonization resulted in slightly lower EPBTs and higher EROI values than the original studies. For mono-Si a mean EPBT of 3.9 years and a mean EROI of 8.6:1 was determined on recalculation. For poly-Si a mean EPBT of 2.9 years and a mean EROI of 9.3 was found on recalculation. The largest recalculations were primarily due to lack of

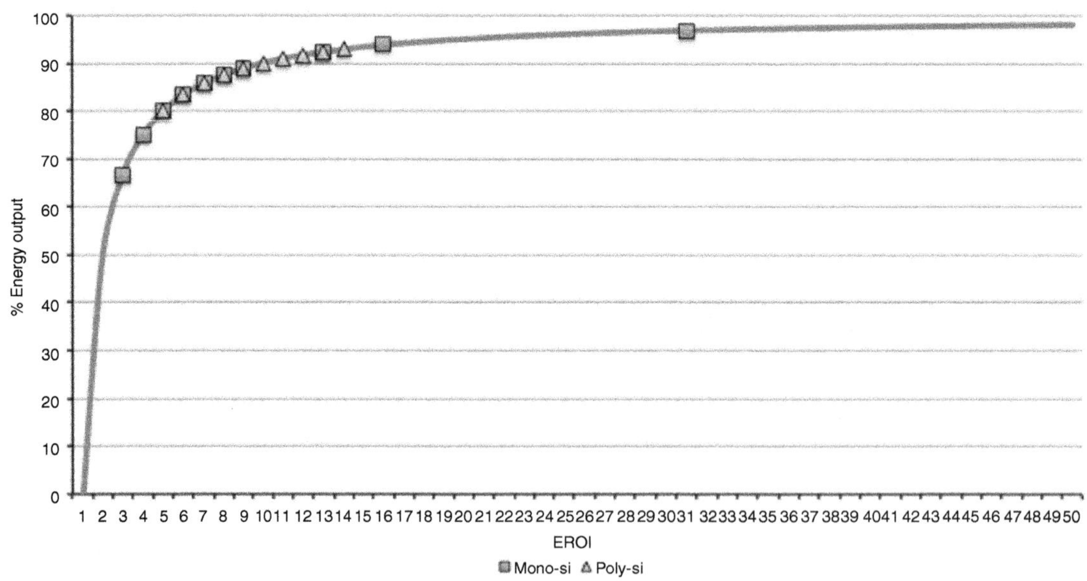

FIGURE 21.6 % Energy Output versus EROI for harmonized studies of mono-Si and poly-Si. *Adapted from Koppelaar RHEM. Solar-PV energy payback and net energy: meta-assessment of study quality, reproducibility, and results harmonization. Renew Sustain Energy Rev 2017;72:1241–55.*

quality correction and harmonizing for large battery systems. Koppelaar also examined the age of systems being analyzed compared with publication dates and found that some authors were using systems that were 6 years, in an extreme case 18 years old. This had a serious impact on values for both EPBT and EROI of the PV modules and systems studied.

Five popular types of PV modules often analyzed are monocrystalline (mono-Si), multicrystalline, or polycrystalline (multi-Si or poly-Si), amorphous silicon (a-Si), CdTe thin-film (CdTe), and CIS thin-film (CIS). Kumar et al. [20] studied these five types of modules in varying environments and systems. As can be expected, they found that EPBT varied greatly from case to case and also that Si volume and methods of production were large factors in determining CED, which along with efficiency were the largest factors for EPBT. Overall they found mono-Si to have the highest lifecycle energy demand and a large range in EPBT due to estimations of silicon purification and crystallization processing. Poly-Si had a much lower EPBT and the thin-film systems (a-Si, CdTe, and CIS) had the lowest EPBTs. Among the thin-films, CIS consumed the most primary energy and the a-Si had the longest EPBT due to lower conversion efficiency. CdTe held the lowest EPBT across systems analyzed. Bhandari et al. [38] found similar results when they examined several LCAs for EPBT and CED. After harmonizing the variables for output, they calculated mean EPBT varying from 1.0 to 4.1 years. CdTe modules ranked lowest, followed by copper indium gallium diselenide, a-Si, poly-Si, and mono-Si with the highest EPBT. They also found that across different types of PV, variation in CED was greater than that in efficiency and performance ratio, concluding that CED, not efficiency, has greater influence on EPBT.

If a low CED is the important factor for a low EPBT, we can assume that new manufacturing technologies and application methods, such as advanced production processes and reducing raw material consumption, especially silicon, would show a reduction in EPBTs over time. Wong et al. [39] studied the difference in mono-Si versus poly-Si PV processing and found that the former had a larger CED due to an additional Czochralski process, which has a significant energy requirement. Although this results in greater conversion efficiency for mono-Si, it is insufficient to lower the EPBT below that of poly-Si. Wong et al. describe another approach to manufacturing a combination of mono-Si and poly-Si into a hybrid c-Si, which eliminates the need for the extra Czochralski process and produces a PV module with greater efficiencies than poly-Si but lower energy costs than mono-Si. They also note that wafer thickness has been greatly reduced without losing efficiency by the use of reflective back-coatings, which increase chances of photon capture.

In the few studies that have applied harmonization to NEA, EPBT, and EROI analyses, a reduction in CED across the industry was found. Gorig and Breyer [40] calculated and compared the CED of different modules and systems over time using LCAs. They employed financial learning curve concepts to determine the energy demand for major PV systems. They weighted both module and system energy demand according to their share in the PV market at the time and found that the CED for all modules and systems, for which there was adequate marketing data, had decreased over time. The modules and systems with higher amounts of Si decreased at faster rates. They showed that energy consumption in PV manufacturing followed the log-linear learning curve and BOS, such as aluminum frames, also reduced in energy consumption. C-Si modules had the highest energy learning rates, which were mainly driven by improved silicon production. Louwen et al. [41] re-assessed the body of PV LCAs over a 40-year period and corroborate Gorig and Breyer's findings of decreasing CED for PV systems and modules. They found that EPBT dropped from around 5 years in 1992 to currently under 1 year for poly-Si and just over 1 year for mono-Si concurrent with rapid growth in capacity. Overall, they found that CED decreased 11.9%–12.6% with each doubling of capacity.

21.3.3 Future Possibilities

The future of PV technology development is very difficult to predict; however, it may be likely to experience a high potential for decrease in energy costs and an increase in efficiencies for the industry. It is expected that EPBTs will therefore decrease and EROIs will likely increase. Of the very little information available to predict these things, it seems to be mostly positive for PV. Overall, new manufacturing technologies and application methods, such as advanced production processes, reducing Si and other raw materials consumption, and increasing material recycling rates are all avenues for improving PV performance and EROI.

As calculated by Gorig and Breyer [40], the learning rate for PV modules is 17% and for PV systems it is 14%. They expect strong development until 2020 and forecast an EROI of 20–60:1 approaching the year 2030. The study by Louwen et al. [41] reinforces the idea that CED will decrease for PV modules in the future. They expect the production of mono-Si to be influenced by a stronger learning rate to poly-Si modules due to the fact that monocrystalline Si is more energy intensive and thus benefits most from energy and material use reduction. Overall, lifecycle energy costs have realized real improvements given developments in terms of material usage. Energy efficiency continues to improve for a range of PV technologies available for economic electricity generation. The most noticeable improvements have occurred for CdTe technology in terms of overall systems improvements [33]. This is without the serious recycling efforts expected in the future of PV manufacturing.

Insight into recycling methods and what they mean for the EPBT and EROI of PV are still being developed. Goe and Gaustad [42] predict that recycling rates are likely to have the most impact for low efficiency modules, especially those with aluminum frames, due to the less complex nature of their composition, but rates might be low due to small returns for customers. More complex high efficiency modules might realize low recycling rates due to the potential high costs of reintegrating them into the manufacturing process. Overall they caution that realized recycling rates are likely to be low without regulation, or mandatory recycling. They estimate, however, that exhaustive recovery of PV materials could have the potential to reduce EPBTs of mounted modules by more than half for mature Si-based and thin-film technologies.

As we can see, the industry is moving at a very fast pace toward increased efficiencies, lower CED, thus lower EPBT, and therefore, we assume, higher EROIs. We can also start to see that there are many assumptions in making industry-wide statements concerning PV technology deployment. There are many variables that are specific to geographies, climates, supply chains, markets, and even politics. In the past, varying methodologies were employed in EROI analyses, mainly due to lack of data and experience. Today the methodology behind EROI analysis is even more important as the systems and technologies themselves are quickly changing together with advancement in deployment and complexity. Although LCAs for PV systems have increased in numbers lately, the data required for comprehensive EROI calculations is still behind in the race for analyses to keep up with technological improvements. Larger, up to date, and transparent datasets would be a great improvement.

References

[1] Dale M, Raugei M, Fthenakis V, Barnhart C: Energy return on investment (EROI) of solar PV: an attempt at reconciliation, *Proc IEEE* 3(7):995–999, 2015.

[2] Hall CAS: Migration and metabolism in a temperate stream ecosystem, *Ecology* 53:585–604, 1972.

[3] Hall CAS, Cleveland CJ, Kaufmann R: *Energy and resource quality: the ecology of the economic process*, John Wiley & Sons, 1986.

[4] Lee R: Kung bushman subsistence: an input-output analysis, Environ cult behav. In Vayada A, editor: *Ecol Stud in Cultural Anthropol*, NewYork, 1969, Natural History Press, pp 47–79.

[5] Georgescu-Roegen N: Energy and economic myths, *South Econ J* 41:347–381, 1975.

[6] Berndt ER. From technocracy to net energy analysis: engineers, economists and recurring energy theories of value. Studies in Energy and the American Economy Discussion Paper No.11, MIT-EL 81-065WP.

[7] Odum HT: *Environment, power, and society*, New York, 1973, Wiley InterScience.

[8] Hall CAS: Energy return on investment: a unifying principle for biology, economics, and sustainability, Springer Nature, Lecture Notes in Energy. Cham, 2017, Springer International Publishing AG.

[9] Hall CAS, Powers R, Schoenberg W: Peak oil, EROI, investments and the economy in an uncertain future. In David P, editor: *Renewable energy systems: environmental and energetic issues*, London, 2008, Elsevier, pp 113–136.

[10] King C, Hall CAS: Relating financial and energy return on investment, *Sustainability* 3(10):1810–1832, 2011, Special Issue on EROI.

[11] Lambert J, Hall CAS, Balogh S, Gupta A, Arnold M: Energy EROI and quality of life, *Energy Policy* 64:153–167, 2014.

[12] Cleveland CJ, Costanza R, Hall CAS, Kaufmann R: Energy and the U.S. economy: a biophysical perspective, *Science* 225:890–897, 1984.

[13] Murphy D, Hall CAS: Year in review—EROI or energy return on (energy) invested, *Ann NY Acad Sci* 1185:102–118, 2010.

[14] Murphy DJ, Hall CAS: Energy return on investment, peak oil, and the end of economic growth, *Ann NY Acad Sci* 1219:52–72, 2011.

[15] Court V, Fizaine F: Long-term estimates of the energy return on investment (EROI) of coal, oil, and gas global productions, *Ecol Econ* 138:145–159, 2017.

[16] Hall CAS, Lambert JG, Balogh SB: EROI of different fuels and the implications for society, *Energy Policy* 64:141–152, 2014.

[17] Gupta AK, Hall CAS: A review of the past and current state of EROI data, *Sustainability* 3:1796–1809, 2011.

[18] Hall CAS: EROI and its implications for long-term prosperity. In Ruth M, editor: *Research methods and application in environmental studies*, Cheltenham, 2015, Edward Elgar, pp 197–224.

[19] Murphy D, Hall CAS, Cleveland C, O'Connor P: Order from chaos: a preliminary protocol for determining EROI for fuels, *Sustainability* 3(10):1888–1907, 2011, Special Issue on EROI.

[20] Kumar A, Das S, Sadhu PK, Pal N: Emerging trend in life cycle assessment of various photo-voltaic systems, *Int J Curr Eng Techn* 4(3), 2014.

[21] Raugei M, Dale M, Barnhart CJ, Fthenakis V: Rebuttal: "comments on 'energy intensities, EROIs (energy returned on invested), and energy payback times of electricity generating power plants'—making clear of quite some confusion", *Energy*:1–4, 2015.

[22] Raugei M, Frischknecht R, Olson C, Sinha P, Heath G. Methodological guidelines on net energy analysis of photovoltaic electrcity. IEA-PVPS Task 12, Report T12-07: 2016, ISBN 978-3-906042-39-8.

[23] ISO 14040. Environmental Management—Life Cycle Assessment—Principles and Framework, 1997.

[24] King CW, Maxwell JP, Donovan A: Comparing world economic and net energy metrics, Part 1: single technology and commodity perspective, *Energies* 8:12949–12974, 2015.

[25] Fthenakis V, Frischknecht R, Raugei M, Kim HC, Alsema E, Held M, et al: *Methodology guidelines on life cycle assessment of photovoltaic electricity*, 2nd ed., International Energy Agency Photovoltaic Power systems Programme, 2011, IEA PVPS Task 12.

[26] Ruagei M, Palmer F, Fthenakis V: The energy return on energy investment (EROI) of photovoltaics: methodology and comparisons with fossil fuel life cycles, *Energy Policy* 45:576–582, 2012.

[27] Prieto P, Hall CAS: *Spain's photovoltaic revolution: the energy return on investment*, New York, 2013, Briefs in Energy, Springer.

[28] Palmer G: Household solar photovoltaics: supplier of marginal abatement, or primary source of low-emission power? *Sustainability* 5(4):1406–1442, 2013.

[29] Weissbach D, Ruprecht GA, Huke K, Czerski S, Hussein Gottlieb A: Energy intensities, EROIs (energy return on invested), and energy payback times of electricity generating power plants, *Energy* 52:210–221, 2013.

[30] Ferroni F, Hopkirk RJ: Energy return on energy invested (ERoEI) for photovoltaic solar systems in regions of moderate insolation, *Energy Policy* 94:336–344, 2016.

[31] Raugei M: Comments on "energy intensities, EROIs (energy returned on invested), and energy payback times of electricity generating power plants"—making clear of quite some confusion, *Energy* 59:781, 2013.

[32] Weissbach D, Rurprecht GA, Huke K, Czerski D, Hussein Gottlieb A: Reply on "comments on 'energy intensities, EROIs (energy returned on invested), and energy payback times of electricity generating power plants'—making clear of quite some confusion", *Energy* 68:1004–1006, 2014.

[33] Leccesi E, Raugei M, Fthenakis V: The energy and environmental performance of ground-mounted photovoltaic systems—a timely update, *Energies* 9(8):622, 2016.

[34] Raugei M, Leccisi E: A comprehensive assessment of the energy performance of the full range of electricity generation technologies deployed in the United Kingdom, *Energy Policy* 90:46–59, 2016.

[35] Frischknecht R, Wyss F, Busser Knopfel S, Lutzkendorf T, Balouktsi M: Cumulative energy demand in LCA: the energy harvested approach, *Int J Lifecycle Assess*, 2015doi: 10.1007/s11367-015-0897-4.

[36] Brand-Correa LI, Brockway PE, Copeland CL, Foxon TJ, Owen A, Taylor PG: Developing an input-output based method to estimate a national-level energy return on investment, *Energies* 10:534, 2017.

[37] Koppelaar RHEM: Solar-PV energy payback and net energy: meta-assessment of study quality, reproducibility, and results harmonization, *Renew Sustain Energy Rev* 72:1241–1255, 2017.

[38] Bhandari KP, Collier JM, Ellingson RJ, Apul DS: Energy payback time (EPBT) and energy return on invested (EROI) of solar photovoltaic systems: a systematic review, *Renew Sustain Energy Rev* 47:133–141, 2015.

[39] Wong JH, Royapoor M, Chan CW: Review of life cycle analyses and embodied energy requirements of single-crystalline and multi-crystalline silicon photovoltaic systems, *Renew Sustain Energy Rev* 58:608–618, 2016.

[40] Gorig M, Breyer C: Energy learning curves of PV systems, *Environ Prog Sustain Energy* 35(3):914–923, 2016.

[41] Louwen A, Sark W, Faaij A, Schropp R: Re-assessment of net energy production and greenhouse gas emissions avoidance after 40 years of photovoltaics development, *Nat Commun* 7:13728, 2016.

[42] Goe M, Gaustad G: Strengthening the case for recycling photovoltaics: an energy payback analysis, *Appl Energy* 120:41–48, 2014.

Further Reading

[43] Lambert J, Hall CAS, Balogh S, Gupta A, Arnold M: Energy, EROI, and quality of life, Energy Policy 64:153–167, 2014.

22

Life Cycle Analysis of Photovoltaics: Strategic Technology Assessment

Vasilis M. Fthenakis
CENTER FOR LIFE CYCLE ANALYSIS, COLUMBIA UNIVERSITY, NEW YORK, NY, UNITED STATES
fthenakis@bnl.gov

22.1 Introduction

Over the last 10 yr, the market for photovoltaics (PVs), as measured by their cumulative installed capacity, has been growing by about 45% per yr. Between 2005 and 2015, global solar-PV capacity increased from about 5–220 GW, and it exceeded 350 GW as of the end of 2017 [1].

The fast growth of the PV markets is largely based on the promise to produce abundant electricity at low cost to the environment. It is therefore important to monitor and report indicators of the environmental performance of PV systems. Such indicators include the energy payback time (EPBT), the energy return on energy investment (EROI), and the global warming potential (GWP). The framework for this type of evaluations is life cycle assessment; it is also called life cycle analysis (LCA), accounting for material and energy flows in all the stages of the system.

This chapter shows the current LCA status for the major currently commercial large-PV technologies, namely: single-crystalline Si (sc-Si), multicrystalline Si (mc-Si), thin-film cadmium telluride (CdTe), and copper indium gallium diselenide (CIGS). Within this variety of technologies, Si-wafer based technologies account for approximately 92% of the total production, while CdTe PV technology represents the largest contributor to nonsilicon based PV systems. As efficiencies and material and energy usage in PV modules is continuously increasing, an up-to-date estimate and reasonable projections of the energy and environmental performance of PV technologies are of key importance for long-term energy strategy decisions. This chapter also presents a prospective LCA accounting for potential improvements in silicon PV technologies, which account for most of the market and have higher environmental footprint than thin-film technologies.

22.2 Life Cycle Analysis Methodology

LCA is a comprehensive framework for quantifying the environmental impacts caused by material and energy flows in each and all the stages of the life cycle of a product or an activity. It describes all the life stages, from "cradle to grave", thus from raw materials

A Comprehensive Guide to Solar Energy Systems. http://dx.doi.org/10.1016/B978-0-12-811479-7.00022-1

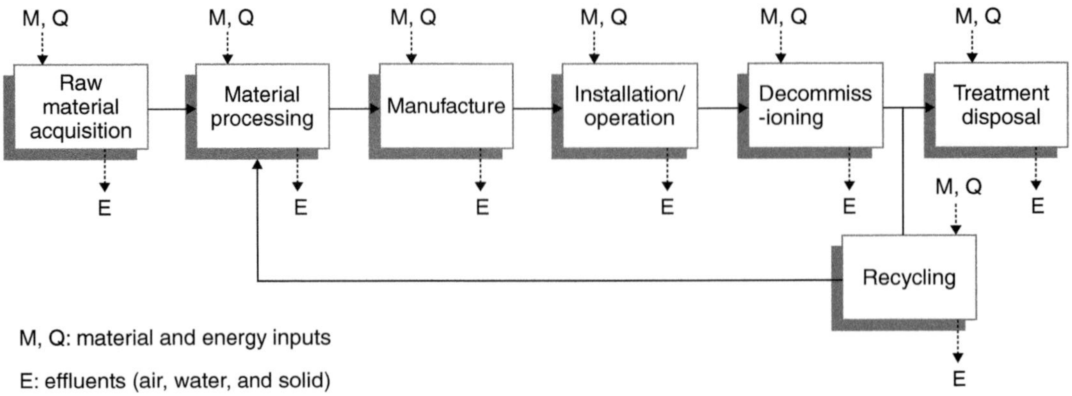

FIGURE 22.1 The Life Cycle Stages of PVs.

extraction to end of life. The cycle typically starts from the mining of materials from the ground and continues with the processing and purification of the materials, manufacturing of the compounds and chemicals used in processing, manufacturing of the product, transport, installation if applicable, use, maintenance, and eventual decommissioning and disposal and/or recycling. To the extent that materials are reused or recycled at the end of their first life into new products, then the framework is extended from "cradle to cradle." This life-cycle for PV is shown in Fig. 22.1.

The life-cycle cumulative energy demand (CED) of a PV system is the total of the (renewable and nonrenewable) primary energy (PE) harvested from the geo-biosphere in order to supply the direct energy (e.g., fuels and electricity) and material (e.g., Si, metals, and glass) inputs used in all its life-cycle stages (excluding the solar energy directly harvested by the system during its operation). Thus,

$$\mathrm{CED}\left[\mathrm{MJ}_{\mathrm{PE\text{-}eq}}\right] = \mathrm{E}_{\mathrm{mat}} + \mathrm{E}_{\mathrm{manuf}} + \mathrm{E}_{\mathrm{trans}} + \mathrm{E}_{\mathrm{inst}} + \mathrm{E}_{\mathrm{EOL}} \tag{22.1}$$

where, E_{mat} [$MJ_{PE\text{-}eq}$]: PE demand to produce materials comprising PV system, E_{manuf} [$MJ_{PE\text{-}eq}$]: PE demand to manufacture PV system, E_{trans} [$MJ_{PE\text{-}eq}$]: PE demand to transport materials used during the life cycle, E_{inst} [$MJ_{PE\text{-}eq}$]: PE demand to install the system, and E_{EOL} [$MJ_{PE\text{-}eq}$]: PE demand for end-of-life management.

The CED of a PV system may be regarded as the energy investment that is required in order to be able to obtain an energy return in the form of PV electricity.

The life-cycle nonrenewable CED is a similar metric in which only the nonrenewable PE harvested is accounted for; details are given in the International Energy Agency (IEA) Photovoltaics Power Systems (PVPS) Task 12 LCA Guidelines report [2].

EPBT is defined as the period required for a renewable energy system to generate the same amount of energy (*in terms of equivalent PE*) that was used to produce (and manage at end-of-life) the system itself.

$$\text{EPBT}\left[\text{yr}\right]=\text{CED}/\left(\left(\text{E}_{\text{agen}}/\eta_{\text{G}}\right)-\text{E}_{\text{O\&M}}\right) \tag{22.2}$$

where E_{mat}, E_{manuf}, E_{trans}, E_{inst}, and E_{EOL} are defined as above; additionally, E_{agen} [MJ_{el} yr^{-1}]: annual electricity generation, $E_{O\&M}$ [$MJ_{PE\text{-}eq}$]: annual PE demand for operation and maintenance, and η_G [MJ_{el} $(MJ_{PE\text{-}eq})^{-1}$]: grid efficiency, i.e., the average life-cycle PE to electricity conversion efficiency at the demand side.

For systems in operation for which records exist, the annual electricity generation (E_{agen}) is taken from the actual records. Otherwise, it would be estimated with the simple equation below (note the units into parentheses]:

$$\text{E}_{\text{agen}}\left[\text{kW h yr}^{-1}\right]=\text{irradiation}\left[\text{kW h m}^{-2}\text{yr}^{-1}\right]\times\text{module efficiency}\left[\text{kW}\left(\text{kWp}\right)^{-1}\text{m}^{-2}\right]\times \\ \text{performance ratio}\left[\text{dimensionless}\right]$$

where, irradiation is the global irradiation on the plane of the PV, module efficiency is the reported by the manufacturer-rated efficiency measured under 1 kW m^{-2} irradiance, and the performance ratio (PR) (also called derate factor) describes the difference between the modules' (DC) rated performance (the product of irradiation and module efficiency) and the actual (AC) electricity generation. It mainly depends upon the kind of installation. Mean annual PR data collected from many installations show PR values of 0.75 from residential and 0.80–0.90 from utility systems.

In general, the PR increases with (1) declining temperature and (2) monitoring the PV systems to detect and rectify defects early. Shading, if any, would have an adverse effect on PR. This is why well-designed, well-ventilated, and large-scale systems have a higher PR.

E_{agen} is then converted into its equivalent PE, based on the efficiency of electricity conversion at the demand side, using the grid mix where the PV plant is being installed. Note that E_{agen} is measured (and calculated) in units of kilowatt hours (kW h), and we first have to convert it to MJ (1 kW h=3.6 MJ) and then use ηG to convert MJ of electricity to MJ of PE ($MJ_{PE\text{-}eq}$). Thus, calculating the primary-energy equivalent of the annual electricity generation (E_{agen}/η_G) requires knowing the life-cycle energy conversion efficiency (η_G) of the country-specific energy-mixture used to generate electricity and produce materials. The average η_G values for the United States of America and Europe are, respectively, approximately 0.30 and 0.31.

Based on the above definition, there are two conceptual approaches to calculate the EPBT of PV power systems:

1. PV as replacement of the energy resources used in the power grid mix. This approach calculates the time needed to compensate for the total (renewable and nonrenewable) PE required during the life cycle of a PV system. The annual electricity generation (E_{agen}) is converted into its equivalent PE, based on the efficiency of electricity conversion at the demand side, using the current average (in attributional LCAs) or

the longer term (in decisional/consequential LCAs) grid mix where the PV plant is being installed.

2. PV as replacement of the nonrenewable energy resources used in the power grid mix. This approach calculates the EPBT by using the nonrenewable PE only; renewable PE is not accounted for, neither on the demand side, nor during the operation phase. This approach calculates the time needed to compensate for the nonrenewable energy required during the life cycle of a PV system. The annual electricity generation (E_{agen}) is likewise converted to PE equivalent considering the nonrenewable PE to electricity conversion efficiency of the average (in attributional LCAs) or the long-term marginal (in decisional/consequential LCAs) grid mix where the PV plant is being installed. The result of using this approach must be identified as nonrenewable energy payback time (NREPBT) to clearly distinguish it from the EPBT derived from the 1st approach. The formula of NREPBT is identical to that of EPBT described above except replacing "PE" with "nonrenewable PE." Accordingly, grid efficiency, ηG, accounts for only nonrenewable PE.

Both EPBT and NREPBT depend on the grid mix; however, excluding the renewable PE makes NREPBT more sensitive to local or regional (e.g., product-specific use of hydropower) conditions, which may not be extrapolated to large global scales. In contrast, EPBT metric with an average large-scale (e.g., EU, US, or World) grid conversion efficiency may not capture the conditions of local or regional climates. The calculated EPBT and NREPBT do not differ significantly in case the power plant mix of a country or region is dominated by nonrenewable power generation. However, because an increasing share of renewable energies is expected in future power grid mixes as well as within the PV supply chain, the two opposing effects of a reduction in the CED of PV and an increase in grid efficiency will require careful consideration, and the numerical values of EPBT or NREPBT may come to vary considerably according to the chosen approach.

Therefore it is important to choose the approach that most accurately describes the system parameters and satisfies the goal of the LCA study.

EROI is defined as the dimensionless ratio of the energy generated over the course of its operating life, over the energy it consumed (i.e., the CED of the system). The electricity generated by PV needs to be converted to PE so that it can be directly compared with CED. Thus EROI is calculated as:

$$\text{EROI}\left[\text{MJ}_{\text{PE-eq}} / \text{MJ}_{\text{PE-eq}}\right] = (\text{T})\left[\left(\text{E}_{\text{agen}} / \eta_{\text{G}}\right) - \text{E}_{\text{O\&M}}\right] / \text{CED} = \text{T} / \text{EPBT}$$

where T is the period of the system operation; both T and EPBT are expressed in years.

EROI and EPBT provide complementary information. EROI looks at the overall energy performance of the PV system over its entire lifetime, whereas, EPBT measures the point in time (t) after which the system is able to provide a net energy return. Further discussion of the EPBT and EROI methodology can be found in the IEA PVPS Task 12 LCA Guidelines, and a discussion of its misrepresentation in a few publications is discussed by Raugei et al. [3–5]

22.2.1 Interpretation and Reporting

The ISO 14040- and 14044-standards provide a framework for LCA. However, this framework leaves the individual practitioner with a range of choices that can affect the validity of the results of an LCA study. Thus, this author initiated and led an IEA PVPS task (Task 12), which developed guidelines to offer guidance for consistency, balance, and quality to enhance the credibility of the findings from PVs LCAs [6]. The guidelines represent a consensus among PV LCA experts in the United States, Europe, and Asia, for assumptions on PV performance, process input and emissions allocation, methods of analysis, and reporting of the results. The latter is of the utmost importance as parameters varying with geographical zones and system boundary conditions can significantly affect the results; accordingly, transparency is essential in comparing product life cycles. As pointed out in the IEA Guidelines, transparency in reporting is of the utmost importance because parameters vary with geographical zones, and a system's boundary conditions and modeling approach can affect the findings significantly. At a minimum, the following parameters should be reported: (1) On-plane irradiation level and location; (2) module-rated efficiency; (3) system's PR; (4) time-frame of data; (5) type of system (e.g., roof-top, ground mount fixed tilt or tracker); (6) expected lifetime and degradation ratio for PV and balance of system (BOS); (7) system's boundaries (whether capital goods, installation, maintenance, disposal, the transportation- and recycling-stages are included for both PV modules and balance-of-system (frame, mounting, cabling, inverter; for utility applications the transformer, site preparation, and maintenance); (8) the place/country/region of production modeled (e.g., average grid, site specific power use (e.g., hydro, coal), and (9) explicit goal of the study (e.g., static or prospective LCA, prototype or commercial production, current performance or expected future development). In addition, a LCA report should identify the following: The LCA method used, especially if is not process-based; the LCA tool (e.g., Simapro, Gabi, etc); databases used [e.g., Ecoinvent, GaBi, Franklin, National Renewable Energy Laboratory (NREL)]; the EPBT calculation method; commercial representativeness of the study (required if the data are from a pilot-scale production), and assumptions for production of major input materials, e.g., solar grade (SoG) silicon, aluminum (primary and/or secondary production).

22.3 Current Photovoltaic Status

22.3.1 Major Technologies

The PV systems comprise PV panels (also called modules) and BOS (mechanical and electrical components such as support and mounting structures, inverters, transformers and cables, as well as system installation, operation and maintenance). The currently dominant PV technologies are: sc-Si, mc-Si, CdTe, and CIGS.

Fig. 22.2 shows the respective flow diagrams for the c-Si and thin film PV systems. After the metallurgical (MG) and SoG Si production stages, mc-Si ingots are cast and sawn into wafers: sc-Si PV cells additionally require an intermediate Czochralski (Cz)

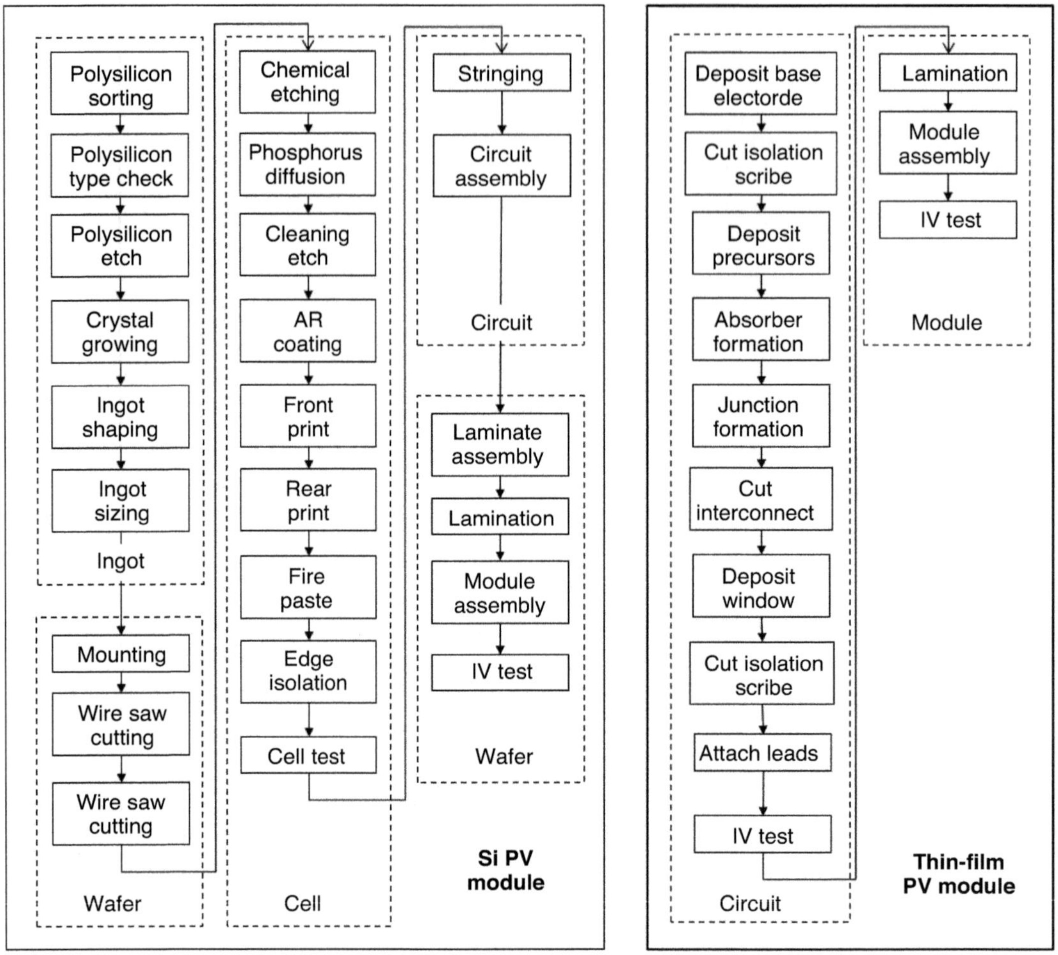

FIGURE 22.2 Process sequence for manufacturing Si and thin-film PV modules.

recrystallization step. Then, the individual PV cells are assembled into framed PV panels, and finally the PV system is completed by the addition of the BOS. There are more steps to arrive at the final stage in c-Si manufacturing than in thin-film PVs (CdTe and CIGS), and a comparatively large amount of energy is required for the production of crystalline silicon [1].

22.3.2 Production Sites and Electricity Mixes

Each analyzed PV system is also classified on the basis of country of production. The c-Si PV production chain is classified into three main producing regions: Europe, China, and the USA, according to the data source used. The sc-Si and mc-Si wafers used in Chinese PV manufacturing are sourced entirely domestically; on the other hand, of those used in United States PV manufacturing, 66% are produced in China and 34%

domestically; and for those used in European PV manufacturing, 89% is produced locally and 11% in China. Regarding CdTe PV panels, the two production countries, as of 2016, are Malaysia and the USA. The main production countries for CIGS PV are Japan and China. All further upstream steps in the Si supply chain are analyzed considering their actual geographical location—for instance, the production of metallurgical grade Si (MG-Si) is divided among the main global producers, i.e., China, Russia, Norway, and the United States.

The individual local updated electricity mixes, used for all PV module manufacturing and for the Si supplying countries, are also considered since they influence the amount of PE ultimately required for each production process, as well as the associated environmental impacts.

22.4 Current Photovoltaic Life Cycle Analysis Results

22.4.1 Fixed-Tilt Ground-Mounted Photovoltaic Systems

Leccisi, Raugei, and Fthenakis have recently published a most of up-to-date LCA of the major PV modules installed in fixed-tilt ground mounting utility-scale systems [7]. This section summarizes the finding of their study. Fig. 22.3 shows the CED of the analyzed PV systems and Fig. 22.4 shows the GWP, both expressed per kW_p—the stacked bars show the individual contributions of the main life cycle stages. The average efficiency for each technology is assumed in accordance with the 2016 status report published by the Fraunhofer Institute for Solar Energy Systems [8]—specifically: 17% for sc-Si PV, 16% for mc-Si, 15.6% for CdTe PV, and 14% for CIGS PV. The reader should be aware that these efficiencies continue to improve and, correspondingly, the environmental indicators will continue to improve as well.

The results clearly show that the most impacting step for c-Si technologies is the energy intensive production of SoG Si and cell manufacturing.

Fig. 22.3 highlights that, per kilowatt (kW_p), c-Si PV systems are about twice as energy-demanding to produce as CdTe PV systems.

Fig. 22.4 illustrates the resulting GWP indicator per kilowatt (kW_p): c-Si PV technologies generally have higher values in comparison with thin film PV panels, and in particular, the lowest GWP values are for CdTe PV.

As shown in Fig. 22.5, the EPBTs of the analyzed PV technologies were found to range from 6 months (for CdTe PV installed in the US South-West) to approximately 2–3 yr (for c-Si PV installed in Central-Northern Europe). Here yr refers to year.

22.5 Technology Roadmaping

As shown in the previous section, among the currently commercial technologies c-Si technologies have the highest environmental impact, and the most impacting steps are silicon feedstock and ingot growth, wafering, and cell processing. Thus to provide a paradigm of

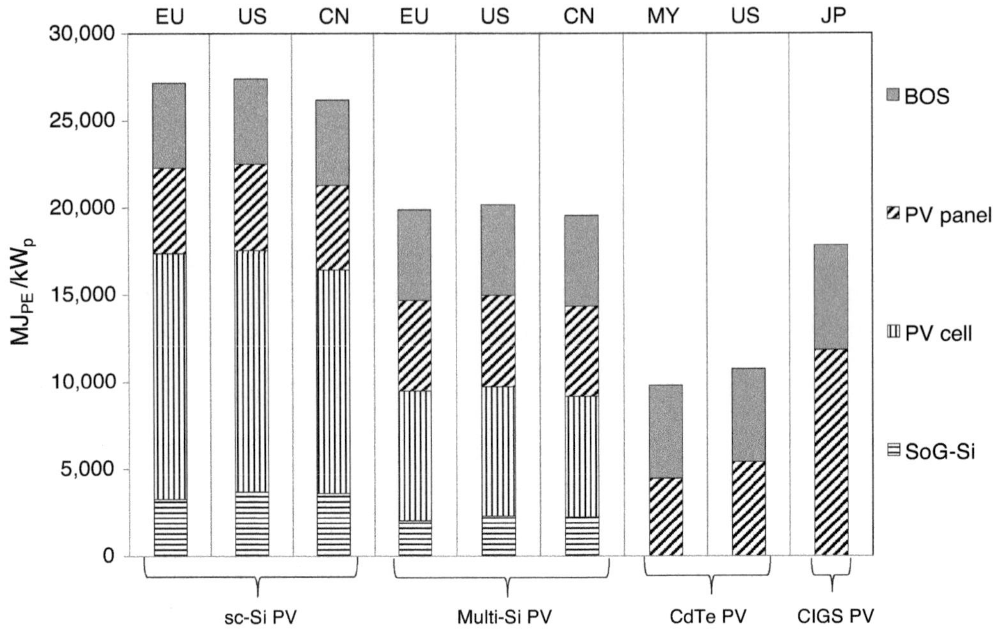

FIGURE 22.3 CED per kWp of major commercial PV systems. *CED*, Cumulative energy demand.

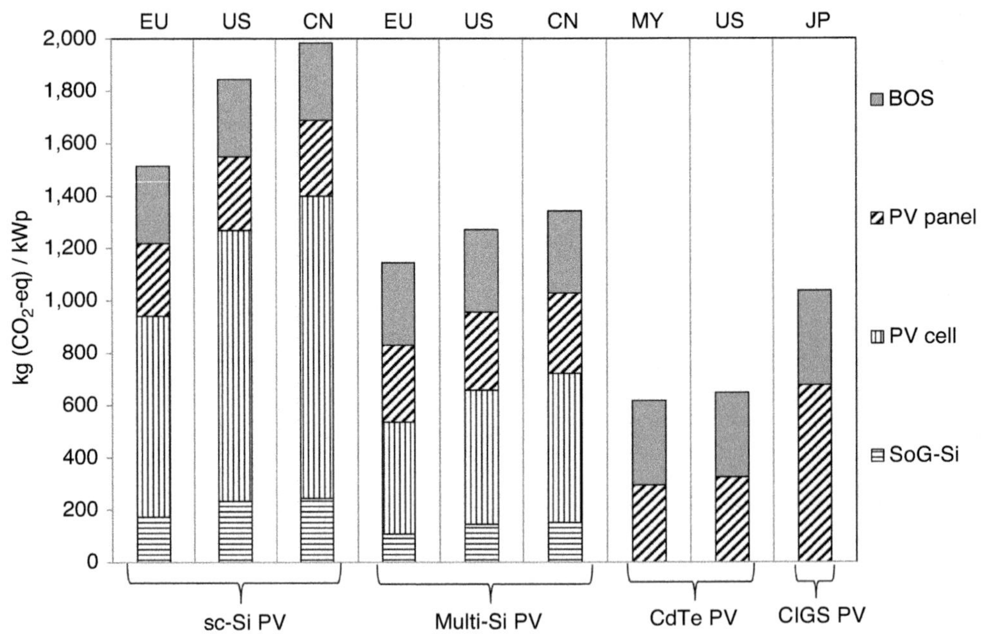

FIGURE 22.4 GWP per kWp of major commercial PV systems. *GWP*, Global warming potential.

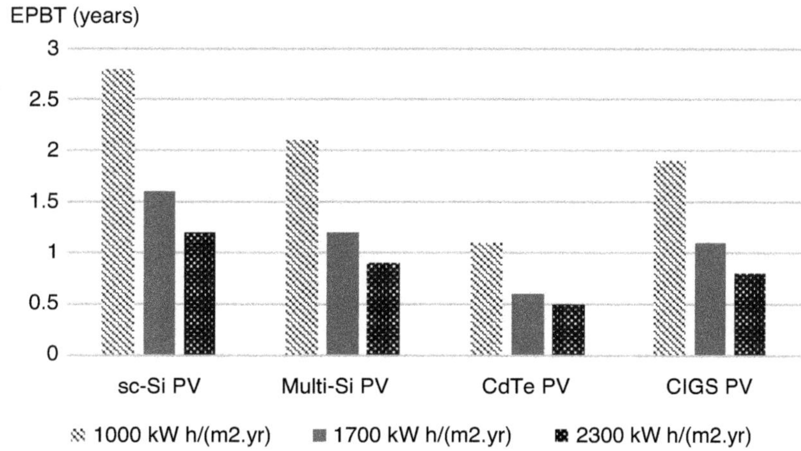

FIGURE 22.5 EPBT of major commercial PV systems for three levels of solar irradiation. *EPBT*, Energy payback time.

LCA used in strategic technology assessment, we will discuss proposed technological improvements in these stages that have the potential to improve the environmental impact of c-Si PV.

22.5.1 Feedstock and Ingot Growth

Today almost all commercially available modules are made from polysilicon using the Siemens process. In this process silicon is converted into crystalline ingots through either Cz crystal growth (for monocrystalline silicon) or directional solidification (DS) (for multicrystalline silicon). As polysilicon accounts for almost 25% of the total module production cost today, some facilities try to reduce cost by process integration with material and energy recycling routes. Additionally, the high cost has triggered development of alternatives to the traditional Siemens route (e.g., the fluidized bed reactor process and even abandonment of poly-Si altogether, as with upgraded metallurgical grade silicon (UMG-Si) [9].

Numbers of ingots in polysilicon production and ingot sizes in Cz production, sizes will likely increase, which increases the throughput and yield of good quality silicon [1].

22.5.2 Wafering

To reduce the required amount of polysilicon the wafer thickness is continuously being reduced, i.e., from 500 μm in 1979 to 180 μm for mc-Si (~140 μm for monocrystalline-Si) in 2016. The kerf loss, the silicon loss due to sawing, has decreased as well, to about 150 μm per wafer today. With further advances in the multiwire slurry saw (MWSS) process it is expected that 120 μm wafers with 120 μm kerf loss may become achievable by 2020 [10], which already reduces silicon consumption by 27% if breakage does not increase and efficiencies can be maintained.

Aside from the kerf loss the MWSS process has other drawbacks: wafer contamination, thickness variation across the wafer, high breakage, and high material consumption. To address those drawbacks alternative wafering methods were proposed, most notably a wire sawing process using a diamond abrasive on the wire. Most machines on the market today are adaptable for diamond wires, and implementing them will increase the throughput by a factor of 2–3. In addition, the kerf is commercially recyclable as opposed to kerf from the slurry-based process, and wire consumption per wafer is much lower [9].

22.5.3 Cell Processing

According to the International Technology Roadmap for Photovoltaics (ITRPV) roadmap as wafers get thinner, so large fraction of cells become rear-contact cells because front-to-rear interconnections and soldering of interconnects induces too much stress on thin wafers. There are three main approaches to rear-contact cells [11]: metal wrap-through (MWT), emitter wrap-through (EWT), and back-junction (BJ). In the first two approaches the emitter is still at the front of the device, but holes are laser drilled through the wafer that transport carriers to the rear, either through the metal contacts (MWT) or the emitter (EWT). The main difference between MWT and EWT is thus that the MWT still has grid lines (but no busbars) on the front surface. In a BJ cell, the emitter is located at the rear surface, typically in an interdigitated fashion with the back surface field (BSF). A BJ cell has the benefit that the contacts can cover almost the whole rear side of the cell, greatly reducing series resistance. All three approaches reduce contact shading, although this is especially true for the EWT and BJ types. Large-area efficiencies of 24.2% have been reached on BJ solar cells, and over 20% for MWT and EWT cells. Efficiencies of 20.2% are reported for interdigitated back contact silicon heterojunction cells (IBC-HIT) but simulations show that efficiencies of up to 26% are possible [9].

22.5.4 Technological Scenarios

Based on the developments described in the previous subsections Mann, de Wild-Scholten, Fthenakis et al. [9] have defined three advanced modules, and described their potential for environmental impact improvements (Table 22.1). Module 1 is a high efficiency scenario based on 120 μm thick MWSS wafers, and an IBC-HIT design with dielectric stack passivation, random pyramidal texture on the front surface and plated copper contacts. Based on the technology road mapping cell approach, module efficiencies are assumed to be 23.5%.

Module 2 is a slight modification of 1: Wafering is now done through light ion implantation, as in Silicon Genesis' PolyMax process. Wafers from this process are stronger mechanically, which is why a higher yield (98% vs. 90%) can be assumed [9].

Module 3 is a low-cost alternative based on UMG-Si feedstock and quasi-monocrystalline wafers. The efficiency for this cell is 20%, based on promising UMG-Si and MWT results and ITRPV projections [8,9].

Table 22.1 Description of the Three Technological Scenarios Under Investigation

Process Step	Module 1	Module 2	Module 3
Feedstock	Poly-Si		UMG-Si
Crystallization	Czochralski ingot growth		Seeded DS
Ingot yield	95%	95%	95%
Good Si out/Si in	85%	85%	85%
Wafering	MWSS	Ion implantation	MWSS
Wafer thickness/μm	120	40	120
Kerf loss	120	0	120
Yield	90%	98%	90%
Cell processing	IBC-HIT design, random pyramidal texture, plated copper contacts		Metal wrap-through design, random texture, front and rear passivation, plated copper contacts
Yield	99%	99.5%	99%
Efficiency	23.5%	22.5%	20%
Module assembly	Frameless encapsulation without lamination foil, with all-rear low-stress interconnection, and high packing density		
Yield	99.5%	99.5%	99.5%
Efficiency	22%	21%	19%

22.6 Prospective Life Cycle Analysis of Future Designs

To get a sense of the environmental impacts of these modules Mann, de-Wild, Fthenakis et al. performed a prospective life cycle assessment [9]. Where LCA is normally performed ex-post, ex-ante (before the event) the application is more challenging because of the need for a consistent comparison with the incumbent technology, defining anticipated future states of the technology under study, and properly addressing data quality and extrapolation and characterizing uncertainty. These challenges are addressed below.

We used as a functional unit one square meter module area of each the three modules introduced in the previous section; this allows for a direct comparison with the current (incumbent) technologies. Our analysis focused only on the cradle-to-gate stages of the modules, excluding the BOS and end-of-life stages.

22.6.1 Data Collection, Modeling, and Inventory Analysis

The life cycle inventory analysis is performed using the Ecoinvent database (version 2.2) [12] for material and energy inputs. For electricity inputs we used the continental European electricity mix (UCTE), which has a conversion efficiency of 0.31. The PE requirement is calculated using the CED method in units of MJ_p.

In the inventory we have used as much original data as possible. As shown in Table 22.2, we used new data from four different companies, and complemented that with data published in literature and from the Ecoinvent database. In two cases, the heterojunction formation step and conductive patterning of the back sheet data were unavailable and modeling was required.

Table 22.2 Primary Energy Data Values and Sources

Process Step	Key Inputs	Value
Feedstock	Poly-Si, Siemens	545 $MJ_p\ kg^{-1}$
	UMG-Si	322 $MJ_p\ kg^{-1}$
Crystallization	CZ-Si electricity use	85.6 $kW\ h\ kg^{-1}$
	Quasi-mono Si electricity use	19.3 $kW\ h\ kg^{-1}$ quasi-mono Si
Wafering	Electricity	Confidential
Wafering (ion implantation)	Electricity requirement for total process	51.4 $kW\ h\ m^{-2}$
Cell processing	Plating of contacts	Confidential
	IBC-HIT deposition	44.5 $MJ_p\ m^{-2}$
Encapsulation	Solar glass (front 3.0 mm, rear 2.8 mm)	14.6 $kg\ m^{-2}$
	Conductive back sheet	66.7 $MJ_p\ m^{-2}$

The interpretation of the results is the final stage of the LCA. In addition, because of the prospective nature of this study, a thorough uncertainty analysis is required.

22.6.2 Uncertainty Analysis

We can distinguish between various sources of uncertainty in LCA [13]; parameter and scenario uncertainties are the most significant in the current study. These parameters can be hard to measure precisely, or precise values might be unavailable, and furthermore, they might be inherently variable. Scenario uncertainty is related to the normative choices made in constructing scenarios. Another source of uncertainty is modeling uncertainty, which comes from the structure of the model. Due to its prospective nature, data are subject to change in the coming years, and the scenarios that we have developed may not represent the actual situation in 2020. An approach to deal with these uncertainties is discussed below.

22.6.2.1 Parameter Uncertainty

A prospective LCA study has two major data quality issues: data representativeness (e.g., scaling up from laboratory scale or temporal fluctuations in PE requirements for material inputs) and data availability (e.g., confidential or nonexistent data, that therefore have to be modeled). The challenge is to quantify these uncertainties in a transparent manner. Commonly, a pedigree matrix approach is used, as in the eco-invent database [14]. Here, the assumption is made that all inputs are accurately described by a lognormal distribution. The variance of this distribution is a measure of the uncertainty of the input, and is found using a pedigree matrix with data quality indicators (DQI) [13,15]. There are five data characteristics on which this variance is based: (1) data reliability, (2) data completeness, (3) temporal correlation, (4) geographic correlation, and (5) further technical correlation.

22.6.2.2 Scenario Uncertainty

The technological scenarios and performance forecasts that we have made are also subject to uncertainty. For instance, cell and module efficiencies might be higher or lower than we predicted, so we will present the EPBT and EROI results on an efficiency interval. We have based our yield assumptions on the values reported in the ITRPV roadmap [12], which are based on expert judgment. We address the yield uncertainty through parameter uncertainty, as described in the previous subsection: industrial expert estimates receive a data quality indicator score of 4 for reliability. Together with the other indicator scores one can find the variance and 95% confidence interval of the yield assumptions, as outlined by Weidema et al. [15]. Other sources of scenario uncertainty include development and adoption of: diamond-coated wire sawing, encapsulation methods, double-sided contact cells, and passivation methods. Finally, crystalline silicon PV modules might be produced by processes in 2020 that are unknown today, or that we did not account for in our scenarios; such uncertainties are discussed by Mann et al. [9].

22.7 Results

22.7.1 Cells and Modules

Based on the life cycle inventory we calculated the cradle-to-gate (factory) PE consumption for production of the three modules, including the 95% confidence interval, as shown in Fig. 22.6A (in PE requirement per square meter of module). To put these values into context: the most recent LCA results for multi- and monocrystalline silicon PV modules are 2150 and 2750 $MJ_p\ m^{-2}$ respectively, so the CED of modules 1, 2 and 3 are only 74% (2030/2750), 55% (1500/2750) and 62% (1330/2150) of the previous generation CEDs.

The CED of module 1 is considerably higher than that of modules 2 and 3. The breakdown in Fig. 22.6B shows that this is almost solely due to the relatively high consumption of monocrystalline silicon. Moving towards a kerfless cutting method reduces the silicon consumption by a factor of five, but at the same time the wafering energy is in-

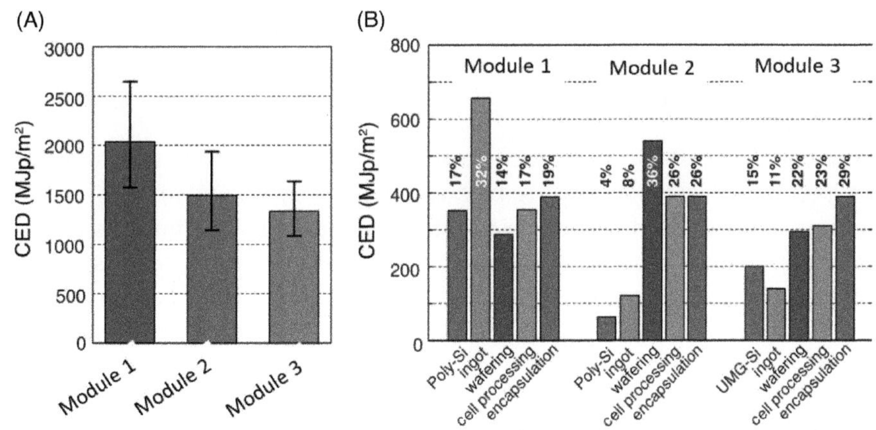

FIGURE 22.6 (A) CED per square meter of cell and (B) breakdown into feedstock, ingot, wafering, cell processing, and encapsulation steps. The percentage shows the contribution of each process step to the total CED [9]. *CED*, Cumulative energy demand.

creased due to the energy intensive ion implantation process. Module 1 and 3 require the same amount of feedstock, but the UMG-Si feedstock and seeded DS reduce the PE requirement for this step by almost a factor of three (although this comes at an efficiency penalty).

The breakdown for the aforementioned recent LCA results on monocrystalline silicon modules into poly-Si, ingot growth and wafering, cell processing, and encapsulation is approximately 500, 1260, 390, and 600 $MJ_p\ m^{-2}$ (or 18, 46, 14, and 22%) respectively [15]. Comparing this to the breakdown shown in Fig. 22.3B we see that in module 1a modest reductions have been achieved in feedstock, crystallization and wafering (owing to thinner wafers), and encapsulation. The cell processing energy requirement has not changed much, most likely because low-temperature processing has offset increases due to the increased overall process complexity. As mentioned before, poly-Si consumption is drastically reduced in module 1b, and module 2 uses low-energy feedstock. Because module 2 is based on quasi-mono wafers, the energy requirement here is reduced further (the corresponding energy requirement for ingot growth and wafering in 2011 in a multi-Si module was ~670 $MJ_p\ m^{-2}$ [16]).

The confidence intervals are quite large: [1570, 2640], [1140, 1930] and [1070, 1630] $MJ_p\ m^{-2}$ for modules 1, 2, and 3 respectively (Fig. 22.2A). This is mainly due to the temporal uncertainty: About half (or even 63% for module 1a) of the PE requirement comes from poly-Si feedstock, ingot growth and wafering, and these production steps have a large potential for reduction in energy intensity by scaling up. We therefore expect the actual CED to decrease towards the lower bound rather than increase.

We also calculated the EPBTs for these modules (Fig. 22.4A). The mean payback time of module 1 is just over 0.6 yr, while 2 and 3 have an EPBT of a little under 0.5 yr. Adding 0.2 yr for BOS brings those EPBT to 0.7–0.8 yr. As shown in Table 22.1 the most recent module EPBT values are 1.6 for mono-Si and 1.2 for multi-Si [7], so a reduction in the EPBT of over a factor of two may be feasible. Correspondingly, the EROI of produced electricity for a lifetime of 30 yr would be about 40.

22.7.2 Balance of System

Although the BOS lies outside of the scope of this investigation, it is a crucial part of any PV energy system and we will therefore briefly discuss its impacts on our results. One can distinguish between two common types of grid-connected PV systems: building-integrated systems (e.g., rooftop panels) and ground-based systems (e.g., large scale centralized power plants). In Table 22.3, we show the EPBT and EROI for both types of system, calculated with data from Fthenakis et al. [2]. The rooftop system consists of a 2.5 kW inverter and a Schletter mounting system. The PV plant uses average mounting structure data from market surveys and inverter data based on a 4.6 MW_p plant in Tucson, USA. According to the IEA LCA guidelines a PR of 0.80 is used for a utility PV plant [6].

To reduce costs associated with the BOS the EUPVTP has identified some research priorities [16]. For the EPBT the most relevant goals are an increase in the inverter lifetime (of

Table 22.3 The Primary Energy Consumption of the BOS Components for Two Types of PV Systems, and the EPBT and EROI on a System Level

	Rooftop			Grount-Mount PV Plant		
Module	**1**	**2**	**3**	**1**	**2**	**3**
Structure, cabling/MJ m^{-2}	225	225	225	645	645	645
Inverter/MJ m^{-2}	704	675	614	254	243	222
Total/MJ m^{-2}	929	900	839	899	888	867
EPBT/yr	0.9	0.7	0.7	0.8	0.7	0.7
EROI (E_{out}/E_{in})	13	15	15	14	16	16

Source: Data are from [2].

over 30 yr by 2025 compared to 15 yr now) and low-cost support structures, cabling and electrical connections. As Table 22.3 shows the inverter accounts for a considerable share of the CED of rooftop systems, and even a modest increase in the inverter lifetime would therefore significantly decrease the system EPBT. The inverter is less important in the case of large-scale PV plants, where the support structure dominates the CED.

Table 22.3 also shows the system EROI values of crystalline silicon PV in 2020 with old BOS data.

22.8 Conclusion

In this chapter, we summarized life-cycle impacts from (1) current vintage of commercial PV technologies and (2) future c-Si modules via a prospective LCA of crystalline silicon PV technologies expected to materialize on or about the year 2020. To do so, we developed three technological scenarios (two based on monocrystalline silicon and one based on quasi-monocrystalline silicon), building on various existing roadmaps. To summarize, increasing the cell efficiency is (after scaling up) the most important lever to reduce energy demand and costs. By using high quality passivation and BJ higher efficiencies can be achieved while simultaneously reducing the wafer thickness, which reduces the embodied energy even further. However, thinner wafers also require novel cell processing and encapsulation schemes, which we accounted for. We forecast that the EPBT of crystalline silicon modules could be reduced to 0.5 yr (0.7 yr when including BOS) provided that planned technological advances will occur. The EROI of PV modules could increase by a factor two to three in the coming years, and lie in the same range as electricity from coal-fired power plants.

This is a prospective LCA, subject to technological improvements taking place, and as such it carries a considerable degree of uncertainty. To address parameter uncertainty, results were provided with a 95% confidence interval, but scenario uncertainty remains. However, since the general tendency of maturing technologies is to become less energy and material intensive and we have used data that are representative of, at best, the situation today, these forecasts can be considered conservative.

Finally, it is important to note that these results cannot easily be compared with other PV technologies, because they too can change over time. Nonetheless, these results show that there is considerable potential to reduce the environmental impact of crystalline silicon PVs while at the same time reducing production costs.

References

[1] Fthenakis VM, Lynn PA: *Electricity form sunlight: systems integration and sustainability*, 2nd ed., Wiley, 2018.

[2] Fthenakis V, Kim HC, Frischknecht R, Raugei M, Sinha P, Stucki M, et al: *Life Cycle Inventories and Life Cycle Assessments of Photovoltaic Systems*, International Energy Agency, Report IEA-PVPS T12-02:2011, 2011.

[3] Raugei M, Fullana-i-Palmer P, Fthenakis V: The energy return on energy investment (EROI) of photovoltaics: methodology and comparisons with fossil fuel life cycles, *Energy Policy* 45:576–587, 2012.

[4] Carbajales-Dale M, Raugei M, Fthenakis V, Barnhart: Energy return on investment (EROI) of solar PV: an attempt at reconciliation, *Proc. IEEE* 103(7):995–999, 2015.

[5] Raugei M, Sgouridis S, Murphy D, Fthenakis V, et al: Energy return on energy invested (ERoEI) for photovoltaic solar systems in regions of moderate insolation: a comprehensive response, *Energy Policy* 102:377–384, 2017.

[6] Fthenakis V, Frischknecht R, Raugei M, Kim HC, Alsema E, Held M, et al: *Methodology guidelines on life cycle assessment of photovoltaic electricity*, 2nd ed., International Energy Agency, Report IEA-PVPS T12-03:2011, 2011.

[7] Leccisi E, Raugei M, Fthenakis V: The energy and environmental performance of ground-mounted photovoltaic systems – a timely update, *Energies, Special Issue on Life Cycle Assessment in Current and Evolving Grids* 9(8):622, 2016.

[8] *Photovoltaics Report*, Freiburg, Germany, 2016, Fraunhofer Institute for Solar Energy Systems.

[9] Mann SA, de Wild-Scholten MJ, Fthenakis VM, van Sark W, Sinke WC: The energy payback time of advanced crystalline silicon PV modules in 2020: a prospective study, *Prog. Photovolt. Res. Appl.* 22:1180–1194, 2014.

[10] Crystalline Silicon Technology Manufacturing Group: *International Technology Roadmap for Photovoltaics: Results 2011*, 3rd ed., Berlin, 2012, SEMI PV Group Europe, Available from: http://www.itrpv.net.

[11] van Kerschaver E, Beaucarne G: Back-contact solar cells: a review, *Prog. Photovolt. Res. Appl.* 14:107–123., 2006doi: 10.1002/pip.657.

[12] Frischknecht R, Jungbluth N, Althaus HJ, Doka G, Dones R, Hischier R, et al. Overview and Methodology. Dübendorf; 2011. Available from: http://www.ecoinvent.org/fileadmin/documents/en/01_OverviewAndMethodology.pdf.

[13] Lloyd SM, Rise R: Characterizing, propagating, and analyzing uncertainty in life-cycle assessment, *J. Ind. Ecol.* 11(1):161–179, 2007.

[14] Weidema BP, Bauer C, Hischier R, Motel C, Nemecek T, Reinhardt J, et al: *Overview and Methodology: Data Quality Guideline for the Ecoinvent Database Version 3. Ecoinvent Report 1(v3)*, St. Gallen, 2012, The Ecoinvent Center.

[15] Weidema BP: Multi-user test of the data quality matrix for product life cycle inventory data, *Int. J. Life Cycle Assess.* 3(5):259–265., 1998doi: 10.1007/BF02979832.

[16] European Photovoltaic Technology Platform. A Strategic Research Agenda for Photovoltaic Solar Energy Technology. 2nd ed. 2011. Available from: http://www.eupvplatform.org/.

PART 6

Economics, Financial Modeling, and Investment in PVs, Growth Trends, and the Future of Solar Energy

23

Materials: Abundance, Purification, and the Energy Cost Associated with the Manufacture of Si, CdTe, and CIGS PV

Ajay Gupta
EROI ENERGY ADVISORS INC., BRAMPTON, ON, CANADA
akg78002@me.com

23.1 Introduction

"We're made of star stuff," Carl Sagan explained decades ago in his iconic television show Cosmos. It is theorized that during the Big Bang the lightest elements—hydrogen, helium, and trace amounts of lithium and beryllium—were formed. As stars form even today, their cores fuse hydrogen into helium, and then when they begin to die they manufacture carbon from fusing helium atoms. Very massive stars begin a further series of reactions forming oxygen through iron (Fe). The most massive stars die in a supernova, which releases enough energy and neutrons to form elements heavier than Fe, such as uranium and gold (Au). Today we have identified 90 such naturally occurring elements and in total 119 elements are known to man. On Earth we have found methods of extracting, refining, and using every material found for our increasingly complex industrial society [1,2]. These materials not only make today's industry possible; they are also what will make tomorrow's industry a reality.

The world is currently in a state of transition and the shift from fossil fuels to renewable technologies will lead to greater consumption of certain essential materials [3–6]. Although much attention is being paid to the decrease in coal mining across the globe, almost no attention is being paid to the implications of growing demand for materials required in renewable energy technologies. Renewable forms of energy and specifically low or zero-carbon forms of energy are being sought to fulfill future needs in the face of depleting fossil fuels and climate change. Photovoltaic (PV) energy is one of these renewable forms of energy and although implications of the economic and energy availability of

A Comprehensive Guide to Solar Energy Systems. http://dx.doi.org/10.1016/B978-0-12-811479-7.00023-3

the technology continues to be heavily evaluated, the energy cost of developing PV is just starting to come to light for certain systems (see Chapter 27). The implications of rapid up-scaling in PV technology deployment to the materials industry, including the energy costs of developing materials necessary for a variety of PV technologies, however, is just beginning to be widely discussed [6].

Our current modern economies are dependent on growth; thus, the same technological and resource dependence that has allowed us to experience a wealth of materials defines our vulnerability within the system. Disruptions in the complex network of global resource chains and industrial development can lead to serious degradation of productivity and where energy production is concerned, even a decrease in social well-being. Recent rapid development of PV energy throughout the world has led to questions of material constraints on it, especially given other profitable industries competing for resources.

23.2 Critical Metals

Material constraints are those that arise due to limited amounts of critical metals required for rapid PV production [7,8]. Critical metals are those that are vulnerable to supply disruptions, where such disruptions would have significant impact to the industry and have a high likelihood of occurrence [9–11]. Metals critical to the rapid upscale of the PV industry are such typically due to the required rapid rates of development, not to any inherent properties of the material itself. Material constraints apply to both primary ores (those extracted from the earth to obtain metal) and secondary ores (those that are derived from primary ores). For example, tellurium (Te)—a secondary ore—is obtained from the impurities of copper (Cu) refinement—Cu being the primary ore. This is not to be confused with secondary production (Cu production from recycling products containing Cu). Another important distinction is that between “resources” and “reserves.” “Known resources” refers to the crustal abundance of an elemental material. It is the total of that material which exists on Earth and typically does not change over time. “Known reserves” or the “reserve base” refers to the subset of resources that can be economically used given current technologies of extraction and processing. Historically, the reserve base of a given material tends to increase over time as new technologies are developed to access deposits more difficult to process. Technology can also increase extraction efficiency. Consumption acts to decrease geologic reserves. Assessing the criticality of a metal toward PV production requires a good understanding of the metal resources, reserves, and supply dynamics; however, the reserves and resources of secondary ores are very uncertain today [12].

Ever since Hubbert predicted, in 1956, a future peak in crude oil production in the United States (US) [13], similar efforts have been made to predict the peak of other nonrenewable resources. It has been suggested that Canada and the United States have already passed peak Cu production [14,15] and that zinc (Zn) might also be approaching global production peak [7,16]. Cu is a primary metal necessary for PV production, and like Zn so are its secondary products. Like the criticality of a material, peak production refers not to the amount of material in reserves, but primarily the rate at which it can be developed for use; however, to determine criticality both are necessary to observe. Unfortunately, resources and reserves of critical metals are very difficult to calculate because they are almost all extracted from mineral deposits as by-products of Zn, Cu, Fe, Au, lead (Pb), nickel (Ni), titanium (Ti), aluminum (Al), platinum (Pt), and tin (Sn) [17]. Therefore, despite their respective importance to PV technologies, secondary ores such as Te, indium (In), germanium (Ge), gallium (Ga), and selenium (Se) are typically of significantly lesser economic value than their primary source ores in a given mining operation. As such, they are rarely part of a company's core business and hence are less likely to be extracted or even have their presence reported [18]. For example, global demand for Te was 200 t in 2009 and at 145 $ kg^{-1} it can be considered valuable, however, it all came as a by-product of millions of tons of Cu and Au mining and so is insignificant to the profits of the companies involved [19]. This makes secondary ores more vulnerable to supply disruptions, which increases their chances of being critical metals to the PV industry—among other high-tech industries such as cellphones, computers, and televisions.

Given the interest in material constraints to industries such as PV production, it is no surprise that recent efforts have been made to quantify future supply disruptions. Sherwood et al. modeled resource criticality in modern economies using agent-based dynamics in 2017 and found that there are vulnerabilities from technological interdependence [20]. Contrary to traditional economic models, their model removed simplifying assumptions of smooth growth, cost-shares in determining production, economic agents that act with perfect information and foresight, and included that the economy consists of representative agents and not a multitude of heterogeneous agents. They modeled technological growth as a complex adaptive system and applied biophysical constraints on macroeconomic growth. They found that with high levels of technological interdependence, removing any resource led to a significant decline in production output. Grandell and Thorenz [21] and Grandell et al. [22] examined the role of critical metals in future markets of renewable energy technologies, including PV. After modeling for a number of assumptions in future energy and economic output and composition, they found that a number of metals could constrain clean energy development in the future. Specific to solar energy were Ag, In, Te, and ruthenium (Re). Notably, Ag demand was modeled to exceed known resources by more than 300% and reserves by almost 450% as it is used in virtually all solar energy technologies and electronics in general, including electric vehicles. In other words, Ag demand is expected to grow significantly in the future of both energy production and consumption technologies. There is some, but very little, literature already available on the subject of material constraints specific to the solar energy industry.

Unfortunately, information on material constraints to PV is relatively scarce and given its novelty, there are more questions than answers at this time. This is due mainly to the uncertainty of the composition of future energy economies, as is today, but was even more so in the past. As we can make better predictions of what the future may look like, the possibility of material constraints will become clearer. In the year 2000, Andersson evaluated the material necessary for large-scale thin-film PV development and found that copper indium gallium selenide (CIGS) technology was limited by In reserves and that cadmium telluride (CdTe) technology was limited by Te reserves [23]. Later, the potential for large-scale manufacture of PV and its metal requirements were estimated in several studies with similar conclusions [24–26]. In response to findings of possibilities in material constraints associated with rapid PV development, Wadia et al. examined extraction costs and supply constraints for 23 materials relevant to semiconductors, including those relevant to PV, and found large differences in material extraction costs over space and time [27]. In 2009, Fthenakis raised the issue of secondary ore constraints to thin-film PV production due to limitations in annual production of primary ores [28]. Then in 2011 Candelise et al. examined the possibility of material constraints, specifically for In and Te in thin-film PV and found that although there was little evidence that production would necessarily be constrained, cost reductions would be [29]. By identifying potential PV and material utilization efficiency improvements in 2012, Fthenakis determined that the Te available from copper refineries is sufficient for several TWs of production by mid century [30]. In 2013, the Critical Materials Institute was founded in the Ames Laboratory as part of the US Department of Energy and identifies Te as a critical material for solar cells to this day [31]. Later in 2014, Houari et al. used a systems dynamic model to evaluate Te availability and thin-film PV growth and found the industry would be less constrained than previously calculated [32]. In 2015 Grandell and Hook identified In, Te, Ge, and Ru as potentially critical to thin-film PV development in the terawatt range [33]. Finally in 2017, Davidsson and Hook concluded that although the scale of the required materials could become problematic for currently commercial technologies, especially thin-film and crystalline silicon (c-Si) PV, the conclusion cannot be drawn that material availability is likely to act as a constraint on total growth of PV [34]. There have been recent attempts at calculating reserves without explicit, accurate, and time sensitive reporting from corporations, however, without any agreement on methods or definitive values as of yet [35,36]. As can be understood from this reading, the problem of calculating the possibility of material constraints concerning secondary ores, given better forecasts of future output or in terms of their respective reserve base, is not settled.

23.3 Material Requirements for PV

There are five different technologies typically discussed in the literature pertaining to materials relevant to the PV industry: mono-crystalline silicon (mono-Si) and multi or poly-crystalline silicon (multi-Si or poly-Si), which are both forms of c-Si PV, and amorphous silicon (a-Si), CIGS and CdTe, which are all forms of thin-film PV. All five technologies have different pathways for production and therefore can carry different footprints based on where they are manufactured [37]; however, the amounts of nonfuel materials required for each technology is generally the same, although it varies widely based on technology. Table 23.1 illustrates some of the materials of interest based on PV technology.

Each technology has its advantages and disadvantages as well as different metal content. C-Si is the most popular and widely deployed PV technology to date. It is most popular in large-scale and rooftop applications. Thin films are generally less efficient than c-Si cells, but they are also less energy intensive to produce and can be applied in

Table 23.1 General Comparison of Metal Content in Different PV Technologies

	c-Si	CIGS	CdTe	a-Si
Al	x			
Cu		x	x	
In		x		
Fe	x			
Pb	x			
Ni	x			
Ag	x			
Zn			x	x

Adapted from [6]. Note that this is not a comprehensive list and many secondary materials are not represented here.

more versatile forms. A breakdown of the current composition of global PV capacity by technology is illustrated in Fig. 23.1 and annual production of PV capacity by technology in Fig. 23.2. Currently, c-Si accounts for about 90% of the current PV deployment by capacity and 66% is comprised of poly-Si. Although poly-Si does not have the high efficiency of mono-Si, it is more cost effective for most applications; however, mono-Si has recently experienced an increase in development. Thin films comprise around 9% of the total deployment by capacity to date and around 6% of total 2016 PV development. Among the thin films, CdTe is the most popular due to its higher efficiencies and low cost of production; CIGS can also become relatively cheap to produce, and a-Si suffer from the lowest performance but can be printed onto flexible surfaces.

As can be understood from Table 23.1 and Figs. 23.1 and 23.2, materials for c-Si solar cells such as Ag and for thin films such as In and Te are very important to the PV industry. This is especially true moving forward into the future with the kind of growth expected from these technologies. Table 23.2 provides the most up to date values for the amounts of certain metals required based on capacity output for the PV industry as a whole. Even within PV technologies, there are ranges of material requirements due to differences in production pathways, countries of origin, and technological developments over time [37]. Each metal is more thoroughly discussed in terms of its part in specific PV technologies in later sections of the chapter, which is summarized in Table 23.2.

23.3.1 Mining and Refining Materials for PV

Mining the Earth for materials is something humans have been doing for thousands of years. Today mining is done mostly by heavy machinery, but it is still integral to industry and can be the backbone of a nation's economy. There are thousands of mines on earth today, ranging from small family-size operations to large-scale mountain top removal. It is important to remember that every ore deposit is different, and so is every mine.

Mining includes searching for, extracting and beneficiating (separating commercially useful ore and mineral content from surrounding materials) from the Earth. There are seven distinct main steps in any mining operation: (1) exploration, (2) design, (3) construction,

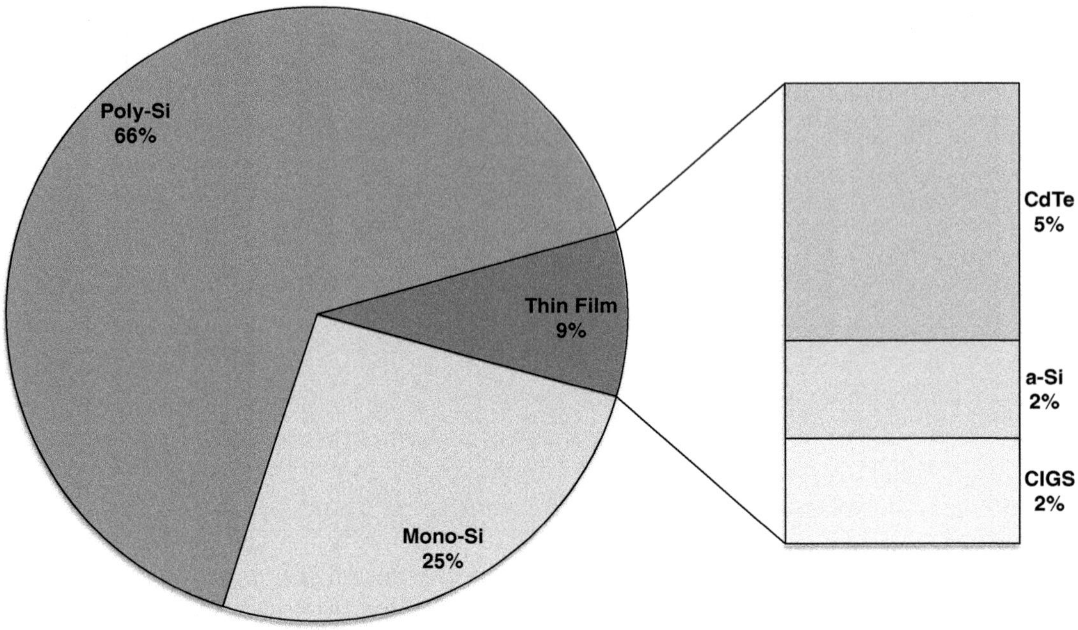

FIGURE 23.1 Current composition of global PV capacity by technology. Calculations based on industry data from [38] and global capacity data from 1997–2016 from [39].

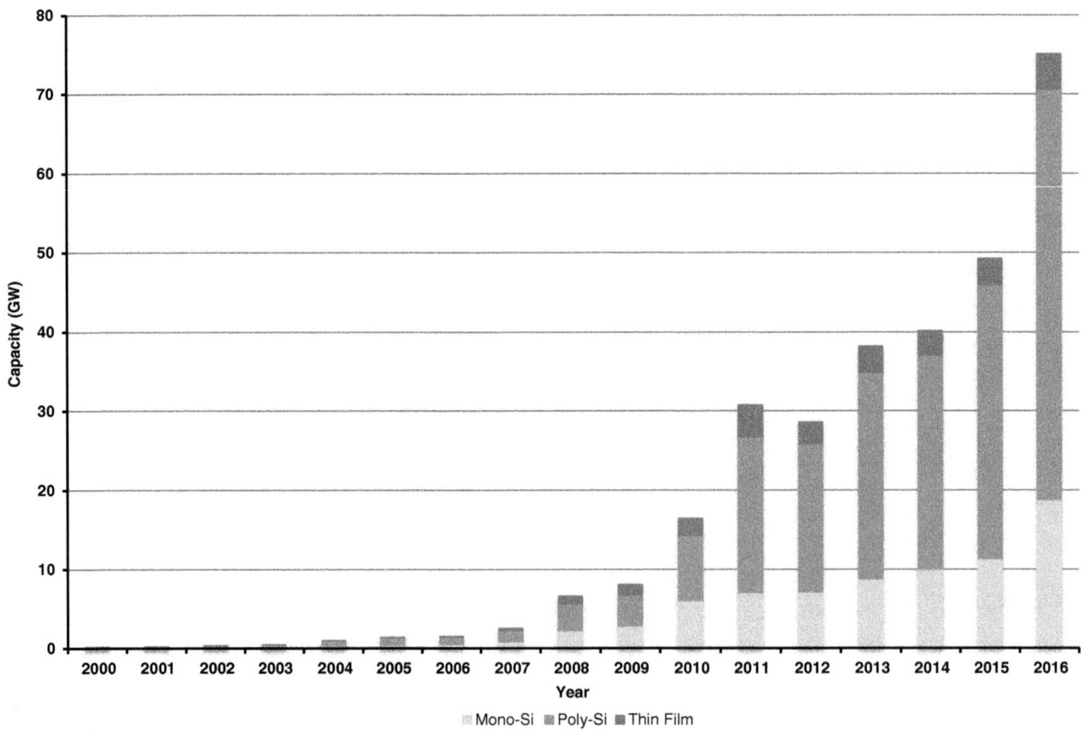

FIGURE 23.2 Annual global capacity additions by technology. Calculations based on industry data from [38] and global capacity data from 1997–2016 from [39].

Table 23.2 Selected Material Requirements for a Range of PV Installments and Their Respective Global Production and Reserves

Metal	PV Requirements/ kg $(MW)^{-1}$	c-Si/kg $(MW)^{-1}$	CIGS/kg $(MW)^{-1}$	CdTe/kg $(MW)^{-1}$	a-Si/kg $(MW)^{-1}$	Mix/kg $(MW)^{-1}$	2016 Production/Mt	Reserves/Mt
Aluminum (AL)	102						57.6	–
Cadmium (Cd)	0.93–83.51		0.93	83.51			0.023	0.5
Copper (Cu)	16.97–2194.1		16.97			2194.1	19.4	720
Gallium (Ga)	0.12–6.17		0.12–6.17				0.00018	0.0065
Germanium (Ge)	4.07				4.07		0.0001605	0.119
Indium (In)	4.5–83.79		4.5–83.79				0.000665	0.011
Lead (Pb)	72.38–269.3	72.38				269.3	4.82	88
Selenium (Se)	0.5–84.41		0.5–84.41				0.0022	0.1
Silicon (Si)	18.4–40.07	33.08–40.07			18.4		7.2	–
Silver (Ag)	5.17–19.2						0.027	0.57
Tellurium (Te)	4.7–90.38			90.38			0.0004	0.025
Tin (Sn)	5.95–463.1		5.95			463.1	0.28	4.7
Zinc(Zn)	29.99						11.9	220

Metal	PV Requirements/ kg $(MW)^{-1}$	c-Si/kg $(MW)^{-1}$	CIGS/kg $(MW)^{-1}$	CdTe/kg $(MW)^{-1}$	a-Si/kg $(MW)^{-1}$	Mix/kg $(MW)^{-1}$	2016 Production/Mt	Reserves/Mt
Aluminum (AL)	102						57.6	–
Cadmium (Cd)	0.93–83.51		0.93	83.51			0.023	0.5
Copper (Cu)	16.97–2194.1		16.97			2194.1	19.4	720
Gallium (Ga)	0.12–6.17		0.12–6.17				0.00018	0.0065
Germanium (Ge)	4.07				4.07		0.0001605	0.119
Indium (In)	4.5–83.79		4.5–83.79				0.000665	0.011
Lead (Pb)	72.38–269.3	72.38				269.3	4.82	88
Selenium (Se)	0.5–84.41		0.5–84.41				0.0022	0.1
Silicon (Si)	18.4–40.07	33.08–40.07			18.4		7.2	–
Silver (Ag)	5.17–19.2						0.027	0.57
Tellurium (Te)	4.7–90.38		4.7	90.38			0.0004	0.025
Tin (Sn)	5.95–463.1		5.95			463.1	0.28	4.7
Zinc(Zn)	29.99						11.9	220

Material requirements adapted from [6,33,37,48,49] and material production and reserves from [33,46,48,49]. These numbers should be regarded as estimates and not be taken at face value. Reserves for Cd, Ga, Ge, and In are from 2014.

(4) extraction, (5) beneficiation, (6) processing, and (7) reclamation [40]. Each step can be different for each mine. Popular types of mines include surface and underground mining, which have both been employed for thousands of years of mining activity. Solution, or "in situ" mining, which uses acids to dissolve metals into solutions and then recover ores from those solutions, is a relatively new process gaining popularity. Physical factors of the ore deposits determine the type and size of a mine, such as nature, location, size, depth, and grade [41]. Surface mining is the most popular because it entails the least amount of cost to developers and can be done relatively inexpensively with heavy machinery and explosives. Underground mines are employed if the deposits are deep and concentrated enough to warrant extensive operations such as the added planning and drilling. Other types of mines include deep-sea mining and asteroid or outer space mining. Although they speak to our need to find new deposits to feed our industries, they have not yet been employed.

Once ore is beneficiated and processed, it must be refined. There are two categories for the process of refining ore: pyrometallurgical processing and hydrometallurgical processing. Pyrometallurgical processing uses physical properties such as smelting and is more widespread. Hydrometallurgical processing uses chemical properties and relies on leaching ores with strong acids and recovering concentrated materials by precipitation or solvent extraction/electrowinning (SX/EW) [42]. Both processes use physics and chemistry to gain higher concentrations of ore throughout many steps that may or may not be needed or economical depending on the nature of the deposit—specifically, the ore grade and concentrations of other desirable by-products.

For pyrometallurigical processing, or smelting, the ore is simply heated in a furnace until the metals physically separate. Electric or "flash" furnaces are used for lower material volumes and higher by-product recovery rates [43]. Impurities can be further removed by blowing gases into the molten form of the ore to achieve even higher concentrations. Since PV modules require extremely pure concentrations of materials, metals must be even further refined "electrolytically," or "electrorefined." During "electrorefinement," the concentrate is electrically dissolved by a specific current into a specific electrolyte and accumulates on a cathode in nearly pure concentrations [44]. Such material is referred to as "cathode grade" material.

For hydrometallurgical processing, several steps must be taken. The ore goes through crushing, grinding, washing, filtering, sorting and sizing, gravity concentrating, leaching, ion exchange, solvent extraction, electrowinning, and precipitation [45]. In other words, the ore is put into solution and then refined through precipitation or SX/EW. Finally, it is electrorefined to cathode grade. The leftover electrolyte often contains other economically retrievable metals that can be further refined.

23.3.2 Aluminum

Al is second to Si as the most abundant metallic element on Earth; however, its commercial industrial application is only about 100 years old because it is very reactive in nature and difficult to separate from other materials. Today, Al is favored for its lightweight malleability, ductility, resistance to corrosion and durability. It can be easily machined and cast. Second to Fe, Al is the most widely used metal in the global economy, including areas of transportation, packaging, construction, consumer durables, electrical transmission lines and machinery. Al is derived from bauxite ore in the Earth, which is an abundant material. Bauxite is converted to alumina, an aluminum oxide, which is further electrolyzed to very high purity concentrations. Al is used in the PV industry for making supports, or the Body of System (BOS), including frames for solar panels, and electricity transmission lines. An amount of 102 kg of Al is consumed per megawatt of overall PV installations [6,37]. The 2016 production of worldwide Al was estimated at 57.6×10^6 t (57.6 Mt) with a yearend capacity of 72.5 Mt [46]. Reserves are not calculated due to the immense global resources of bauxite, which are sufficient to meet world Al demand well into the future. Al is also important to PV production, as it is the primary ore for Ga.

23.3.3 Gallium

Ga is a secondary by-product from aluminum oxide production from bauxite, although some is produced from Zn-processing residues. It is used in producing integrated circuits and optoelectronic devices (mostly laser diodes and light-emitting diodes). For the most part, specific methods for refining Ga to semiconductor grade are proprietary information. Some likely processes involved include fractional distillation, electrolysis, extracting, vacuum distillation, fractional crystallization, zone melting, and single crystal growth. The exact process would depend upon Ga concentrations in the alumina being refined. Ga is most relevant to the PV industry in terms of its application in thin-film PV technology, specifically CIGS. Amounts of between 0.12–6.17 kg of high-grade Ga are consumed per megawatt of CIGS installations [6,33]. The 2016 worldwide production of low-grade Ga was estimated at 375 t and high-grade at 180 t. Year-end low-grade capacity was estimated to be at 730 t and high-grade at 320 t [46]. As only a very small portion of the Ga present in bauxite and zinc ores is recoverable and factors controlling recovery are proprietary, United States Geological Survey (USGS) does not estimate Ga reserves, however, Grendall and Hook estimate them to be 6.5×10^3 t (6.5 kt) [33]. The Ga contained in world resources of bauxite is estimated to exceed 1 Mt and the USGS postulates that a considerable quantity could be contained in world Zn resources, however only a very small percentage is techno-economically recoverable [46].

23.3.4 Copper

Cu has played an important role in human development for thousands of years. It is favored for its properties of ductility, malleability, resistance to corrosion, and thermal and electrical conductivity. Cu is an invaluable industrial metal mostly used in building construction, building wiring, telecommunications, and electronics. It can also be found in transportation, industrial machinery, and general consumer products. Usually associated with sulfur in natural deposits, it is currently produced by a variety of multistage processes in many countries. Generally, it begins with the mining and concentrating of low-grade ores (<0.5% Cu) containing copper sulfide minerals and then smelting and electrolytic refining to produce cathode grade Cu (>99.99% pure Cu). Hydrometallurgical Cu processing is increasing in popularity. Given that Cu is the second-best-known conductor of electricity, it is used in electricity transmission, distribution, and to some degree in every PV technology today, however, it is explicitly important for CIGS PV. Amounts as low as 16.97 kg of Cu are consumed per megawatt of CIGS installation. If one assumes a mix of solar technologies consisting of 80% c-Si, 10% a-Si, and 5% CdTe and CIGS each, then up to 2.19 t of Cu are consumed per megawatt of mix installed [6]. In 2016 worldwide production of Cu was estimated at 19.4 Mt with reserves at 720 Mt, while known world resources are currently estimated at 2.1×10^9 t (2.1 Gt), plus another estimated undiscovered 3.5 Gt [46]. Cu is also important to PV production as it is the primary ore for Se and Te.

23.3.5 Selenium

Compared to other materials, Se is relatively new to industry. Growth in consumption was driven by applications in rubber compounding, steel alloying, and Se rectifiers. By 1970 its major use was as a photoconductor in paper copiers. Today Se is used to decolorize green tint in different glasses, in architectural plate glass as an insolating agent, in catalysts to enhance selective oxidation, plating solutions, blasting caps, corrosion resistance, electrolytic production of manganese, and in Cu, Pb and steel alloys to improve machinability. It is also an essential micronutrient found in fertilizer and used as a dietary supplement for humans and livestock. Se is recovered from the electrolytic refinement of Cu. It accumulates in the electrolyte or "anode slimes" after Cu electrolysis has been performed. In general, Se is concentrated using very strong acids or roasting and leaching, followed by hydrometallurgical processing and precipitation. Se is an integral semiconductor for producing CIGS modules. Amounts ranging between 0.5 and 84.41 kg of Se are consumed per megawatt of CIGS installed [6,33]. In 2016 worldwide production of Se was estimated at 2.2 kt with reserves at 100 kt [46].

23.3.6 Tellurium

Tellurium is also relatively new to industry and also a relatively rare element. It was used in the chemical industry as a vulcanizing agent and accelerator in the processing of rubber and a catalyst for synthetic fiber production. Today it used as an alloying additive in steel, copper, lead, cast iron, and malleable iron. Like Se, Te is recovered from Cu anode slimes but in much lower concentrations. It is also recovered from effluent in the process of refining Pb. General steps in refining Te include separation, roasting, leaching, cementation, electrolysis, vacuum distillation, and zone refinement. One of the major uses for high-grade Te today is as a semiconductor in the development of CdTe thin-film PV modules. Up to 90.38 kg of Te is consumed per megawatt of CdTe installed [6,33]. In 2016 worldwide production of Te was estimated at 400 t with reserves at 2.5 kt [46].

23.3.7 Silicon

Si is the most abundant element on Earth. Si in the form of silicates constitutes more than 25% of the Earth's crust. Today the uses of Si include the production of Al and Al alloys, various uses in the chemical industry and the electronics industry. Relatively small quantities of Si are processed into high-purity Si for use in the semiconductor industry as chips for computers and the solar energy industry as photovoltaic cells. Si is typically sourced from silica in various natural forms, such as quartzite. It is generally produced in submerged-arc electric furnaces and then further refined by the Siemens, modified Siemens, Fluidized Bed Reactor or Czochralski process. From quartz sand, Si is refined into metallurgical grade Si and from there it is further refined into electronic grade Si, which is of the highest purity. Historically, electronic grade Si was mostly supplied to the computer industry for making integrated circuits and other electronic components. Since the purity required for producing c-Si and a-Si modules is lower than that for integrated circuits, the PV industry relied on the "off-grade" Si that was not suitable for the electronics industry. Today, the PV industry has grown to the size that a small proportion of solar-grade Si comes from "off-grade" sources and most of it now comes from sources specific to PV production. Amounts in the area of 33.08 kg of Si are consumed per megawatt of p-Si installed, 40.07 kg $(MW)^{-1}$ for m-Si and 18.4 kg $(MW)^{-1}$ for a-Si [6,37]. It is possible that solar grade Si be sourced from recycled waste in the future, unfortunately the present economics are not in favor of it. The USGS reports that an estimated 7.2 Mt of Si was produced in 2016; however, that includes ferrosilicon, which is not a pure Si metal [46]. They provide no reserves or resources data due to its abundance. Despite this abundance in material, glass production capacity is a concern and Burrows and Fthenakis predict that thousands of new float-glass plants will have to be built to meet industry needs over the next 20 years [47].

23.3.8 Silver

The history of Ag in human civilization is thousands of years old as it has been used for utensils, jewellery, and even currency. Ag exhibits high ductility, electrical conductivity, malleability, and reflectivity. In fact, of all metals, Ag has the whitest color, the highest optical reflectivity, and highest thermal and electrical conductivity. Today it can be found in electronics, coins, jewellery and silverware, photography, antimicrobial applications, clothing, pharmaceuticals, plastics, batteries, bearings, brazing, soldering, catalytic converters, electroplating, inks, mirrors, water purification, wood treatment, and solar photovoltaics. The chemical characteristics of Ag are similar to Cu and it can even substitute Cu on an atomic level in most minerals formed in the ground. Ag can be a principle product for a mine, or it can be obtained as a by-product of Pb, Zn, Cu, and Au mining. Once extracted, Ag containing material is beneficiated and refined by smelting and leaching. To produce solar-grade Ag, the concentrate also undergoes electrolytic Cu refining. Ag can be found in all forms of PV technology depending upon the design. It is used in c-Si for metallization of the module and in general for its excellent electrical conductivity. Cu can substitute for it in some applications. Amounts ranging from 5.17 to 19.2 kg of Ag are consumed per megawatt of overall PV installations [6,21]. In 2016 an estimated 27 kt of Ag was produced worldwide and reserves were estimated at 570 kt [46].

23.3.9 Zinc

After Fe, Al, and Cu, Zn is the most mined metal in the world. Zinc sulfide, or sphalerite is the primary mineral ore for Zn production and it is never found in its elemental form. It is mostly consumed as a metal for coatings to protect iron and steel from rusting, alloys such as bronze and brass, Zn based dye casting alloys and rolled Zn; however, it also has applications in rubber production and medicines. It is required for proper growth and development in plants and animals and is the second most common trace metal found in the human body after Fe. Zn is typically refined through electrolysis. Although Zn is not directly used in PV modules, it is necessary for PV in that it is the primary metal for Cd, Ge, and In, which may not otherwise be produced economically. It is estimated that 29.99 kg of Zn is required for every megawatt of overall PV installation [6]. In 2016 worldwide production was estimated at 11.9 Mt with reserves of 220 Mt [46].

23.3.10 Cadmium

Cd is also a relatively new industrial metal. It is soft, malleable, and ductile. Industrial applications include rechargeable batteries, pigments, coatings, plating, and stabilizers for plastics. Since Cd shares certain chemical properties with Zn, it can substitute Zn in sphalerite formations. Zn:Cd ratios in sphalerite range from about 200:1 to 400:1 [46]. Cd is generally recovered as a by-product from the process of electrolytic refinement or smelting of Zn. It is concentrated through a SX/EW process, although it can also be further refined using vacuum distillation or zone refinement. One of the major uses for high-grade Cd today is in the production of CdTe thin-film PV modules, and is also used in the production of CdS that forms a heterojunction in CIGS manufacturing as well. Amounts as low as 0.93 kg of Cd are consumed per megawatt of CIGS installed, compared to about 83 kg consumed per megawatt of CdTe installed [6,33]. In 2016 worldwide production of Cd was estimated at 23 kt [46] and reserves are not calculated by the USGS, although Grendall and Hook estimate them to be 500 kt [33].

23.3.11 Germanium

Ge is another metal that is relatively new to industry. It is hard, has a metallic luster, the same crystal structure as a diamond and is brittle. Ge is also a semiconductor and once seemed very promising as a transistor material. Unfortunately, among other deficiencies, it proved difficult to concentrate to the necessary purities. It was replaced by very-high quality Si in transistors but was then used for fiber optics, infrared systems, and polymerization catalysts. Today, the energy costs associated with semiconductor grade Si are being reevaluated and Ge is once again considered as a good option. Like Cd and In, Ge is a by-product of Zn production and occurs as germanite, which is associated with Zn ores. Concentration of Ge is very energy intense and typically consists of precipitation, electrorefining or leaching with solvent extraction, and then zone refinement. Today, Ge is mostly of interest in a-Si production. It is estimated that 4.07 kg of Ge is consumed per megawatt of a-Si installment [33,48]. In 2016 worldwide production was estimated at 160.5 t [46] and although the USGS do not have data on Ge reserves, Frenzel et al. estimate the reserves to be around 119 kt [49].

23.3.12 Indium

Like Ge and Cd, In is a relatively new industrial metal and is derived from Zn production. Applications for In include fusible alloys, solders and electronics, semiconductors and thin films for liquid crystal displays—which is its main application today. Typically, the residues from Zn smelting and roasting are leached and concentrated into a precipitate that includes precipitation. Process steps for purification include recovery from Zn leaching, roasting, concentration, solvent extraction, back extraction, plate immersion, anode casting, electrorefinement, vacuum distillation, and zone refinement. The last three steps can be repeated several times until the metal is of sufficient purity for application in CIGS technology. Amounts ranging between 4.5 and 83.79 kg of In are consumed per megawatt of CIGS installed [6,33]. In 2016 worldwide production of Se was estimated at 655 t [46]. The reserves are not calculated by the USGS although Grendall and Hook estimate them to be 11 kt [33].

23.4 Energy Costs of Materials

This section considers the energy costs of developing the materials necessary to construct PV systems. Since the PV systems are an energy delivery system, insight into the energy required to develop the materials, especially as a trend over time, is of concern (see Chapter 26). If the literature on energy requirements for PV systems is small, that for material constraints is meager, and synthesis for the energy requirements for materials in PV systems is inadequate. The role of energy in the processes by which materials are obtained and prepared for industry is rarely mentioned in the past [50–52]. Little work is available on the energy requirements to generate the elements from primary ores and is scarce concerning that for secondary ores important to advanced technologies, such as PV.

The extraction of metals from geological deposits on Earth and their concentration for modern industrial utility requires a significant amount of energy. In 1986, Hall et al. [53] provided a thorough historic examination into resource materials and their energy requirements, though unfortunately their data is for the 1970s and previous years. Subsequent data has been comparatively inconsistent, although improving due to Life-cycle Analysis standards, and usually held or withheld by private interests. In 2009, Fthenakis et al. did give some estimates for the energy use in mining and smelting of minor elements [26]. Later, Hall et al.'s work was somewhat updated by Gupta and Hall in 2012 [54] where the authors also raised the issue of rising energy costs as ore grades are degraded and the impact of such increases on the overall efficiency of PV systems as an energy delivery system. That question was further explored in the recent works of Koppelaar and Koppelaar [55] who evaluate the impact of ore grade and depth on the energy inputs to Cu mining and development, and also by Fizaine and Court [56] who developed a model for estimating the impacts of material depletion on renewable energy in general.

As can be understood from this chapter so far, Cu is integral to the future of the PV industry because it is required in PV systems themselves and in the infrastructure required to transport and consume the energy produced. It also has a useful amount of scientific information associated with it given its history and prevalence in industry. As is the case for most resource extractions, as Cu is mined, the grade of the ore tends to decline overtime [53–60]. The economic rationale is to first produce metals from the most profitable, easily accessible and thus less costly deposits. These tend to be of higher concentration and closer to the surface. As the best resources are consumed first, the operation tends toward moving deeper into the ground to lower grade materials. Therefore, it is observed that as one depletes a metal deposit, it reduces in concentration and thus the more material must be handled and the more energy must be expended per unit of metal obtained [54]. This relation was recently observed and updated for Cu by Koppelaar and Kopelaar [55] and 34 metals by Court and Fizaine [56]. The opposing factor to increasing energy costs given ore degradation is process efficiency, or improved technology in general.

As discussed it is very difficult to generate good data for mining metals, as no two mines are the same. Differences include those in the deposits, both in quantity and quality of the ore, the geography, and even the technology at hand. A time-series industry-wide view can therefore be very useful. It is important to observe how the energy costs associated with mining and metal extraction and production relate to the rest of the economy over time. Fig. 23.3 illustrates the growth in energy consumption of the industry as it relates to the energy consumption of the total economy.

Global mining and quarrying includes all upstream activities pertaining to mineral extraction and beneficiation applied to metals and nonmetals, excluding fossil fuels. Energy consumption for nonmetallic minerals, nonferrous metals, and iron and steel refer to downstream activities, including final energy consumption, which refers to the entire global economy. We can see that, especially in the twenty-first century, the energy consumption for the materials industry is growing faster than that for the global economy as a whole. This is especially true for upstream activity (mining and quarrying), which is sensitive to ore grades and where the most material is handled. One reason for this increase is the extraction and processing of increasing volumes of ore. Another reason is the increasing energy required to obtain more remote and less concentrated deposits. In other words, there could be an increase in quantity and decrease in quality.

In the United States, during the year 2000, 92% of metals obtained were through surface mining [56]. Of the energy consumed approximately 35% was electrical, 32% fuel oil, and 33% coal, gas, and gasoline [40]. Electricity is mostly used for ventilation, water pumping, crushing, and grinding. Diesel fuel is used mostly in hauling and transportation in general. Typically, electricity is the major source of energy consumed in an underground mine due to the ventilation needs, where surface mining uses mostly diesel fuel for digging and hauling. It is estimated that on average two-thirds of the energy consumption in

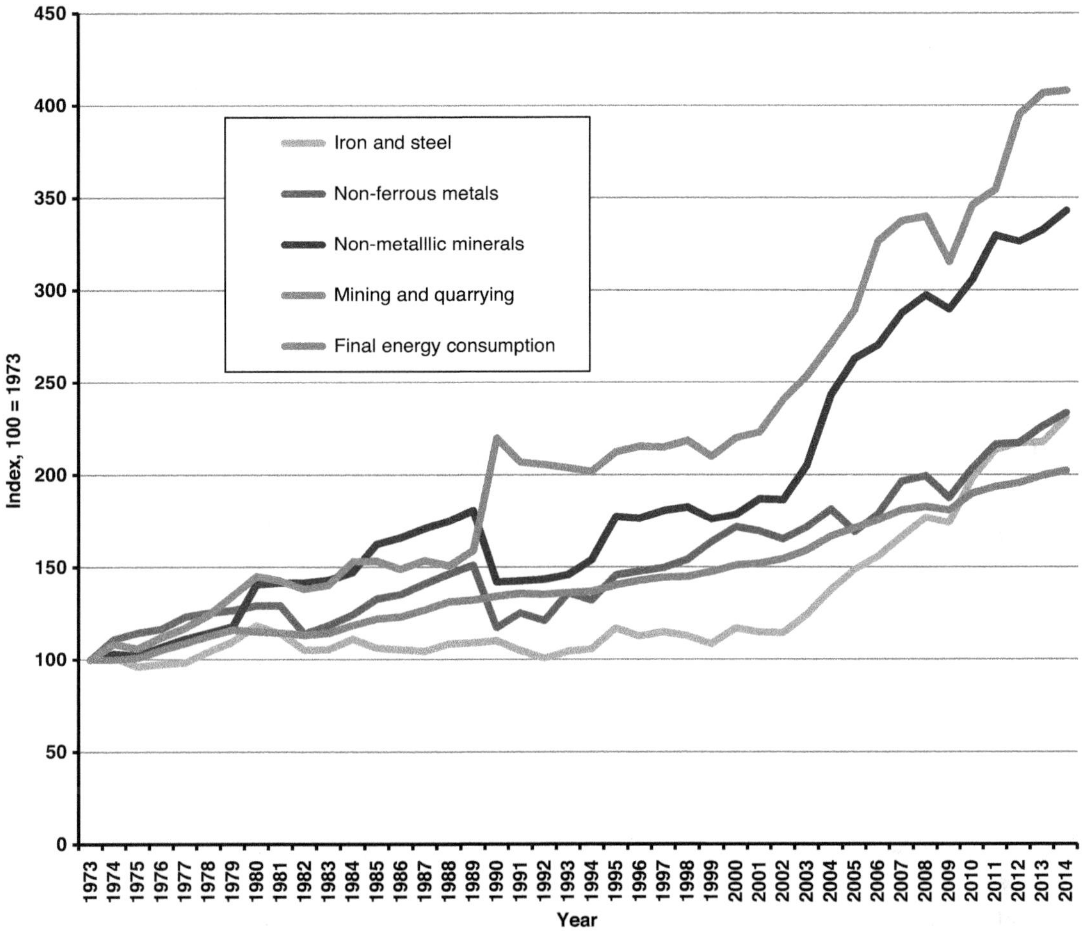

FIGURE 23.3 Indexed growth in the global mining sector versus total energy consumption [61]. Mining and quarrying activities do not include fossil fuel extraction.

materials processing is due to the crushing and grinding process [41,54]. Ultimately, ore grade defines the final energy cost of a mining operation, but given the numerous additional processes in pyrometallurgical processing, it is estimated that the energy cost can be up to 325 times higher compared to hydrometallurgical beneficiation [40].

Table 23.3 lists some of the energy costs per phase of Cu mining and refining. As can be understood from the table, the energy cost is highly variable based upon the process pathway chosen for production for a given mine. In their study, Koppelaar and Koppelaar [55] found that ore grades had a significant impact on both diesel and electricity consumption, while depth only influenced electricity use with little impact from variation. They also found that ore grades and depth impact energy consumption more significantly in underground than surface mines. Mines that use only leaching or in situ recovery showed no significant impact from ore grade or depth variation.

Table 23.3 Summary of Energy Costs Associated with Cu Mining and Refining Processes by Ore Grade

Process	Energy Cost/MJ(kg)$^{-1}$	Ore Grade
Mining	13.90	1.00
	6.30	1.00
	13.70	0.50
	27.40	0.25
Mining/100 m depth	4.60	0.55
	2.90	–
Crushing	2.50	1.00
	0.90	1.00
	1.80	0.50
	3.60	0.25
Grinding	6.4–10.6	1.00
	11.9–21.1	0.50
	23.9–42.2	0.25
Floatation, regrinding, and tailings disposal	2.00	1.00
	4.10	0.50
	8.10	0.25
Smelting	19.9–45.3	–
	10.30	–
	9.3–12.8	–
	11.30	–
Electrorefining	5.90	–
	3.10	–
Heap leaching	1.10	1.00
	2.10	0.50
	4.20	0.25
Solvent extraction	4.60	–
Electrowinning	8.50	–
Heap leaching and SW/EW	45.50	1.00
Heap leaching including embodied energy	103.00	1.00
	179.00	0.50
	322.00	0.25
Beneficiating and smelting including embodied energy	72.13	1.00
	131–244	0.50
	249–474	0.25
Site transport	5.3–8	0.75
	7.20	–
Product transport	0.28	

Adapted from [55].

The idea of handling large quantities of materials and its relation to energy costs can be used to describe another trend in materials production. Table 23.3 implies that there should be a difference between the energy costs of primary ores, which require crushing and grinding, and secondary ores, which are typically electrorefined out of waste streams from primary ore beneficiation. In general, the energy cost for annual primary ore is much higher than that of its secondary ores. Fig. 23.4 illustrates the breakdown of the global downstream mining costs of materials, both primary and secondary in the year 2014. The secondary ores are all represented within the 3% "other nonferrous metals," including some primary metals.

Fizaine and Court found that sectors of global metal extraction and production represent a significant share of total global final energy consumption. They reported a range of about 7%–10% from literature and themselves calculated around 10% for 2011 [56]. Uncertainty around their calculation is due to differing years of energy cost reporting and that the method of energy cost accounting for coproduction of metals may differ from one

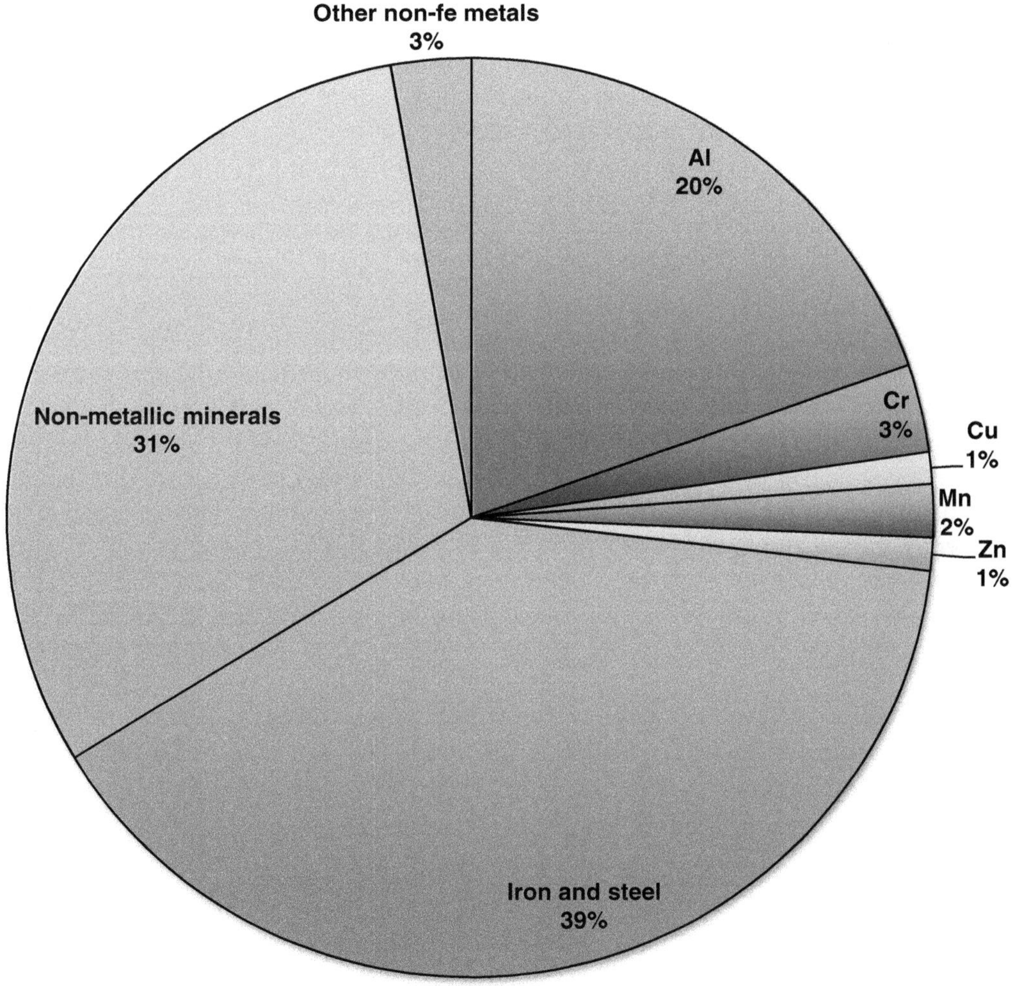

FIGURE 23.4 Downstream energy consumption costs by type of metal in 2014. Energy consumption of individual metals referenced from [56,62] and industry data referenced from [61]. Thirty-five ores, including some primary ores, secondary ores, platinum group metals, and rare earth metals are represented with the "other non-Fe metals."

study to another. For 2014, the global metal mining industry represented about 14.5% of total primary energy finally consumed and about 28% of the energy consumed in total mining [56,61,62]. A summary of the estimated energy cost per elemental material is available in Table 23.4. As illustrated, the highest energy cost among metals produced in 2014 was for Al by far. Al is also the most widely recycled metal on Earth.

Just as primary ore energy costs tend to be higher than that for secondary ores, primary materials can also have a much higher energy cost than secondary, or recycled, materials. In general, it is accepted that to increase life cycle efficiency and lower energy costs for PV systems, a serious effort is going to have to be made for the recycling of electronics sometime in the future. In fact, from 1900 to 2010, global human-made material stocks accumulating in buildings, infrastructure, and machinery increased 23 fold and required half of all the annual global resource use to be improved, maintained, or operated [63]. Despite efforts to improve recycling rates, only 12% of materials inflows to stocks are from secondary sources. There are avenues for hope in this regard but the situation is complex.

The impact of recycling PV panels and electronics to the energy costs of PV production is already beginning to be studied. This work could inform material and design choices in the future to minimize life cycle energy costs and maximize material throughput. Pathways for recycling PV and manufacturing waste have been studied for some time [64] and preliminary results show that exhaustive recovery of PV materials has the potential to reduce energy costs by more than half for mature c-Si and thin-film technologies [65]. However, although intuition would suggest that cheaper low-efficiency modules be discarded and expensive energy intense modules recycled, the complex materials might be too costly to disassemble and so there may be no incentive for producers to use them. Also, although Al frames are easily recyclable, they also increase energy intensity and must be disassembled. Ultimately, producer take-back, if not mandated will continue to be dictated by economics, but when regulated may provide incentives for design efforts that ease recycling.

Currently, of the most important metals utilized in the PV industry, only Al and Cu experience any significant recycling efforts from industry or the public [46]. No secondary ores are recycled in any great quantities. Although there is plenty of room for much needed improvement in these rates, especially from an energy consumption point of view, recycling rates are not yet increasing significantly. Other metals that are expected to require significant secondary recovery and could be called critical include In, Te, and Ag. Economic factors that influence extraction and processing have led to significant wastage in the PV industry and it is becoming important to consider mine wastes as a major contributor to future supplies. Possibly, the immense and potentially growing amounts of pollution due to the mining required for materials extraction will lead to the public demanding their clean up [66,67]. Perhaps this would provide incentive for recovery from waste. It is possible that with concerted efforts to recycle and employ secondary recovery technology and policy that In supplies might last well into the twenty-first century [35]. To avoid the issue of critical metals, at some point at least, the PV industry will have to either encourage public recycling or find incentives to improve material recovery and secondary production on their own [30].

Table 23.4 Estimates of the Energy Cost Associated With the Production of Different Metals in 2014 [56,62]

Metal	Energy Cost of Production/GJ t^{-1}	2014 Production/t	Total Energy Cost/GJ	Share of Energy Consumption for Metals Production/%
Aluminum	212	50,500,000	10,706,000,000	66.1953
Antimony	13	158,000	2,054,000	0.0127
Arsenic	28	36,400	1,019,200	0.0063
Beryllium	457.2	290	132,588	0.0008
Bismuth	56.4	13,600	767,040	0.0047
Cadmium	110	22,400	2,464,000	0.0152
Cerium	354	24,000	8,496,000	0.0525
Chromium	64	26,400,000	1,689,600,000	10.4468
Cobalt	322	123,000	39,606,000	0.2449
Copper	33	18,500,000	610,500,000	3.7747
Gadolinium	2,162	7,500	16,215,000	0.1003
Gallium	12,660	435	5,507,100	0.0341
Germanium	2,215	173	383,195	0.0024
Gold	68,400	2,990	204,516,000	1.2645
Hafnium	633	90	56,970	0.0004
Indium	2,875	844	2,426,500	0.0150
Iridium	2,100	3	6,300	0.0000
Lanthanium	219	12,500	2,737,500	0.0169
Lead	20	4,870,000	97,400,000	0.6022
Lithium	433	33,000	14,289,000	0.0883
Magnesium	437.3	1,055,000	461,351,500	2.8525
Manganese	56.9	17,800,000	1,012,820,000	6.2623
Mercury	409	2,350	961,150	0.0059
Molybdenum	148	281,000	41,588,000	0.2571
Neodymium	392	75,900	29,752,800	0.1840
Nickel	114	2,450,000	279,300,000	1.7269
Palladium	5,500	193	1,061,500	0.0066
Platinum	270,500	147	39,763,500	0.2459
Praseodymium	220	2,500	550,000	0.0034
Rhenium	171	44.7	7,643.7	0.0000
Rhodium	14,200	25	355,000	0.0022
Silver	1,582	26,800	42,397,600	0.2621
Tantalum	1,755	1,200	2,106,000	0.0130
Tin	207	286,000	59,202,000	0.3660
Titanium	430	194,000	83,420,000	0.5158
Tungsten	357	86,800	30,987,600	0.1916
Vanadium	517	82,700	42,755,900	0.2644
Yttrium	756	600	453,600	0.0028
Zinc	48	13,300,000	638,400,000	3.9472
Zirconium	1,371.5	1,420	1,947,530	0.0120

Uncertainty around these numbers is due to variation in years of reporting, accounting for coproduction, and variability in processing methods.

Options to relieve this stress on the PV industry include increasing energy efficiency in production at rates that offset the effects of decreasing ore grades, or improving the ore grades themselves. This can be achieved through greater recycling efforts, switching materials to those that are not energy intensive or constrained, or finding new resources of high-quality materials. Interestingly, it is the third option to increase ore grades—finding new resources—that seems to be receiving the most investment. Currently, there is a push for plans and technology being developed for NASA to optically mine asteroids for materials [68]. In other words, using lenses in outer space to concentrate the sun's energy to such use as to mine different materials from asteroids. Another colossal endeavor to find new discoveries is that for deep-sea mining of the ocean floor [69]. The other two options, including recycling, are currently deemed uneconomical.

23.5 Conclusion

It is not just we that are made of "star stuff," but our economy as well. Materials are vital to the economy, especially when it comes to those parts of our industries that we depend upon for our energy, such as the PV industry. Not just the availability, but also the energy costs of developing such materials are very important. It is currently uncertain if we are going to realize disruptions in very large scales of PV development from metals critical to the industry, whether from depletion, competition from other industries, or uneconomic energy costs. Assessing precisely how much material we have, especially secondary ores, and also the energy costs associated with them is complicated as every mining operation is different. Currently we are experiencing a rapid-upscale in global PV capacity installations and there are some concerns over whether or not we have the materials or the production capacity to keep up. Also, although PV modules are becoming more efficient at producing electricity, the energy costs for securing the materials for development is increasing. So far the efficiency gains seem to be outperforming the increased costs, but this dynamic needs more study. One of our most critical and urgent needs if we are to achieve a better understanding of a future involving a continued rapid-upscale of PV electricity, is better information about material availability in the long term and the energy costs associated with that.

References

[1] Greenfield A, Graedel TE: The omnivorous diet of modern technology, *Res Cons Recyc* 74:1–7, 2013.

[2] Graedel TE, Erdmann L: Will metal scarcity impede routine industrial use? *Mat Res Soc Bull* 37:325–331, 2012.

[3] Kleijn R, Van Der Voet E, Kramer GJ, Van Oers L, Van der Giesen C: Metal requirements of low-carbon power generation, *Energy* 36(9):5640–5648, 2011.

[4] Vidal O, Bruno G, Arndt N: Metals for a low-carbon society, *Nat Geosci* 6 Oct:894–896, 2013.

[5] Ali SH, Giurco D, Arndt N, Nickless E, Brown G, Demetriades A, et al: Mineral supply for sustainable development requires resource governance, *Nature* 543(Mar 16):367–372, 2017.

[6] The World Bank: *The growing role of minerals and metals for a low carbon future*, Washington D.C, 2017, World Bank Publications, The World Bank Group.

[7] Andersson BA: *Material constraints on technology evolution: the case of scarce metals and emerging energy technologies*, Goteborg, Sweden, 2001, Thesis for the Degree of Doctor of Philosophy, Department of Physical Resource Theory, Chalmers University of Technology and Goteborg University.

[8] Cranstone DA: *A history of mining and mineral exploration in Canada and outlook for the future*, Ottawa, Canada, 2002, Department of Natural Resources.

[9] N.R.C.: *Minerals, critical minerals and the US economy*, Washington D.C., US, 2007, The National Academies Press.

[10] Erdmann L, Graedel TE: Criticality of non-fuel minerals: a review of major approaches and analyses, *Environ Sci Technol* 45:7620–7630, 2011.

[11] Graedel TE, Reck BK: Six years of criticality assessments: what have we learned so far? *J Ind Ecol* 20:692–699, 2015.

[12] Mudd GM, Jowitt SM, Werner TT: The world's by-product and critical metal resources part I: uncertainties, current reporting practices, implications and grounds for optimism, *Ore Geol Rev* 86:924–938, 2017.

[13] Hubbert MK: *Nuclear energy and the fossil fuels (presented before the spring meeting of the Southern District Division of Production, American Petroleum Institute, San Antonio, Texas, March 8, 1956)*, Houston, US, 1956, Shell Development Company, Exploration and Production Research Division.

[14] Rosa RN, Rosa DRN: Exergy cost of extracting mineral resources. In Miguel AF, Heitor Reis, Rosa RN, editors: *Evora, Portugal: Proceedings of the 3rd International Energy, Exergy and Environment Symposium*, CGE-Evora Geophysics Centre, 2007.

[15] Gordon RB, Bertram M, Graedel TE: Metal stocks and sustainability, *Proc Natl Acad Sci* 103(5):1209–1214, 2006.

[16] Laherrere J: *Peaks in Argentina Latin America and the world*, Washington D.C., US, 2010, ASPO Conference.

[17] Nassar NT, Graedel TE, Harper EM: By-product metals are technologically essential but have problematic supply, *Sci Adv*:1–10, 2015.

[18] Willis P, Chapman A, Fryer A: *Study of by-products of copper, lead zinc and nickel*, Oakdene Hollind for International Lead and Zinc Study Group, International Nickel Study Group & International Copper Study Group, 2012.

[19] Jones N: *A scarcity of metals is hindering green technologies*, Yale Environment 360, Yale School of Forestry and Environmental Studies, 2013.

[20] Sherwood J, Ditta A, Haney B, Haarsma L, Dale M: Resource criticality in modern economies: agent-based model demonstrates vulnerabilities from technological interdependence, *Biophys Econ Resour Qual* 2:9, 2017.

[21] Grandell L, Thorenz A: Silver supply risk analysis for the solar sector, *Renew Energy* 69:157–165, 2014.

[22] Grandell L, Lehtila A, Kivinen M, Koljonen T, Khilman S, Lauri LA: Role of critical metals in the future markets of clean energy, *Renew Energy* 95:53–62, 2016.

[23] Andersson BA: Materials availability for large-scale thin-film photovoltaics, *Prog Photovolt Res Appl* 8:61–76, 2000.

[24] Keshner MS, Arya R: *Study of potential cost reductions resulting from super large- scale manufacturing of PV modules: final subcontract report*, Golden, US, 2004, National Renewable Energy Laboratory.

[25] Feltrin A, Freundlich A: Material considerations for terawatt level deployment of photovoltaics, *Renew Energy* 33:180–185, 2008.

[26] FthenakisV, Wang W, Kim HC: Life cycle inventory analysis of the production of metals used in photovoltaics, *Renew Sustain Energy Rev* 13:493–517, 2009.

[27] Wadia C, Alivisatos AP, Kammen DM: Materials availability expands the opportunity for large-scale photovoltaics deployment, *Environ Sci Technol* 43(6):2072–2077, 2009.

[28] Fthenakis V: Sustainability of photovoltaics: the case for thin-film solar cells, *Renew Sust Energy Rev* 13(9):2746–2750, 2009.

[29] Candelise C, Speirs JF, Gross RJK: Materials availability for thin film (TF) PV technologies development: a real concern? *Renew Sust Energy Rev* 15(9):4972–4981, 2011.

[30] Fthenakis VM: Sustainability metrics for extending thin-film photovoltaics to terawatt levels, *MRS Bull* 37(4):425–430, 2012.

[31] CMI. The Critical Materials Institute, an energy innovation hub: What is the problem? The Ames Laboratory, U.S. Department of Energy (CMI). Available from: https://cmi.ameslab.gov/what-CMI-does.

[32] Houari Y, Speirs J, Candelise C, Gross R: A system dynamics model of tellurium availability for CdTe PV, *Prog Photovolt: Res Appl* 22:129–146, 2014.

[33] Grandell L, Hook M: Assessing rare metal availability challenges for solar energy technologies, *Sustainability* 7:11818–11837, 2015.

[34] Davidsson S, Hook M: Material requirements and availability for multi-terawatt deployment of photovoltaics, *Energy Policy* 108:574–582, 2017.

[35] Werner TT, Mudd GM, Jowitt SM: The world's by-product and critical metal resources part II: A method for quantifying the resources of rarely reported metals, *Ore Geol Rev* 80:658–675, 2017.

[36] Mueller SR, Wager PA, Turner DA, Shaw PJ, Williams ID: A framework for evaluating the accessibility of raw materials from end-of-life products and the Earth's crust, *Waste Manag*, 2017.

[37] Frischknecht R, Itten R, Sinha P, Wild-Scholten M, Zhang J, Fthenakis V, et al: Life cycle inventories and life cycle assessment of photovoltaic systems, *Int Energy Agency (IEA) PVPS Task 12, Report T12-04:2015.*, 2015.

[38] Fraunhofer ISE. Photovoltaics report. Freiburg, Germany: Fraunhofer Institute for Solar Energy Systems, ISE, PSE AG; 12 July 2017. Available from: www.ise.fraunhofer.de.

[39] British Petroleum: *BP statistical review of world energy June 2017*, British Petroleum Group, June 2017, Available from: http://www.bp.com/content/dam/bp/en/corporate/pdf/energy-economics/statistical-review-2017/bp-statistical-review-of-world-energy full-report.pdf.

[40] BCS Inc. Mining industry of the future: energy and environmental profile of the U.S. mining industry. Washington D.C., US: US Department of Energy, Office of Energy Efficiency and Renewable Energy.

[41] USDOI: *Mining and quarrying trends*, Washington D.C., US, 2012, US Geological Survey.

[42] Vignes A: *Extractive metallurgy: metallurgical reaction processes*, London, England, 2011, ISTE, Wiley.

[43] Habashi F: *Textbook of pyrometallurgy*, Sainte-Foy, Canada, 2002, Metallurgie Extractive Quebec.

[44] Watt A, Philip A: *Electroplating and electrorefining of metals*, New York, US, 2006, Merchant Books.

[45] Habashi F: *A textbook of pyrometallurgy*, Sainte-Foy, Canada, 1993, Metallurgie Extractive Quebec.

[46] U.S.G.S.: *Commodity statistics and information*, U.S, 2017 January, Department of Interior, U.S. Geological Survey, Mineral Resources Program (USGS), minerals.usgs.gov/minerals/pubs/commodity/index.html.

[47] Fthenakis V, Burrows K: Glass needs for a growing photovoltaics industry, *Solar Energy Materials and Solar Cells*, 2015 January.

[48] Andersson BA, Azar C, Holmberg J, Karlsson S: Material constraints for thin-film solar cells, *Energy* 23(5):407–411, 1998.

[49] Frenzel M, Ketris MP, Gutzmer J: On the geological availability of germanium, *Miner Deposita*:471–486, 2014, Springer, Heidelberg, Berlin, Germany.

[50] Hannon B, Ruth M, Delucia E: A physical view of sustainability, *Ecol Econ* 8:253–268, 1993.

[51] Hall CAS, Lindenberger D, Kummel R, Kroeger T, Eichorn W: The need to reintegrate the natural sciences with economics, *BioScience* 51(8):663–673, 2001.

[52] Ayerse RU: *Information, entropy, and progress: a new evolutionary paradigm*, New York, US, 1994, American Institute of Physics.

[53] Hall CAS, Cleveland C, Kaufmann R: *Energy and resource quality: the ecology of the economic process*, New York, US, 1986, Wiley & Sons, Inc, 221-228.

[54] Gupta AK, Hall CAS: Energy costs of materials associated with the exponential growth of thin-film photovoltaic systems. In Ginley DS, Cahen D, editors: *Fundamentals of materials for energy and environmental sustainability*, Warendale, US and Cambridge, England, 2012, Cambridge Univ. Press, Materials Research Society.

[55] Koppelaar RHEM, Koppelaar H: The ore grade and depth influence on copper energy inputs, *Biophys Econ Resour Qual* 1:11, 2016.

[56] Fizaine F, Court V: Renewable electricity producing technologies and metal depletion: a sensitivity analysis using the EROI, *Ecol Econ* 110:106–118, 2015.

[57] Page NJ, Creasey SC: Ore grade, metal production, and energy, *J Res US Geol Surv* 3:9–13, 1975.

[58] Mudd G, Diesendorf M: Sustainability of uranium mining and milling: toward quantifying resources and eco-efficiency, *Environ Sci Technol* 42(7):2624–2630, 2008.

[59] Norgate T, Janahashi S: Low grade ores – smelt, leach, or concentrate? *Miner Eng* 23:65–73, 2010.

[60] Memary R, GiurcoD, Mudd G, Mason L: Life cycle assessment: a time-series analysis of copper, *J Clean Prod* 33:97–108, 2012.

[61] IEA. Statistics: energy balance flows, world final consumption 2014. OECD/IEA (IEA); 2017. Available from: https://www.iea.org/Sankey/#?c=World&s=Final%20consumption

[62] MMTA. Metals information page. Minor Metals Trade Association (MMTA); 2017. Available from: https://mmta.co.uk/metals

[63] Krausmann F, Wiedenhofer D, Lauk C, Haas W, Tanikawa H, Fishman T, Miatto A, Schandl H, Haberl H: Global socioeconomic material stocks rise 23-fold over the 20th century and require half of annual resource use, *Proc Nat Acad Sci*, 2017 February.

[64] Fthenakis VM: End-of-life management and recycling of PV modules, *Energy Policy* 28(14):1051–1058, 2000.

[65] Goe M, Gaustad G: Strengthening the case for recycling photovoltaics: an energy payback analysis, *Appl Energy* 120:41–48, 2014.

[66] Ives M: *Boom in mining rare earths poses mounting toxic risks*, Yale Environment 360, Yale School of Forestry and Environmental Studies, 2013 January 28.

[67] Wilt J. Why we need to clean up mining if we want a renewable energy economy. DeSmog Canada (Wilt J); July 20, 2017. Available from: https://www.desmog.ca/print/11963

[68] Sercel J: *Optical mining of asteroids, moons, and planets to enable sustainable human exploration and space industrialization*, Loura Hall, 2017, National Aeronautics and Space Administration, Trans-Astra Corp (Srcel J); April 6, 2017. Available from: https://www.nasa.gov/directorates/spacetech/niac/2017_Phase_I_Phase_II/Sustainable_Human_Exploration.

[69] Trager R. Countries poised to roll out deep sea mining in new "gold rush." Chemistry World, News, The Royal Society of Chemistry (Trager R); March 7, 2017. Available from: https://www.chemistryworld.com/news/countries-poised-to-roll-out-deep-sea-mining-in-new-gold-rush/2500509.article

24

Global Growth Trends and the Future of Solar Power: Leading Countries, Segments, and Their Prospects

James Watson, Kristina Thoring, Alyssa Pek
SOLARPOWER EUROPE, BRUSSELS, BELGIUM
j.watson@solarpowereurope.org

24.1 Introduction

Solar photovoltaic (PV) (from here on called solar) power has seen enormous growth across the globe in the past decade. In 2005, solar power deployment was limited to a global total of around 5 GWs; whereas today, just some 12 years later, the global total, from Chile to China, is 306.5 GW [1]. The story of this phenomenal growth can be attributed to two main factors. First, the drive for clean energy, built on regulatory frameworks that were designed to support early adopters across the world, though ostensibly in Europe. Second, this was supplemented by a fall in the cost of solar systems of around 75%, in less than 10 years, according to the International Renewable Energy Agency [2].

The past 10 years have been revolutionary in terms of solar deployment, but the future is even brighter with forecasts from SolarPower Europe predicting that there could be almost 1 TW of solar power production installed globally by 2021 [1]. Again, this is due to political and economic factors that are combining to offer solar power to a wide spectrum of the global population, thus cementing its place in the modern power generation arena.

This chapter examines the growth of solar power across regions and countries, from the early 2000s until the current period. It will identify the countries that initially developed solar power, and those countries where solar power is a growing industry. We will also assess the likely future path of solar power in the coming years and offer some predictions as to how and where solar will grow. This will entail an analysis of the drivers of deployment in the past, present, and future. Finally, the chapter will also examine the breakdown of the types of deployment by segment, offering views on their future development.

A Comprehensive Guide to Solar Energy Systems. http://dx.doi.org/10.1016/B978-0-12-811479-7.00024-5

24.2 Solar Growth Trends

Solar power was not a major contributor to energy production anywhere in the world until the early 21st century. At that stage, it began to grow in Europe supported by the creation of a target for renewable energy which is discussed in a European Commission White Paper [3]. A small target of 12% for the contribution by renewable sources of energy to the European Union's gross inland energy consumption by 2010 was set and countries, notably Germany, began developing the means to produce solar power on a larger scale. Nevertheless, the market remained small, being just 5.25 GWs in 2007 [4]. The main change in the fortune of solar power came in 2009 with the completion of the clean and renewable energy (CARE) package of the European Commission [5]. This regulatory initiative paved the way for the growth in solar power so that in 5 years it becomes a dominant feature of global markets.

The CARE package contained a directive on renewable energy, which set a goal of 20% of total energy to be derived from renewable resources by 2020 [5]. This ultimately created the regulatory certainty that investors needed to begin a large-scale interest in solar power. It is true that the Spanish market began to grow even before the CARE package was signed and sealed. In 2008, Spain had the largest solar market in the world [1]; this was driven by a generous feed-in-tariff scheme, which ultimately came undone after only a year. It is also true that Germany had support in place for solar before the CARE package; however, this did initially remain as a relatively small-scale industry and certainly not large enough to drive a major solar power deployment.

In 2007, solar was still a relatively novel concept. In the rest of the world with relatively small capacities installed, even in places, such as the United States and Japan and the global market was only 8.9 GW [4]. By 2012, the global market had reached 101.3 GW. This was not evenly spread globally, in fact 80% of solar installations were in Europe [1]. The solar boom in Europe, driven by regulatory measures and generous public support schemes, saw markets, such as Germany and Italy emerge as the solar leaders in the early part of the 2010s. The Energiewende in Germany placed a premium on renewable energy sources, with both solar and wind deployment growing rapidly. Germany soon became the largest solar power market in the world, a position it would hold for a number of years. The evolution of total global installed Solar PV power capacity is shown in Fig. 24.1.

The solar boom in Europe had the fortuitous effect of influencing other countries around the world. In China, the growth of solar in Europe had been well documented and the government decided to declare that solar power would be a strategic industry [6]. The burgeoning demand in Europe encouraged the Chinese government and entrepreneurs to invest heavily in solar manufacturing: mainly in cells, wafers, and panels, and later in inverters. This allowed a huge economy of scale effect to begin, delivering the cost reduction seen in solar power over the past decade [7]. Coupled with the leading research and development being undertaken in Europe, the prices of solar systems began to fall and begin the process of taking solar power from a niche interest, to a mass consumer product.

Nevertheless, the momentum of deployment began to slowdown in Europe (Fig. 24.2). Adverse economic conditions made it difficult for many countries to maintain the support schemes and some, such as Spain, the Czech Republic, Bulgaria, and Romania, applied retroactive measures to solar investments, seriously limiting the value of the assets held by owners. This began to inject uncertainty into the market, which was compounded in 2012 with the introduction of preliminary antidumping measures on solar panels and cells imported from China.

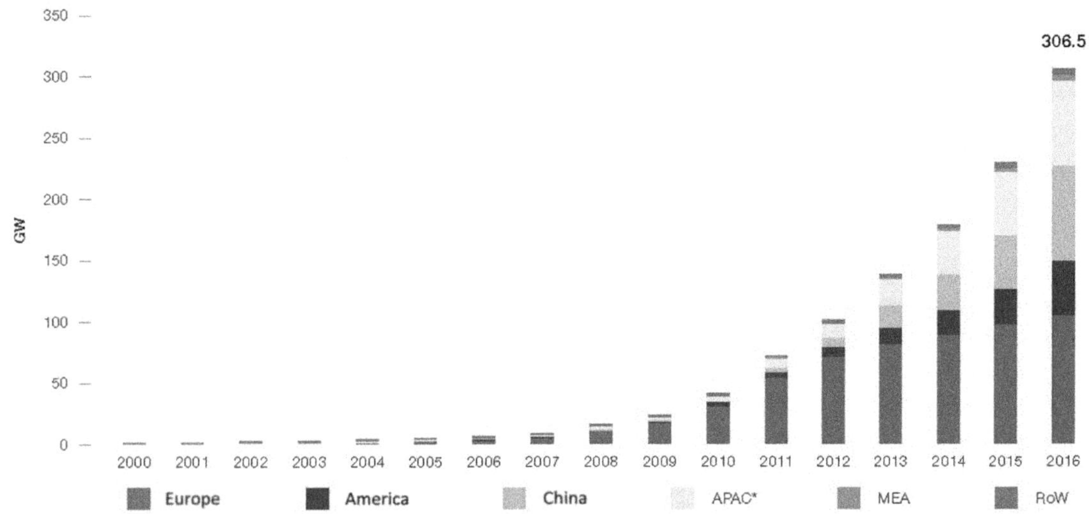

FIGURE 24.1 Evolution of total global installed solar photovoltaic (PV) capacity 2000–2016. *SolarPower Europe, 2017, p. 4.*

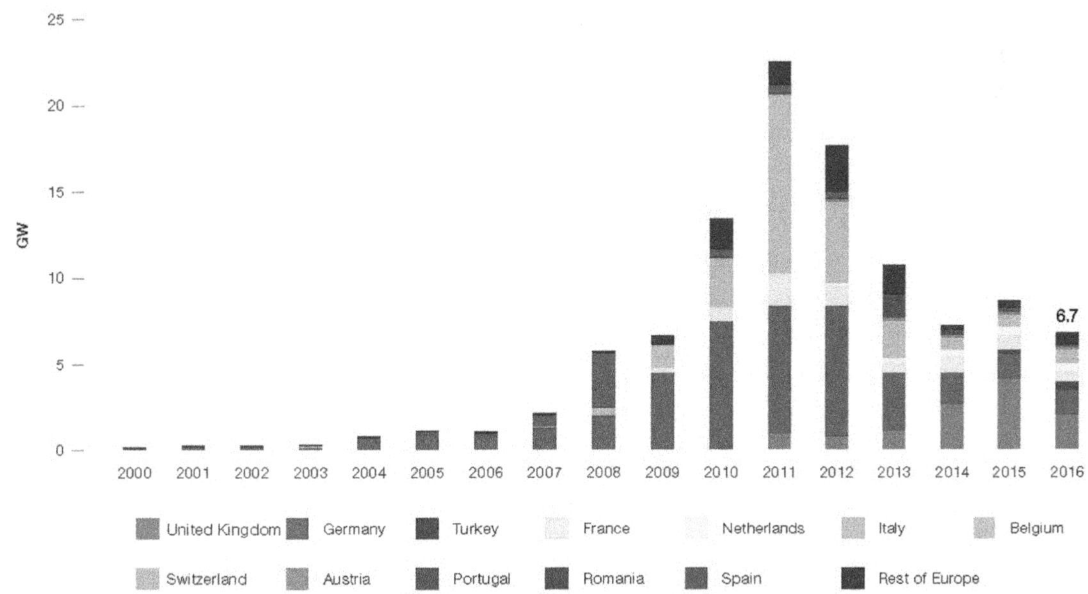

FIGURE 24.2 European solar PV connections 2000–2016 for selected countries. *SolarPower Europe, 2017, p. 16.*

The combination of these factors introduced a steep decline in the rate of installation of solar power grid connections in Europe; falling from a high of 22 GW installed in 2011 to just 7 GW 3 years later in 2014 [1]. This naturally had a huge impact on the sector in Europe, with numerous companies falling into bankruptcy and adding more pressure on the perception of the stability of the solar market in Europe. A powerful group of companies had also formed in 2011, the so-called Magritte Group, with a clear agenda to slow the deployment of renewables in Europe, and solar in particular.

As mentioned earlier, the leading national market in Europe has been Germany, with over 40 GW of solar power installed, which is more than double the second placed market, Italy, who have just under 20 GW of installed capacity. The United Kingdom is Europe's next largest market, with over 11 GW installed by the end of 2016 [1]. No other markets have over 10 GW-installed capacity in Europe. Thus three markets make up about 70% of Europe's installed capacity, proving that there is much more room for solar power growth in Europe despite the slowdown in recent years. Much of Central and Eastern Europe, which has favorable solar conditions, remains virtually a virgin solar power territory. Southern Europe also has huge potential, with Spain, Greece, and Italy all representing countries with unfulfilled solar promise. While the rate of deployment in Europe has flagged, there remains reason for optimism that the market can be reignited given the large number of countries that have yet to embrace solar.

While Europe plunged into a solar slowdown, the rest of the world began to experience a major solar boom. Asia began to awake its interest in solar power and levels of deployment slowly began to rise. China turned its huge industrial machinery not only into serving the then dominant European market, but also began to deploy high levels of solar power on an annual basis. Given the availability of the technology, space, and an enormous need to clean up air quality, solar became the power generation technology of choice. China now has the honor of hosting the world's largest solar park, the Longyangxia Dam Solar Park, which comprises 850 MW of solar—enough to power 200 000 homes. This record site has cost around 6×10^9 Yuan and has been under construction since 2013 [8].

The 13th Solar Energy Development Five Year Plan (2016–20) was adopted by the National Energy Administration in December 2016 establishing the goal to install at least 105 GW by 2020 [9]. In 2013, China was already responsible for about 20% of global installations of solar power [1]. This grew through 2017 when China, with an installed capacity of about 40 GW in 2017 represented half of the world's annual solar power deployment [1].

China was certainly not alone in driving the Asia Pacific regions solar power deployment. After the Fukishima disaster of 2011 in Japan, the government decided to move to less volatile energy sources to supply its electricity needs. This meant a major investment in solar power, and the government duly went about setting up support schemes and provisions for solar power to grow in Japan [10]. The fruit of this labor saw solar deployment move rapidly into double-digit growth, with around 10 GW of installation annually from 2012 through 2016 [1]. This has made Japan one of the leading solar power markets in the world with an installed capacity of 11% of the global total.

In global terms, the deployment of solar power has gone through a major shift. As previously outlined, the initial major development of solar power was in Europe, and through 2015, Europe remained the globally dominant region for solar deployment. In 2015, Europe still represented over half of the total deployment of solar power globally. Additionally, in 2016, Europe became the first region to reach the landmark of 100 GW of installed capacity. This was a major achievement, but was quickly equaled and surpassed by the Asia Pacific region, which went on to record a total solar deployment of over 140 GW by the end of 2016 [1]. Asia is now the powerhouse of solar power deployment, and this is a trend that is set to continue.

In 2016, India joined the group of countries turning to solar power. Prime Minister Modi launched the International Solar Alliance (ISA) in 2015 at the United Nations Conference of the Parties (COP) 21 meeting in Paris. The ISA brings together over 120 countries that commit to making solar power the core of their energy supply [11]. Having taken this bold step with but a handful of gigawatts installed in India, the necessity for action to back political deeds became apparent. India quickly set itself a target of reaching 100 GW of installed solar by 2022 [12]. This would require a rate of deployment close to 20 GW per year on average, but the plan clearly identified the need for a ramped-up approach. Adding 5 GW in 2016, India doubled its installed capacity; the expectation is that this market will indeed be the next major boom in the solar sector [13]. For this to happen, much depends on getting the system to support such deployment in India, but Prime Minister Modi is already actively pursuing matters, such as interstate electricity supply and speeding up application processes. The only danger comes from the use of local content requirements, which threaten to add instability to the market for investors.

Outside of Asia and Europe, there have been small but growing solar power markets. The Paris Climate Change Agreement has set in place a framework for solar power to be deployed in many more markets across Africa, the Middle East, Central Asia, and the Americas. In the Americas, there have been a number of countries where solar has seen recent significant growth.

Not least of these is the United States, which has seen a dramatic increase in the deployment of solar power in the past 4 years. This has been spurred on by the Obama Presidency, and in 2016 the United States became the second largest annual market for solar power globally with an installed capacity of over 14 GW; the rate of deployment almost doubled in comparison to 2015 [1]. This was driven by the Investment Tax Credit (ITC) that offered an incentive of around 30% of the cost of a solar system for consumers. With the price of solar already falling, this incentive scheme gave critical weight to see the United States become an important global leader in solar power. The solar sector has become one of the most dynamic in the US energy business; it is estimated that more people work in the solar sector than in the combined gas, oil, and coal sectors in the United States today [14]. The dynamism of the US market has also resulted in the solar sector trail blazing with new business models, such as power purchasing agreements (PPAs) directly between solar developers and large corporates, third party ownership schemes, and net metering [15].

Outside of the United States, other markets in the Americas have shown potential. Chile has seen the lowest PPA award of US$ 29.1 $(\mathrm{MW\,h})^{-1}$ and installed close to 1 GW of solar in 2016, while Mexico also has seen growth with over 1 GW of projects awarded in 2016 [1].

Other areas of the world have also seen solar growth; the Middle East and Africa countries are starting to look to solar to fill the energy gap in a cheap and efficient way. There are no landmark developments in these regions yet, but it is likely that Africa and the Middle East will soon also experience high levels of solar power penetration.

The global market has witnessed an explosion in solar deployment since the 2000s. In just 10 years, the global market has developed from roughly 20 GW installed in 2007 to over 306.5 GW in 2016 [1]. All regions of the world now play their role in solar, a onetime European phenomenon is now embraced globally. Fig. 24.3 shows the global breakdown of the top solar PV markets by the end of 2016. This trend is set to progress as the reduction in cost of solar power systems, which is expected to persist, continues to drive political support in most of the world. With global commitments, such as the Paris Climate Agreement holding the majority of countries together in addressing climate change; it is clear that the future of global solar power deployment is indeed very bright.

24.3 Future Market Growth Potential

With the continuing fall in the price of solar systems, solar electricity generation is set to become the world's cheapest energy source. The IEA forecast that by 2050, solar will indeed be the leading generation source for electricity across the globe [16]. However, price alone does not indicate whether a technology is embraced, this has been demonstrated very effectively in the earlier section with regard to Europe. It is the combination of regulation and price that determine the success of any electricity generation technology [1]. Despite Herculean attempts to develop markets in energy, the reality is that regulation

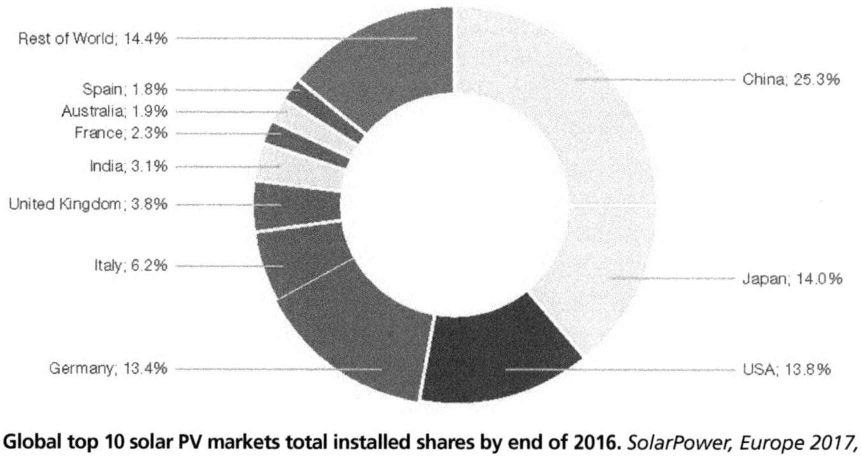

FIGURE 24.3 Global top 10 solar PV markets total installed shares by end of 2016. *SolarPower, Europe 2017, p. 6.*

of those markets and the support policies around specific generation technologies still have greater impact than the simple cost of producing energy.

Given that regulation is so important in energy, it is not surprising that when one considers what comes next in solar, one must examine the known solar potential in a country against the energy mix bias and the known regulatory environment. This makes the art of predicting solar deployment quite difficult. Although one thing that can be said with confidence, is that solar will continue to decrease in cost and that it is likely to be taken up by more and more countries, due to international initiatives, such as the Paris Climate Agreement and also through clean energy plans, such as the one currently being designed in the European Union [17]. SolarPower Europe has devised numerous forward-looking assessments in its annual Global Market Outlook, and in the most recent 2017 edition, a forecast has been made that perhaps 1 TW of solar will be installed by the end of 2021 [1].

Assessing the potential of the global solar market is a good place to begin an analysis of where solar power deployment could be heading in the short term, as well as the long term. This can then be used to locate where national markets could contribute to the overall forecast of the global total. These predictions are shown in Fig. 24.4. To begin, the forecast of SolarPower Europe is for 935.5 GW of solar to be installed globally by 2021, under optimal conditions [1]. These conditions include continued price decreases and political support for solar being kept at high levels. Taking these assumptions into account, this growth would represent an almost tripling of global solar installations over the next 5 years; a rate that is perhaps a slowdown on the last 5 years, which has seen a sevenfold

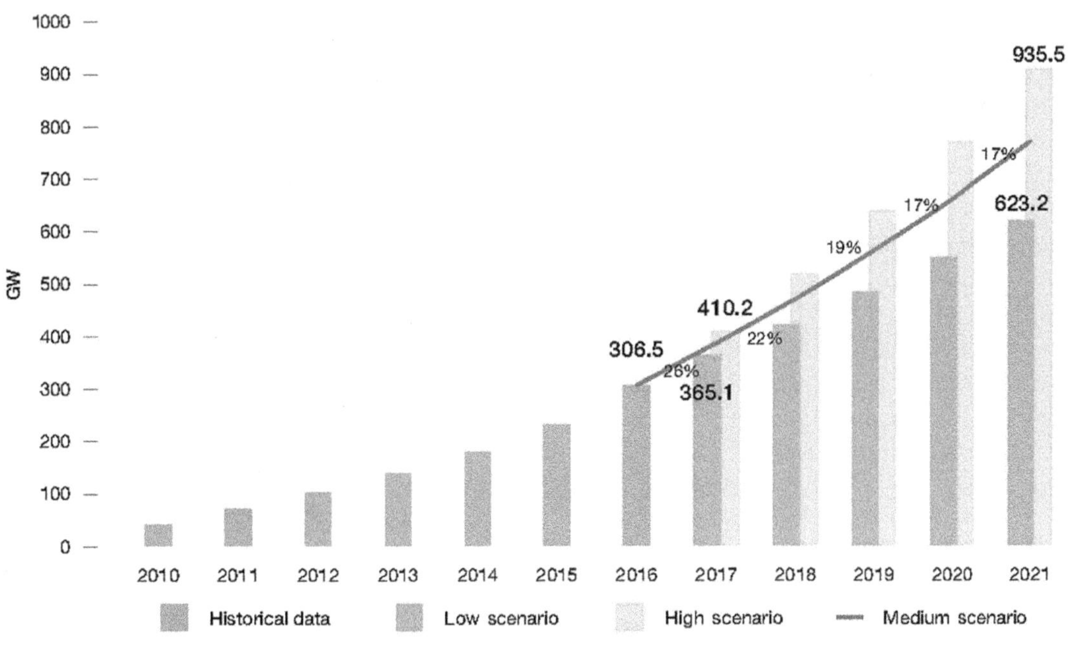

FIGURE 24.4 World total solar PV market scenarios 2017–21. *SolarPower Europe, 2017, p.16.*

increase in solar power deployment [1]. Nevertheless, as the volumes are now much larger, this represents a global solar revolution in terms of deployment. Currently, solar power represents less than 1% of global electricity demand, but with such predicted deployment of figures this would quickly grow [18].

The countries that would drive this wave of deployment are already using solar power; these include China and India. These two countries are likely to dominate the global solar markets in the future. India fully believes that it will be able to install 175 GW of solar by the mid-2020s [12]. This type of initiative will see large amounts of the near 600 GW of solar installed, which are needed to hit the forecast of SolarPower Europe. Other regions are also likely to come to the forefront. The Middle East, Saudi Arabia, Jordan, the United Arab Emirates, to name but a few, are all main lining solar power in the coming years [1]. The Central Asian Republics are also expanding their solar interests—with Uzbekistan looking to build its electricity system largely on solar power and Kazakhstan also rapidly increasing deployment [19]. Africa will likewise see a boom in solar power, from Dakar to Djibouti and from Cape Town to Cairo, most African countries have huge need for energy and no constraints from existing electricity systems. This is the perfect combination for low cost solar power to come into its own in Africa. Thus, it can be expected that 100 GW will be installed in this region in the coming few years [1]. Finally, Latin America is also on the verge of a big solar power push: Brazil, Chile, Mexico, and many others are all embracing their natural climatic advantages and the cost effectiveness of solar power.

Given the political environment in much of the world being very favorable to solar power, it is possible to imagine that in just a few years' time, more than 1 TW of solar will be installed.

As for the early pioneer markets, Europe, Japan, and the United States, it is not entirely clear what will happen at this stage. All have huge potential, as mentioned earlier in the chapter, Europe has many virgin solar power territories. It is highly plausible that as many EU countries scramble to meet their 2020 Renewable Energy Directive targets, solar power will experience a second wave in the EU. This phenomenon may already be currently witnessed in countries, such as the Netherlands, which is now holding large solar-power auctions and has been identified as a country far behind on reaching its renewable energy target for 2020 [20]. Poland, Hungary, and Romania are all solar power dwarves at the moment, yet have favorable natural conditions and may need to act to hit the 2020 targets. If this is a trend that is continued, it is not hard to imagine that Europe too could pass the next milestone—200 GW—within the next few years. Outside of the EU, Turkey is also stirring, although some of the domestic policies may limit the full potential of solar power, it is not unimaginable that more than 5 GW will be installed in the coming 5 years, which is the current government target [1]. The Ukraine, Belarus, and Russia itself likewise have solar programs that could further add significant numbers of solar deployment. This is not to mention that the original Western European cradle countries can also continue on a solar path with Germany and France committed to large auctions through to 2022.

In Japan and the United States, it is similarly clear that solar will continue to be deployed. In Japan, the danger is that the government reverts to nuclear as the Fukishima disaster diminishes in the minds of many people in Japan. Support schemes have already been seriously degraded and the tender system may not allow Japan to reach the levels of deployment seen in the past 5 years. In the United States, President Donald Trump has implemented trade measures as of January 2018 on solar panels coming from China, Malaysia, Canada, and Mexico. These tariffs will be a big hit to the US solar industry, with significant impact across the entire solar value chain leading to a decrease in demand, manufacturing and loss of jobs. Due to these trade measures, it will be hard to predict the United States role in the coming 5 years; it could be anything from close to 100 GW to less than 10 GW.

Fig. 24.5 outlines the forecasts of the top solar PV markets globally based on both total capacity/MW and political prospects. Together, the combined outputs of these nations

	2016 **Total Capacity** (MW)	2021 **Total Capacity Medium Scenario by 2021** (MW)	2017–21 **New Capacity** (MW)	2017–21 **Compound Annual Growth Rate** (%)	**Political support prospects**
China	77 921	197 921	120 000	20	
United States	42 362	112 262	69 900	22	
India	9548	75 898	66 350	51	
Japan	42 947	72 547	29 600	11	
Mexico	505	10 505	10 000	84	
Australia	5843	13 623	7780	18	
Korea, Republic of (South Korea)	4921	12 121	7250	20	
Brazil	112	6492	6380	125	
Pakistan	910	5310	4400	42	
Taiwan (Republic of China)	320	4020	3700	66	
United Arab Emirates	219	3769	3550	77	
Chile	1675	4925	3250	24	
Thailand	2167	5267	3100	19	
Algeria	85	3085	3000	105	
Philippines	892	3632	2740	32	
Canada	2671	5357	2686	15	
South Africa	1470	4125	2655	23	
Saudi Arabia	4	2309	2305	257	
Egypt	26	2076	2050	140	
Jordan	367	2097	1730	42	

FIGURE 24.5 Top global PV markets "prospects." Top global markets does not include European countries. For top European markets, see Fig. 19. *SolarPower, Europe 2017, p. 11.*

will result in a major deployment of solar power across the world in the coming 5 years. Whether we reach 1 TW of installed solar, the rate of deployment will undoubtedly accelerate. Therefore, solar power is set to become a growing part of the electricity system of all countries in the world and is on track to meet the IEAs prediction that by 2050 solar will be the world's number one energy generation source.

24.4 Segmental Growth

In terms of the segments that can be identified in solar, there are three distinct types of installation: utility scale (solar farms), industrial/commercial, and household. The latter two are almost all entirely roof top based and thus some commentators often make a split between rooftop and ground mounted solar power [1]. For the purposes of this chapter, the delineation will be made between rooftop and ground mounted utility scale for simplicity of expression.

To date, a common model of solar development in most countries has been to start with large-scale ground mounted solar and to slowly develop the rooftop segment alongside the utility scale solar [1]. The global leading solar markets all started in this way: China, the United States, Germany, and the United Kingdom for easy examples. The trend that can be discerned is interesting, as in these markets the utility scale solar develops through stimulation normally caused by incentive schemes, and once the large-scale solar is up and running, rooftop solar begins to speed up in deployment.

There are many potential reasons for this: the land for utility scale can be consumed and then developers move on to larger roofs (applicable more in Europe); incentives for large scale solar are diminished, as the market takes off, while incentives remain for rooftops; perception of success of ground mounted projects leads to more acceptance and more consumers looking for their own solar; and so on. What is clear is that land constraints certainly lead to rooftop markets, with small countries, such as Slovakia, Ireland, Belgium, and Switzerland having almost 100% of their solar on roofs.

In Europe, we can observe that where retroactive measures have been brought into play in a solar market, the expressed trend of moving from ground mounted solar into rooftops does not happen. Romania, Bulgaria and Spain remain markets with less than 20% of their deployed solar on roofs, against the market trend of close to 70% of solar being on European roofs [1]. This certainly suggests that governments that seek to slow down or kill fledgling solar markets can do so with regressive and destructive policies, designed purely to limit solar power. This highlights once again the crucial role that regulation plays in solar deployment, as the economics of solar in Spain suggest that for much of Southern Spain solar is already at grid parity [21]. Therefore, if economics alone were to drive solar, you would expect much more than the 5 GW that have been installed to date in Spain.

Returning to the segmentation, Europe has become a rooftop market and most commentators expect this to continue [1]. The fact remains that in markets like the United Kingdom, which have experienced a large boom in ground mounted systems, the rooftop market remains wide open for exploitation. It is reckoned that a city, such as London, has solar installed on fewer than 0.5% of the city's over 3 million homes [22]. The opportunity is thus great. Italy and Germany today enjoy markets with over 70% rooftop solar, which is very different from their starting points [1]. France remains balanced, but with the tender system expected to deliver up to 15 GW in the next 6 years, this could be a market where utility scale solar will gain dominance.

Globally, with the dominance of utility scale markets there is a situation where ground mounted solar is in fact much more prevalent than roof top. In 2016, 21.6 GW of solar was installed on roofs, while 55 GW was installed as utility scale projects [1]. This is more than double the amount of rooftop solar. Driven by uncapped subsidy schemes in the three largest markets: China, the United States, and Japan [1], this has helped to increase the utility scale solar power installations fraction of the global market to 72% in 2016.

This is a trend that is likely to continue, as countries are now beginning their solar programs with tender systems rather than feed-in-tariffs, but the same process is evident. In India, for example, tenders are being used to drive cost effective deployment of ground mounted solar power, but the government retains an ambitious target of 40% of its 100 GW program to eventually be on roof tops. This is a model that is being embraced across the world: Chile, Mexico, UAE, Turkey, etc. All are looking to tender driven solar energy projects with a secondary development in rooftops.

SolarPower Europe forecasts that by 2020 the gap will be narrowing between the rate of deployment of ground mounted and rooftop solar, and that by 2021 the trend will reverse [1]. This trend is demonstrated in Fig. 24.6. In the high scenario, it is envisaged that 82.6 GW of rooftop solar could be deployed against 79.4 GW of ground mounted [1]. This can be explained by the expected strong rooftop market growth in Europe, the United States, Japan, and China, not to mention India, by the turn of the decade. As many of these markets move into tenders and controlled ground mounted deployment, new business

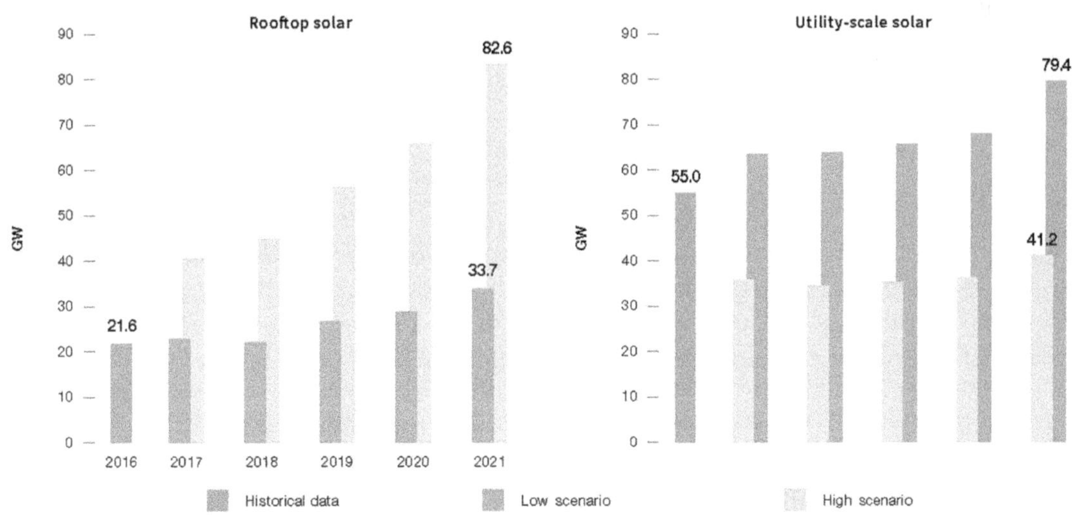

FIGURE 24.6 Scenarios for global PV rooftop and utilities segments development 2016–21. *SolarPower Europe, 2017, p. 21.*

models, such as self-consumption will drive large and small consumers to embrace solar. Government policy is also more likely to favor rooftop solar, even in unlikely solar markets, such as Poland rooftop incentives remain available today [23]. Governments see solar as a natural roof top technology, and with a new range of building integrated solar products likely to be reaching the market at scale in the next 5 years, the last few remaining criticisms of the aesthetics of solar will be gone [1].

Alongside the development of solar on traditional terrain, solar is also beginning to be deployed in more unconventional ways. Floating solar is becoming a new opportunity for solar developers using water as a host for new solar parks. In just 10 years, the value of the market is expected to jump from $\$13.8 \times 10^6$ in 2015 to $\$2.7 \times 10^9$ by 2025 [24]. In Europe, the largest floating solar panel array is installed on London's Queen Elizabeth II reservoir and consist of over 23 000 panels [25]. This development clearly offers another route for the development of solar, and another important market.

The technology is not limited to calm water conditions. In the Netherlands, in 2017, the feasibility of floating panels in rough water conditions is being tested [26]. If these tests are positive it could be imaginable that small island countries, such as Malta, that have huge solar potential, but limited land space to install arrays, could find the perfect way to harness the energy of the sun for their societies. Clearly the development of floating solar is in its early stages, but the potential for the market to grow in the coming years is already well recognized.

Ground mounted solar dominates the global deployment activities today, but it is likely that more markets will follow the lead of Europe and therefore in the 2020s, we will witness the emergence of the domination of roof top systems across the world. This is, of course, subject to the caveat that governments embrace solar power, rather than trying to kill it as demonstrated in countries, such as Spain, Romania, and Bulgaria. At the present time, utility scale solar offers good return on investment at the utility scale, much higher than that which can be achieved on rooftop solar [27]. This is in part due to scale, but as new business models, such as self-consumption, grow across the world the role of rooftop solar is set to equally grow and offer good return on investment for developers.

24.5 Industrial Growth

Solar is also offering the world a huge bonanza in clean jobs. IRENA estimates that more than three million jobs existed in solar in 2016 [28]. This is the largest employer of any renewable technology and of any traditional energy source. In terms of the upstream growth in solar power, it is clear that different regions of the world dominate different segments of the value chain. It is also clear that China has a strong desire to dominate most of the segments, as well as associated sectors, such as battery storage [29].

Currently, Europe is leading in raw materials, particularly the production of polysilicon, although China is developing its own providers [30]. The market for modules is dominated by China, although notable companies continue to compete in specific segments, such as thin film panels. In terms of inverter production, China and Europe share the leadership at the present time [30]. It is in the balance of systems (trackers, mounting, cables, etc.) that domestic markets usually dominate the scene, as it is not so cheap to ship steel frames around the world.

This is good news for societies and governments. Utilizing this global value chain correctly allows societies to experience solar at the lowest cost, where interference in the value chain occurs, the increase in cost is always passed on to the end consumer. Mixing local production with imported at scale products keeps costs down and supports the drive for more solar power, as governments can expect large levels of deployment at levels of cost close to or at grid parity. Importantly, in relation to jobs, the key segments almost always remain in local hands. For example, in Europe a recent study identified that more than 50% of the upstream jobs were in the balance of systems, while overall 84% of jobs were in the downstream segment, construction, development, engineering, and operations and maintenance [31]. The results of this study are outlined in Fig. 24.7.

This means that solar power provides local jobs and helps societies transition away from fossil fuels with new clean jobs to replace old redundant ones. President Obama noted in 2014 that solar would be put in place by American hands with jobs that could not be exported overseas [32]. This was a strong recognition from the former US President of the power of solar energy to deliver jobs and growth for local communities. It applies across the world and where there is a strong market for solar power, there is also a strong market for jobs.

The trends in the solar industrial sector are likely to continue, the only question is how far can China go in dominating each segment? This race bodes well for consumers all around the world and is part of the reason that commentators still expect the cost of solar systems to drop in the coming years [33].

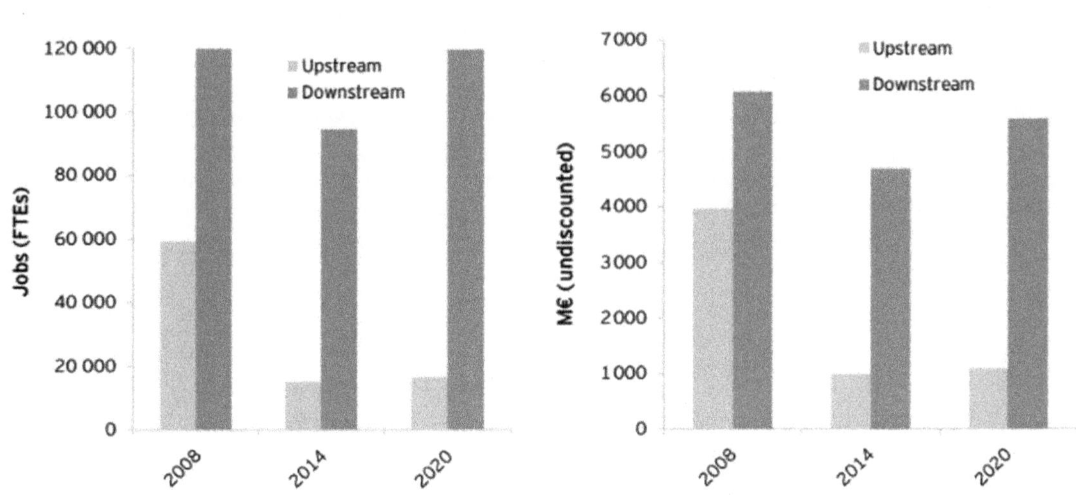

FIGURE 24.7 Job support and GVA creation upstream and downstream activities. *EY, 2015.*

24.6 Conclusions

The future for solar is very bright, rates of installation are expected to continue to grow through the next 5 years and beyond. Long-term forecasts show that solar will be at the core of the electricity system by the mid-century. The European solar phenomenon has now been exported globally, and solar is likely to play a role in almost every countries energy mix. Its versatility and reliability mean that it will be embraced by governments across the globe; no other energy generation technology enjoys such public support [1].

Solar will also become an increasingly rooftop phenomenon as the sector matures, but ground mounted will continue playing a major role for the foreseeable future. Solar will also bring jobs and growth to countries, and as such offers governments and citizens the perfect means to benefit from the energy transition. As this chapter has discussed, solar provides cheap energy and the opportunity to develop local employment. With these factors combining, it is likely that solar will indeed reach its full potential in the coming decade.

References

[1] SolarPower Europe: *Global Market Outlook For Solar Power: 2017–2021*, Brussels, 2017, SolarPower Europe, Available from: http://www.solarpowereurope.org/reports/global-market-outlook-2017/.

[2] IRENA, 2015. Renewable Power Generation Costs in 2014. Available from: http://www.irena.org/DocumentDownloads/Publications/IRENA_RE_Power_Costs_2014_report.pdf.

[3] European Commission, Energy for the Future: Renewable Sources of Energy. Brussels, 1997, European Union.

[4] SolarPower Europe, Global Market Outlook for Solar Power: 2008–2012. Brussels, 2008, SolarPower Europe.

[5] European Commission. Directive 2009/28/EC of the European Parliament and of the Council; 2009. Available from: http://eur-lex.europa.eu/legal-content/EN/ALL/?uri=CELEX: 32009L0028.

[6] Ball J, Reicher D, Sun X, Pollock C: *The New Solar System: China's Evolving Solar Industry and its Implications for Competitive Solar Power in the United States and the World*, Stanford, 2017, Stanford University.

[7] NREL. Renewable Energy Policy in China: Overview; Renewable Energy in ChinaNREL International Programs (Fact Sheet); 2004. Available from: http://www.nrel.gov/docs/fy04osti/35786.pdf.

[8] Phillips T. China Builds World's Biggest Solar Farm in Journey to Become Green Superpower; 2017. Available from: https://www.theguardian.com/environment/2017/jan/19/china-builds-worlds-biggest-solar-farm-in-journey-to-become-green-superpower.

[9] NEA. China 13th Solar Energy Development Five Year Plan (2016–2020); 2016. Available from: http://zfxxgk.nea.gov.cn/auto87/201612/t20161216_2358.htm.

[10] Abe S. After Fukushima, Japan Beginning to See the Light in Solar Energy; 2013. Available from: https://www.theguardian.com/world/2013/jun/18/japan-solar-energy-fukushima-nuclear-renewable-abe.

[11] Neslen A. India Unveils Global Solar Alliance of 120 Countries at Paris Climate Summit; 2015. Available from: https://www.theguardian.com/environment/2015/nov/30/india-set-to-unveil-global-solar-alliance-of-120-countries-at-paris-climate-summit.

[12] NITI. Report on the Expert Group on 175GW RE by 2022; 2015. Available from: http://niti.gov.in/writereaddata/files/writereaddata/files/document_publication/report-175-GW-RE.pdf.

[13] Kenning T. India Added 5.5 GW Solar in FY2016/17; 2017. Available from: https://www.pv-tech.org/news/india-added-5.5gw-solar-in-fy2016-17.

[14] *U.S. Energy and Employment Report*, Washington, DC, 2017, US Department of Energy, United States Government.

[15] SEIA. Model Leases and PPAs; 2017. Available from: http://www.seia.org/research-resources/model-leases-ppas.

[16] IEA. How Solar Energy Could be the Largest Source of Electricity by Mid-Century; 2014. Available from: https://www.iea.org/newsroom/news/2014/september/how-solar-energy-could-be-the-largest-source-of-electricity-by-mid-century.html.

[17] European Commission. Commission Proposes New Rules for Consumer Centred Clean Energy Transition. Available from: https://ec.europa.eu/energy/en/news/commission-proposes-new-rules-consumer-centred-clean-energy-transition.

[18] Solar Power Europe: *Global Market Report For Solar Power: 2016–2020*, Brussels, 2016, SolarPower Europe.

[19] Watson J. Kazakhstan: Opportunity for Our Members; 2016. Available from: http://www.solarpowereurope.org/newsletter-archives/may-2016/our-news/kazakhstan-opportunity-for-our-members/.

[20] European Commission. Renewable Energy Progress Report; 2017. Available from: https://ec.europa.eu/commission/sites/beta-political/files/report-renewable-energy_en.pdf.

[21] Masson G, Werner C, Gerlach A, Macé.: *A Framework Model for Post-Subsidy PV Market Forecast*, Brussels, 2017, Becquerel Institute in collaboration with Chris Werner Energy Consulting, Gerlach Consulting GmbH.

[22] Greater London Authority: *The London Plan: Spatial Development Strategy for Greater London*, London, 2011, Greater London Authority.

[23] Bellini E. Poland to Hold Auction for Solar and Renewables on June 29; 2017. Available from: https://www.pv-magazine.com/2017/06/06/poland-to-hold-auction-for-solar-and-renewables-on-june-29/.

[24] Grand View Research. Floating Solar Panels Market Size & Trend Analysis, By Product (Tracking Floating Solar Panels, Stationery Floating Solar Panels), By Region (U.S., Europe, Asia Pacific, Japan, Central & South America, Rest of World), And Segment Forecasts, 2014-2025; 2017. Available from: http://www.grandviewresearch.com/industry-analysis/floating-solar-panels-market.

[25] Lightsource. Lightsource Begins Work on Europe's Largest Floating Solar Project; 2016. Available from: https://www.lightsource-re.com/2016/02/15/lightsource-begins-work-on-europes-largest-floating-solar-project/.

[26] Osborne M. Floating Solar Pilot Projects in the Netherlands Set Sail; 2017. Available from: https://www.pv-tech.org/news/floating-solar-pilot-projects-in-the-netherlands-disembark.

[27] EY. Capturing the Sun: The Economics of Solar Investment; 2016. Available from: http://www.ey.com/Publication/vwLUAssets/EY-capturing-the-sun-the-economics-of-solar-investment/$FILE/EY-capturing-the-sun-the-economics-of-solar-investment.pdf.

[28] IRENA. Renewable Energy and Jobs Annual Review 2017; 2017. Available from: https://www.irena.org/DocumentDownloads/Publications/IRENA_RE_Jobs_Annual_Review_2017.pdf.

[29] IEEFA. China's Global Renewables Expansion; 2017. Available from: http://ieefa.org/wp-content/uploads/2017/01/Chinas-Global-Renewable-Energy-Expansion_January-2017.pdf.

[30] SolarPower Europe. An Industrial Competitiveness Strategy for the Solar Value Chain in Europe; 2017. Available from: http://www.solarpowereurope.org/index.php?eID=tx_nawsecuredl&u=0&g=0&t=1500040567&hash=67d1b38be3522ec6bf2fe65b2cb5badfd037907c&file=fileadmin/user_upload/documents/Industrial_Strategy/An_Industrial_Competitiveness_Strategy_for_Solar_in_Europe_2017.pdf.

[31] E.Y: *Solar Power Making a Major Contribution to Jobs and Wealth in Europe But the Full Potential is Not Yet Realised*, Brussels, 2015, SolarPower Europe, Available from: http://www.solarpowereurope.org/index.php?id=324&ADMCMD_cooluri=1.

[32] The White House Office of the Press Secretary. FACT SHEET: President Obama Announces Commitments and Executive Actions to Advance Solar Deployment and Energy Efficiency; 2014. Available from: https://obamawhitehouse.archives.gov/the-press-office/2014/05/09/fact-sheet-president-obama-announces-commitments-and-executive-actions-a.

[33] Bloomberg New Energy Finance. New Energy Outlook 2017; 2017. Available from: https://about.bnef.com/new-energy-outlook/.

25

Optimal Renewable Energy Systems: Minimizing the Cost of Intermittent Sources and Energy Storage

David Timmons

UNIVERSITY OF MASSACHUSETTS BOSTON, BOSTON, MA, UNITED STATES

David.Timmons@umb.edu

25.1 Introduction

Carbon dioxide is a by-product of burning any fossil fuel, and is the main greenhouse gas responsible for climate change. Controlling climate change will require greatly reducing fossil fuel use. And regardless of climate change, society will eventually adopt renewable energy, as fossil fuels have finite availability and are only created over geologic time. Thus the question is not whether society will adopt renewable energy, but when: carbon-free renewable energy sources must eventually be used in all global economies. This chapter describes economic principles that should govern renewable energy choices, with a focus on accommodating renewable energy intermittency. A cost-effectiveness approach is used, describing the least costly way to achieve up to 100% of energy from intermittent and dispatchable renewable energy sources plus energy storage.

Electricity is the form of energy produced by the major renewable energy sources, including solar photovoltaics (PV), wind power, and hydropower, suggesting that electricity will be the predominant form of energy in a renewable energy economy. Electric motors are also much more efficient than internal combustion engines, and electrification of the transportation sector will facilitate the transition to renewable energy. Building heating and cooling can be provided by electric heat pumps, a renewable source of thermal energy when coupled with a renewable electricity source. Energy in the form of electricity can be used for most applications (with some exceptions, including jet fuel and shipping fuel). This chapter thus focuses on electricity supply, though the same principles can be applied to supplying thermal loads or transportation fuels from renewable sources.

A key characteristic of ambient energy sources including solar and wind energy is that they are not dispatchable, or not available on demand. Most ambient energy sources have seasonal as well as daily fluctuations, which vary by energy source and location. Using

A Comprehensive Guide to Solar Energy Systems. http://dx.doi.org/10.1016/B978-0-12-811479-7.00025-7

renewable sources to meet electricity demand thus presents challenges not encountered with using fossil fuels. In addition to providing enough total electricity from renewable sources, supply must always meet demand. The intermittency problem may suggest the use of renewable energy portfolios that include a variety of renewable sources and energy storage.

While biomass, hydroelectric, and geothermal energy can provide stored energy comparable to fossil fuels, these sources are not sufficient to meet electricity demand in many regions. Future renewable energy systems may not have an equivalent of today's "base load" power plants, which produce energy continuously. Rather, a variety of energy sources plus energy storage and demand management will be used to provide a continuous flow of energy services. This chapter describes an economic approach to designing such a system.

A core concept of modern economics is that optimal solutions are identified not by average costs, but by marginal costs—not how much it costs to produce solar energy on average, but how much it costs to obtain *another* unit of solar energy. This is a key concept for minimizing the cost of a renewable energy system. And given the intermittency issue with ambient energy sources, temporal problems are prominent—how much it costs to obtain *another* unit of solar energy *now*. As shown below, least-cost renewable energy systems are designed by minimizing the cost of providing energy at critical times with the most challenging ambient conditions—with little sun, water, or wind. Accommodating such critical times drives the design of energy infrastructure and total energy cost.

This chapter focuses on costs of renewable energy for society, ignoring any taxes or subsidies that may change these costs and affect the cost-minimizing decisions of businesses and homeowners. It also considers only direct costs for energy sources, ignoring any external costs, for example, any environmental costs related to hydropower development. Such costs should of course be considered in developing a renewable energy system, but vary greatly by site. And policies to optimize renewable energy systems are another important aspect of the renewable energy transition not considered in this chapter.

25.2 Renewable Energy Microeconomic Considerations

Costs for energy sources are often expressed as the levelized cost of energy (LCOE), or the cost per unit of energy based on amortized capital cost, assumed project life, present value of operating costs, and energy production. For example, in a renewable energy system with no change in annual output, LCOE is the amortized capital cost plus operating cost divided by annual energy output:

$$LCOE = \frac{K\left(\dfrac{r(1+r)^T}{(1+r)^T - 1}\right) + C}{(P.8760.CF)} \tag{25.1}$$

where K is capital cost, C is annual operating cost, r is an annual interest rate, T is the useful life of the project in years, P is the maximum power output, 8760 is the number of hours in a year, and CF is the capacity factor. CF represents the portion of maximum energy output actually obtained over the course of the year, perhaps 20% in the case of solar PV (given nights and cloudy days, and depending on location). Based on Eq. (25.1), greater CF clearly reduces LCOE: for given capital and operating costs, energy cost falls with more energy production.

Compared to fossil fuels, renewable energy sources often require large capital investments, as shown in Table 25.1. A large portion of fossil-fuel LCOE is for ongoing fuel purchases. While solar, wind, hydro, and geothermal energy sources have no fuel costs, significant capital expenditure is required.

Marginal costs for renewable energy are typically rising in any particular place. If we build the first increments of hydropower, wind, solar, and geothermal power generation on the most-accessible, least-costly, and most-productive sites, producing additional energy will be more expensive. At the minimum total cost, all energy alternatives should have the same marginal cost. For example, if new solar electricity were available for \$100 $(\text{MW h})^{-1}$ and wind electricity for \$60 $(\text{MW h})^{-1}$, wind power would be the less expensive choice. But investing in a wind-power installation means the next wind installation will be more expensive (marginal cost rises, if the least expensive options are used first). To minimize total energy cost, wind power should be chosen as long as it less expensive than solar (depending on temporal availability, as discussed later). If the marginal costs of energy sources differ, possibilities for substitution have not been fully exploited, and total cost is not minimized. The equimarginal principle states that minimum total cost occurs when all marginal costs are equal.

Table 25.1 Capital Cost of Renewable and Nonrenewable Electricity Sources

	Nominal Capacity/ MW	Capital Cost/10^6\$ MW^{-1}	Assumed Capacity Factor	Capital Cost[a]/10^6\$ (expected MW)$^{-1}$
Natural gas: combined cycle[b]	702	0.98	0.90	1.09
Hydroelectric: conventional[c]	500	3.02	0.75	4.03
Coal: ultra supercritical[c]	650	3.64	0.90	4.04
Biomass: bubbling fluidized bed[b]	50	4.99	0.90	5.54
Wind: onshore[b]	100	1.88	0.30	6.26
Nuclear: advanced[b]	2234	5.95	0.90	6.61
Solar: photovoltaic tracking[b]	150	2.53	0.25	10.14
Wind: offshore[c]	400	6.42	0.35	18.33
Solar: thermal electric[c]	100	5.22	0.25	20.88

[a]For comparing sources with different capacity factors, a million dollars per expected megawatt is defined as (10^6 \$ MW^{-1})/(capacity factor), or the capital cost needed to produce the same amount of electricity as one MW of capacity running continuously.
[b]Adapted from EIA [2].
[c]Adapted from EIA [1], adjusted for inflation to 2016.

Fig. 25.1 illustrates the equimarginal principle, with schematic representations of marginal cost (MC) curves illustrative of a renewable energy supply in a hypothetical region. Hydropower (panel A) is the least expensive renewable source for initial quantities, though marginal costs rise steeply as more difficult hydropower sites are used. Wind power (panel B) is somewhat more expensive than hydropower, but available in greater quantities with moderate increases in marginal cost. Solar PV energy (panel C) has the greatest cost for initial quantities, but additional quantities near the same cost are virtually unlimited. Panel D shows an aggregate renewable energy marginal cost curve, a combination of curves A–C. Of course in practice, actual renewable energy supply functions will vary.

In Fig. 25.1, panel D, the intersection of the aggregate supply curve and the market demand curve (D) identifies the quantity of energy produced in this hypothetical market. As in any market, the marginal cost of the last resource needed to meet demand effectively determines the market price of all resources used. Solar PV is the most expensive but most abundant renewable energy resource in this example, as in many parts of the world, and effectively the cost of solar PV sets the price for all renewable energy. In some regions, offshore wind or geothermal energy might have a similar supply curve and play a similar role.

The dashed horizontal line across panels A–D in Fig. 25.1 shows that the marginal costs of all renewable energy sources used are equal. Though initial quantities of hydropower are less expensive than solar PV, the least-cost energy supply is provided by a combination of high-priced hydropower and wind, as well as solar (as in this example all three sources are needed to provide the required quantity of energy).

In addition to minimizing cost of energy production, energy should be used more efficiently whenever this can be accomplished less expensively than producing it: the marginal cost of energy efficiency should equal the marginal cost of renewable energy [3]. This is shown graphically in Fig. 25.1, panel E, with the marginal cost of energy efficiency rising as the quantity of energy used is reduced (cost is rising from right to left on the graph).

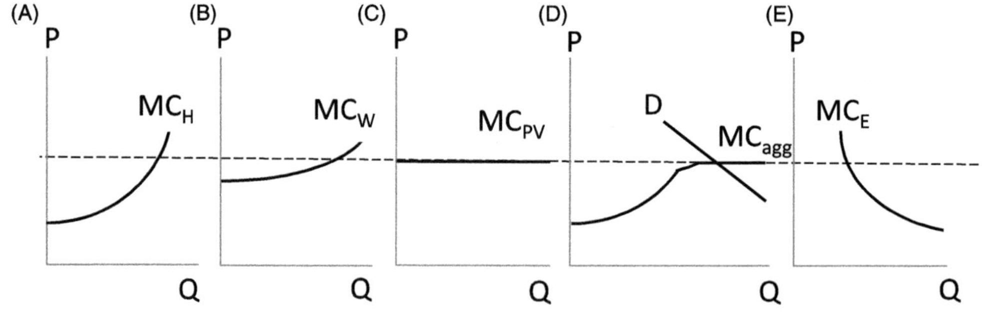

FIGURE 25.1 Equality of renewable energy marginal costs (MC) and marginal cost of energy efficiency.
(A) Hydropower: marginal cost increases sharply. (B) Wind power: marginal cost increases moderately. (C) Solar PV: almost constant marginal cost. (D) Aggregate renewable supply with demand. (E) Efficiency: can reduce much energy use at MC of renewables.

The marginal costs of energy sources may vary greatly by location. In the United States, for example, a given solar PV panel produces more electricity in the sunny Southwest than in the rainy Northwest, and the cost of solar PV is thus lower in the Southwest (as indicated by Eq. (25.1), as the capacity factor is greater in the Southwest). The spatial equimarginal principle is that marginal costs of renewable energy from sources in different locations should be equal, where marginal cost includes transmission from point of production to point of use. Again this is substitution principle: if importing solar energy from a sunnier region is less expensive than producing locally, total cost can be reduced by substituting imported energy.

Depending on differences in production and transmission costs, this could imply that renewable energy costs are minimized with large, centralized production facilities, which could be at great distances from users. Yet such an energy system might not meet energy resiliency criteria, and optimal systems should balance energy cost and resiliency. Just as many countries now subsidize domestic food production for greater food security, domestic and localized energy production may be attractive despite greater costs.

The variability of individual renewable energy sources is mitigated by having multiple sources in different geographic locations. For example, the mean wind speed across an area is less variable than in any particular place. Increasing distance between generation sites typically reduces the covariance between sites, providing a more constant flow of energy [4].

For electrical energy, supply must always equal demand—unlike in other markets, temporary electricity shortages and surpluses can create severe problems. Yet the supply of ambient energy varies both systematically (e.g., seasonally) and randomly (e.g., daily), creating a significant challenge for matching supply and demand. Approaches to intermittency include having a diversity of energy sources (which may have complementary variability), having a renewable energy portfolio that includes dispatchable sources such as biomass and hydropower, oversizing generation, and storing energy. All of these approaches have costs that must be considered to minimize the total cost of a renewable system, the problem to which we now turn.

25.3 Economic Theory of Renewable Energy Intermittency

As described in Section 25.2, the equimarginal principle suggests that to minimize total cost of obtaining energy, marginal costs of different renewable sources should be equal. But the problem is considerably more complex when temporal dimensions are considered. Ambient energy sources are unable to be delivered on demand, or are nondispatchable—we cannot control when the sun shines or the wind blows. This may be the most prominent issue in renewable energy economics: marginal costs of delivering renewable energy vary over time. Moreover, in an optimal renewable energy system marginal costs of different energy sources are not necessarily equal at every moment in time, but rather are equal only at specific critical times.

As shown in Section 25.2, most renewable energy sources are capital intensive, and operating costs can be small after initial capital expenditures are made. In economic terms, costs for such energy sources are mostly fixed. For example, after incurring the cost to install a solar panel, there is virtually no operating cost. The capital cost is the same regardless of the how much sunlight the panel captures, and regardless of whether the energy produced is actually useful at any given moment. The economic question then becomes how much capital expenditure is necessary given anticipated production and anticipated electrical demand. Extending this question to a portfolio of energy sources, the objective is to minimize the total expense (including capital and operating expenses) for energy production facilities that meet all demand conditions under all conditions of ambient energy availability.

It is more expensive to collect renewable energy when ambient energy is limited. In the Vermont example presented in Section 25.4 below, average daily solar insolation during the period of November–February is only 60% of the other eight months of the year. The sun shines every day, and it is possible to collect some solar energy even on a cloudy winter day. But meeting demand is more expensive when sunshine is limited, since more solar panels are required to produce a given amount of electricity. Other intermittent renewable energy sources have similar seasonal as well as daily variation. Total costs for renewable energy systems are driven not by average conditions, but by fixed capital investments required for critical times when the renewable energy supply is low and/or the electricity demand is high. The marginal cost of an energy resource represents the expenditure (mostly capital expenditure) needed to deliver more energy at such a time.

Energy storage is an alternative to producing energy at each moment [5]. With storage, energy produced now can be used later. Many energy storage technologies are available today [6,7]. Electrical storage options include batteries and capacitors; pumped water storage, a hydropower variation; compressed air storage, for example, in underground caverns; and energy stored as electrolytic hydrogen. Also, storage of thermal energy figures prominently in some simulations of renewable energy systems [8,9]. Cost of energy storage is in addition to production cost, and the financial characteristics of energy storage are similar to those of intermittent renewables: most costs are for fixed capital expenditures, and operating costs tend to be small.

Biomass represents solar energy that has already been stored by plants, and which can be released on demand. Biomass can thus provide a complement to intermittent sources like solar energy, with biomass combustion providing energy during times of insufficient insolation. But in most industrialized countries, biomass (the predominant energy source before the industrial revolution) is available in relatively small quantities compared to current energy consumption. Fossil fuels represent similar ancient biomass energy, which can be burned on demand. Though the focus here is on renewable energy, one application for any residual fossil fuel use is to provide stored energy at times of low ambient energy availability. Total fossil fuel use is constrained to a quantity consistent with meeting climate-change mitigation goals (a quantity that must eventually approach zero).

For renewable energy systems, the equimarginal principle can be modified as follows: for an energy system that can meet energy demand at each moment in time, marginal costs should be equal for energy sources plus storage options at critical times of limited energy availability. For example, on a still night with no wind or solar energy available, the marginal cost of obtaining more hydropower should equal the marginal cost of wind or solar energy plus required storage.

Formally, the objective is to minimize total cost *TC* for renewable energy. For simplicity, here we consider only solar energy *s* and wind energy *w* on days 1 and 2, but the proof can be extended to any number of energy sources and energy storage options operating over any number of time periods. The function *f* describes the capital and operating costs of obtaining energy quantities *s* and *w*. The marginal cost of obtaining energy depends on the renewable source (*s* or *w*) and the day (1 or 2), given daily fluctuations in ambient energy. The objective function is thus:

$$\text{Minimize}\, TC = f(s_1, w_1, s_2, w_2)$$

subject to constraints that total energy production exceeds demand *d* on each day:

$$\begin{aligned} s_1 + w_1 &\geq d_1 \\ s_2 + w_2 &\geq d_2 \end{aligned}$$

and to nonnegativity constraints, as energy quantities cannot be negative:

$$s_1, w_1, s_2, w_2 \geq 0 \tag{25.2}$$

The Lagrangian method can be used to describe the optimum solution, where the Greek letter lambda (λ) is the Lagrangian multiplier [10]:

$$\mathrm{L} = f(s_1, w_1, s_2, w_2) + \lambda_1(d_1 - s_1 - w_1) + \lambda_2(d_2 - s_2 - w_2) \tag{25.3}$$

Since the objective function (Eq. (25.2)) includes inequality constraints, the Kuhn-Tucker conditions (Eqs. (25.4)–(25.7)) are necessary for the least-cost solution ([10]; [eq. 13.17, p. 410]):

$$s_1, w_1, s_2, w_2, \lambda_1, \lambda_2 \geq 0 \tag{25.4}$$

This condition reiterates the nonnegativity requirement for the energy quantities, and extends it to the Lagrangian multipliers. In this type of optimization problem, λ has a specific interpretation: it is the marginal cost of slightly tightening the associated constraint [10], a cost which cannot be negative.

$$\begin{aligned} \mathrm{L}_{\lambda 1} \leq 0 \;&\rightarrow\; d_1 - s_1 - w_1 \leq 0 \;\rightarrow\; d_1 \leq s_1 + w_1 \\ \mathrm{L}_{\lambda 2} \leq 0 \;&\rightarrow\; d_2 - s_2 - w_2 \leq 0 \;\rightarrow\; d_2 \leq s_2 + w_2 \end{aligned} \tag{25.5}$$

Subscripts denote partial first derivatives of the Lagrangian equation (e.g., $L_{\lambda 1} = \partial L / \partial \lambda_1$).

These conditions reflect the demand constraints, that total energy production on each day must be equal to or greater than demand.

$$\begin{aligned} L_{s1} \geq 0 &\rightarrow f_{s1} - \lambda_1 \geq 0 \rightarrow f_{s1} \geq \lambda_1 \\ L_{w1} \geq 0 &\rightarrow f_{w1} - \lambda_1 \geq 0 \rightarrow f_{w1} \geq \lambda_1 \\ L_{s2} \geq 0 &\rightarrow f_{s2} - \lambda_2 \geq 0 \rightarrow f_{s2} \geq \lambda_2 \\ L_{w2} \geq 0 &\rightarrow f_{w2} - \lambda_2 \geq 0 \rightarrow f_{w2} \geq \lambda_2 \end{aligned} \tag{25.6}$$

These are similar to the classic first-order conditions for a Lagrangian equation with equality constraints, and reflect the marginal costs of energy on each day. For example, f_{s1} is the marginal cost of solar energy on day 1. By themselves, these conditions do not require that marginal costs be equal.

$$\begin{aligned} s_1 L_{s1} = 0 &\rightarrow s_1(f_{s1} - \lambda_1) = 0 \\ w_1 L_{w1} = 0 &\rightarrow w_1(f_{w1} - \lambda_1) = 0 \\ s_2 L_{s2} = 0 &\rightarrow s_2(f_{s2} - \lambda_2) = 0 \\ w_2 L_{w2} = 0 &\rightarrow w_2(f_{w2} - \lambda_2) = 0 \end{aligned} \tag{25.7}$$

This is the first of two so-called complementary slackness conditions. For solar energy on day 1, for example, either $s_1 = 0$ (no solar energy is used in the solution), or $(f_{s1} - \lambda_1) = 0$ (which requires $f_{s1} = \lambda_1$, making the value of the second term zero). Note that if a λ_i is zero, these conditions require that either the respective s_i or w_i is zero (no solar or wind is used), or that f_{si} or f_{wi} is zero (that marginal cost is zero). For a feasible solution, this cannot hold in all four equations, that is, some energy with a nonzero cost must be provided. Either λ_1 or λ_2 must take a positive value in the solution.

$$\begin{aligned} \lambda_1 L_{\lambda 1} = 0 &\rightarrow \lambda_1(d_1 - s_1 - w_1) = 0 \\ \lambda_2 L_{\lambda 2} = 0 &\rightarrow \lambda_2(d_2 - s_2 - w_2) = 0 \end{aligned} \tag{25.8}$$

The final conditions are again for complementary slackness: on day 1, either $\lambda_1 = 0$ or $d_1 = s_1 + w_1$ (so that supply exactly equals demand, and the second term is zero). This condition is key to demonstrating equality of marginal costs. For example, when $d_1 - s_1 - w_1 = 0$, or when the demand constraint is binding on a critical day, λ_1 can take a value greater than zero, that is, if just enough solar panels have been deployed to meet demand on this day.

In Eq. (25.7) for solar energy on day 1, a positive value of λ_1 allows s_1 to take a positive value (the optimum solution uses solar energy on this day) and the marginal cost of solar to have a nonzero value ($f_{s1} > 0$), but only if $f_{s1} = \lambda_1$. Similarly, a positive value of λ_1 allows w_1 to be positive in the second line of Eq. (25.7), and $f_{w1} = \lambda_1$. And if both f_{s1} and f_{w1} are equal to λ_1, then f_{s1} and f_{w1} are equal to each other ($\lambda_1 = f_{s1} = f_{w1}$): the marginal costs of solar and wind energy are the same. When demand constraints bind, that is, on critical days when the total energy supply equals demand without exceeding it, the marginal cost of obtaining renewable energy is equal for each renewable energy source and for stored energy. If this were not the case, that is, if marginal energy costs were different on critical days, it would be possible to reduce total cost with a solution that more fully exploited the lower-cost resource.

Though this proof is simplified, it can be expanded to include any number of energy sources and energy storage, and to include any number of time periods, for example, the 365 days in a year or 8760 hours in a year. The results are always the same: at critical times when demand constraints bind, the marginal costs of all energy options are equal, and the equimarginal principle holds.

Though Eqs. (25.2) through (25.8) demonstrate how the equimarginal principle applies to designing a renewable energy system, the Lagrangian approach does not necessarily represent an easy empirical way to identify critical days or their associated marginal costs. And the Kuhn-Tucker conditions are necessary (they must hold at the optimum solution) but not sufficient (they may also hold in a solution that is not the global minimum). The following section turns to an empirical example where optimization software is used to identify the optimum portfolio of investments in a renewable energy system. As we will see, the equimarginal principle holds for the critical days.

25.4 Economics of Renewable Energy Intermittency: Empirical Example from Vermont

Vermont is small state in the northeastern, New England region of the United States, with a 2016 population of 625 000. The state is mostly rural, with its largest metropolitan area, Burlington, having 216 000 residents [11]. Vermont contains both small and large-scale examples of most renewable energy sources, including solar PV, wind energy, hydroelectric energy, and electricity generation from woody biomass. As of 2017, the state had a goal of using 90% renewable energy by 2050, a goal endorsed by recent administrations from both major political parties [12]. For illustration, a 100% renewable electricity goal is considered, with all electricity sourced from within the state (though in reality, some out-of-state sources are likely). Daily electricity demand data are taken from the New England grid that serves Vermont [13], and scaled to Vermont based on population.

The Vermont example is based on case studies of actual solar PV, wind power, hydropower, and biomass plants in Vermont, in addition to a pumped-hydro energy storage facility in nearby Massachusetts. Energy data from actual generating facilities are used to illustrate real possibilities for integration of renewable energy sources that are already proven on a limited scale. But as specific project costs vary greatly, and in some cases costs have declined sharply in recent years, current cost data are used to represent US national averages.

The solar PV component of the example is based on a 2016 installation of 2.7 MW capacity, in Williston, Vermont. The project is owned by Vermont's largest electric utility, Green Mountain Power. It produces an estimated 8186 MW h a^{-1} (where a refers to year) (19.9% capacity factor) from 21 417 solar PV panels on land area of approximately 11 ha [14]. Daily production is estimated from 2005 solar insolation data at the nearby Burlington airport [15]. The estimated cost of \$1375 $(kW_{AC})^{-1}$ is based on the midpoint of the Lazard [16] range, yielding an expected cost for a new project of \$6.5 $\times 10^6$ (actual cost was approximately \$12 $\times 10^6$) [17].

While this is currently Vermont's largest solar PV installation, there is no obvious upper limit on the number of projects of this scale that could be built, given that it occupies only 11 ha of a largely rural state.

The wind energy component is based on the Kingdom Community Wind project in Lowell, Vermont, constructed in 2012, and also owned by Green Mountain Power. The project includes 21 Vestas turbines of 3 MW each, for a maximum output of 63 MW [18]. Capital cost of \$1475 MW^{-1} is based on the midpoint of the Lazard [16] range, resulting in an expected project cost today of \$93 $\times$ 10^6 (actual cost in 2012 was \$150 $\times$ 10^6) [19]. The project generates approximately 186 000 MW h a^{-1}, for a 33.7% capacity factor. Green Mountain Power provided monthly energy forecast data along with 2016 daily wind speed data at the site, from which daily energy production was estimated. While the Kingdom Community Wind mountaintop site is not unusual in Vermont, there is a limit to the number of comparable sites that could feasibly be developed for wind power, given the needs for road and power access, for reliable wind conditions, and for public acceptance of these prominent features in the landscape.

The model hydropower project is a privately owned project at a former mill on the Walloomsac River, near Bennington, Vermont. The 0.36 MW capacity plant started generating in 2015, and produces 1640 MW h a^{-1} [20], for a 52% capacity factor. The site has approximately 4.7 m of head and captures flows of up to 9.8 m^3 s^{-1} [21]. Head is provided by an old dam with a small impoundment, but here it is modeled as a run-of-river plant, as the plant is licensed to operate in this way (and for this example it also simpler to consider energy production and storage separately). Walloomsac River flow data are from a nearby monitoring station maintained by the US Geological Survey [22]. The capital cost estimate is \$3.2 $\times$ 10^6 per megawatt ([2]; 2013 estimate adjusted for inflation), resulting in estimated cost of \$1.2 $\times$ 10^6 (actual cost on this site was \$2.5 $\times$ 10^6) [20]. Though this is a small hydropower project, it is representative of the ubiquity of similar old mill sites around Vermont that could potentially be developed (as compared to new sites for large hydropower, which are few within the state).

The 50 MW McNeil Generating Station in Burlington, Vermont (owned jointly by three Vermont utilities) provides an example of an electricity generating plant burning woody biomass. Constructed in 1984, the plant consumes 76 t (where *t* refers to metric tons) of wood chips per hour at full load [23]. In 2015 the plant generated 290 000 MW h of electricity [24] for a 66% capacity factor, though this has varied greatly over the years. Given fuel requirements, operating costs are greater for biomass plants than for the other energy sources modeled here. Cost estimates are from adjacent Massachusetts [25]. Though biomass is attractive as a dispatchable renewable energy source, the supply of biomass is clearly finite. The upper limit of 1.6 $\times$ 10^6 t of woody biomass per year is from a previous study of Vermont biomass ([26]; moderate scenario). If fully exploited for electricity generation, this would provide 23% of Vermont's current electricity demand. New biomass plants have been proposed in Vermont but have been controversial, due to concerns about both air quality and sustainable forest management.

While many forms of energy storage are technically feasible, pumped-hydro storage has one of the longest histories and lowest costs [16]. When excess electricity is available in the grid, it is used to pump water from a lower reservoir to an upper reservoir. When more electricity is needed, the operation is reversed, with water falling from the upper to the lower reservoir and generating electricity. The Northfield Mountain pumped hydro plant lies a few miles south of the Vermont border in Northfield, Massachusetts. Constructed in 1972, the plant pumps water from the Connecticut river to a manmade reservoir that holds 21×10^6 m^3 of water at an elevation 240 m above the river. The plant generates 1143 MW at full capacity [27], with an assumed storage efficiency of 81% [16]: for every 100 MW h put into the system, 81 MW h are later delivered back to the grid. Lazard's [16] midpoint estimate for constructing a new plant of this type is \$122 $(\text{MW h})^{-1}$, or \$$2.3 \times 10^6$ MW^{-1} capacity, given other Lazard assumptions. This implies a capital cost of \$$2.6 \times 10^9$ for a facility like Northfield Mountain built today. Half of the capital investment is assumed to be for megawatt capacity (turbine-pumps, generators, etc.) and half for megawatt-hours of storage (the upper reservoir). Capacity and storage investments are modeled separately. Finding additional sites for pumped storage on the scale of the Northfield Mountain project could clearly be challenging, but smaller facilities are also feasible—Lazard [16] models a 100 MW facility, just 9% of the capacity at Northfield Mountain.

In addition to the parameters described above, some simplifying assumptions are used to estimate supply functions for each of the sources, as an illustration of the possible cost of expanding use of these resources. For solar PV, a capacity sufficient to supply the whole state (not including energy storage cost) is assumed to be available for a marginal cost of 10% more than the base cost given above (i.e., marginal cost is almost constant); for wind, 50% more than the base cost (marginal cost rises moderately), and for hydroelectric energy, 100% more (marginal cost rises sharply). These increases in marginal costs correspond roughly to Fig. 25.1 in Section 25.2. An actual energy plan for the state would consider marginal cost increases for each source in more detail, and would likely result in nonlinear supply functions.

Though electricity supply must equal demand in each moment, this model uses a simplifying assumption that supply is sufficient for each of the 365 days in a year. In most cases the one-day interval is likely to be reasonably representative of an actual cost minimization problem, given that within-day supply-demand discrepancies can be met with multi-day energy storage included in the optimum generating and storage portfolio.

The model incorporates daily data on insolation, wind generation, and water flow for the case study sites, as described earlier. The cost of obtaining energy from a source varies inversely with the available ambient energy; for example, solar power is less expensive to obtain on days with abundant sunshine. LCOE_D is defined as the levelized cost of energy, if the ambient conditions on a particular day were to prevail on every day for the duration of an energy project. LCOE_D is thus a measure of how expensive it is to obtain a given energy source on a particular day. LCOE_D is calculated using the following equation, similar to Eq. (25.1):

$$LCOE_D = \frac{K\left(\frac{r(1+r)^T}{(1+r)^T-1}\right)+C}{(MWh_D.365)} \tag{25.9}$$

where MWh_D is the daily megawatt hours of energy output given ambient conditions on a particular day, and all other terms are as defined for Eq. (25.1).

Fig. 25.2 shows the daily levelized cost of energy ($LCOE_D$) of producing solar, hydroelectric, and wind energy at the Vermont sites modeled, given each day's ambient energy availability.

In this example, solar PV energy and hydro/wind energy are complementary, with solar providing less expensive energy in the summer and hydro/wind energy being less expensive

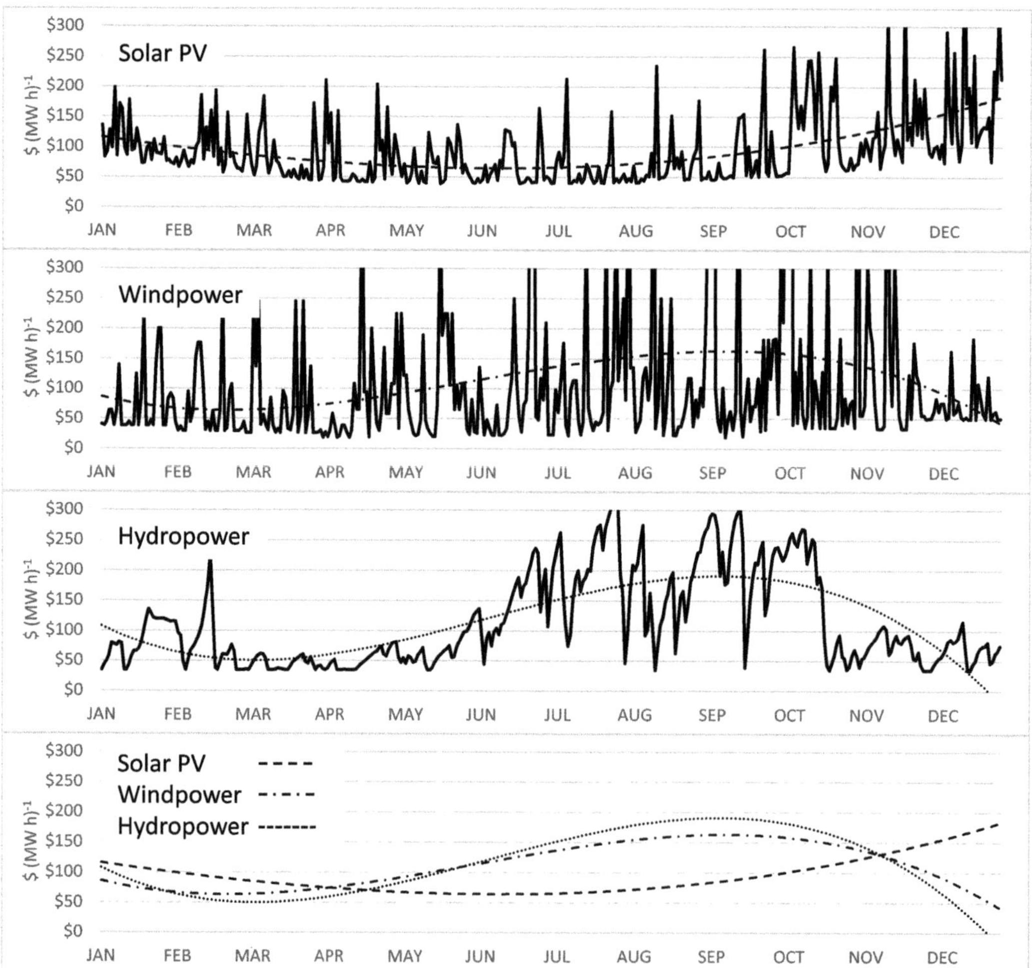

FIGURE 25.2 $LCOE_D$ **for solar, wind power, and hydropower in Vermont example.**

in the winter. On the other hand, hydropower and wind energy costs are positively correlated, with hydro and wind energy being potential substitutes. Price spikes on individual days (due to low ambient energy availability) are not necessarily a large concern, since energy from previous, less expensive days can be stored for such days. The objective is to balance different energy sources and storage costs to arrive at the minimum total cost for the year.

Optimization software[a] can be used to identify solutions for constrained optimization problems such as the one described in Section 25.3. For this example, the optimization software is set to pick combinations of energy production capacity from solar PV, wind, hydropower, and biomass energy, and well as energy storage capacity and storage quantity. Capacity for each intermittent source results in production estimates by source for each of the 365 days in a year, given ambient conditions for each day. If biomass investments have been chosen and demand exceeds ambient supply on a particular day, biomass energy may be used. Similarly, including storage investments allows any excess energy to be stored and consumed later, if demand should exceed supply.

Production capacity choices result in capital expenditures, which are amortized at an interest rate of 8%. Annual operating expenses are added to arrive at total annual expense. The optimization software attempts to minimize total annual expense, subject to the constraint that energy supplied is greater than or equal to demand on each of the 365 days in the year. Biomass use is also constrained to the total annual biomass availability. The optimization software iteratively picks combinations of renewable energy sources and storage, until total cost cannot be further reduced.

Results for this example are shown in Table 25.2. Given the assumptions earlier, and the weather patterns observed for single representative years (which vary by energy source), the minimum-cost solution includes 42% of Vermont's electricity from solar PV, 45% wind power, 13% hydropower, and no biomass, plus an energy storage plant of approximately 41% of the size of the Northfield Mountain case-study site.

While this solution reflects the equimarginal principle, this result is not obvious from looking at any particular day, where marginal costs may differ. Though on a day when marginal costs differ, supply could be shifted from higher to lower marginal cost sources to reduce costs, this may result in failing to meet the supply constraint on other critical or near-critical days. Thus all critical days must be considered as a group. For example, considering for simplicity only the nondispatchable sources (scenario 4 in Table 25.3), the five days where supply is most limited have less than 6% excess capacity. For the total energy produced on these five days, there is less than a 5% difference between the lowest and highest marginal costs. The equimarginal principle thus holds, approximately, when all the critical and near-critical days are considered together.

[a] Finding a global minimum cost from among the many combinations of different energy sources is a nontrivial mathematical problem. For this example, Microsoft *Excel's Solver* add-in is used. Note that results may be sensitive to *Solver* settings, for example, the length of time in which Solver is allowed to search for lower-cost combinations. Other software including *Mathematica*, *Matlab*, *GAMS*, and *Stata* can perform similar optimization routines. An optimization result is not necessarily unique, and is not guaranteed to be a global minimum—results should always be checked for consistency and reasonableness.

Table 25.2 Renewable Energy Cost Minimization Example: Vermont

	Solar PV	Hydropower	Wind	Biomass	Storage Flow[a]	Storage Stock[b]	System
Optimal capacity/MW	1567	219	1022	0	471	45 864	
Case study plant size/MW	4.7	0.4	63.0	50.0	1143		
Case study plants needed	334	607	16	0	0.41		
Annual generation/TW h	2.8	0.9	3.0	0	0.4		6.7
Portion of total generation/%	42%	13%	45%	0%			100%
Capacity factor	20%	47%	33%	na			
Total investment/10^6\$	\$2,210	\$774	\$1,630	\$0	\$811	\$553	\$5,978
Portion of total investment/%	37%	13%	27%	0%	14%	9%	100%
Plant life/years	30	40	25	30	40	40	
Interest rate/%	8.0%	8.0%	8.0%	8.0%	8.0%	8.0%	
Levelized capital cost/10^6 \$	\$196	\$65	\$153	\$0	\$68	\$46	\$528
Operating cost/10^6 \$	\$16	\$4	\$38	\$0	\$1	\$1	\$61
Total annual cost/10^6 \$	\$212.8	\$68.9	\$191.0	\$0	\$69	\$47	\$589
LCOE/\$ (MWh)$^{-1}$	\$76	\$77	\$64	na	\$270		\$111

[a]Storage flow is the maximum storage input or output power in MW.
[b]Storage stock is the maximum quantity of energy storage in MW h.

Table 25.3 Levelized Cost of Energy (LCOE) for Alternative Production Scenarios

		Energy Facilities Included						
	Scenario	Solar PV	Hydro-power	Wind	Biomass	Storage	LCOE/\$ (MW h)$^{-1}$	Increase From Base/%
1	All sources	X	X	X	X	X	111	na
2	No storage	X	X	X	X		177	59
3	No biomass	X	X	X		X	111	0
4	No biomass or storage	X	X	X			252	128
5	No wind	X	X		X	X	134	21
6	No hydro	X		X	X	X	116	5
7	No solar		X	X	X	X	149	34
8	Wind and storage			X		X	151	37
9	Hydro and storage		X			X	368	233
10	Solar and storage	X				X	167	51

Several other points are worth noting about the solution in the Vermont example. Here the intermittency problem has been solved chiefly with a diversity of energy sources rather than with energy storage. A balanced portfolio of energy generating sources was chosen to accommodate ambient conditions and energy demand on each day of the year. Use of energy storage is minimal, since storage adds cost without producing any energy—in difficult conditions, even a relatively costly energy source can be less expensive than the alternative of investing in both energy generation and storage. And given the cost of energy storage, in this case it is used only for the short term, not for accommodating seasonal energy variability: at the optimal solution, energy storage is sufficient for just 3.1 days of average demand. In this example hydropower contributed only 13% of the energy to the optimal solution, but only run-of-river hydropower was included, and hydropower with reservoir storage would likely play a more prominent role.

Biomass energy is available, but was not used in the minimum-cost solution. However, the value of waste heat from biomass electricity plants was excluded, and utilizing this could make biomass energy significantly more economical. In another test, baseload biomass electricity (running at 90% capacity factor) was forced into the solution. LCOE for all sources rose only slightly, to \$112 $(\mathrm{MW\,h})^{-1}$, but LCOE for biomass alone was \$107 $(\mathrm{MW\,h})^{-1}$—greater than for solar, wind, or hydropower. In a renewable energy system, having some biomass capacity would likely have advantages for accommodating unforeseen production losses, and so on.

All else equal, greater utilization and associated greater capacity factor decrease LCOE for an energy source (Eq. (25.1)). But for minimizing the cost of a portfolio of resources that accommodates all demand conditions, some oversizing of energy production resources can be optimal. For the minimum-cost portfolio in this example, the utilization rate is 79%, that is, only this portion of the electricity generated is actually needed to meet demand. While this might appear to represent waste when considering individual sources, for the system it is apparently optimal to modestly oversize generating capacity, that is, for the year as a whole this is less expensive than other alternatives. In some systems, valuable uses for this intermittently available "waste" energy might be found.

While this example does not include a fossil-fuel component, from the Lagrangian approach (Section 25.3), its characteristics are known: the marginal cost of fossil fuel will be equal to all other alternatives on the critical days. This implies that any fossil-fuel facilities used in a renewable energy system will have relative large capacity, low capacity factor, and greater LCOE than when used in a baseload application.

At \$111 $(\mathrm{MW\,h})^{-1}$, total cost for the 100% renewable scenario is significantly greater than the LCOE for the principle solar, wind, and hydro sources alone, as shown in Table 25.2. This is a cost of renewable energy intermittency: meeting demand on critical days with relatively high-cost energy sources and energy storage increases the overall LCOE.

In this example the LCOE likely exceeds present wholesale costs for electricity, but this example includes no value for existing infrastructure, no interstate energy flows, no subsidies, and so on, so the \$111 $(\mathrm{MW\,h})^{-1}$ figure is not strictly comparable to current market prices. And a number of factors are likely to reduce the cost of a 100% renewable energy scenario in the long run, as discussed in Section 25.5.

Table 25.3 presents the LCOE results for testing of alternative scenarios. First the model is run as shown in Table 25.2, sequentially omitting one renewable resource—how much would LCOE increase if hydropower were excluded, for example, or if biomass energy were excluded? Next, scenarios are evaluated that rely entirely on a single energy source plus energy storage. All of these scenarios are feasible, that is, it is possible to supply enough energy to meet demand with any scenario shown in Table 25.3 (assuming a sufficient number of solar, hydropower, wind energy, and storage sites). Every scenario shown is at least as expensive as the base scenario in Table 25.2 (since limiting options cannot reduce costs), but in some cases constrained scenarios are not much more expensive than the base scenario. In particular, the scenario omitting biomass is the same cost as the optimum solution (since the optimum does not include biomass), and omitting run-of-river hydropower (without storage) is only slightly more expensive than the optimum solution.

To assess the effect of different ambient conditions than represented by these sample years, periods of time longer than one year can be modeled. Also, statistical distributions of ambient energy patterns can be used to generate hypothetical weather patterns using Monte Carlo analysis, allowing analysis of hundreds or thousands of weather-condition combinations [28]. While an engineered solution would also incorporate extra capacity to account for unusual conditions, plant maintenance, and so on, legacy fossil-fuel plants will likely provide such back-up capacity for some time.

25.5 Extensions and Conclusions

The Vermont example in Section 25.4 is intended to illustrate that a 100% renewable energy scenario is feasible, and to describe a method to estimate its cost. Yet there are several elements of the example that are unrealistically restrictive, and which offer possibilities for long-term cost reduction.

Costs of renewable energy sources are not static, and have been declining sharply for some sources [29]. Cost reductions are particularly likely for solar energy, given developments described elsewhere in this volume. Based on the equimarginal principle, a cost reduction for any of the energy sources or for energy storage will reduce total system LCOE.

One simplifying assumption for the example is that all of Vermont's electricity be sourced from within the state. Vermont already imports large-scale hydropower from Quebec, a dispatchable source given the size of its reservoirs, and Vermont is likely to import more Canadian hydropower in the future [30]. In the longer term, cross-country transmission lines may provide paths to import solar energy from the US Southwest and wind energy from the Great Plains, which may be both less expensive and less variable than solar and wind energy produced in Vermont, in spite of the transmission costs [31].

The simplified Vermont example considers only storage of energy in the form of electricity. But for cooling applications it may be less expensive to store energy in the form of ice, and for heating applications in the form of hot water or another heated mass. While electricity may ultimately be used to generate the cooling and heating, with thermal energy storage the electricity need not be produced at the moment the cooling or heating is required, relieving the need to produce electricity under the most expensive conditions.

Similarly, the simplified example includes only large electricity-generating biomass plants in which at least two-thirds of the potential energy in biomass is lost as waste heat. Cogenerating heat and electricity in smaller, more widely distributed biomass plants would allow use of this waste heat and relieve the need to produce electricity for heat pumps. This would likely reduce the cost of biomass energy, and lower-cost biomass could enter the optimum portfolio, reducing its cost.

Carbon neutrality in the transportation sector suggests a large increase in the number of electric cars [32]. While these represent a new electrical load, every electric car includes a large battery pack, providing opportunities for electricity storage. At a minimum, there could be some flexibility in charging times, assuming charging stations are provided at both homes and businesses (or wherever cars may be parked). Cars should be charged when energy is most available—currently during the night, though this might not be the case with variable ambient energy. And possibilities go beyond shifting charging times. Many cars will have battery capacities greater than required for daily use. For example, a car with a 300 km range might only be driven 50 km on a normal day. With appropriate incentives, an owner might agree to relinquish part of the battery capacity and its associated range, at least on some days. A utility could then make use of this relinquished capacity to bolster supply, a capacity that could be substantial given the potential number of electric cars.

As described in Section 25.2, the equimarginal principle implies that the marginal cost of energy conservation should equal the marginal cost of energy production. From the example in Section 25.4 it can be seen that the value of energy conservation on a critical day could be very high indeed. Much conservation could be obtained for less than the cost of energy production [33], and appropriate conservation investments could greatly reduce total costs.

The Vermont example in Section 25.4 also treats daily electricity demand as immutable, which is another oversimplification. Substantial possibilities for load shifting exist, if appropriate incentives are provided. For example at the author's home, the electric utility offers off-peak energy priced at approximately half the rate of peak hours (given the different marginal costs of obtaining energy during these times). Peak hours are 6:00 until 23:00 h Mondays through Fridays, or approximately half of the hours in a week. Since subscribing to this rate, the author's household has shifted the great majority of its electricity use to off-peak hours, with about 80% of total usage now being off peak. The electric water heater is on a timer that heats the tank at night. The dishwasher is only run after 23:00 h and on the weekend. The clothes washer and dryer have normally been used only on weekends in any case. There are two electric cars that charge at night. A heat pump which provides back-up heat (the primary source being wood) runs more during the night, when it is colder outside and there is no solar gain (and with thermal storage, the heat pump could run entirely at night). All related controls have default off-peak settings, but can be overridden with the push of a button (accounting for a portion of the 20% peak-hour usage), so very little convenience is lost. This same concept could be applied in a renewable energy system, where instead of fixed peak and off-peak hours, prices would vary with ambient energy availability. Customers could get signals for high and low-rate times, perhaps with a forecast several days ahead. With appropriate controls some devices could automatically operate during times of greater ambient energy availability. Such approaches would greatly simplify the task of matching energy demand with intermittent energy supply.

The problem of climate change requires a shift to carbon-neutral energy sources across the world. While there are a few examples of predominantly renewable energy grids (e.g., Iceland, Costa Rica) most countries still rely heavily on fossil fuels for energy. And even where use of renewable sources has greatly expanded (e.g., Germany, Denmark), there is still sufficient fossil-fuel generation to avoid most of the challenges of ambient energy variability. But a complete change to renewable energy is clearly feasible with today's technology. The guiding economic principle for minimizing the cost of a renewable energy transition is equality of marginal costs for different energy sources, for stored energy, and for energy conservation. Given the temporal nature of the renewable energy supply problem, equimarginal costs should occur at specific critical times of reduced ambient energy availability and/or increased energy demand. In the Vermont example, seasonal ambient energy variability was most cost effectively addressed with source diversity, and daily variability addressed with minimal energy storage. This conclusion would not hold in all places—an area with a consistent low-cost renewable source might rely on that source exclusively, for example, hydropower in a wet area, geothermal energy in a volcanic zone, or solar power in place with consistent year-round sun. While the costs of complete reliance on renewable energy may initially appear to be greater than current energy cost, there are numerous avenues by which system costs can be reduced through innovative approaches. And the climate-change related costs of not transitioning to carbon-neutral energy sources are likely to be far greater still.

References

[1] EIA, Updated capital cost estimates for utility scale electricity generating plants. 2013, Energy Information Administration: Washington D.C.

[2] EIA, Capital cost estimates for utility scale electricity generating plants. 2016, U.S. Energy Information Administration: Washington D.C.

[3] Timmons D, et al: Decarbonizing residential building energy: a cost-effective approach, Energy Policy 92:382–392, 2016.

[4] Hart EK, Stoutenburg ED, Jacobson MZ: The potential of intermittent renewables to meet electric power demand: current methods and emerging analytical techniques, Proc. IEEE 100(2):322–334, 2012.

[5] Denholm, P. et al., The role of energy storage with renewable electricity generation. National Renewable Energy Laboratory: Golden, Colorado, USA, 2010.

[6] Beaudin M, et al: Energy storage for mitigating the variability of renewable electricity sources: an updated review, Energy Sustain. Dev 14(4):302–314, 2010.

[7] Díaz-González F, et al: A review of energy storage technologies for wind power applications, Renew Sust Energy Rev 16(4):2154–2171, 2012.

[8] Palzer A, Henning H-M: A comprehensive model for the German electricity and heat sector in a future energy system with a dominant contribution from renewable energy technologies–Part II: results, Renew Sust Energy Rev 30:1019–1034, 2014.

[9] Jacobson MZ, et al: Low-cost solution to the grid reliability problem with 100% penetration of intermittent wind, water, and solar for all purposes, Proc Nat Acad Sci 112(49):15060–15065, 2015.

[10] Chiang A, Wainwright K: Fundamental methods of mathematical economics, 4th ed., New York, 2005, McGraw-Hill.

[11] American Fact Finder. United States metropolitan statistical areas. 2015 February 19, 2017)]. Available from: https://factfinder.census.gov/faces/tableservices/jsf/pages/productview.xhtml?src=CF.

[12] Ring, W., Scott sticks with renewable energy goal, Burlington, Vermont: Burlington Free Press, 2017.

[13] ISO New England. Hourly real time system demand. 2017 February 19, 2017. Available from: https://www.iso-ne.com/isoexpress/web/reports/load-and-demand.

[14] State of Vermont, GMP Solar-Williston, Certificate of Public Good, Final Order docket no. 8562, Montpelier, Vermont: Public Service Board, Editor. 2016.

[15] NOAA. NSRDB solar, location 726170. 2005 (January 19, 2017). Available from: ftp://ftp.ncdc.noaa.gov/pub/data/nsrdb-solar/solar-only/.

[16] Lazard, Lazard's levelized cost of storage analysis--version 1.0. 2015.

[17] Green Mountain Power, Personal communication, D. Timmons, Editor; 2017.

[18] State of Vermont, GMP Lowell wind, Certificate of Public Good, Final Order docket no. 7628, Montpelier, Vermont: Public Service Board, Editor; 2011.

[19] Green Mountain Power, Vermont Public Service Board Rules in Favor of Green Mountain Power's Lowell Wind Project; 2011.

[20] Damon, E., Seeking new life for an old North Bennington Dam, in Bennington Banner. Bennington, Vermont; 2015.

[21] State of Vermont, Water quality certification: Vermont tissue mill hydroelectric project, Montpelier, Vermont: D.o.E.C. Agency of Natural Resources, Watershed Management Division, Editor; 2013.

[22] USGS. Water Watch: USGS 01334000 Walloomsac River near North Bennington, Vermont. 2016 January 19, 2017.

[23] Burlington Electric Department. Joseph C. McNeil Generating Station. 2017 January 19, 2017)]. Available from: https://www.burlingtonelectric.com/about-us/what-we-do/joseph-c-mcneil-generating-station.

[24] Burlington Electric Department. McNeil Station Joint Ownership Operating Committee Meeting Notes. 2016 January 19, 2017)]. Available from: https://www.burlingtonelectric.com/sites/default/files/2016-03-mcneil-mins.pdf.

[25] Timmons, D., et al., Energy from forest biomass: Potential economic impacts in Massachusetts, in Massachusetts Sustainable Forest Bioenergy Initiative. Massachusetts Division of Energy Resources and Department of Conservation and Recreation; 2007.

[26] BERC, Vermont Wood Fuel Supply Study: 2010 Update, Biomass Energy Resource Center: Burlington, Vermont; 2011.

[27] FirstLight. Northfield Mountain Pumped Storage Project: Relicensing Website 2017 January 19, 2017. Available from: http://www.northfieldrelicensing.com/Pages/Northfield.aspx.

[28] Henning HM, Palzer A: A comprehensive model for the German electricity and heat sector in a future energy system with a dominant contribution from renewable energy technologies—Part I: methodology, Renew Sust Energy Rev 30:1003–1018, 2014.

[29] Barbose, G, Darghouth, N, Tracking the Sun IX: The Installed Price of Residential and Non-Residential Photovoltaic Systems in the United States. SunShot Initiative, U.S. Department of Energy and Lawrence Berkeley National Laboratory; 2016.

[30] Baird, J.B., VT weighs turbo-charge from Hydro-Québec, Burlington, Vermont: Burlington Free Press; 2015.

[31] Fthenakis V, Mason JE, Zweibel K: The technical, geographical, and economic feasibility for solar energy to supply the energy needs of the US, Energy Policy 37(2):387–399, 2009.

[32] Jacobson MZ, et al: 100% clean and renewable wind, water, and sunlight (WWS) all-sector energy roadmaps for the 50 United States, Energy Environ Sci 8:2093–2117, 2015.

[33] McKinsey and Company, Impact of the financial crisis on carbon economics: version 2.1 of the global greenhouse gas abatement cost curve; 2010.

Index

CPI Antony Rowe
Chippenham, UK
2018-12-24 13:43